AF773946

Asymmetric Organo-Metal Catalysis

Asymmetric Organo-Metal Catalysis

Concepts, Principles, and Applications

Liu-Zhu Gong

WILEY-VCH

Author

Prof. Liu-Zhu Gong
University of Science and Technology of China
Department of Chemistry
230026 Hefei
China

Cover Design: Wiley
Cover Image: © Valentin Valkov/Getty Images

All books published by **WILEY-VCH** are carefully produced. Nevertheless, authors, editors, and publisher do not warrant the information contained in these books, including this book, to be free of errors. Readers are advised to keep in mind that statements, data, illustrations, procedural details or other items may inadvertently be inaccurate.

Library of Congress Card No.: applied for

British Library Cataloguing-in-Publication Data
A catalogue record for this book is available from the British Library.

Bibliographic information published by the Deutsche Nationalbibliothek
The Deutsche Nationalbibliothek lists this publication in the Deutsche Nationalbibliografie; detailed bibliographic data are available on the Internet at <http://dnb.d-nb.de>.

© 2022 WILEY-VCH GmbH, Boschstr. 12, 69469 Weinheim, Germany

All rights reserved (including those of translation into other languages). No part of this book may be reproduced in any form – by photoprinting, microfilm, or any other means – nor transmitted or translated into a machine language without written permission from the publishers. Registered names, trademarks, etc. used in this book, even when not specifically marked as such, are not to be considered unprotected by law.

Print ISBN: 978-3-527-34592-2
ePDF ISBN: 978-3-527-34596-0
ePub ISBN: 978-3-527-34594-6
oBook ISBN: 978-3-527-34593-9

Typesetting Straive, Chennai, India
Printing and Binding CPI Group (UK) Ltd, Croydon, CR0 4YY

Printed on acid-free paper

C092401_081021

Contents

Preface

The alarming thalidomide tragedy of the 1960s has immediately led to a worldwide revisit of enantiomers in the medicinal chemistry and pharmaceutical industries, as the biological systems would recognize each enantiomer for either targeted activity or inactivity or severe side effect. Of all the synthetic strategies developed to distinguish enantiomers in the intervening years, the catalytic process induced by a chiral molecule, namely asymmetric catalysis, is undoubtedly the most efficient approach to access enantiomerically pure compounds. In this context, enzymatic transformations are the most natural way, albeit with limited tolerance of reaction conditions; transition metal complexes are of stably increasing importance and popularity in both fundamental research and industrial applications; organocatalysis employs purely small organic molecules as the catalysts to offer a convenient and green solution to asymmetric synthesis.

In the pursuit of ideal synthesis, many have experienced the shortage of effective chiral catalysts in regards to bond activation or stereochemical control. So, instead of de novo design and synthesis of new chiral catalysts, why don't try catalysts blending: *Combined Catalysis*? Incompatibilities notwithstanding, there are vast opportunities in combining distinct catalysts in a single operation for improving reaction efficiency or offering a unique solution to challenging transformations. This book focuses on the concept of *Organo/Metal Combined Catalysis*, the proof of concept, and thoroughgoing discussions of the reactions achieved, as the number of publications on this field has been increasing exponentially during the last decades. *Organo/Metal Combined Catalysis* dates back to the early 2000s when a chiral organocatalyst was exploited to activate the nucleophiles, thus controlling the stereoselectivity of an allylic alkylation with π-allyl palladium species, in the most desirable cooperative manner. Interestingly, one catalyst of a compatible catalytic system may serve as a relay shuttle for other catalytic cycles to enable orthogonal bond-forming reactions, which has been named as relay catalysis (also known as cascade, domino, or tandem catalysis). Once the incompatibility issue between catalysts has developed, the *Organo/Metal Combined Catalysis* fells into a category of sequential catalysis, a compromise relay catalysis featuring stepwise catalysts addition.

This book covers the whole array of organo/metal combined catalytic systems achieved to date, including 10 chapters: general introduction (Chapter 1); phase-transfer-catalyst and metals (Chapter 2); enamine and metals (Chapter 3);

iminium and metals (Chapter 4); Brønsted acid and transition metal cooperative catalysis (Chapter 5); Brønsted acid and metal relay catalysis (Chapter 6); Lewis base and Lewis acid cooperative catalysis (Chapter 7); Lewis base and transition metal cooperative catalysis (Chapter 8); chiral organocatalyst and transition metal-based photoredox catalyst (Chapter 9); applications in total synthesis (Chapter 10). All knowledgeable authors, Dr. Zhi-Yong Han, Dr. Jie Yu, Dr. Chang Guo, Dr. Jin Song, Dr. Dian-Feng Chen, and Dr. Pu-Sheng Wang, have significantly contributed to the development of *Organo/Metal Combined Catalysis*, and can provide invaluable perspectives.

This book could serve as an excellent textbook for graduate students and a great handbook for researchers and other practitioners at all levels of asymmetric synthesis.

1

Why Is Organo/Metal Combined Catalysis Necessary?

1.1 Introduction

Molecular chirality has played an important role in a broad scope of fields, including synthetic chemistry, drug discovery, biological system, and materials science and will continue to exert a great impact on physical science. Such unparalleled significance of chirality leads to increasing demand for efficient asymmetric protocols to build up chiral structures.

Chiral resolution is the oldest way to isolate optically pure chiral molecules from the racemic form. Chiral pool- and auxiliary-induced asymmetric synthesis has frequently been synthetic strategies of choice to create chiral elements in organic synthesis [1]. Although chiral auxiliary-induced asymmetric synthesis has been prevalently applied to the asymmetric synthesis of natural products and pharmaceutically significant substances, and thus held the historical impact on synthetic chemistry [2], the installation and removal of chiral auxiliary basically require additional reaction steps to thereby attenuate the synthetic efficiency.

Asymmetric catalysis has globally been accepted as the most efficient concept to stereoselectively build up molecular chirality. Since the advent of asymmetric cyclopropanation and hydrogenation catalyzed by chiral copper and rhodium complexes, respectively [3, 4], asymmetric metal catalysis has continuously been the central focus of asymmetric synthesis. The versatility and robustness of metals in the activation of a wide spectrum of chemical bonds, even those with high bond energy, have rendered many families of asymmetric transformations to be accessed by either Lewis acid or transition metal catalysis [5, 6].

The control of stereochemistry in asymmetric metal catalysis principally relies on the chiral ligand and to a large degree on the ligand acceleration [7]. The stereochemical control events involved in the transition metal catalysis might be one or some of the typical elementary reactions including chiral ligand coordination, oxidative addition, insertion, and reductive elimination. The oxidative addition occurs more easily with an electronically richer and low-valent metal to increase the oxidation state and coordination number of the metal center; therefore the ligand coordination facilitates this reaction. The global and long-standing interest in the design and development of chiral ligands has culminated in the explosive appearance of privileged ligands [8], which actually propel the proliferation of elegant and practical

Asymmetric Organo-Metal Catalysis: Concepts, Principles, and Applications, First Edition. Liu-Zhu Gong.
© 2022 WILEY-VCH GmbH. Published 2022 by WILEY-VCH GmbH.

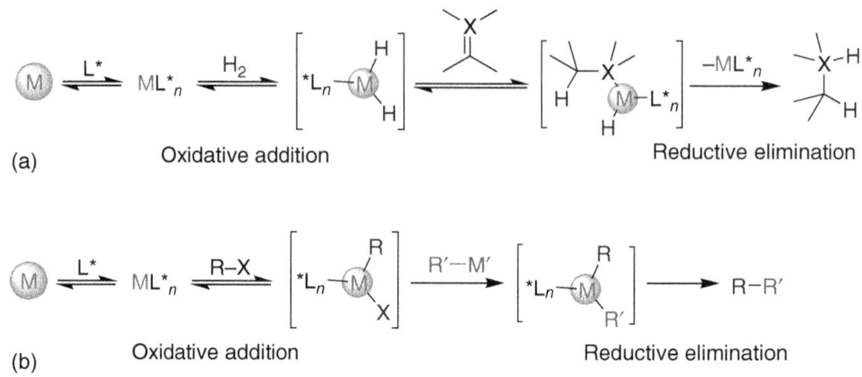

Figure 1.1 Transition metal-catalyzed reactions initiated with oxidative addition. (a) Hydrogenation. (b) Cross coupling.

asymmetric processes commencing with the oxidative addition, for example, asymmetric hydrogenation and cross-coupling reaction (Figure 1.1).

High-valent transition metals have also been found to enable a tremendous number of organic reactions. In contrast to abundantly available chiral ligands for asymmetric catalysis beginning with the oxidative addition, which undergoes with low oxidation state metals, rather fewer chiral ligands are compatible with high-valent metal catalysis and reactions undergoing under oxidation conditions to pose a great challenge to the control of stereoselectivity. For example, although the high-valent metal-catalyzed transformations commencing with nucleometallation (Eq. (1), Figure 1.2), aryl and allylic C–H activation (Eqs. (2) and (3)), have been well established, a very limited number of chiral ligands can enable highly enantioselective variants, in particular, those using molecular oxygen as the terminal oxidant [9]. So far, chiral Lewis acids are successful representatives among massive asymmetric high-valent metal catalysis [10]. As such, a new concept to break the conventional wisdom that relies on the chiral ligand to control the stereochemistry of transition metal catalysis is greatly desirable.

Asymmetric organocatalysis represents an important tool, independent, and conceptually distinct from metal catalysis, to build up molecular chirality [11, 12]. The typical principles in organocatalysis for the activation of chemical bonds cover a broad scope of concepts, including amine catalysis by enamine raising highest occupied molecular orbital (HOMO) and iminium lowering lowest unoccupied molecular orbital (LUMO), Brønsted acid catalysis by hydrogen-bonding interaction or protonation, NHC catalysis via umpolung of aldehyde, Lewis base catalysis by nucleophilic addition to either carboxylic acid derivatives or electron-deficient carbon–carbon double bonds to form reactive enolate or acylammonium species, and phase transfer catalysis by using ammonium and phosphonium to form ion pairs with anionic nucleophiles [13]. Such versatile principles in the activation of chemical bonds and structural diversity of organocatalysts have enabled the explosive appearance of fundamentally novel asymmetric reactions and processes featured by environmentally benign, atom, and step economies. Nevertheless, the

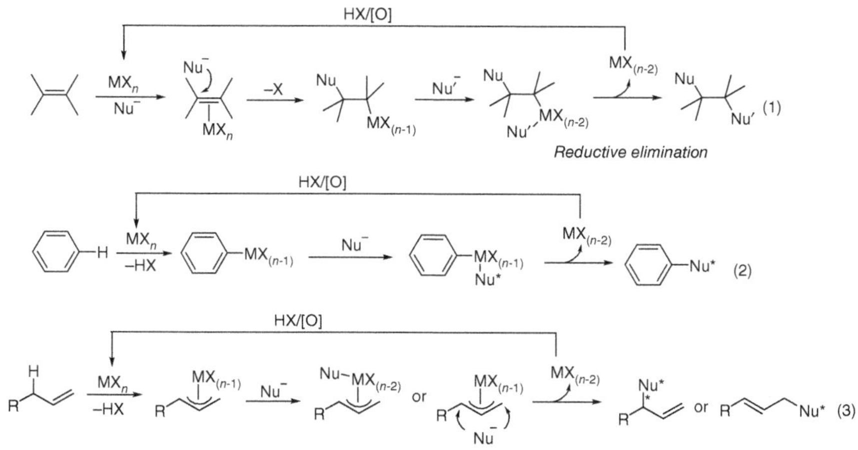

Figure 1.2 Representatives of high-valent metal catalysis.

complete dependence upon the interactions between a highly active functionality and the organocatalyst (Table 1.1) poses the organocatalysis essential constraints to activate relatively inactive chemical bonds and unfunctionalized substrates.

1.2 Early Stage of Organo/Metal Combined Catalysis and General Principles

The combination of asymmetric organocatalysis and metal catalysis integrates the catalytic activity of metals and organocatalysts, hence allows the simultaneous or sequential occurrence of multiply bond-breaking and forming events in stereochemical control to provide much more diverse ranges of concepts or principles capable of enabling unconventional enantioselective transformations that are toughly accessed by the individual catalyst [14]. Very early reports on the asymmetric organo/metal combined catalysis describe a Pd-catalyzed asymmetric allylic alkylation of an imino ester, in concert with chiral phase transfer catalysis [15]. Gong and coworkers found that the use of a cinchona alkaloid-derived ammonium bromide **4a** that Corey developed [16] as the chiral phase transfer catalyst, in combination with an achiral palladium complex of triphenylphosphine, is able to enable the reaction to deliver 59% ee [15]. Takemoto identified that the electron density of the trivalent phosphorus ligand exerts considerable impact on the reaction performance and the highest enantioselectivity of 94% ee was obtained with **4b** that Lygo introduced [17] in the presence of triphenyl phosphite ligand. In both the cases, the respective and synergistic action of the palladium complex and chiral PTC on the allylic ester **1** and nucleophile **2** renders the reaction to proceed more efficiently via a transition state **TS-1** and allows the stereochemical control to be accessed by chiral phase transfer catalyst, alone (Figure 1.3). This strategy indicates that the stereoselection of metal-catalyzed reactions can be controlled without chiral ligand, instead, by a co-organocatalyst, thus

Table 1.1 Typical activation modes in organocatalysis (OC).

Concepts of OC	Active intermediates	Applicable substrates	Reaction types
Enamine catalysis	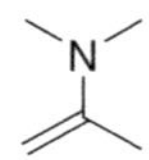	Enolizable aldehydes and ketones	Aldol reaction Mannich reaction Michael addition
Iminium catalysis	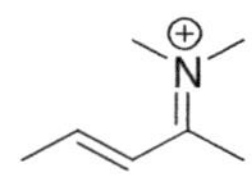	Enal and enones	Michael addition Diels–Alder reaction Friedel–Crafts reaction
Brønsted acid catalysis	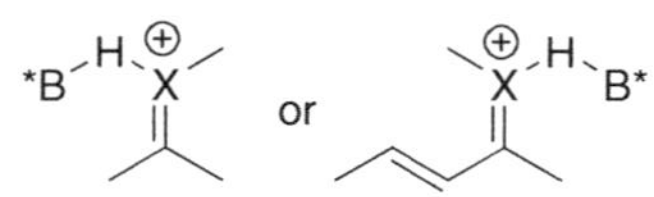	Aldehydes, ketones, and imines, enal, and enones	Reduction Friedel–Crafts reaction Michael addition
NHC catalysis	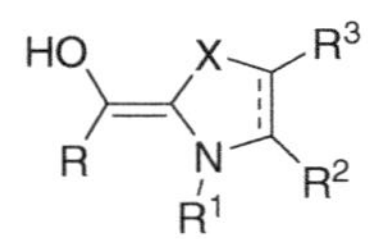	Aldehydes	Annulation Benzoin reaction Stetter reaction
Lewis base catalysis	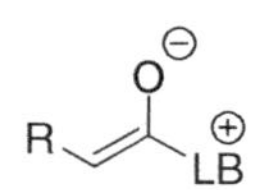	Ketene, acylhalides, anhydride, and other analogues	[2+2] cyclization [4+2] cyclization Kinetic resolution
	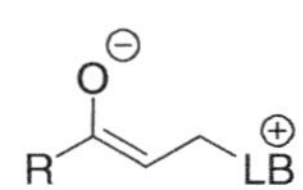	Enals, enones, and α,β-unsaturated esters	[3+2] cycloaddition Baylis–Hillman reaction
Phase transfer catalysis	$R_4\overset{\oplus}{N}\ \overset{\ominus}{C}(R')_3$ or $R_4\overset{\oplus}{P}\ \overset{\ominus}{C}(R')_3$	Acidic nucleophiles	Alkylation Aldol reaction Mannich reaction Michael addition

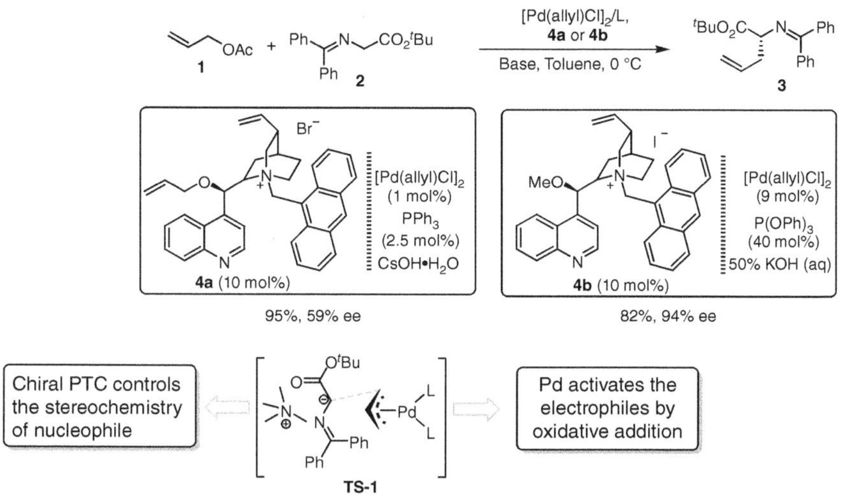

Figure 1.3 Pd and PTC cooperative catalysis.

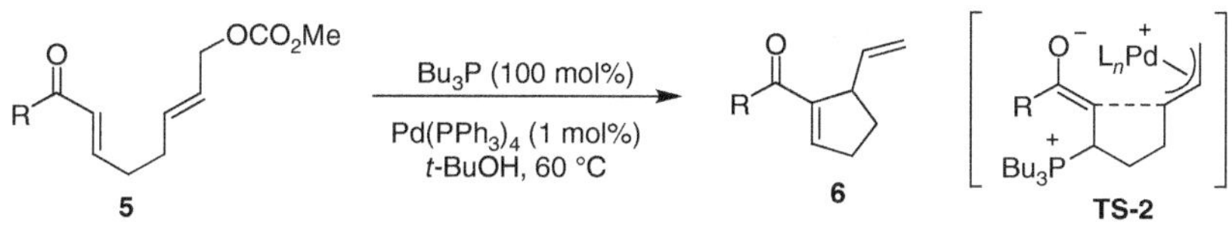

Figure 1.4 Pd and phosphine cooperative catalysis.

opens up a window to seek unconventional modes to address issues of the stereochemical control encountered in the asymmetric metal catalysis.

The cooperative catalysis of transition metal and Lewis base was first showcased by Krische and coworkers [18]. Tributylphosphine undergoes Rauhut–Currier type addition [19] with the enone moiety of **5** to generate a transient enolate and simultaneously, the palladium complex reacts with the allylic carbonate part to give a π-allylpalladium species. A subsequent intramolecular substitution occurs via a transition state **TS-2**, and followed by elimination of tributylphosphine, to yield the final product **6** (Figure 1.4). The perfect integration of the Lewis base and palladium catalysis offers a unique activation mode to make the reaction that is otherwise unable to proceed possible.

The cooperative catalysis of Lewis base and Lewis acid to drive an asymmetric [2+2] annulation of acetyl chloride **7** and imino esters **8** was reported by Lectka and coworkers [20]. The chelation of indium triflate to the imino ester **8** enhances the reactivity of the imine functionality. Thus, the nonmetal-coordinated zwitterionic enolate **Int-1** generated from the acetyl chloride **7** with BQ **10** and base **11** is able to undergo an enantioselective Mannich-type reaction with an In(III) cocatalyst-bound imino ester **Int-2** to form an intermediate **Int-3**. Finally, an intramolecular amide bond formation delivers β-lactams **9** and regenerates the catalyst (Figure 1.5).

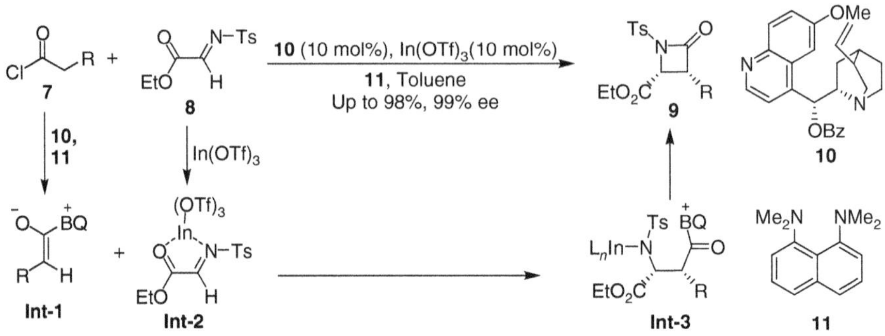

Figure 1.5 Lewis acid and Lewis base cooperative catalysis.

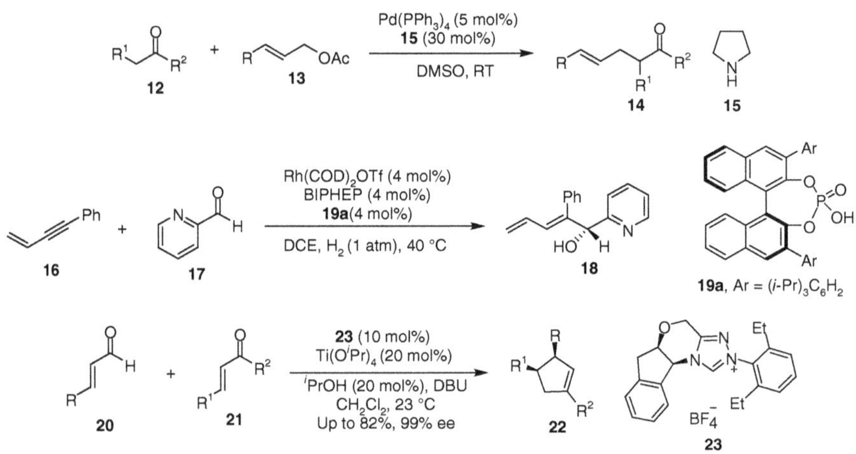

Figure 1.6 Typical organo/metal combined catalysis in the early stage. Source: Modified from Arndtsen and Gong [25].

In the same period, the enamine and palladium [21], Brønsted acids and transition metals [22, 23], NHC and metal complexes [24], and some other combined catalyst systems were successively reported, providing unusual fundamental bond activation modes that turn out to be versatile platforms to allow for the proliferation of unprecedented transformations (Figure 1.6) [25].

Most organocatalysts contain heteroatoms, which basically coordinate to metal catalysts to change the catalytic activity and to result in the "self-quenching" in some cases. As a consequence, the compatibility of metals and organocatalysts turns out to be the key to success in the organo/metal combined catalysis. On the other hand, the synergistic effect among the components of the combined systems is actually most desirable. Based on the activation modes and reaction pathway, the asymmetric organocatalysis combined with metal catalysis generally consists of cooperative catalysis, relay catalysis (also known as cascade, demino, and tandem catalysis), and sequential catalysis as well (Figure 1.7) [26].

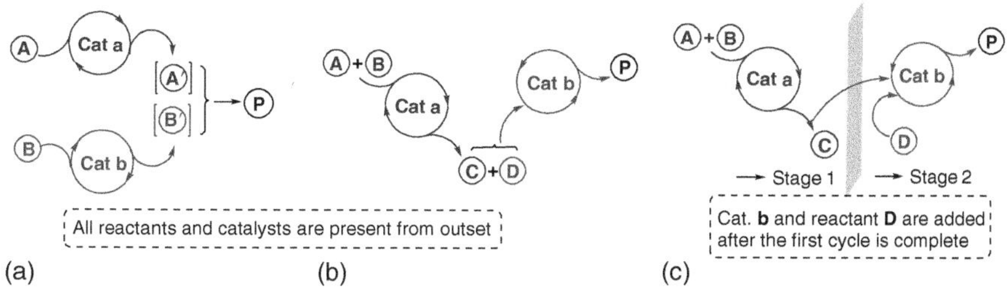

Figure 1.7 Categories of asymmetric organocatalysis combined with metal catalysis. (a) Cooperative catalysis. (b) Relay catalysis. (c) Sequential catalysis. Source: Chen et al. [26].

1.3 Organo/Metal Cooperative Catalysis

1.3.1 Control of Stereochemistry

Cooperative catalysis refers to a catalytic process, which is initiated via the simultaneous and respective activation of two or more substrates, functionalities, or chemical bonds enabled by two or more individual metal and organocatalysts (Figure 1.7a). The cooperative catalysis actually provides more possibilities to control the stereochemistry of an asymmetric transformation by tuning the chirality of each individual catalyst. For example, if the metal only works efficiently for breaking or assembling the chemical bonds, but is unable to control the stereochemistry by varying chiral ligands, the chiral organocatalysis concepts can be adapted to conquer the stereochemical control issue (Figure 1.8a). In case no appropriate organocatalysts are able to efficiently induce the stereoselectivity, chiral metal complexes would stand in for addressing the stereochemical issue while the organocatalyst solely acts to activate

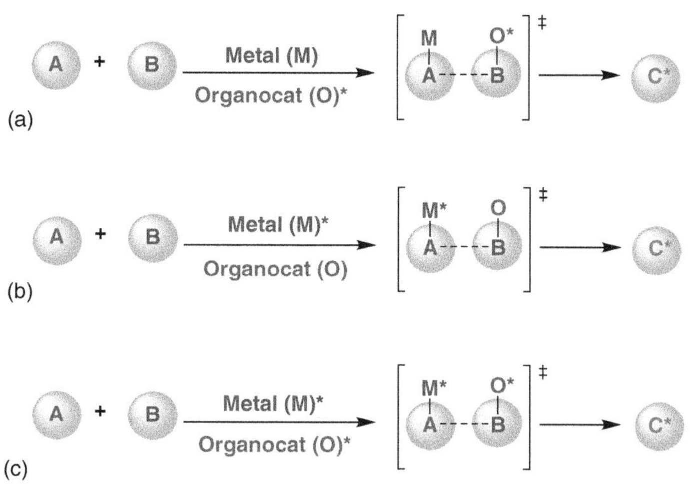

Figure 1.8 Strategies of stereochemical control in cooperative catalysis. (a) Combine achiral metal complex with chiral organocatalyst. (b) Combine chiral metal complex with achiral organocatalyst. (c) Combine chiral metal complex with chiral organocatalyst. Source: Han et al. [27].

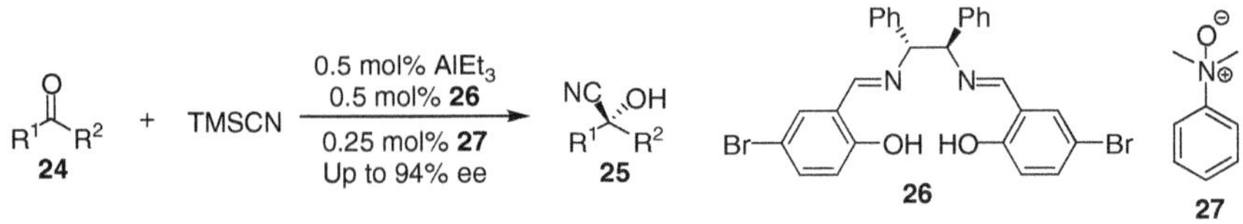

Figure 1.9 Chiral Lewis acid and achiral Lewis base cooperative catalysis. Source: Chen et al. [28].

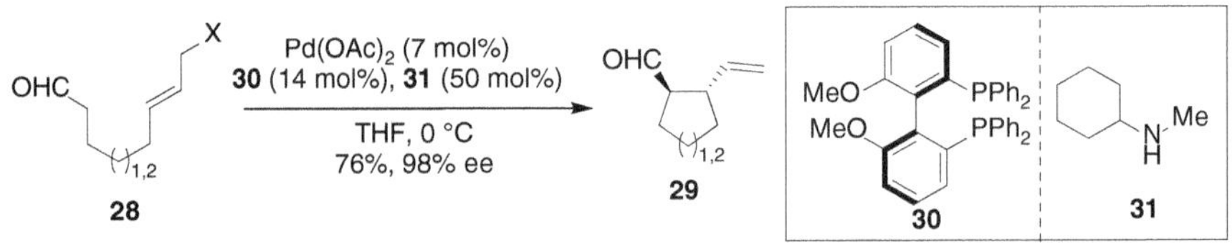

Figure 1.10 Asymmetric allylation enabled by chiral palladium complex and achiral amine. Source: Bihelovic et al. [29].

the substrates (Figure 1.8b). If neither of the concepts offers high stereoselectivity, both chiral metal complex and chiral organocatalyst can be applied to synergistically control the stereoselectivity (Figure 1.8c) [27].

The combination of achiral metal catalysts and chiral organocatalysts appeared at the beginning of this field [15]. In the same period, one of the earliest examples describing merging the chiral metal complex and achiral organocatalyst was introduced by Feng and coworkers who identified that the combination of an aluminum complex of chiral salen **26** and an *N*-oxide Lewis base **27** was able to afford a highly efficient and enantioselective cyanosilylation of ketones **24** (Figure 1.9) [28].

Saicic and coworkers described the combined use of a chiral palladium complex of (*R*)-Ph-MeOBiPHEP **30** and an achiral amine **31** to establish a highly enantioselective intramolecular α-allylation of aldehydes **28** (Figure 1.10) [29].

A binary catalyst made up of a chiral phosphoric acid and an iridium complex **34** adorned with Noyori-diamine ligand allows a highly efficient asymmetric hydrogenation of acyclic imines **32** to give chiral amine **33** in almost perfect levels of enantioselectivity. In this case, the matched chirality between the chiral diamine ligand of the iridium complex **34** and the chiral phosphoric acid (*R*)-**19a** offers high stereoselectivity (Figure 1.11) [30].

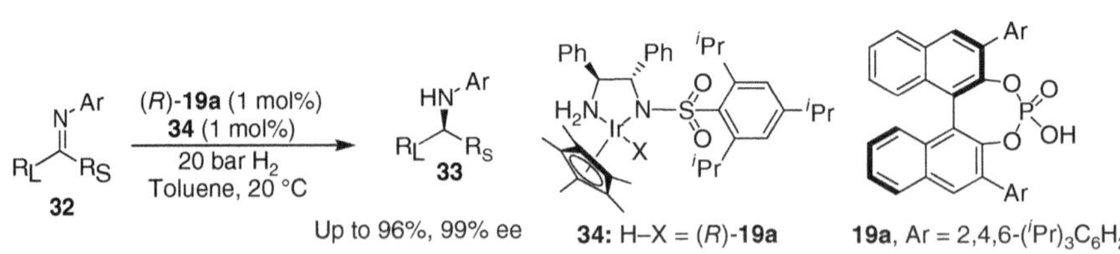

Figure 1.11 Asymmetric hydrogenation enabled by chiral iridium complex and phosphoric acid. Source: Modified from Li et al. [30].

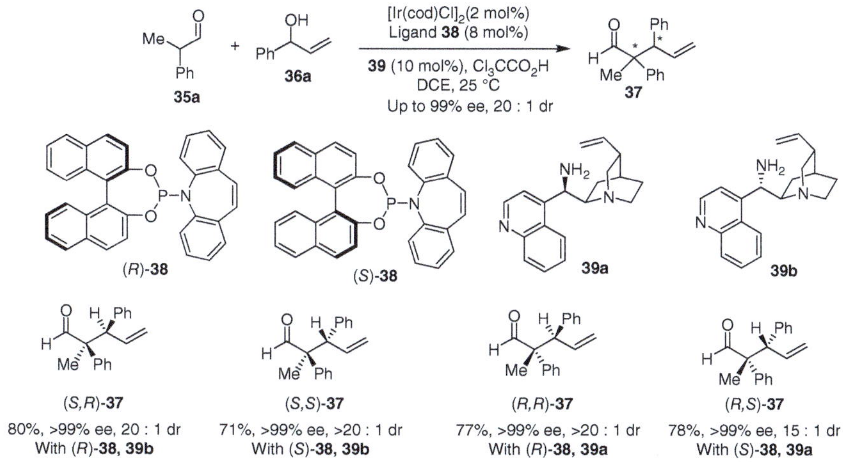

Figure 1.12 Stereodivergent cooperative catalysis. Source: Modified from Krautwald et al. [31].

In the cases of combining chiral metal and chiral organocatalysts, if both of them are robust enough to control the stereochemistry, the cooperative catalysis will be able to establish the stereodivergent construction of chiral structures by simply varying the stereochemistry of each individual chiral catalyst. The proof of concept in this field was demonstrated by Carriera who first reported a stereodivergent allylic alkylation between α-enolizable aldehyde **35a** and an allylic alcohol **36a** based on the asymmetric cooperative catalysis of iridium complex of chiral phosphoramidite **38** and chiral primary amine **39**, allowing access to four stereomers by altering the stereochemistry of the chiral ligand and chiral amine (Figure 1.12) [31]. Afterward, the cooperative catalysis of chiral rhodium complex and chiral amine, chiral iridium complex and chiral Lewis base, as well as the iridium complex and chiral NHC carbene was successively discovered by different research groups for the establishment of stereodivergent processes [32].

1.3.2 Cooperative Activation of Chemical Bonds

Simultaneous action of both the metal and organocatalyst on the same substrate actually provides a unique tool to activate a chemical bond and thereby makes the reaction that may not work upon, being promoted by either of individual catalysts probably occur. The feasibility and robustness of this strategy have been implicit in the activation of relatively inactive chemical bonds. For instance, both the allylic C—N bond of allyl amines **40** and C—O bond of allylic alcohols **41** can be activated by protonation as shown in **Int-4** and hydrogen-bonding interaction as shown in **Int-5** with the chiral phosphoric acid, respectively, to facilitate the oxidative addition of the palladium complex to give the π-allylpalladium phosphate complex intermediate, which smoothly undergoes the enantioselective substitution reaction with the enamine **Int-6** via a transition state **TS-4** to give α-allylic aldehydes **42** (Figure 1.13) [33]. It is worthy to mention that neither the allyl amine nor the allylic

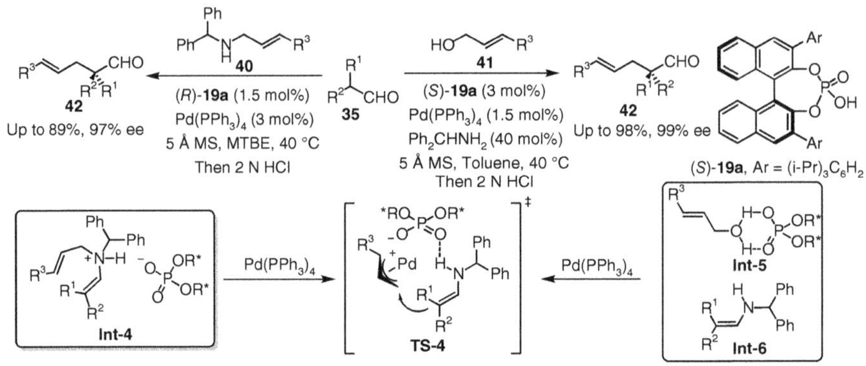

Figure 1.13 Allylation of aldehydes by cooperative catalysis of Pd, chiral Brønsted acid, and amine. Source: Mukherjee et al. [33].

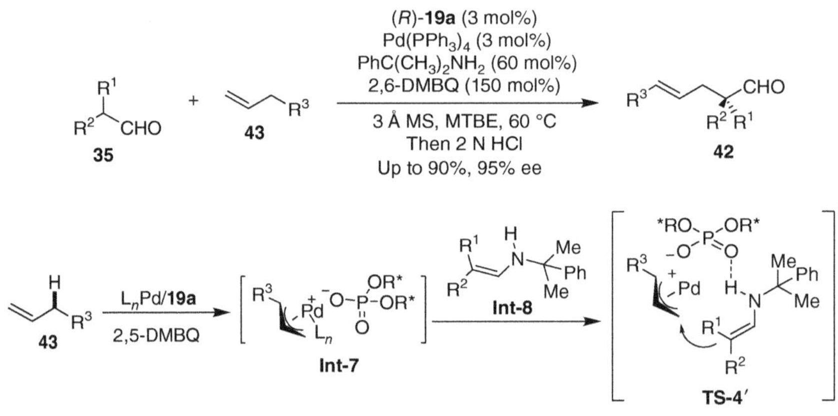

Figure 1.14 Asymmetric allylic C–H functionalization by cooperative catalysis.

alcohol can undergo the oxidative addition without the assistance of the Brønsted acid [34].

Even more remarkably, the Brønsted acid can facilitate the palladium-mediated allylic C–H cleavage. Thus, the combination of the palladium complex, chiral Brønsted acid, and primary amine allows an enantioselective oxidative α-allylation of aldehydes **35** with α-olefins **43** [35]. The Pd-mediated allylic C–H oxidation generates the π-allylpalladium phosphate complex **Int-7** that undergoes asymmetric allylation with enamine **Int-8** via the transition state **TS-4′** similar to List's model **TS-4** to smoothly give the chiral aldehydes **42** in excellent enantioselectivity (Figure 1.14).

Adoption of visible light photocatalysis to combine with chiral organocatalysis leads to another versatile concept for the development of asymmetric transformations that are unable to work based on traditional bond-activation and assembly strategies [36]. MacMillan and coworker first described the integration of a photocatalyst **48** and a chiral amine catalyst **47a** to establish an efficient α-alkylation of aldehydes **44** with α-bromocarbonyls **45** (Figure 1.15) [37]. Subsequently, merging the

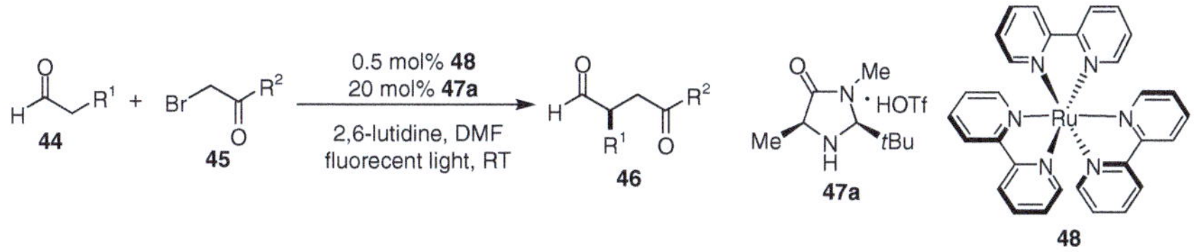

Figure 1.15 Asymmetric organocatalysis combined with photocatalysis. Source: Modified from Nicewicz and MacMillan [37].

ruthenium(II) or iridium(III) visible photocatalysts and chiral Brønsted acids [38], NHC carbene [39], or other organocatalysts [40] were disclosed to promote asymmetric transformations that proceed via transient radical intermediates in excellent stereochemical control.

The organo/metal combined catalysts aforementioned are typical representatives in the field of hybrid cooperative catalysis. Each of them consists of a large family of chiral catalyst systems that provide plentiful synergistic activation of chemical bonds and many possibilities to induce the stereoselectivity, allowing for the creation of new asymmetric transformations. A detailed description of this field will be presented in Chapters 2–9.

1.4 Organo/Metal Relay and Sequential Catalysis

Relay catalysis is defined as a cascade process in which two or more sequential bond-forming events are independently promoted by two or more catalysts in a cascade manner. Based on the substrates, either the linear ligation (Figure 1.16a) or annulation (Figure 1.16b) can be accessed by this catalytic strategy.

The sequential catalysis generally describes a one-pot reaction promoted by different catalysts, which are chemically incompatible, and therefore parts of the catalysts or, in some cases, reagents must be added after the previous catalytic reaction is complete (Figure 1.7c) [26]. Both relay and sequential catalysis can provide general platforms to design new protocols for the enantioselective construction of molecular complexity from readily available starting materials, featured by avoiding the additional workup of either isolable or transient intermediates to thereby reduce labor, save time, and minimize waste. This book will be focusing on discussion of cooperative and relay catalysis, both of which are chemically compatible. Thus, sequential catalytic processes that proceed based on artificial operation, will be excluded from this book except the very original ones that have inspired the following developments.

Rovis coined the name of relay catalysis defining a cascade reaction constituted of elementary steps promoted by the individual catalyst, respectively, and reported a nucleophilic carbene and HOAt (**53**) relay catalysis for an amidation of α-reducible aldehydes **49** and amines **50** to produce α-reduced amides **51** [41]. The aldehyde **49** undergoes a redox reaction with carbene to yield an acyl azolium **Int-8**, which then condenses with **53** to give an even more reactive intermediate **Int-9** capable of participating in the subsequent amidation reaction (Figure 1.17).

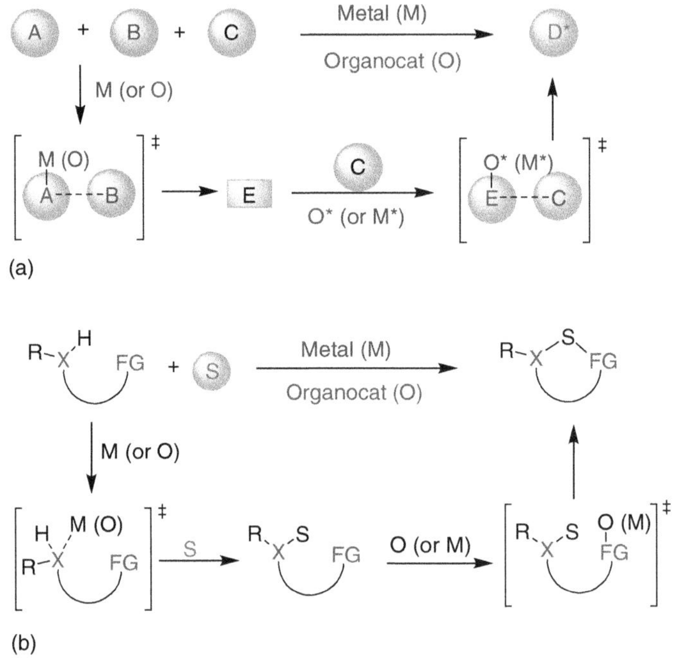

Figure 1.16 General principal of organo/metal relay catalysis. (a) Intermolecular relay catalysis. (b) Relay catalysis for annulation. Source: Han et al. [27].

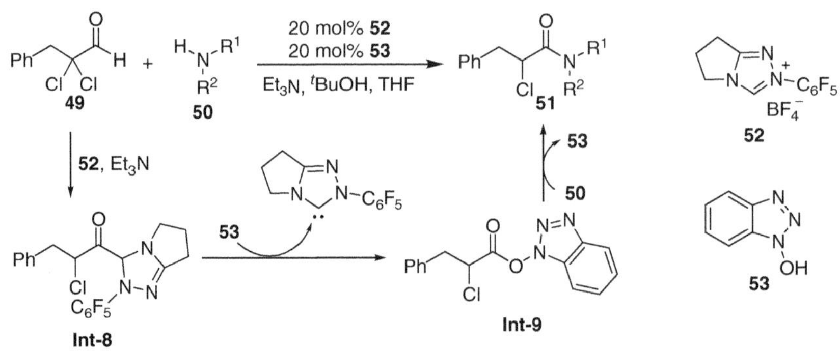

Figure 1.17 Proof of concept of relay catalysis.

One of the earliest organo/metal relay catalysis describes cascade hydroformylation and asymmetric aldol reaction (Figure 1.18). The styrene derivatives **54** prefer to undergo branch-selective hydroformylation catalyzed by rhodium complex of **57** to generate α-branched aldehydes, which then undergo a proline-catalyzed asymmetric direct aldol reaction to give β-hydroxyketones **56**. In contrast, α-alkenes **58** favorably undergo the linear-selective hydroformylation to give enolizable aldehydes, which then participate in the subsequent proline-catalyzed cross-aldol reaction with another aldehyde **59** [42].

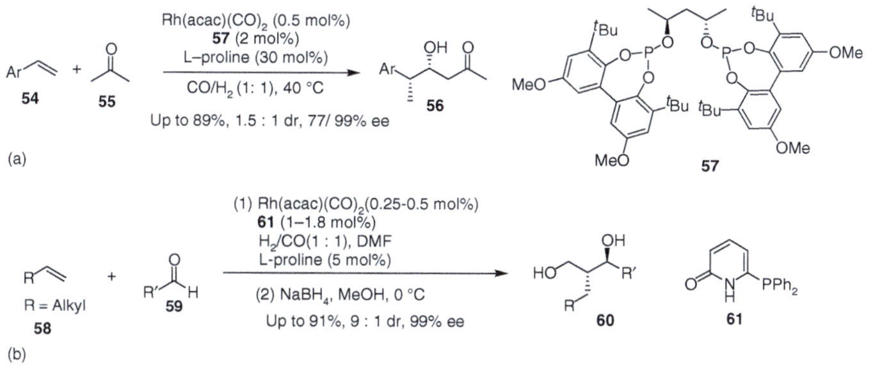

Figure 1.18 Relay catalytic cascade hydroformylation and aldol reaction. (a) Eilbracht's work. (b) Breit's work.

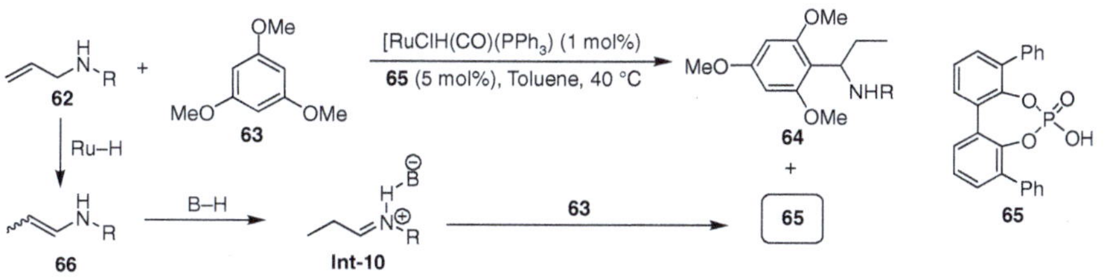

Figure 1.19 Ru/Brønsted relay catalysis. Source: Modified from Sorimachi and Terada [43].

Terada and coworker established a one-pot tandem isomerization/carbon–carbon bond-forming sequence of allylamines **62** and **63** enabled by the relay catalysis of a ruthenium complex and Brønsted acid **65**. The readily available allylamines **62** initially undergoes a double-bond migration catalyzed by ruthenium complex to give an enamine **66**, which then participates in a Brønsted acid-catalyzed addition reaction with 1,3,5-trimethoxybenzene (**63**) via the intermediate **Int-10** to yield **64** and regenerate phosphoric acid **65** (Figure 1.19) [43].

Soon after these events, hydroamination of alkynes and asymmetric transfer hydrogenation cascade processes, enabling the direct transformation of a linear functionality, the carbon–carbon triple bond, to a stereogenic center, were established by relay catalysis of gold complex and chiral phosphoric acid [44]. Gong and coworkers elaborated a gold complex/chiral phosphoric acid relay catalytic transformation to directly convert the propargylic anilines **67** to a number of chiral tetrahydroquinolines **69** in excellent yields and with high levels of enantioselectivity (Figure 1.20a) [44]. Coincidently, Liu and Che also found a cascade intermolecular hydroamination and transfer hydrogenation reaction of alkynes **71** and anilines **72** rendered by gold and chiral phosphoric acid relay catalysis to furnish chiral secondary amine products **73** in excellent yields and enantioselectivities (Figure 1.20b) [44].

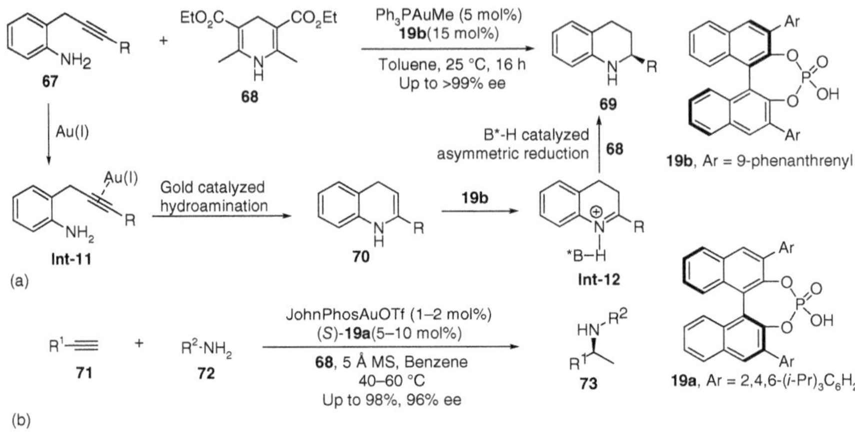

Figure 1.20 Gold/chiral Brønsted acid relay catalysis. Source: Modified from Liu and Che [40].

Dixon combined a gold(I)-catalyzed intramolecular hydroalkoxylation of 3-butylnoic acids **74** to generate furan-2(3*H*)-one intermediates **77**, a chiral phosphoric acid-mediated condensation with tryptamine **75** to form *N*-acyliminium phosphate **Int-13**, and *N*-acyliminium cyclization, to accomplish a one-pot process enabling the direct transformation of the 3-butylnoic acids **74** and the tryptamine **75** into chiral *N*-heterocyclic compounds **76** in great yields and stereoselectivities (Figure 1.21) [45].

In the same year, You and coworkers reported a relay catalytic cascade cross-metathesis and asymmetric Friedel–Crafts alkylation between indole derivatives **78** and unsaturated carbonyls **79** [46]. The presence of the Grubbs catalyst is sufficient to drive cascade reaction but in the racemic version [47]. The addition of the chiral phosphoric acid **19a** as co-catalyst allows the Friedel–Crafts alkylation to proceed much faster and thus offers high stereoselectivity (Figure 1.22).

The earliest combination of iminium activation mode and transition metal for sequential catalysis was described by MacMillan et al. The Grubbs II catalyst **81b**,

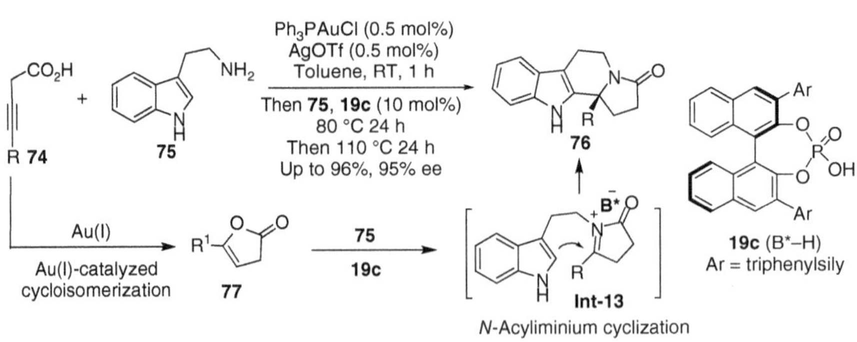

Figure 1.21 Gold/chiral Brønsted acid sequential catalysis. Source: Modified from Yang et al. [45].

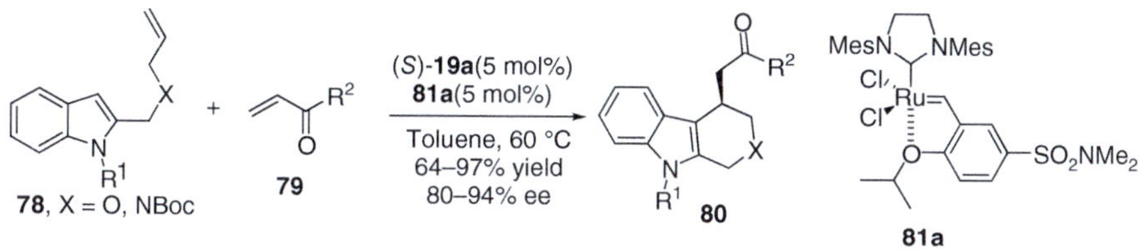

Figure 1.22 Relay catalytic cascade metathesis and Friedel–Crafts reaction.

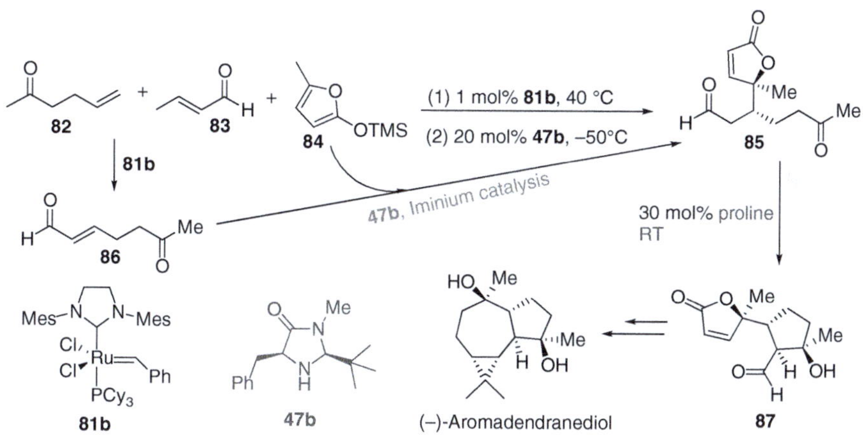

Figure 1.23 Sequential metathesis/Michael addition/aldol reaction. Source: Modified from Simmons et al. [48].

5-hexene-2-one **82**, and crotonaldehyde **83** are combined at 40 °C to undergo the metathesis to give a keto-enol **86**. The subsequent asymmetric Michael addition of trimethylsilyloxyfuran **84–86** catalyzed by imidazolidinone **47b** proceeds at −50°C to generate **85**. Finally, the intramolecular aldol reaction of **85** catalyzed by L-proline allows the production of **87** in overall 64% yield and with 5/1 dr and 95% ee. This sequential one-pot process enabled by the triple catalysis provides the key intermediate **87** to access natural product (−)-aromadendranediol (Figure 1.23) [48].

As aforementioned and also implied in the working models of organocatalysis (Table 1.1), it is basically hard for organocatalysis to activate relatively inert chemical bonds. For example, hydrogen is highly inert for pure organocatalysts and unable to participate in the organocatalytic process. However, the adoption of transition metal catalysis in the relay catalytic process can allow asymmetric hydrogenation to occur by using the chiral organocatalyst to control the stereochemistry. Zhou and coworkers combined chiral ruthenium complex and chiral Brønsted acid to establish highly enantioselective hydrogenation of quinoxalines [49]. The Ru(II)-catalyzed hydrogenation of quinoxaline **88** initially proceeds to generate dihydroquinoxaline **90**, which undergoes either a Ru(II)-catalyzed non-enantioselective hydrogenation or a chiral phosphoric acid (*S*)-**19a**-catalyzed asymmetric disproportionation to give **89** (Figure 1.24). Since the chiral phosphoric acid-catalyzed self-transfer hydrogenation

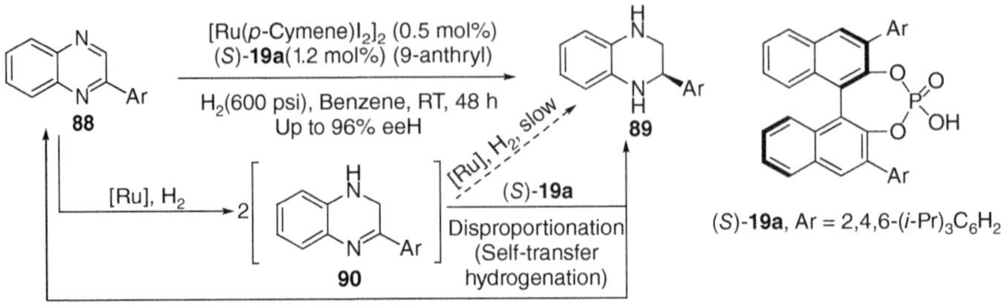

Figure 1.24 Relay catalytic hydrogenation.

is much faster than the Ru-catalyzed process, high levels of enantioselectivity are achieved for the entire relay catalytic hydrogenation reaction.

The relay catalytic systems listed before are typical representatives and actually provide the general reaction modes that are amenable to design new cascade processes. Since these events, hybrid multi-catalyst relay catalysis has gradually been blooming. Increasing attention has been paid to this field, leading to a large number of nonclassical asymmetric cascade transformations that will be highlighted and discussed in detail in the next chapters, in particular in Chapters 3, 4, and 6.

1.5 Conclusion

Since the advent of asymmetric metal catalysis, the incorporation of chiral ligands to the central metal has dominantly been the reliable strategy to address the stereoselective issue encountered in catalytic asymmetric reactions. The organo/metal combined catalysis changes and even actually refutes such conventional wisdom. The preponderance of activation modes and stereochemical options allocates unparalleled capacity to the organo/metal cooperative catalysis for the creation of asymmetric reactions. The sequential occurrence of multiply bond-breaking and forming events in stereoselective manner featured in relay catalysis propels the emergence of nonclassical cascade reactions. The last decades have indeed witnessed exciting progress in asymmetric organocatalysis combined with metal catalysis. It is not to overstate that the organo/metal combined catalysis is gradually altering the status quo of asymmetric catalysis, to some degree, and will continue to exert an essential impact on asymmetric catalysis.

References

1 (a) Ager, D. (ed.) (2005). *Handbook of Chiral Chemicals*, 2e. Boca Raton, FL: CRC Press, Taylor & Francis. (b) Christmann, M. and Bräse, S. (eds.) (2007). *Asymmetric Synthesis – The Essentials*. Weinheim: Wiley-VCH.

2 (a) Hanessian, S. (1983). *Total Synthesis of Natural Products: The Chiron Approach*. Oxford: Pergamon Press. (b) Roos, G. (ed.) (2014). *Key Chiral Auxiliary Applications*, 2e. Academic Press.

3 (a) Nozaki, H., Moriuti, S., Takaya, H. et al. (1966). *Tetrahedron Lett.* 7: 5239–5244. (b) Nozaki, H., Takaya, H., Moriuti, S. et al. (1968). *Tetrahedron* 24: 3655–3669.
4 (a) Knowles, W.S. and Sabacky, M.J. (1968). *J. Chem. Soc., Chem. Commun.*: 1445–1446. (b) Horner, L., Siegel, H., and Büthe, H. (1968). *Angew. Chem. Int. Ed.* 7: 942–942.
5 (a) Noyori, R. (1994). *Asymmetric Catalysis in the Organic Synthesis*. New York: Wiley. (b) Blaser, H.U. and Schmidt, E. (2004). *Asymmetric Catalysis on Industrial Scale: Challenges, Approaches and Solutions*. Weinheim: Wiley-VCH.
6 Jacobson, E.N., Pfaltz, A., and Yamamoto, H. (eds.) (1999). *Comprehensive Asymmetric Catalysis*. Heidelberg: Springer-Verlag.
7 (a) Sharpless, K.B. (1988). *J. Am. Chem. Soc.* 110: 1968–1970. (b) Sharpless, K.B. (1995). *Angew. Chem. Int. Ed. Engl.* 34: 1059–1070.
8 Zhou, Q.L. (ed.) (2011). *Privileged Ligands and Catalysts*. Weinheim: Wiley-VCH.
9 (a) McDonald, R.I., Liu, G.S., and Stahl, S.S. (2011). *Chem. Rev.* 111: 2981. (b) Newton, C.G., Wang, S.G., Oliveira, C.C. et al. (2017). *Chem. Rev.* 117: 8908. (c) Saint-Denis, T.G., Zhu, R.Y., Chen, G. et al. (2018). *Science* 359: 759. (d) Wang, P.S. and Gong, L.Z. (2020). *Acc. Chem. Res.* 53: 2841–2854.
10 (a)Yamamoto, H. (ed.) (2000). *Lewis Acids in Organic Synthesis*. Weinheim: Wiley-VCH. (b) Mlynarski, J. (ed.) (2017). *Chiral Lewis Acids in Organic Synthesis*. Weinheim: Wiley-VCH.
11 MacMillan, D.W.C. (2008). *Nature* 455: 304.
12 List, B. and Yang, J.W. (2006). *Science* 313: 1584.
13 (a) Berkessel, A. and Gröger, H. (2005). *Asymmetric Oranocatalysis – From Biomimetic Concepts to Application in Asymmetric Synthesis*. Weinheim: Wiley-VCH. (b) Dalko, P.I. (2007). *Enantioselective Organocatalysis: Reactions and Experimental Procedures*. Weinheim: Wiley-VCH.
14 (a) Shao, Z.H. and Zhang, H.B. (2009). *Chem. Soc. Rev.* 38: 2745. (b) Du, Z.T. and Shao, Z.H. (2013). *Chem. Soc. Rev.* 42: 1337. (c) Allen, A.E. and MacMillan, D.W.C. (2012). *Chem. Sci.* 3: 633.
15 (a) Chen, G.S., Deng, Y.J., Gong, L.Z. et al. (2001). *Tetrahedron-Asymmetry* 12: 1567–1571. (b) Nakoji, M., Kanayama, T., Okino, T. et al. (2001). *Org. Lett.* 3: 3329–3331.
16 Corey, E.J., Xu, F., and Noe, M.C. (1997). *J. Am. Chem. Soc.* 119: 12414–12415.
17 Lygo, B. and Wainwright, P.G. (1997). *Tetrahedron Lett.* 38: 8595–8598.
18 Jellerichs, B.G., Kong, J.R., and Krische, M.J. (2003). *J. Am. Chem. Soc.* 125: 7758–7759.
19 Rauhut, M.M. and Currier, H. (1963). Repreparation of dialkyl 2-methylense galutarates (American Cyanamid Co.). US Patent 3,074,999; Chem. Abstr. 58, 11224a.
20 France, S., Wack, H., Hafez, A.M. et al. (2002). *Org. Lett.* 4: 1603–1605.
21 Ibrahem, I. and Córdova, A. (2006). *Angew. Chem. Int. Ed.* 45: 1952–1956.
22 Komanduri, V. and Krische, M.J. (2006). *J. Am. Chem. Soc.* 128: 16448–16449.
23 Mukherjee, S. and List, B. (2007). *J. Am. Chem. Soc.* 129: 11336–11337.

24 Cardinal-David, B., Raup, D.E.A., and Scheidt, K.A. (2010). *J. Am. Chem. Soc.* 132: 5345–5347.

25 Arndtsen, B.A. and Gong, L.Z. (eds.) (2020). *Topics in Current Chemistry Collections: Asymmetric Organocatalysis Combined with Metal Catalysis*. Springer.

26 Chen, D.F., Han, Z.Y., Zhou, X.L. et al. (2014). *Acc. Chem. Res.* 47: 2365.

27 Han, Z.Y., Wang, C., and Gong, L.Z. (eds.) (2012). *Science of Synthesis: Asymmetric Organocatalysis*, (ed. K. Maruoka), vol. 2, 697. Stuttgart: Georg Thieme Verlag.

28 (a) Chen, F., Feng, X., Qin, B. et al. (2003). *Org. Lett.* 5: 949–952. (b) Chen, F.X., Zhou, H., Liu, X. et al. (2004). *Chem. Eur. J.* 10: 4790–4797.

29 (a) Bihelovic, F., Matovic, R., Vulovic, B. et al. (2007). *Org. Lett.* 9: 5063–5066. (b) Vulovic, B., Bihelovic, F., Matovic, R. et al. (2009). *Tetrahedron* 65: 10485–10494.

30 Li, C., Wang, C., Villa-Marcos, B. et al. (2008). *J. Am. Chem. Soc.* 130: 14450–14451.

31 Krautwald, S., Sarlah, D., Schafroth, M.A. et al. (2013). *Science* 340: 1065–1068.

32 (a) Cruz, F.A. and Dong, V.M. (2017). *J. Am. Chem. Soc.* 139: 1029–1032. (b) Jiang, X., Beiger, J.J., and Hartwig, J.F. (2017). *J. Am. Chem. Soc.* 139: 87–90. (c) Singha, S., Serrano, E., Mondal, S. et al. (2020). *Nat. Catal.* 3: 48–54.

33 (a) Mukherjee, S. and List, B. (2007). *J. Am. Chem. Soc.* 129: 11336–11337. (b) Jiang, G. and List, B. (2011). *Angew. Chem. Int. Ed.* 50: 9471–9474.

34 Murahashi, S., Makabe, Y., and Kunita, K. (1988). *J. Org. Chem.* 53: 4489–4495.

35 Wang, P.S., Lin, H.C., Zhai, Y.J. et al. (2014). *Angew. Chem. Int. Ed.* 53: 12218–12221.

36 Prier, C.K., Rankic, D.A., and MacMillan, D.W. (2013). *Chem. Rev.* 113: 5322–5363.

37 Nicewicz, D.A. and MacMillan, D.W. (2008). *Science* 322: 77–80.

38 Rono, L.J., Yayla, H.G., Wang, D.Y. et al. (2013). *J. Am. Chem. Soc.* 135: 17735–17738.

39 DiRocco, D.A. and Rovis, T. (2012). *J. Am. Chem. Soc.* 134: 8094.

40 Liu, Y.-Y., Liu, J., Lu, L.-Q. et al. (2019). *Top. Curr. Chem.* 377: 37.

41 Vora, H.U. and Rovis, T. (2007). *J. Am. Chem. Soc.* 129: 13796.

42 (a) Abillard, O. and Breit, O. (2007). *Adv. Synth. Catal.* 349: 1891. (b) Chercheja, S. and Eilbracht, P. (2007). *Adv. Synth. Catal.* 349: 1897.

43 Sorimachi, K. and Terada, M. (2008). *J. Am. Chem. Soc.* 130: 14452.

44 (a) Han, Z.Y., Xiao, H., Chen, X.H. et al. (2009). *J. Am. Chem. Soc.* 131: 9182–9183. (b) Liu, X.Y. and Che, C.M. (2009). *Org. Lett.* 11: 4204–4207.

45 Yang, T., Campbell, L., and Dixon, D.J. (2007). *J. Am. Chem. Soc.* 129: 12070–12071.

46 Cai, Q., Zhao, Z.A., and You, S.L. (2009). *Angew. Chem. Int. Ed.* 48: 7428–7431.

47 Chen, J.R., Li, C.F., An, X.L. et al. (2008). *Angew. Chem. Int. Ed.* 47: 2489–2492.

48 Simmons, B., Walji, A.M., and MacMillan, D.W.C. (2009). *Angew. Chem. Int. Ed.* 48: 4349–4353.

49 Chen, Q.A., Wang, D.S., Zhou, Y.G. et al. (2011). *J. Am. Chem. Soc.* 133: 6126–6129.

2

Metal/Phase-Transfer Catalyst Combined Catalysis

2.1 Introduction

Phase-transfer catalysis (PTC) has been recognized as one of the most useful methodologies in both academia and industry, as it features mild reaction conditions, simple procedures, low cost, environmentally benign reagents and solvents, and the practicability for large-scale synthesis [1–4]. In general, PTC accelerates a reaction by facilitating the migration of a reactant from one phase into another where the reaction occurs. As a simple and classic example of PTC, the reaction of 1-chlorooctane and aqueous sodium cyanide is shown in Scheme 2.1 [3]. Although heating the reaction mixture under reflux with vigorously stirring for one or two days gives no apparent product, the addition of a catalytic amount of phosphonium salt drastically accelerates the reaction to complete in several hours. In the catalytic process, phosphonium cation not only transfers the cyanide anion into the organic phase but also increases the nucleophilicity of the CN^- due to the large ionic radius of Q^+, which reduces the electrostatic interaction between the ion pair. Employing chiral ammonium or phosphonium salts allows for enantioselective transformations, including the industrial synthesis of amino acids via the asymmetric alkylation of glycine iminoesters with a chiral phase-transfer catalyst [5, 6].

Transition metal catalysts are crucial for modern chemistry considering their enormous diversity in lowering the activation barrier for chemical reactions and the huge number of reactions catalyzed by transition metal compounds. The combination of transition metal catalysis and PTC offers more opportunities and advantages compared to conventional catalytic strategies. In this chapter, we will briefly introduce the early stage of transition metal/PTC combined catalysis and focus on the asymmetric transformations enabled by the binary catalyst system.

2.1.1 Early Racemic Examples: PTC and Transition Metal Co-catalyzed Reactions

One of the great challenges in catalysis is to integrate the advantages of heterogeneous and homogeneous catalysis. The former features high efficiency, easy separation of the catalyst from the product, but normally suffers from unsatisfactory selectivity. The latter usually represents high selectivity and mild reaction

Asymmetric Organo-Metal Catalysis: Concepts, Principles, and Applications, First Edition. Liu-Zhu Gong.
© 2022 WILEY-VCH GmbH. Published 2022 by WILEY-VCH GmbH.

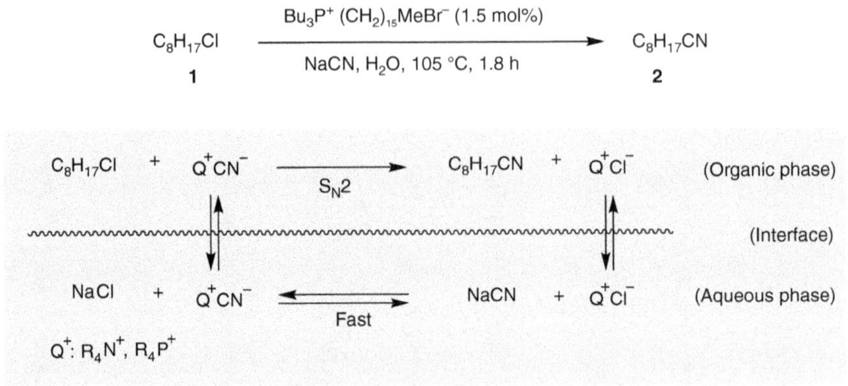

Scheme 2.1 Example of phase-transfer catalysis. MTBE, methyl *tert*-butyl ether; DCM, dichloromethane; TBAB, tetrabutylammonium bromide. Source: Modified from Starks [3].

conditions, but catalyst and product exist in the same phase and difficult to separate. In principle, a two-phase system might allow for good separation and integrates advantages of heterogeneous and homogeneous catalysis.

Initial attempts of applying PTC to transition metal-catalyzed reactions appeared in the 1970s. Alper [7] and Cassar [8] independently reported the catalytic carbonylation of benzyl bromide performed in a liquid–liquid biphase system and cooperatively catalyzed by a cobalt carbonyl and benzyltriethylammonium chloride. The reaction is believed to proceed following the sequence in Scheme 2.2. Cobalt tetracarbonyl anion **Int-1** is generated from dicobalt octacarbonyl and sodium hydroxide and transported into the organic phase by forming a lipophilic intermediate **Int-2** with the phase-transfer catalyst [9]. The reaction of the anion **Int-1** with benzyl bromide **3** gives an intermediate **Int-3**, which undergoes a ligand migration to form the acyl complex **Int-4** and a subsequent clearage of the cobalt carbon bond would afford the carboxylate anion **4** in the aqueous phase. The overall reaction features very mild conditions and a convenient work-up procedure. Besides carbonylation reactions, hydrogenation, coupling, and other reactions enabled by transition metal/PTC cooperative catalysis have also been reported [10–13].

2.2 Asymmetric Metal/Phase-Transfer Catalyst Combined Catalysis

The combination of PTC and transition metal catalysis offers more options and additional stereocontrol strategies in asymmetric catalysis considering that either or both of them can be chiral to impart asymmetric induction individually or synergistically. Depending on the type of phase-transfer catalyst employed, the PTC and metal combined catalysis can be categorized into metal and cationic PTC cooperative catalysis, metal and anionic PTC relay catalysis, PTC combined with reactive metallic anions, and metal combined with onium ligands (Scheme 2.3). In the metal and cationic PTC

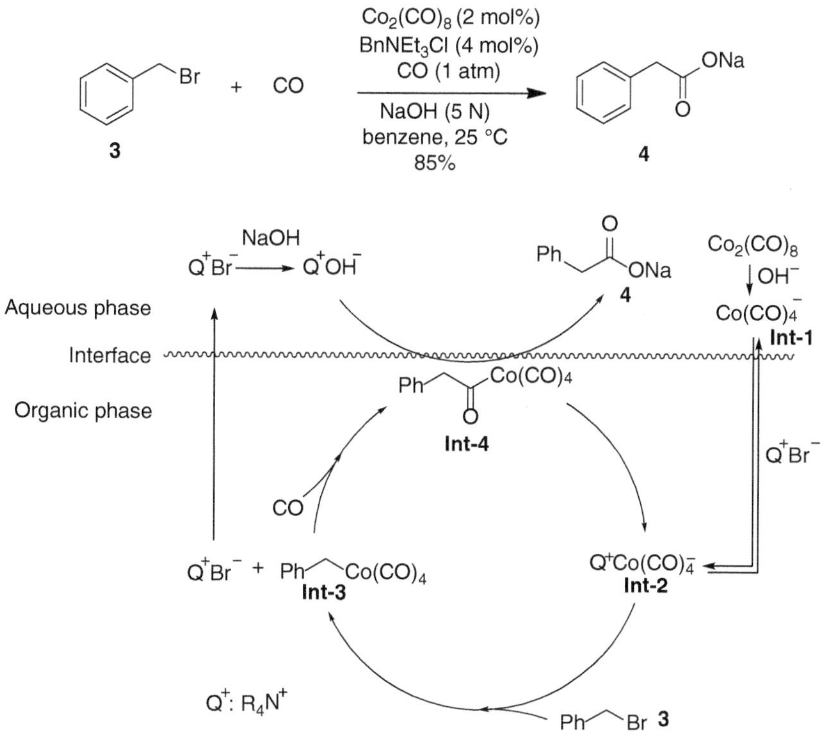

Scheme 2.2 Cobalt/PTC co-catalyzed carbonylation of benzyl halides.

cooperative catalysis, the nucleophile can be transformed into the reactive ion pair Nu^-Q^+ upon exposing to a PTC and a base, and then reacts with the other intermediate, generated from a metal-catalyzed elementary reaction (Scheme 2.3a). In the metal and anionic PTC relay catalysis, a chiral anion is typically employed to transfer the insoluble substrate into the solution as a chiral ion pair, which then participates in a metal-catalyzed process to form the final product (Scheme 2.3b). It is feasible to harness either chiral transition metal catalyst or chiral PTC to control the stereochemistry of these reactions. If two types of chiral catalysts are employed, the additional stereocontrol element is provided for the reaction, and thus a matched or mismatched effect should be observed. In the third model, the PTC catalyst is used to transfer a reactive metallic anion, such as permanganate, generating a chiral ion-pair complex (Scheme 2.3c). The last strategy applies chiral bifunctional ligands bearing onium ion moieties for metal-catalyzed transformations (Scheme 2.3d). Besides chiral skeletons, these bifunctional ligands could also utilize external chiral counter anions to control stereochemistry. In nature, the validity of the last two combined catalysis concepts relies on the chiral ion-paired complex to act as a real catalyst (Scheme 2.3c,d). In this regard, these strategies could also be considered as sole catalyst promotion. As such, these two concepts will be mentioned very briefly at the end of this chapter.

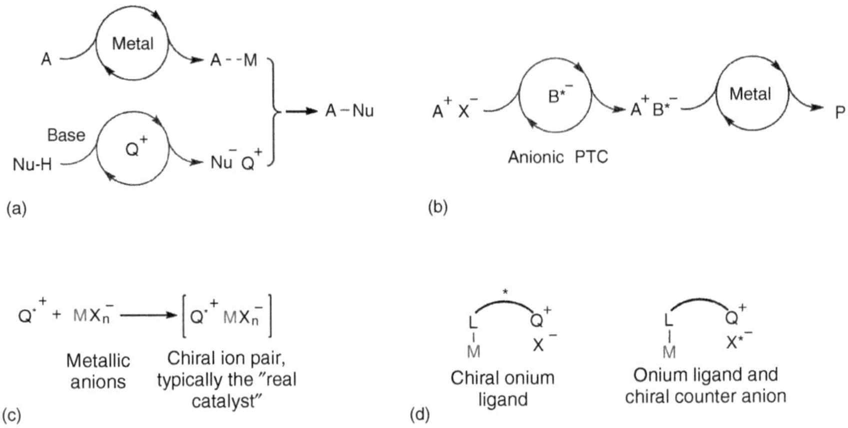

Scheme 2.3 Strategies of metal/PTC combined catalysis. (a) Metal/PTC combined catalysis. (b) Metal/anionic PTC combined catalysis. (c) PTC and reactive metallic anions. (d) Metal and onium ligand.

2.2.1 Combination of Cationic PTC and Transition Metal in Asymmetric Catalysis

In 2001, the first examples of PTC/metal co-catalyzed asymmetric reaction was established independently by Gong and Mi [14] and Takemoto's group [15]. They successively developed an enantioselective allylic alkylation of iminoesters **5** with simple allyl esters **6** by the cooperative catalysis of chiral PTC and palladium complex. Gong and Mi identified a cinchona alkaloid-derived ammonium bromide **PTC-1** as the chiral phase-transfer catalyst, in combination with an achiral palladium complex of Ph_3P, to enable the reaction delivering 59% ee (Scheme 2.4). Either (*S*)- or (*R*)-2,2′-bis(diphenylphosphino)-1,1′-binaphthyl (BINAP) was investigated in combination with **PTC-1** to reveal if the synergistic effect existed on the stereochemical control. The use of (*S*)-BINAP as the ligand with **PTC-1** gave the product with only 20% ee and with the (*R*)-enantiomer predominating. In contrast, the combination of (*R*)-BINAP with PTC-**1** allowed the reaction to give 60% ee, but the (*S*)-enantiomer predominated, the similar yield and ee were also obtained from the reaction with PTC **1** in the presence of PPh_3. These results indicate that (*S*)-BINAP and PTC-**1** are mismatched but no synergistic stereochemical control occurs even if the matched chiral ligand is used.

A significant improvement in the stereoselectivity of the Pd/chiral PTC cooperatively catalyzed allylic allylation of iminoesters was achieved by Takemoto and coworkers (Scheme 2.5) [15]. The electron density of the trivalent phosphorus ligand exerts a considerable impact on the reaction performance. Electronically rich phosphine ligands, such as 1,2-bis(diphenylphosphino)ethane (DPPE) and tributylphosphine, enable the reaction to give higher yield but lower enantioselectivity than relatively electronically poor ones, such as triphenylphosphine and triphenyl phosphite. In particular, the highest enantioselectivity was obtained in the presence of triphenyl phosphite under identical conditions, albeit with the sacrifice of the catalytic efficiency. Ultimately, the combination of the palladium complex

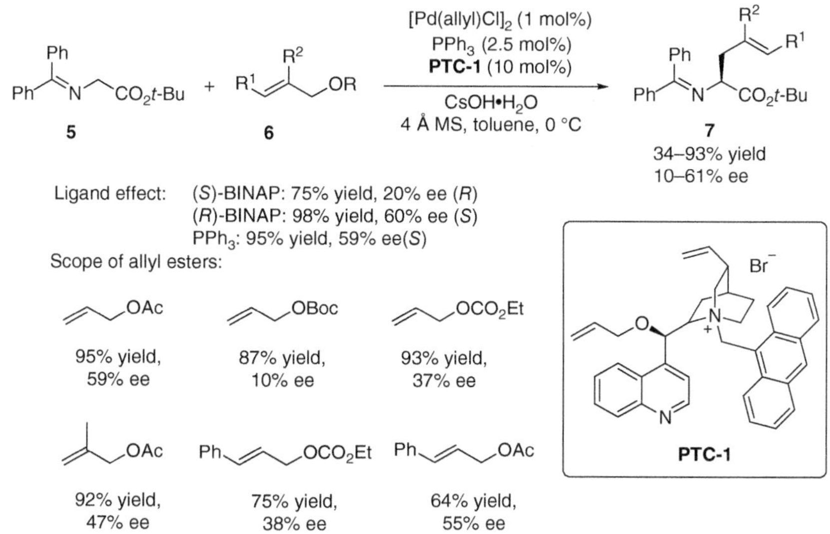

Scheme 2.4 Pd/chiral PTC catalyzed asymmetric allylation of iminoesters.

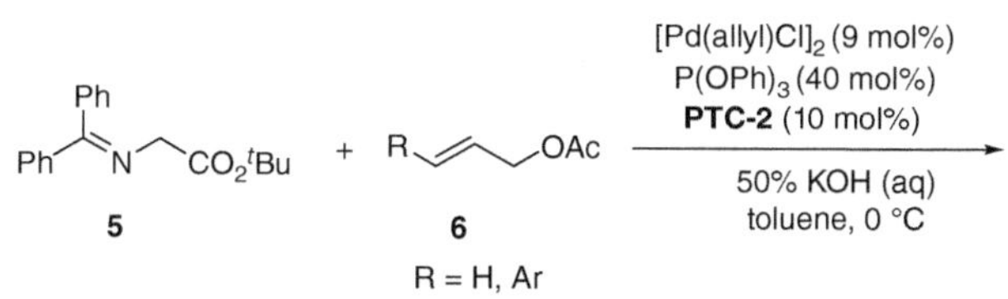

Effect of the achiral phosphine ligand (using KOH solid as the base)

Entry	Ligand (mol %)	Yield (%)	ee (%)
1	DPPE (8)	74	24
2	n-Bu_3P (16)	69	4
3	Ph_3P (16)	32	59
4	$(PhO)_3P$ (16)	19	82

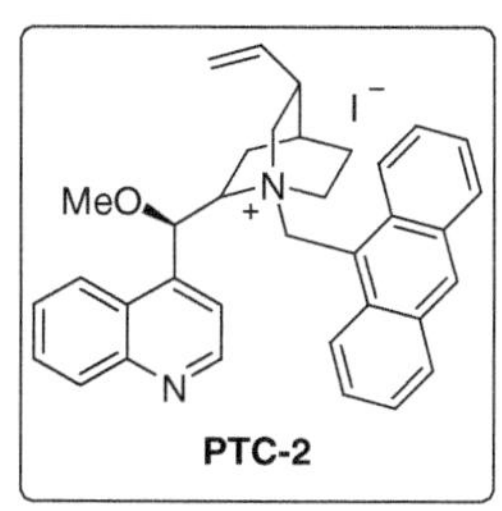

Representative products:

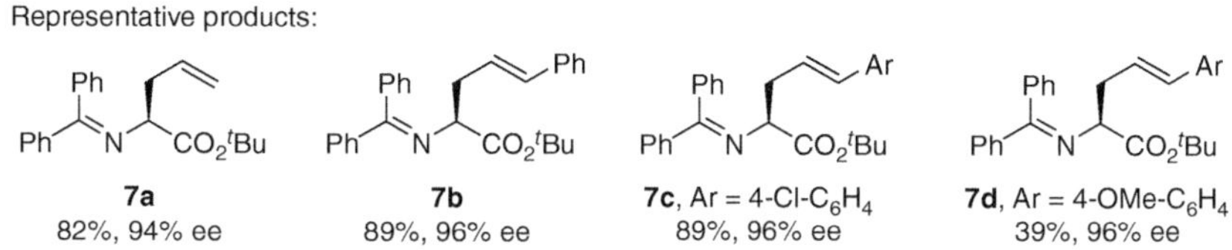

Scheme 2.5 Asymmetric allylic allylation of iminoesters catalyzed by chiral PTC and palladium complex of triphenyl phosphite.

of triphenyl phosphite and *O*-methyl chinconidinium **PTC-2** appeared to be the best catalyst system and delivered good yields and excellent enantioselectivities. Employing 50% KOH in an aqueous solution to supercede KOH solid rendered the reaction to offer higher enantioselectivity. In addition, the asymmetric allylation reaction with racemic branched allyl ester **8** gave the same linear-selective product with almost identical enantioselectivity (93% ee).

The palladium-catalyzed asymmetric allylic alkylation of racemic **8** was also investigated in presence of triphenyl phosphite and MOP, respectively, by using **PTC-2** as the co-catalyst (Scheme 2.6a) [16]. The palladium complex of triphenyl phosphite favored the generation of linear product **7b** but still gave considerable amounts of branched selectivity. As Hayashi reported, the MeO–MOP was convinced to promote an inner attack of nucleophiles to the π-allylpalladium(II) complex generated from **8**, producing the terminal alkene **9** as the major product [17]. However, in contrast to Hayashi's observation, the use of MOP to replace triphenyl phosphite led to the predominant generation of linear product **7b** in 70% yield and with 90% ee, accompanying with small amounts of branched product **9** with 85% ee. From the mechanistic point of view, the π-allylpalladium(II) complex **Int-5** couples with a chiral ion pair formed from the iminoester **5** and **PTC-2** to give product **7**, stereoselectively

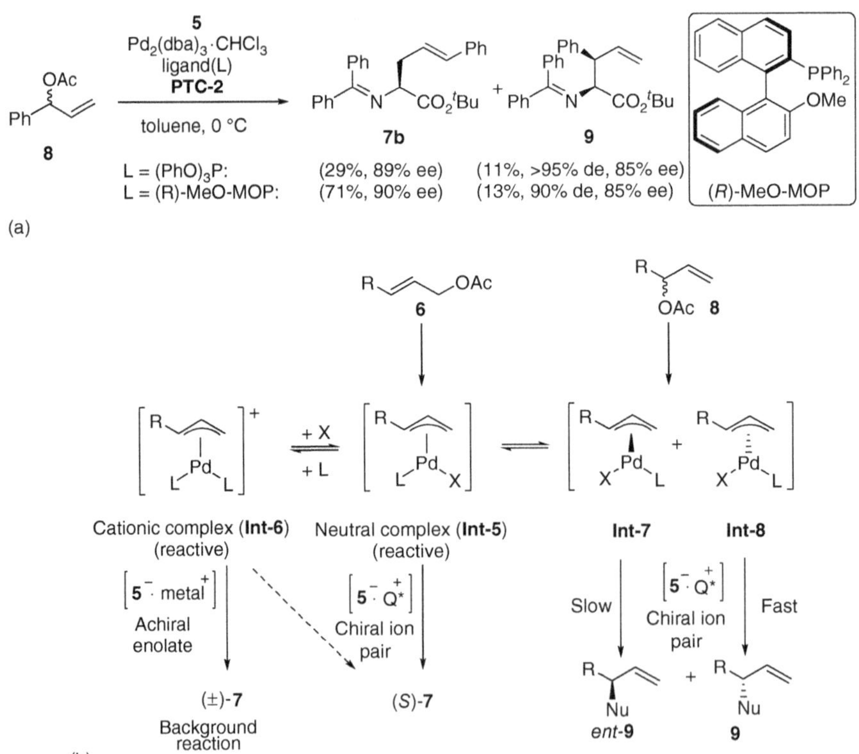

Scheme 2.6 Mechanism studies on the chiral PTC/Pd co-catalyzed asymmetric allylation of iminoesters. Source: Modified from Nakoji et al. [16].

(Scheme 2.6b). The cationic π-allylpalladium complex is more reactive than the neutral complex **Int-5** [18]. In the cases involving more σ-donating ligands, such as DPPE and n-Bu_3P, more active complex **Int-6** might be formed dominantly, which would be able to react with the achiral enolate and to give the racemic product. In contrast, when the more π-accepting ligand such as $(PhO)_3P$ is used, complex **A** is formed exclusively and the racemic background reaction can be significantly inhibited. On the other hand, the branched allyl ester **8** undergoes oxidative addition with Pd to initially generate π-allylpalladium complexes **Int-7** and **Int-8**, respectively. The stronger trans influence of $(PhO)_3P$ and MOP principally enables the nucleophilic attack to favorably occur on the carbon trans to the ligand, yielding branched product. However, the nucleophilic attack is still a rather slow process due to the severe steric hindrance, and the π-allylpalladium complexes **Int-7** and **Int-8** could isomerize rapidly to complex **Int-5** via exchange of the coordination site of L and X, leading to the formation of the linear product as the major product. Regardless of either pathway via that the allylic alkylation event proceeds, the chiral ion pair generated from **5** and the **PTC-2** can effectively control the stereochemistry to give the product **7b** or **9** with high stereoselectivity.

Although palladium catalyst prefers the linear-selective allylic alkylation product, other transition metals, including Ir [19–22], Mo [23, 24], and W [25, 26] as well tend to drive the allylic alkylation to occur at the more substituted terminus of the allylic substrate to give the branched product. In 2003, Takemoto and coworkers reported their endeavors to investigate the iridium complex/PTC cooperatively catalyzed asymmetric allylic alkylation of iminoesters [27]. The combination **PTC-2** and iridium complex of triphenyl phosphite exclusively provided the branched products **9′** and **9** even if the linear cinnamyl alcohol derived allyl esters were employed as the substrates. However, the yield and stereoselectivity of **9′** remained moderate regardless of alternating the leaving group of allylic substrate. On the other hand, the use of bidentate chiral ligand **L1**, without the addition of PTC, considerably enhanced the enantioselectivity of the major product to up to 97% ee (Scheme 2.7).

Umpolung [28] addition of imines, in which imines serve as nucleophiles, offers a unique approach to chiral amines [29–31]. In 2015, Deng's group successfully realized the first asymmetric umpolung addition of imines with various Michael acceptors using chiral ammonium salt PTC [32]. Inspired by this work, Han and coworkers established an Ir/PTC cooperatively catalyzed asymmetric umpolung allylic substitution of simple α-iminoesters with allyl acetates, providing facile access to chiral α-quaternary amino acid derivatives bearing two vicinal stereocenters (Scheme 2.8) [33]. 1,1′-bi-2-naphthol (BINOL)-based phosphite **L1** with a sulfur atom at the alkyl chain, developed by Takemoto's group, proved to be the optimal ligand for the reaction. The existence of a PTC as the co-catalyst was found to be crucial for the reaction. Through a thorough screening of the PTC, including a number of chiral PTCs such as **PTC-4** and **PTC-5**, achiral tetrabutylammonium bromide (**PTC-3**) was found to be the best option. Without the phase-transfer catalyst, the yield of the reaction decreased severely, and diminished enantioselectivity was observed. The reaction generally offers good yields, high diastereo- and enantioselectivities employing aromatic allyl acetates. Alkyl allyl acetate,

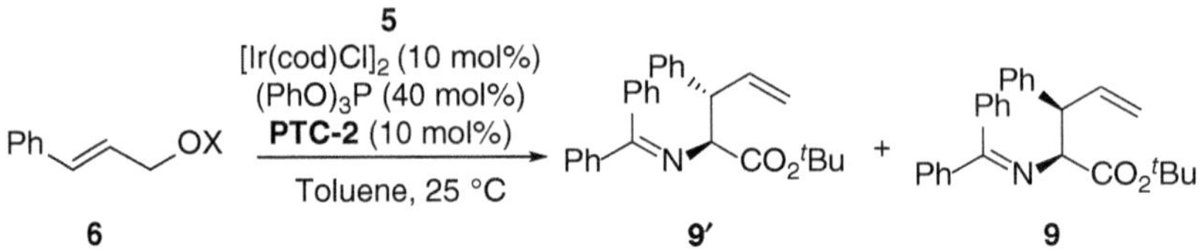

X	Yield (%)	dr (**9′/9**)	ee of **9′**(%)
Ac (**6a**)	31	77/23	49
Bz (**6b**)	40	75/25	46
CO_2Me (**6c**)	15	80/20	29
CO_2Ph (**6d**)	23	83/17	57
$PO(OEt)_2$(**6d**)	47	83/17	26

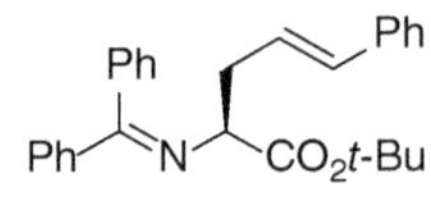

7b, linear product not observed

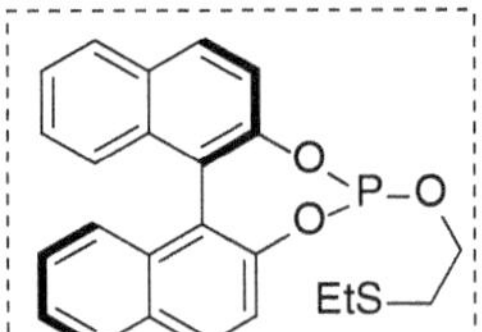

L1

Scheme 2.7 Chiral PTC/achiral iridium co-catalyzed asymmetric allylation of iminoesters.

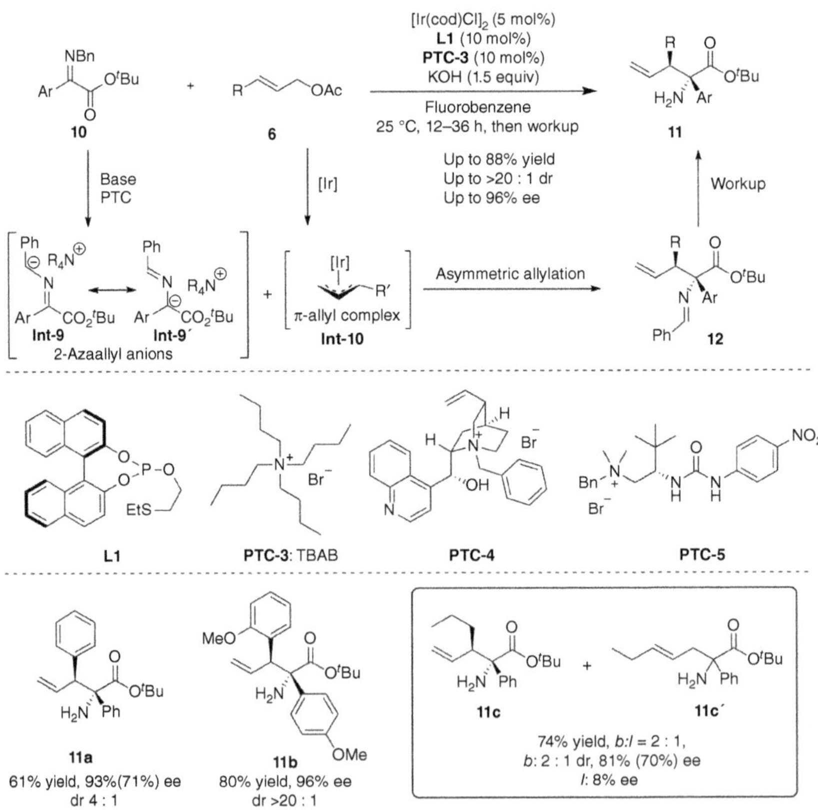

Scheme 2.8 Ir/PTC cooperatively catalyzed asymmetric umpolung allylic substitution of simple α-iminoesters with allyl acetates. Source: Modified from Su et al. [33].

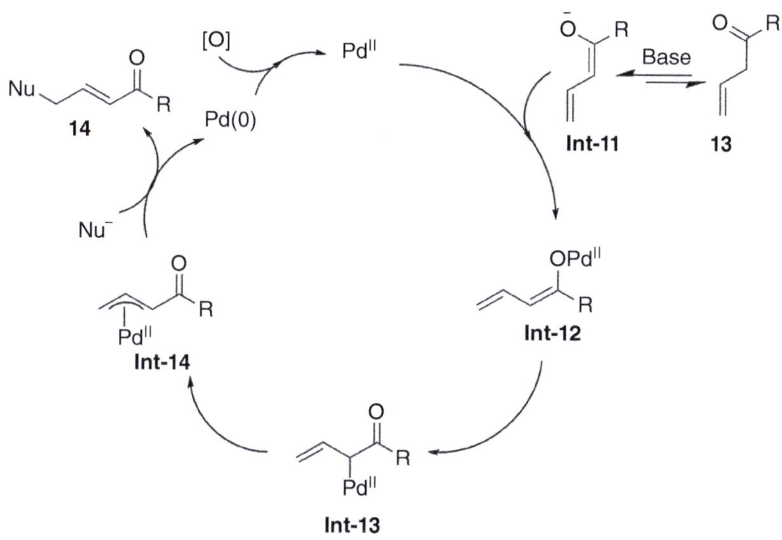

Scheme 2.9 Pd(II)-mediated electrophilic vinylogous reaction.

(*E*)-pent-2-en-1-yl acetate, delivers the desired amino acid derivative with a 2 : 1 b:l ratio and with high ee for the major product.

Stahl found that the enolizable ketone can undergo a Pd(II)-catalyzed dehydrogenation to generate the corresponding α,β-enone by forming Pd(II)-enolate species in the presence of a base and then undergoing the subsequent β-hydride elimination [34]. Chen and coworkers proposed that a similar transmetalation between palladium acetate and dienolate **Int-11** generated from the deprotonation of allyl ketone **13** with a base gave an intermediate **Int-12**. The intermediate **Int-12** then undergoes 1,3-Pd migration to form η^1-allyl palladium **Int-13**, which then tautomerizes into a π-allyl palladium intermediate **Int-14**. The allylic substitution of the intermediate **Int-14** with a nucleophile would occur in a direct vinylogous umpolung manner to yield **14** (Scheme 2.9).

Based on this mechanistic hypothesis, Chen and coworkers established a highly enantioselective oxidative coupling between oxindoles **15** and allyl ketones **13** enabled by the cooperative catalysis of palladium acetate and chiral PTC (Scheme 2.10) [35]. Either L-amino-acid-derived phosphonium salt **PTC-6** or **PTC-7** was identified as the best co-catalyst to offer the highest enantioselectivity. A broad range of 3-substituted oxindoles underwent the reaction smoothly, generally giving the multifunctional products bearing an all-carbon quaternary stereocenter with excellent enantioselectivities. 3-Indolinone-2-carboxylic and 2-cyclopentanonecarboxylic esters were also suitable for the reaction, leading to the products **16c** and **16d** with moderate and high enantioselectivities offered by **PTC-6** and **PTC-7**, respectively. Dienol silyl ether **18**, holding similar reactivity to the dienolate, is also able to undergo oxidative allylic alkylation reaction in lieu of the allyl ketone **13** to provide **19** with excellent enantioselectivity. The chiral anion pair **Int-15** generated from **15** and the phase-transfer catalyst is believed to undergo

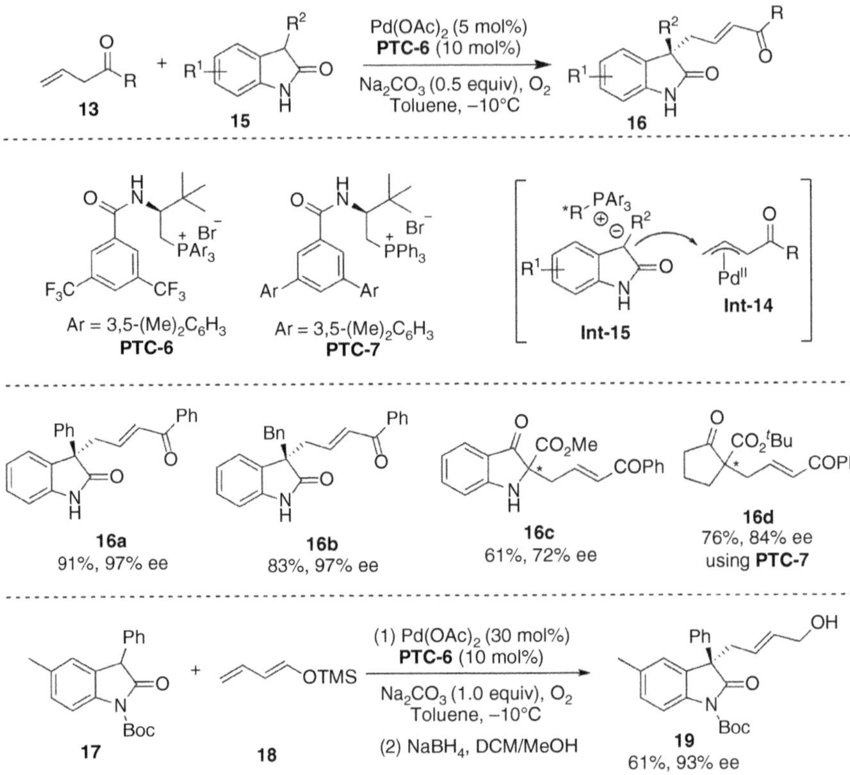

Scheme 2.10 Asymmetric vinylogous umpolung reaction. Source: Modified from Ran et al. [35].

the enantioselective coupling with the π-allyl palladium intermediate **Int-14** to yield the chiral products **16**.

Either ammonium or phosphonium salt PTC-catalyzed enantioselective transformations usually utilize enolizable pronucleophiles under basic conditions and control the stereochemistry via chiral ion pairs. Only a limited number of examples describe nonenolizable, but still stabilized anionic carbon nucleophiles in asymmetric phase-transfer catalysis [36–39]. Recently, Maruoka and coworkers developed an enantioselective alkynylation of isatin derivatives using chiral ammonium bromide **PTC-8** and silver acetate binary catalyst system (Scheme 2.11) [40]. The complexation of AgOAc and the chiral quaternary ammonium salt **PTC-8** generates a chiral ion pair **23**, which can form a chiral silver alkynylide **24** with alkyne **20** in the presence of a base. The chiral ion pair **24** then reacts with isatin **21** to give **25**, which finally undergoes protonation to yield product **22.**

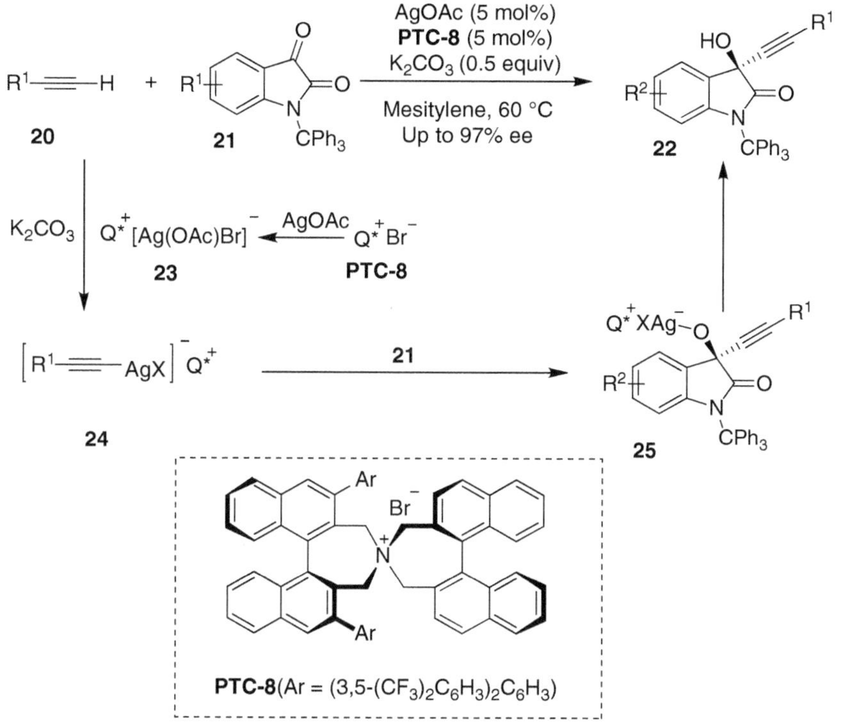

Scheme 2.11 PTC/Ag(I) cooperatively catalyzed enantioselective alkynylation of isatin derivatives.

2.2.2 Combination of Anionic PTC and Transition Metal in Asymmetric Catalysis

In contrast to cationic PTC, anionic PTCs can transfer cations into the reaction phase. As early as the 1980s, tetraphenylborate anion and its sderivatives were identified to transfer cations in several reactions. For example, tetrakis[3,5-bis(trifluoromethyl)phenyl] borate was reported to catalyze the diazotization of anilines in a dichloromethane/aqueous sulfuric acid two-phase system [41, 42]. The resulting diazonium group could further undergo one-pot transformations, such as bromination and hydroxylation reactions. However, the establishment of asymmetric anionic PTC did not have much success until the 2010s [43]. In 2011, Toste reported an asymmetric electrophilic fluorination reaction using an insoluble cationic electrophilic fluorination reagent (Selectfluor) with BINOL-derived phosphate as the chiral anionic phase-transfer catalyst [44]. Since then, chiral anionic PTC has enabled a number of highly enantioselective transformations [45–47].

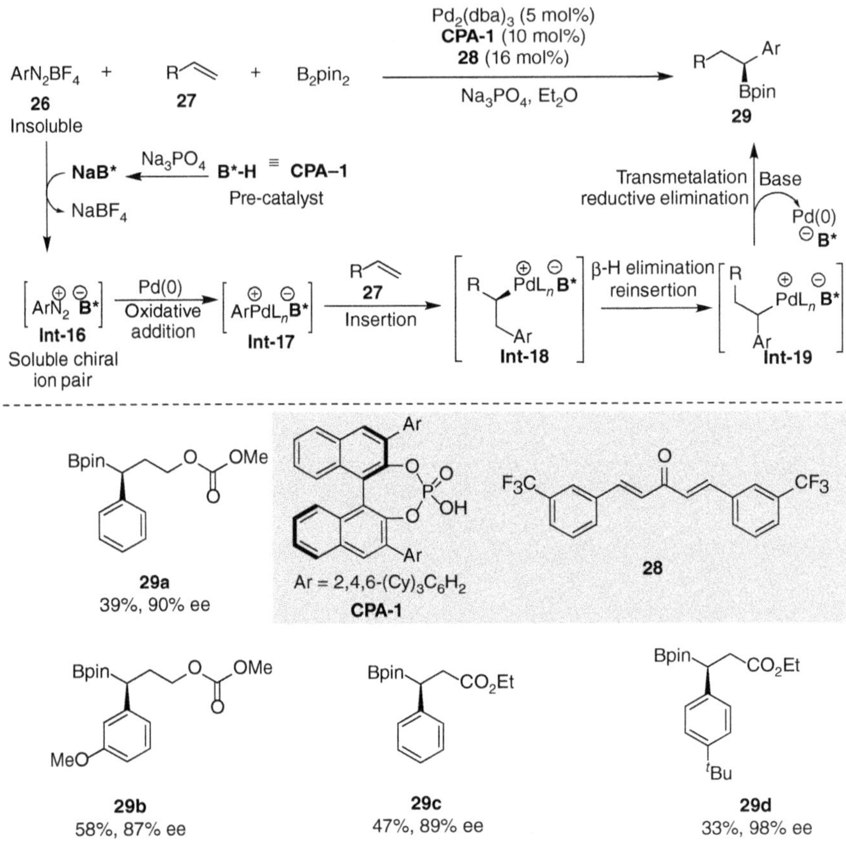

Scheme 2.12 Enantioselective 1,1-arylborylation of alkenes.

In 2015, Toste reported an enantioselective 1,1-arylborylation of alkenes rendered by chiral anionic PTC and palladium cooperative catalysis (Scheme 2.12) [48]. By employing chiral phosphoric acid (**CPA)-1** to generate the anionic PTC and **28** as a ligand for the palladium catalyst, a three-component reaction of aryl diazoniums **26**, alkenes **27**, and pinacol diborate provides chiral benzylic boronates **29** with high to excellent enantioselectivities. In this reaction, the insoluble aryldiazonium salt **26** undergoes a metathesis reaction with sodium phosphate to provide soluble chiral ion pair **Int-16**, which upon undergoing oxidative addition could generate Pd(II) species **Int-17** bearing a chiral counter anion. The complex **Int-17** would undergo alkene insertion to give a complex **Int-18**, which could be transformed into a complex **Int-19** through a β-H elimination/reinsertion sequence. After transmetalation and reductive elimination, **Int-19** could be converted into the chiral borons **29** and release both catalysts.

Inspired by Toste's work, Gong and coworkers developed a highly selective multicomponent carbonyl allylation reaction of 1,3-butadienes, aryldiazonium tetrafluoroborates, and aldehydes (Scheme 2.13) [49]. The Pd(II) complex **Int-17**, generated from the oxidative addition of Pd(0) complex to the soluble chiral iron

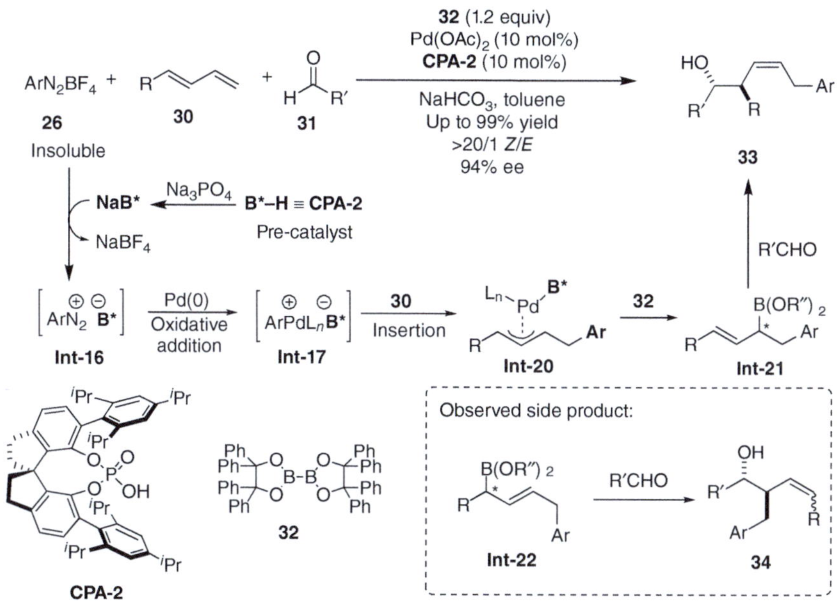

Scheme 2.13 Enantioselective multicomponent carbonyl allylation reaction of 1,3-butadienes, aryldiazonium tetrafluoroborates, and aldehydes.

pair **Int-16**, undergoes an insertion reaction with diene **30** to give a chiral π-allyl Pd complex **Int-20**, which reacts with diboron **26** to give chiral allylboron **Int-21**. **Int-21** would undergo allylation reaction with aldehydes to give enantioenriched homoallylic alcohol **33**. A side product **34**, probably generated from allylboron **Int-22**, was also observed. The 1,1′-spirobiindane-7,7′-diol (SPINOL)-derived phosphoric acid (**CPA-2**) offered the highest enantioselectivity and the use of bulky diboron **32** was the key to inhibit the generation of **34** and provided almost perfect *Z*/*E* selectivity [50].

By virtue of the similar strategy, the same group established an enantioselective three-component carbonyl allylation reaction of 1,3-butadienes, alkynyl bromides, and aldehydes to produce densely functionalized homoallylic alcohols (Scheme 2.14) [51]. Chiral silver phosphate **37** was identified as the best anion PTC and appeared to be slightly more effective for the stereocontrol than the corresponding phosphoric acid. The combination of **37** and palladium acetate, together with the presence of bulky diboron **32** allowed the multicomponent reaction to give excellent *Z*/*E*- and enantioselectivity. In this process, the formation of the chiral Pd(II) species **Int-23** through a metathesis process driven by the precipitation of AgBr is believed to be the key intermediate that is responsible for the stereocontrol of the whole reaction. As such, the use of Ag_2CO_3 is crucial for the reaction to proceed and also for the regeneration of the anionic chiral PTC.

Toste and coworkers created an enantioselective 1,1-diarylation of acrylates **27** by means of Pd and anionic PTC cooperative catalysis (Scheme 2.15) [52]. The reaction proceeds through a pathway similar to the enantioselective alkene

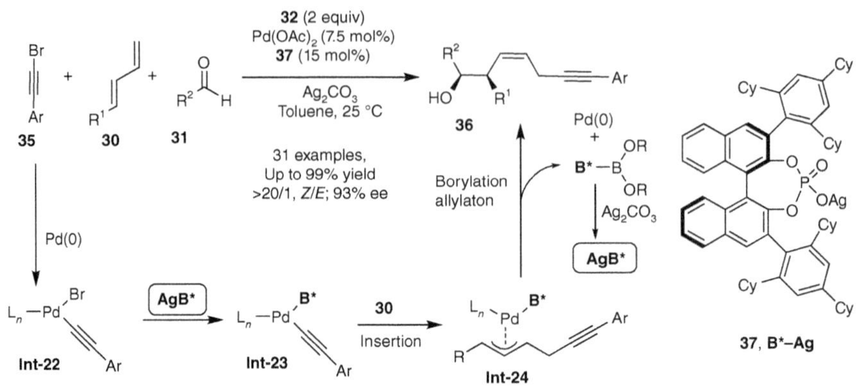

Scheme 2.14 Enantioselective three-component carbonyl allylation reaction of 1,3-butadienes, alkynyl bromides, and aldehydes.

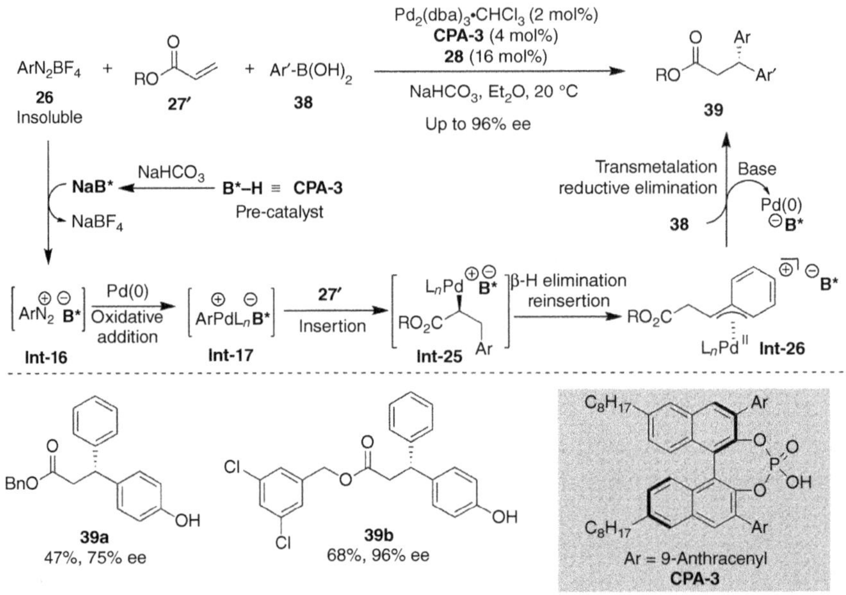

Scheme 2.15 Enantioselective 1,1-diarylation of acrylates. Source: Modified from Yamamoto et al. [52].

1,1-arylborylation reaction described by the same group [48]. After the aryl palladium phosphate **Int-17** undergoes a Heck-type insertion reaction with the acrylate and followed by the β-elimination and reinsertion sequential process to generate a key intermediate **Int-26** stabilized as a π-benzyl species. The intermediate **Int-26** undergoes transmetalation with arylboronic acid and reductive elimination to generate **39. CPA-3** appeared to be the superior catalyst and allows the reaction to give excellent enantioselectivity. However, the enantioselectivity is highly sensitive to the electronic and steric nature as well as the position of substituents on the benzyl acrylate substrate. The use of a multidimensional modeling technique to investigate the effect of the chiral anionic phase-transfer (CAPT) catalyst and the acrylate suggests that the attractive noncovalent interactions between the two components are controlling elements in the enantiodetermining step.

The groups of Toste and Sunoj developed an asymmetric Heck–Matsuda reaction of cyclic alkenes and aryl diazonium salts by taking advantage of palladium and anionic PTC cooperative catalysis (Scheme 2.16) [53, 54]. The catalytic cycle includes the formation of ion-pair **Int-16** through the anion transfer process, oxidative addition with the Pd(0) catalyst to give aryl palladium species **Int-17**, migratory insertion to the cyclic alkene followed by β-H elimination to afford the final product. For some cyclic alkenes, a considerable amount of byproduct caused by double bond isomerization of the substrate could be observed when **CPA-4** was applied. BINAM-derived phosphoric acids (BDPAs), such as **CPA-5** and **CPA-6**, were found to be capable of inhibiting the formation of the byproduct. This work represents the first example of chiral counterion-controlled enantioselective Heck reactions and offers an alternative to those employing chiral ligands to control the stereochemistry. A comprehensive density functional theory (DFT) study indicated that the critical step controlling the enantioselectivity should be the migratory insertion of the Pd-aryl intermediate bearing a chiral counter anion to the cyclic alkene [54]. A modified structure of the intermediate **Int-17′** was suggested based on the calculation results, in which the chiral catalyst BDPA is dispositioned as a counterion bound to sodium bicarbonate, rather than the palladium.

2.3 Conclusion

As highlighted in this chapter, the progress in the transition metal/PTC combined catalysis, especially the asymmetric variants, has illustrated attractive characteristics and potential of this concept. The incorporation of PTCs, either cationic or anionic, into transition metal catalysis not only facilitates the transformations but also offers a supplementary and reliable option for the stereocontrol. In addition, chiral Lewis acid/onium bifunctional catalysts [55–60], chiral ammonium/phosphine hybrid ligands [61–67], and the combination of PTC with metallic anions [68–72], which act closely related to metal/PTC cooperative catalysis in terms of fundamental principle, have emerged for the creation of the new transformations. Despite the remarkable achievements made in the last two decades, when compared to other subfields of metal/organo combined catalysis, the concept of asymmetric metal/PTC combined catalysis, so far is still in its infancy stage. Formidable challenges, as well as opportunities on the other hand, remain in this area. First, metals in the binary system for asymmetric catalysis are limited to palladium and iridium, and the combination of PTC with other metals, such as rhodium and nickel, remains unexplored. Second, the metal/PTC combined catalysis in the relay or sequential manner remains undeveloped, although some effects have been made toward this field [73]. Finally, besides simple cations and anions, PTCs are also known to transfer many other

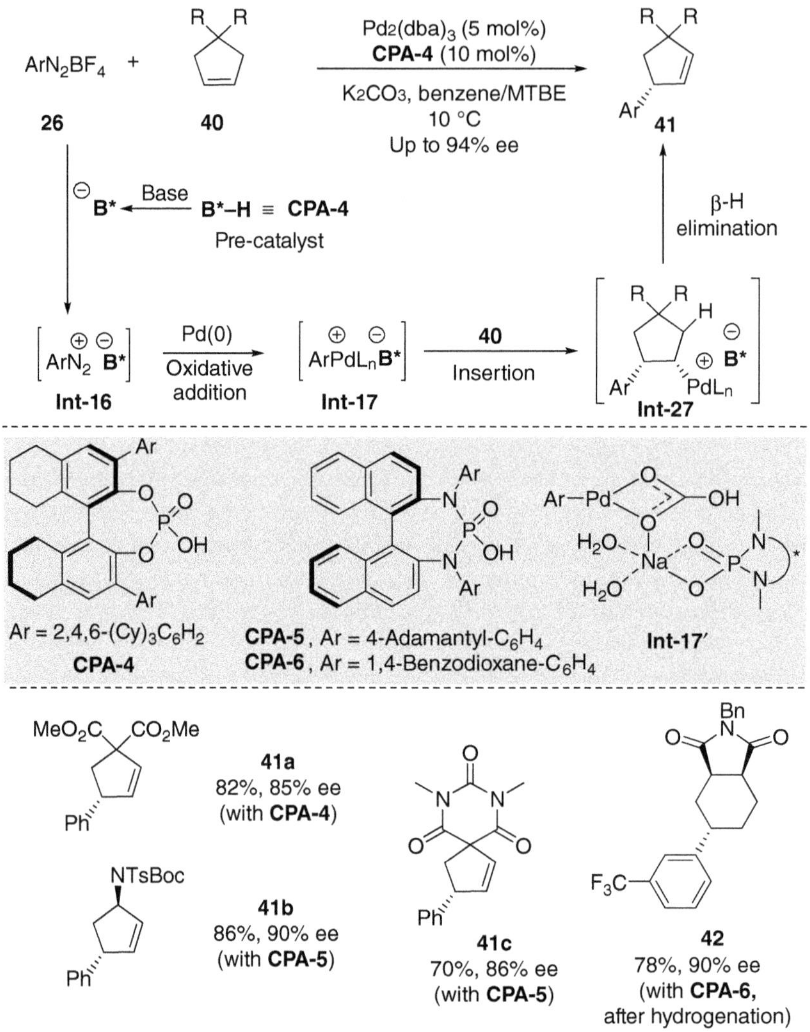

Scheme 2.16 Pd/anionic PTC cooperatively catalyzed asymmetric Heck–Matsuda reaction of cyclic alkenes and aryl diazonium salts. Sources: Avila et al. [53], Reddi et al. [54].

species such as acids [74], formaldehyde [75], radical anions [76], and ammonia [77]. These features actually offer additional dimensions for the development of asymmetric transformations by virtue of metal and PTC cooperative catalysis.

References

1 Makosza, M. (2000). *Pure Appl. Chem.* 72: 1399–1403.
2 Makosza, M. (1966). *Tetrahedron Lett.* 7: 4621–4624.
3 Starks, C.M. (1971). *J. Am. Chem. Soc.* 93: 195–199.

4 Starks, C.M., Liotta, C.L., and Halpern, M.E. (1994). *Phase-Transfer Catalysis*. New York: Chapman & Hall.
5 Ooi, T. and Maruoka, K. (2007). *Angew. Chem. Int. Ed.* 46: 4222–4266.
6 Shirakawa, S. and Maruoka, K. (2013). *Angew. Chem. Int. Ed.* 52: 4312–4348.
7 Alper, H. and Abbayes, H.D. (1977). *J. Organomet. Chem.* 134: C11–C14.
8 Cassar, L. and Foa, M. (1977). *J. Organomet. Chem.* 134: C15–C16.
9 Alper, H., Des Abbayes, H., and Des Roches, D. (1976). *J. Organomet. Chem.* 121: C31–C34.
10 Bar, R., Sasson, Y., and Blum, J. (1982). *J. Mol. Catal.* 16: 175–180.
11 Zoran, A., Sasson, Y., and Blum, J. (1984). *J. Mol. Catal.* 27: 349–353.
12 Jeffery, T. (1985). *Tetrahedron Lett.* 26: 2667–2670.
13 Li, P. and Alper, H. (1986). *J. Org. Chem.* 51: 4354–4356.
14 Chen, G.S., Deng, Y.J., Gong, L.Z. et al. (2001). *Tetrahedron: Asymmetry* 12: 1567–1571.
15 Nakoji, M., Kanayama, T., Okino, T., and Takemoto, Y. (2001). *Org. Lett.* 3: 3329–3331.
16 Nakoji, M., Kanayama, T., Okino, T., and Takemoto, Y. (2002). *J. Org. Chem.* 67: 7418–7423.
17 Hayashi, T., Kawatsura, M., and Uozumi, Y. (1998). *J. Am. Chem. Soc.* 120: 1681–1687.
18 Frost, C.G., Howarth, J., and Williams, J.M.J. (1992). *Tetrahedron: Asymmetry* 3: 1089–1122.
19 Janssen, J.P. and Helmchen, G. (1997). *Tetrahedron Lett.* 38: 8025–8026.
20 Fuji, K., Kinoshita, N., Kawabata, T., and Tanaka, K. (1999). *Chem. Commun.*: 2289–2290.
21 Qu, J. and Helmchen, G. (2017). *Acc. Chem. Res.* 50: 2539–2555.
22 Cheng, Q., Tu, H.F., Zheng, C. et al. (2019). *Chem. Rev.* 119: 1855–1969.
23 Trost, B.M., Hildbrand, S., and Dogra, K. (1999). *J. Am. Chem. Soc.* 121: 10416–10417.
24 Larhed, M., Kaiser, N.F.K., Bremberg, U. et al. (2000). *Angew. Chem. Int. Ed.* 39: 3596–3598.
25 Lloyd-Jones, G.C. and Pfaltz, A. (1995). *Angew. Chem. Int. Ed.* 34: 462–464.
26 Lehmann, J. and Lloyd-Jones, G.C. (1995). *Tetrahedron* 51: 8863–8874.
27 Kanayama, T., Yoshida, K., Miyabe, H. et al. (2003). *J. Org. Chem.* 68: 6197–6201.
28 Seebach, D. and Enders, D. (1975). *Angew. Chem. Int. Ed.* 14: 15–32.
29 Guo, C., Song, J., and Gong, L.Z. (2013). *Org. Lett.* 15: 2676–2679.
30 Tian, L., Hu, X.-Q., Li, Y.-H., and Xu, P.-F. (2013). *Chem. Commun.* 49: 7213–7215.
31 Waser, M. and Novacek, J. (2015). *Angew. Chem. Int. Ed.* 54: 14228–14231.
32 Wu, Y., Hu, L., Li, Z., and Deng, L. (2015). *Nature* 523: 445–450.
33 Su, Y.-L., Li, Y.-H., Chen, Y.-G., and Han, Z.-Y. (2017). *Chem. Commun.* 53: 1985–1988.
34 Diao, T.N. and Stahl, S.S. (2011). *J. Am. Chem. Soc.* 133: 14566–14569.
35 Ran, G.Y., Yang, X.X., Yue, J.F. et al. (2019). *Angew. Chem. Int. Ed.* 58: 9210–9214.

36 Ooi, T., Takahashi, M., Doda, K., and Maruoka, K. (2002). *J. Am. Chem. Soc.* 124: 7640–7641.
37 Liu, X., Xu, C., Wang, M., and Liu, Q. (2015). *Chem. Rev.* 115: 683–730.
38 Yang, X., Wu, T., Phipps, R.J., and Toste, F.D. (2015). *Chem. Rev.* 115: 826–870.
39 Noda, H., Kumagai, N., and Shibasaki, M. (2018). *Asian J. Org. Chem.* 7: 599–612.
40 Paria, S., Lee, H.-J., and Maruoka, K. (2019). *ACS Catal.* 9: 2395–2399.
41 Iwamoto, H., Sonoda, T., and Kobayashi, H. (1984). *J. Fluorine Chem.* 24: 535–537.
42 Iwamoto, H., Sonoda, T., and Kobayashi, H. (1983). *Tetrahedron Lett.* 24: 4703–4706.
43 Carter, C., Fletcher, S., and Nelson, A. (2003). *Tetrahedron: Asymmetry* 14: 1995–2004.
44 Rauniyar, V., Lackner, A.D., Hamilton, G.L., and Toste, F.D. (2011). *Science* 334: 1681–1684.
45 Phipps, R.J., Hamilton, G.L., and Toste, F.D. (2012). *Nat. Chem.* 4: 603–614.
46 Neel, A.J., Milo, A., Sigman, M.S., and Toste, F.D. (2016). *J. Am. Chem. Soc.* 138: 3863–3875.
47 Zi, W.W., Wang, Y.M., and Toste, F.D. (2014). *J. Am. Chem. Soc.* 136: 12864–12867.
48 Nelson, H.M., Williams, B.D., Miro, J., and Toste, F.D. (2015). *J. Am. Chem. Soc.* 137: 3213–3216.
49 Tao, Z.-L., Adili, A., Shen, H.-C. et al. (2016). *Angew. Chem. Int. Ed.* 55: 4322–4326.
50 Hoffmann, R.W. (1988). *Pure Appl. Chem.* 60: 123.
51 Shen, H.-C., Wang, P.-S., Tao, Z.-L. et al. (2017). *Adv. Synth. Catal.* 359: 2383–2389.
52 Yamamoto, E., Hilton, M.J., Orlandi, M. et al. (2016). *J. Am. Chem. Soc.* 138: 15877–15880.
53 Avila, C.M., Patel, J.S., Reddi, Y. et al. (2017). *Angew. Chem. Int. Ed.* 56: 5806–5811.
54 Reddi, Y., Tsai, C.C., Avila, C.M. et al. (2019). *J. Am. Chem. Soc.* 141: 998–1009.
55 Sawamura, M. and Ito, Y. (1992). *Chem. Rev.* 92: 857–871.
56 Kull, T. and Peters, R. (2008). *Angew. Chem. Int. Ed.* 47: 5461–5464.
57 Kull, T., Cabrera, J., and Peters, R. (2010). *Chem. Eur. J.* 16: 9132–9139.
58 Mechler, M. and Peters, R. (2015). *Angew. Chem. Int. Ed.* 54: 10303–10307.
59 Brodbeck, D., Broghammer, F., Meisner, J. et al. (2017). *Angew. Chem. Int. Ed.* 56: 4056–4060.
60 Schmid, J., Junge, T., Lang, J. et al. (2019). *Angew. Chem. Int. Ed.* 58: 5447–5451.
61 Ohmatsu, K., Ito, M., Kunieda, T., and Ooi, T. (2012). *Nat. Chem.* 4: 473.
62 Ohmatsu, K., Ito, M., Kunieda, T., and Ooi, T. (2013). *J. Am. Chem. Soc.* 135: 590–593.
63 Ohmatsu, K., Hara, Y., and Ooi, T. (2014). *Chem. Sci.* 5: 3645–3650.
64 Ohmatsu, K., Imagawa, N., and Ooi, T. (2014). *Nat. Chem.* 6: 47.
65 Ohmatsu, K., Ito, M., and Ooi, T. (2014). *Chem. Commun.* 50: 4554–4557.

66 Ohmatsu, K., Kawai, S., Imagawa, N., and Ooi, T. (2014). *ACS Catal.* 4: 4304–4306.

67 van Leeuwen, P.W.N.M. (2008). *Supramolecular Catalysis*. Weinheim: Wiley-VCH.

68 Brown, R.C.D. and Keily, J.F. (2001). *Angew. Chem. Int. Ed.* 40: 4496–4498.

69 Bhunnoo, R.A., Hu, Y.L., Laine, D.I., and Brown, R.C.D. (2002). *Angew. Chem. Int. Ed.* 41: 3479–3480.

70 Wang, C., Zong, L.L., and Tan, C.H. (2015). *J. Am. Chem. Soc.* 137: 10677–10682.

71 Ye, X., Moeljadi, A.M., Chin, K.F. et al. (2016). *Angew. Chem. Int. Ed.* 55: 7101–7105.

72 Zong, L., Wang, C., Moeljadi, A.M.P. et al. (2016). *Nat. Commun.* 7: 13455.

73 Fang, G., Zheng, C., Cao, D. et al. (2019). *Tetrahedron* 75: 2706–2716.

74 Cousseau, J., Gouin, L., Jones, L.V. et al. (1973). *J. Chem. Soc., Faraday Trans. 2*: 69.

75 Sharf, V.Z., Litvin, E.F., Kasyrnova, K.A., and Afans'ev, V.A. (1985). *Izv. Akad. Nauk SSSR, Ser. Khim.* 34: 1463.

76 Endo, T., Saotome, Y., and Okawara, M. (1984). *J. Am. Chem. Soc.* 106: 1124–1125.

77 Li, S., Cen, R., Zeng, Z., and Liang, Z. (1987). *Huaxue Tongbao* 6: 28.

3

Enamine-Metal Combined Catalysis

3.1 Introduction: Combined Enamine Activation and Metal Catalysis

Enamine, formed from the condensation of an amine with an aldehyde or a ketone, is a versatile nucleophile capable of undergoing a wide scope of bond-forming transformations [1]. The dynamic interconversion between the enamine and the starting materials allocates a large chemical space of amines to work as organocatalysts capable of enabling nucleophilic reactions of enolized aldehydes and ketones, now known as enamine catalysis (Scheme 3.1). Since the seminal discovery of proline-catalyzed direct aldol and Mannich reactions [2], asymmetric enamine catalysis has been a general concept of high atom/step economy and sustainability in asymmetric synthesis [3]. Over the past decades, a huge library of chiral primary and secondary amines has been convinced to be efficient organocatalysts featuring enamine activation to enable a tremendous number of asymmetric transformations [4].

Upon enamine activation, the α-carbon of a carbonyl substrate becomes highly nucleophilic that can be trapped by some transient organometallic electrophiles, which are generated from elementary organometallic reactions or Lewis acid-mediated transformations, in addition to typical organic electrophiles (Scheme 3.2) [5]. As a consequence, the orthogonal combination of enamine activation and transition metal catalysis can provide fundamentally new reactivity capable of rejuvenating both strategies and accessing otherwise inaccessible reaction pathways [6]. This chapter illustrates the general concept of integrating the enamine activation and transition metal catalysis for the development of asymmetric transformations.

3.2 Catalytic Asymmetric α-Allylation of Carbonyls

3.2.1 Oxidative Addition-Initiated Allylic Alkylation

Transition metal-catalyzed allylic alkylation is a fundamental cross-coupling reaction with widespread applications in organic synthesis [7]. A transition metal

Asymmetric Organo-Metal Catalysis: Concepts, Principles, and Applications, First Edition. Liu-Zhu Gong.
© 2022 WILEY-VCH GmbH. Published 2022 by WILEY-VCH GmbH.

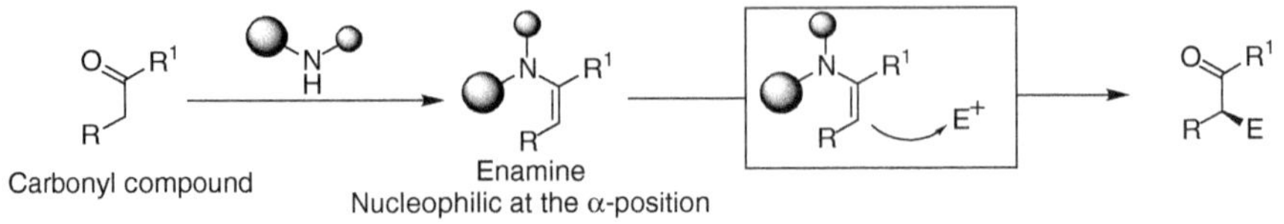

Scheme 3.1 Enamine catalysis to active carbonyl compounds.

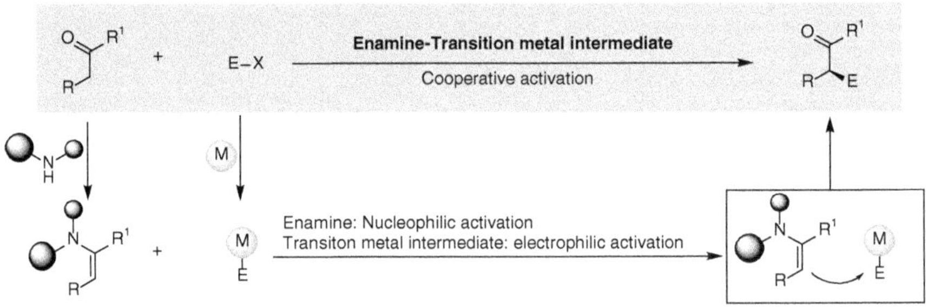

Scheme 3.2 General concept of enamine and metal cooperative catalysis.

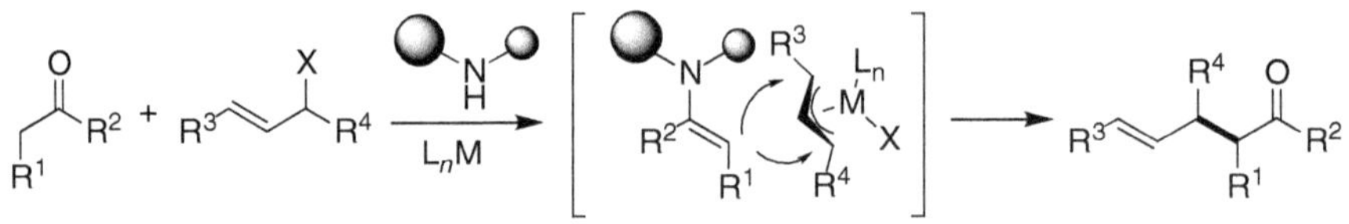

Scheme 3.3 Catalytic asymmetric allylic alkylation of carbonyl compound. Source: Refs [5, 7].

complex activates an allyl substrate through oxidative addition to generate a π-allyl-metal complex that can be captured by a wide scope of nucleophiles. The coupling event between the enamine and π-allyl-metal complex provides a general concept for the development of asymmetric allylic alkylation reaction of aldehydes and ketones by amine and transition metal combined catalysis (Scheme 3.3) [5, 7]. More importantly, because either the chiral ligand or amine catalyst can control the stereoselectivity individually or in a concerted manner, the combined catalysis would provide extra possibilities to induce the stereoselectivity [8].

Córdova and coworker first elaborated the cooperative catalysis of enamine and palladium complex to drive a direct intermolecular α-allylation of aldehydes or ketones with allyl acetates (Scheme 3.4) [9]. In this catalytic process, the nucleophilic enamine **5**, catalytically generated from the condensation of the carbonyl compound **1** and the amine catalyst **4**, undergoes the allylic alkylation with the π-allyl-Pd(II) complex **6** generated via the oxidative addition of the allylic acetate **2** to the Pd(0) complex to forge the C—C bond, together with the regeneration of Pd(0) catalyst. Subsequent hydrolysis of the iminium intermediate **7** leads to the formation of final product **3** and releases the amine catalyst **4**. The hallmark of this process is that the chemical bond-breaking and forming events occur simultaneously and in a concerted manner.

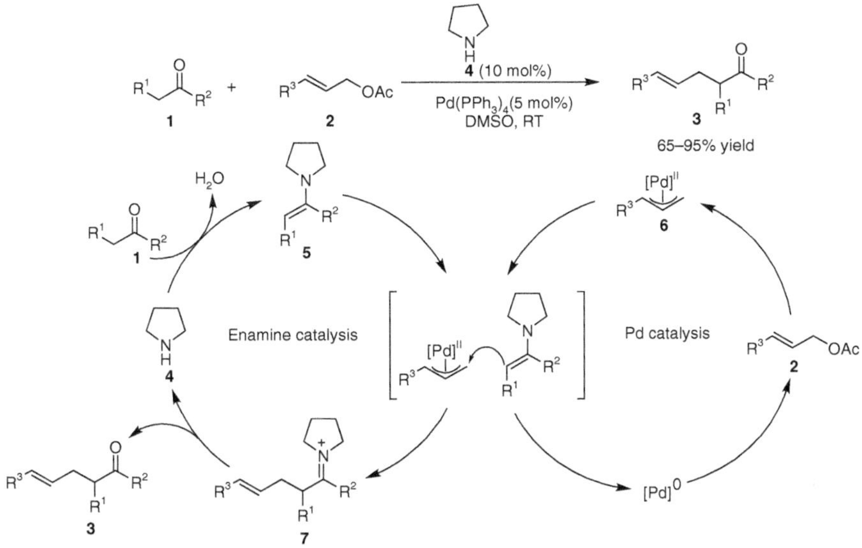

Scheme 3.4 Cooperative activation by amine catalyst and palladium catalyst. Source: Ibrahem and Córdova [9].

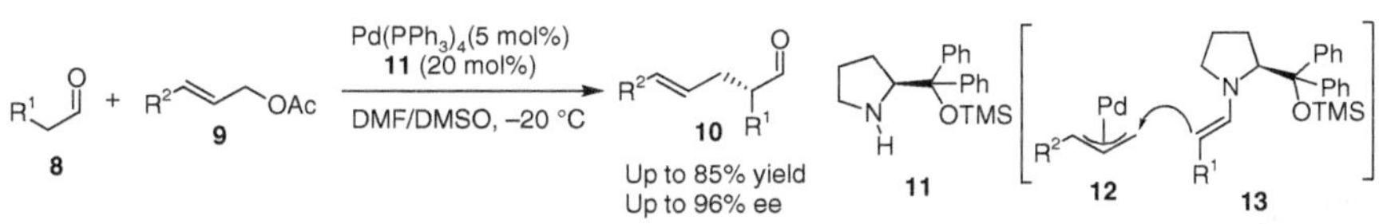

Scheme 3.5 Enantioselective α-allylic alkylation of aldehydes.

The same group expanded the concept to an enantioselective α-allylation of aldehydes **8** by using chiral amine and achiral $Pd(PPh_3)_4$ as the combined catalyst [10]. A prolinol derivative **11** turns out to be a superior organocatalyst and enables the reaction to give high yields and enantioselectivities. The chiral amine solely controls the stereoselectivity by the enantioselective addition of the enamine **13** to the π-allyl-Pd(II) species **12** (Scheme 3.5).

The use of chiral primary amino acid **17** as the co-catalyst enables a direct α-allylation of α-branched aldehydes **14** with an allyl ester **15** synergistically catalyzed by a palladium complex to give the α-allyl aldehydes with up to 87% ee (Scheme 3.6a) [11a]. Moreover, an asymmetric α-allylation of α-branched aldehydes with allyl alcohols **18** orchestrated by a similarly combined catalyst system proceeds successfully to give various α-allylated aldehydes **19** with excellent control of enantioselectivity (Scheme 3.6b) [11b].

The synergistic catalysis of chiral palladium complex and achiral amine for the asymmetric intramolecular α-allylic alkylation of aldehydes was first introduced by Saicic and coworkers [12]. Surprisingly, chiral amine catalyst fails to efficiently induce chirality to the intramolecular allylic alkylation of aldehydes **21**. Instead, a chiral palladium complex, in combination with an achiral amine, delivers notable

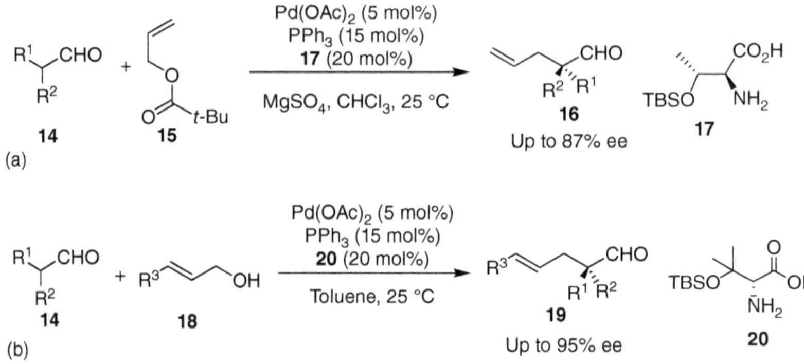

Scheme 3.6 Direct α-allylation of α-branched aldehydes. (a) α-Allylation of α-branched aldehyde with allyl ester. (b) α-Allylation of α-branched aldehyde with allyl alcohol. Source: (a, b) Modified from Yoshida et al. [11].

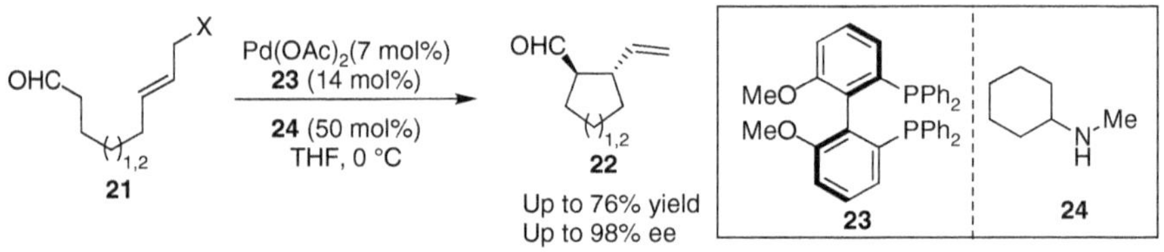

Scheme 3.7 Asymmetric intramolecular α-allylic alkylation of aldehydes.

stereoselectivity. In particular, the combination of *N*-methylcyclohexanamine (**24**) and chiral palladium complex of (*R*)-Ph-MeOBiPHEP (**23**) allows the reaction to afford cyclic products **22** with high enantioselectivities (Scheme 3.7).

Since both chiral metal complexes and chiral amine catalysts can impact the stereochemistry during the bond-forming process, alternating the configuration of the ligand **28** or amine catalyst **29** would be able to tune the stereochemistry of the products. Carreira and coworkers first initiated the concept of stereodivergent synthesis of chiral molecules by using asymmetric α-allylation of branched aldehydes **25** as a model reaction enabled by the cooperative catalysis of chiral iridium complex and chiral amine (Scheme 3.8) [13]. Chiral primary amine catalysts derived from cinchona alkaloids and iridium complexes of the phosphine-olefin ligands are both able to efficiently control the stereochemistry of the allylic alkylation alone; therefore, the orthogonal combination of two diastereomeric amines **29A** or **29B** with iridium complexes of either enantiomer of the chiral ligand **28** led to a fully stereodivergent reaction in excellent levels of yields and stereoselectivity.

The Sunoj group performed a mechanistic study of the stereodivergent cooperative catalysis by using density functional theory calculation [14]. Based on their calculations, the stereochemically unique products were determined by altering the Ir-(P, olefin) and enamine catalysts in the form of an enamine and an Ir–π-allyl intermediate. The multipoint contact analysis of the stereocontrolling transformation revealed that the weak interactions in the transition state are responsible for chiral induction. The geometric disposition of the quinolone ring of the cinchona

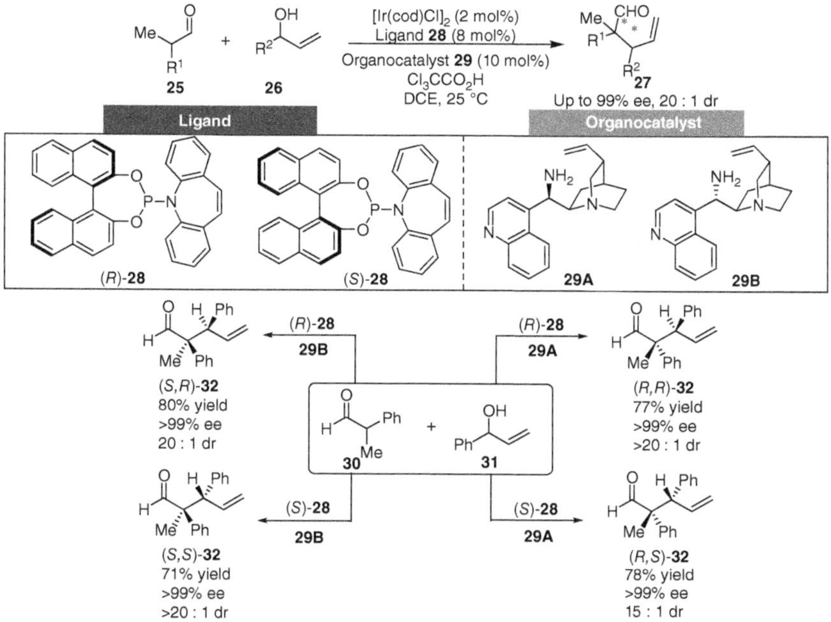

Scheme 3.8 Stereodivergent synthesis by asymmetric cooperative catalysis. Source: Modified from Krautwald et al. [13].

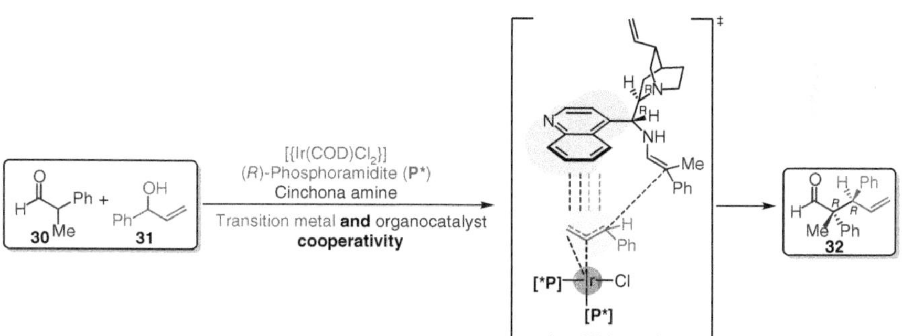

Scheme 3.9 Mechanistic insights on the stereodivergent cooperative catalysis.

amine is noted as participating in a π-stacking interaction with the phenyl ring of the Ir–π-allyl moiety to play a crucial role in relative stabilization of the stereo controlling transition states (Scheme 3.9).

Such a stereodivergent cooperative catalysis is highly general and applicable to similar protocols involving different substrates. Following the proof of concept [13], asymmetric α-allylation reactions of linear aldehydes [15a], α-amido- and -hydroxyacetaldehydes [15b] with allylic alcohols were all accomplished in a stereodivergent manner by cooperative catalysis of iridium complexes and prolinol derivatives (Scheme 3.10a). A fully stereodivergent total synthesis of natural product Δ^9-tetrahydrocannabinol **38** and all other stereomers have been

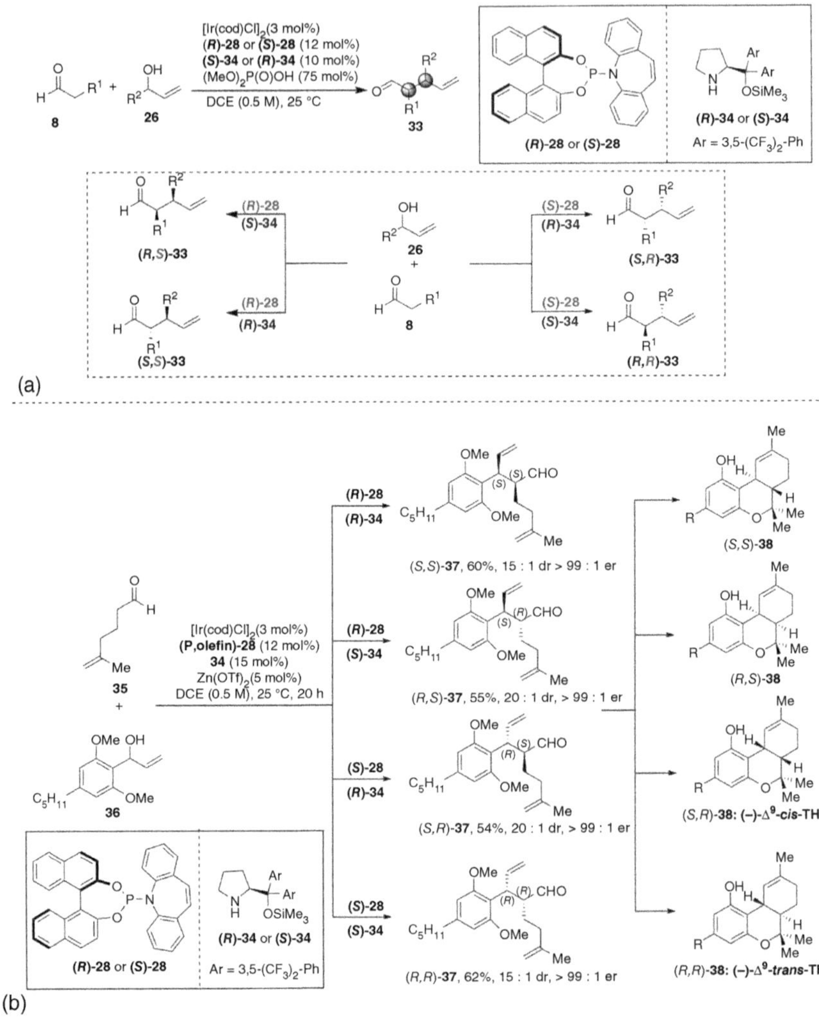

Scheme 3.10 Stereodivergent dual catalytic α-allylation of linear aldehydes. (a) Stereodivergent dual catalytic α-allylation of aldehydes. (b) Stereodivergent preparation of all stereoisomers of Δ^9-tetrahydrocannabinols. Source: Modified from Zhang et al. [16].

accomplished commencing with the stereodivergent α-allylation of aldehydes [15c] (Scheme 3.10b).

A catalytic asymmetric [4+2] annulation of vinyl amino alcohols with aldehydes was developed by iridium complex and amine cooperative catalysis (Scheme 3.11) [16]. Two catalytic cycles act in concert to enable this cycloaddition. The iridium catalyst reacts with vinyl amino alcohol **39** to form a π-allyliridium dipole intermediate **43**, whereas the achiral amine catalyst **42** condenses with the aldehyde **40** to generate enamine species **42**. Then, the allylic alkylation between **42** and **43** gives aldehyde **44**, followed by condensation and pyrindium chlorochromate (PCC) oxidation, leading to products **40** with high levels of enantioselectivity.

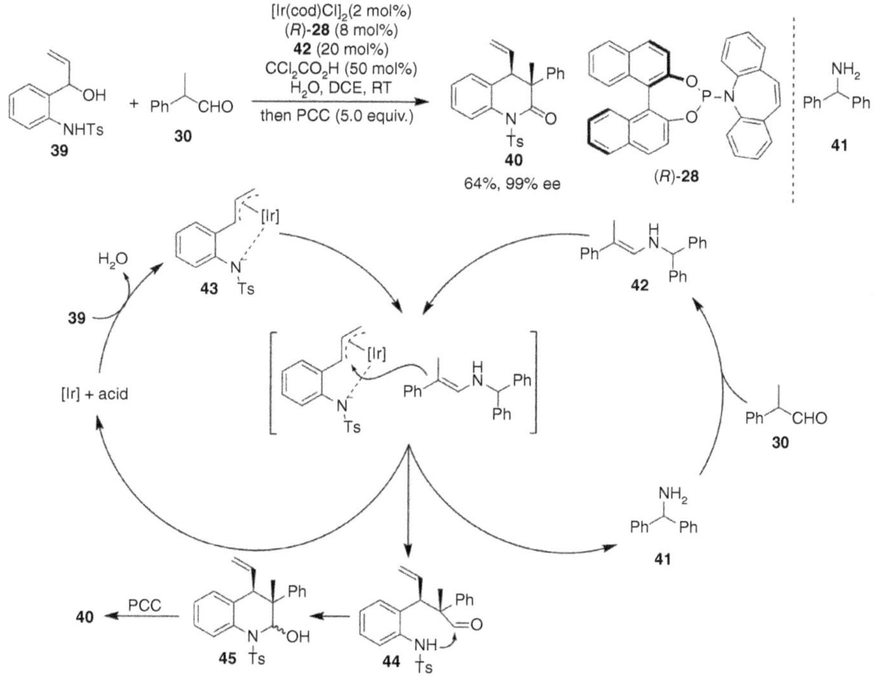

Scheme 3.11 Asymmetric [4+2] annulation by iridium/amine cooperative catalysis.

The combination of dienamine activation with chiral transition metal catalysis has also been successfully applied in γ-allylic alkylation of α,β-unsaturated aldehydes. The enamine **50** generated from aldehydes **46** and prolinol trimethylsilyl either **11** undergoes the asymmetric γ-allylation alkylation with allylic alcohols **47** or allyl esters **52** co-catalyzed by chiral iridium or palladium complex and phosphoric acid **48** (Scheme 3.12). The iridium complex of Carreira ligand **28** combined with (*R*)-**11** allows the reaction to give branched products **49** in excellent diastereo- and enantioselectivities (Scheme 3.12a) whereas the palladium complex of achiral ligand **54**, in concert with (*S*)-**11**, affords linear products **53** with excellent stereoselectivities (>20 : 1 dr, up to 99% ee) (Scheme 3.12b) [17].

The furfural **57** condenses with chiral amine **61** to generate a trienamine **62**, which can undergo Michael addition to nitroalkenes [18a]. The allylic alkylation of trienamine **62** with π-allylpalladium intermediate **63** generated from allylic alcohol **58** occurs at the benzylic position of the furfural **57**. Because the reactive site of the trienamine **62** locates remote to the chiral center, the combination of chiral amine **61** and achiral palladium catalyst is unable to offer high enantioselectivity. A chiral phosphoramidite **60** derived from TADDOL was found to show a considerable synergistic effect with the chiral amine **61** on stereochemical control, leading to high levels of enantioselectivity (Scheme 3.13) [18b].

By using synergistic catalysis of achiral amine and chiral palladium complex, Zhang and coworkers created an asymmetric allylic alkylation reaction of allylic amines **65** and enolizable ketones **64** (Scheme 3.14a). Stoichiometric amounts of

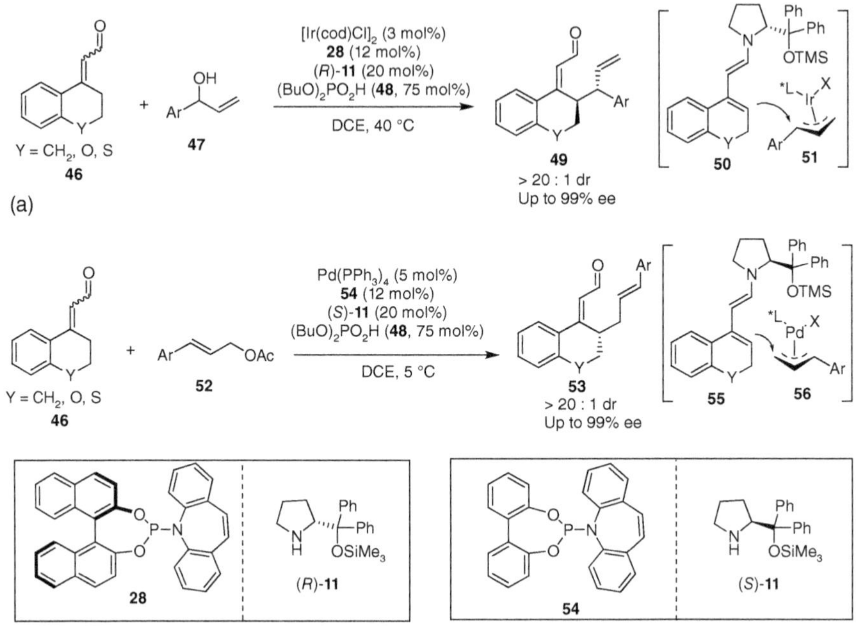

Scheme 3.12 γ-Allylation of cyclic α,β-unsaturated aldehydes. (a) γ-Allylation to form branched product. (b) γ-Allylation to form linear product. Source: (b) Næsborg et al. [17].

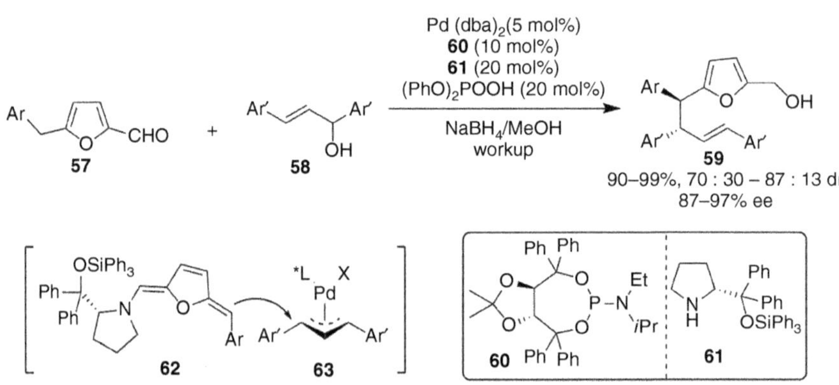

Scheme 3.13 Asymmetric benzylic allylation of furan derivatives enabled by trienamine/Pd cooperative catalysis. Source: Modified from Su et al. [18].

pyrrolidine are required to ensure reaction efficiency. Besides, the protic solvent is necessary to activate the C—N bond of the *N*-allylpyrrolidine by hydrogen bonding interaction and thereby facilitates the oxidative addition for the generation of a π-allyl-Pd(II) complex [19a]. The ferrocene-based phosphinooxazoline ligand **67** turns out to be optimal for the stereochemical control. The enamine concurrently

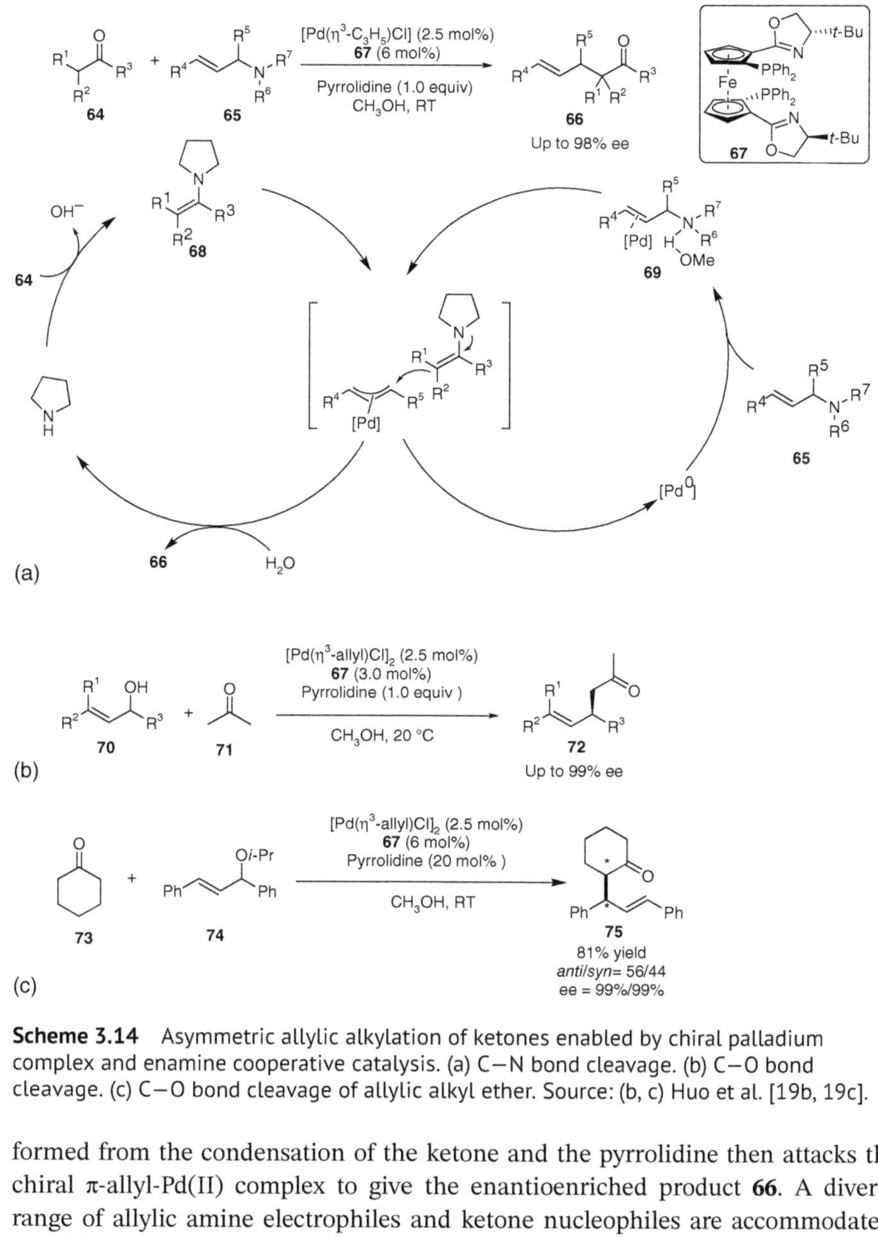

Scheme 3.14 Asymmetric allylic alkylation of ketones enabled by chiral palladium complex and enamine cooperative catalysis. (a) C—N bond cleavage. (b) C—O bond cleavage. (c) C—O bond cleavage of allylic alkyl ether. Source: (b, c) Huo et al. [19b, 19c].

formed from the condensation of the ketone and the pyrrolidine then attacks the chiral π-allyl-Pd(II) complex to give the enantioenriched product **66**. A diverse range of allylic amine electrophiles and ketone nucleophiles are accommodated. In addition, asymmetric allylic alkylation reactions of simple ketones with allylic alcohols **70** and an allyl isopropyl ether **74** driven by a similar binary catalyst system both work well with excellent levels of enantioselectivity (Scheme 3.14b,c) [19b, c].

Chiral primary amine **79** in combination with the palladium complex can synergistically catalyze an asymmetric allylic alkylation of β-ketoesters **76** with allylic alcohols **77** (Scheme 3.15) [20]. The chiral primary amine **79** is unique to activate

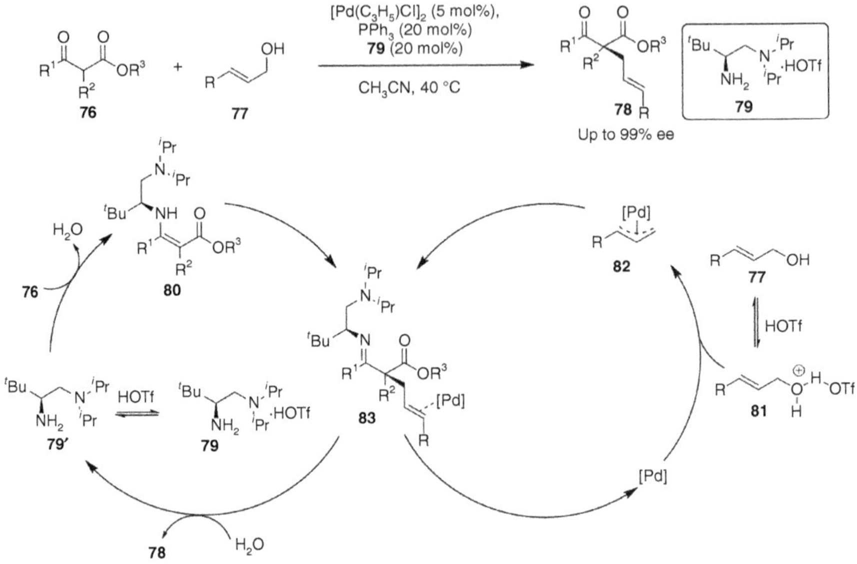

Scheme 3.15 Asymmetric allylic alkylation of β-keto esters cooperatively catalyzed by chiral primary amine and palladium complex. Source: Zhou et al. [20a].

β-ketoesters by rapidly and reversibly forming enamines **80**, which are highly reactive toward the π-allyl-Pd(II) **82** to undergo the allylic alkylation reaction and yield allylation products **78** in almost perfect stereochemical control.

3.2.2 Metal Hydride-Initiated Allylic Alkylation

The oxidative addition of low-valent transition metals to Brønsted acids transforms protons into metal hydrides. The hydrometallation occurs between these metal hydrides and allene, dienes, or propyne derivatives to directly access π-allylmetallic complexes [21]. The resultant π-allyl-metal species are reactive electrophiles to undergo substitution reactions with nucleophiles. Therefore, the combination of low-valent transition metal complexes with Brønsted acids and amine would be able to catalyze transformations of alkynes, allenes, and dienes with enolizable carbonyl compounds to build up structural complexity with high atom economy (Scheme 3.16).

In 2012, Dixon reported a dual amine and palladium catalysis in an enantioselective allene carbocyclization reaction [22]. Initially, a rapid condensation of secondary amine **86** with aldehyde **84** gives an iminium intermediate **87**. Then, the Pd(0) species removes the acidic α-proton of iminium ion via oxidative addition and thereby gives an enamine and a hydrido palladium complex **89**. A subsequent hydropalladation of the allene moiety generates a reactive π-allyl complex **90**, which is then poised to undergo an intramolecular allylic alkylation with the enamine moiety (Scheme 3.17).

Luo and coworkers disclosed a chiral primary amine/achiral palladium complex cooperatively catalyzed enantioselective addition of β-ketocarbonyls **91** to allenes **92**

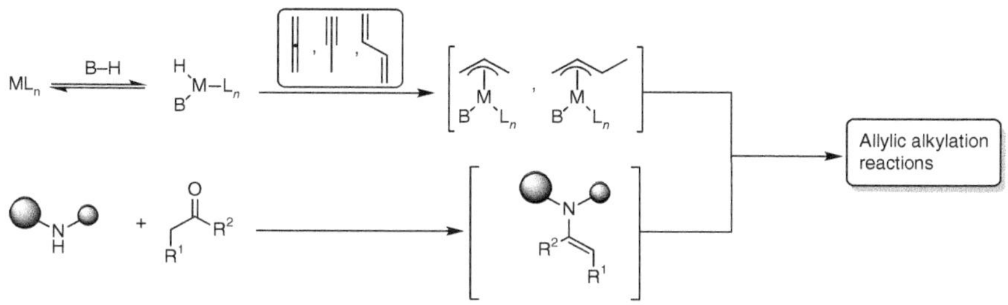

Scheme 3.16 Cooperative catalytic asymmetric allylic alkylation via hydrometallation.

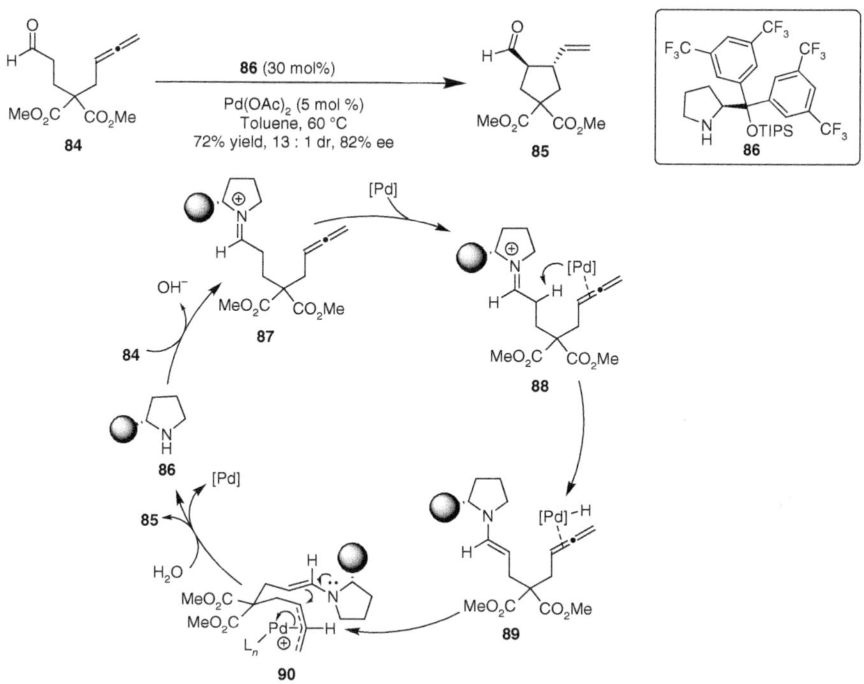

Scheme 3.17 Dual amine/palladium catalysis in enantioselective allene carbocyclization reaction.

(Scheme 3.18) [23]. Mechanistic studies suggest that the hydrido palladium complex undergoes a hydrocarbonation with allenes **92** to afford the key π-allyl-palladium species **96**, which then react with the enamine intermediates **95** to generate the allylation intermediates **97**, and followed by hydrolysis, leading to the allylation products **93**. The synergistic nature of the dual catalysis has been revealed by control experiments that the reaction does not work in the absence of either the chiral primary amine catalyst or palladium complex.

Cruz and Dong established a stereodivergent allylic alkylation reaction between α-branched aldehydes and alkynes based on rhodium and enamine cooperative catalysis [24]. The rhodium hydride **102** generated from the oxidative addition of an Rh(I) complex to the Brønsted acid first undergoes an insertion reaction with

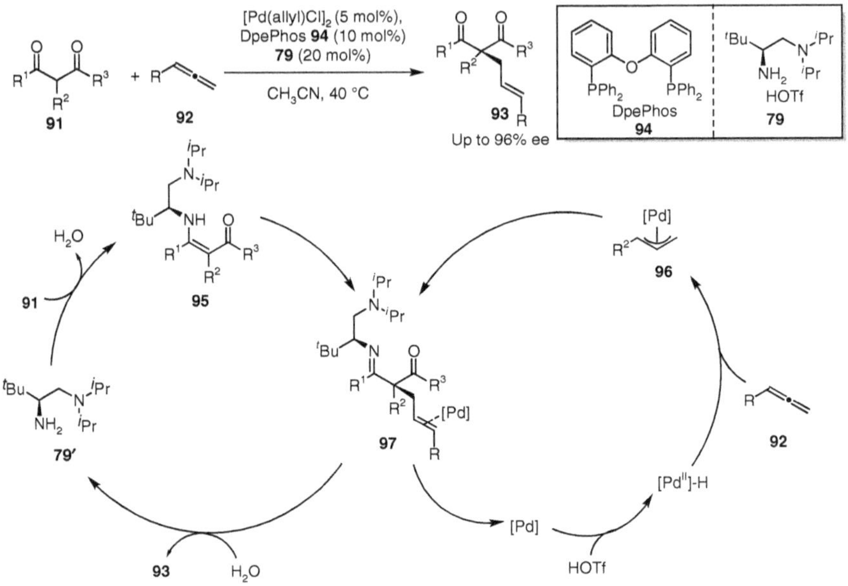

Scheme 3.18 Enamine/palladium synergistically catalyzed-terminal addition to allenes. Source: Modified from Zhou et al. [23].

acetylenes **99** to give a vinylrhodium intermediate **103**, which then undergoes an β-hydride elimination to generate an allene **104** and to regenerate the rhodium hydride **102**. The insertion of allene **104** to Rh-hydride gives an Rh-π-allyl species **105**, which is captured by an *in situ* generated enamine **106** to give an allylation product **100**. The chiral amine **101** turns out to be the most efficient organocatalyst and provides excellent stereoselectivities. The diastereoselectivity is switchable and all of the possible diastereomeric products can be obtained with high regio-, diastereo-, and enantioselectivities by a simply alternating combination of chiral rhodium complexes and amine catalysts (Scheme 3.19).

3.2.3 Lewis Acid-Mediated S_N1 or S_N2 Reaction

In 2010, the Cozzi group described a stereoselective substitution reaction of allylic alcohols **108** with aldehydes **8** cooperatively promoted by a Lewis acid and the MacMillan catalyst **110** [25]. The stabilized carbocation generated from the $InBr_3$-mediated dehydroxylation process of allylic alcohols undergoes an S_N1-type substitution with the enamine formed from imidazolidinone **110** and the aldehyde **8** to provide allylic alkylation products **109** in good yields and with excellent stereoselectivity (Scheme 3.20).

Bandini group established an enantioselective intramolecular α-allylic alkylation of aldehydes by gold/enamine cooperative catalysis [26]. The gold complex acts as a π-Lewis acid that coordinates with the double bond of allylic alcohol moiety of **111** to facilitate the nucleophilic addition of the enamine moiety simultaneously generated from the aldehyde and amine catalyst **113**. The subsequent β-hydroxy elimination releases the S_N2'-type product **112** and the gold catalyst (Scheme 3.21).

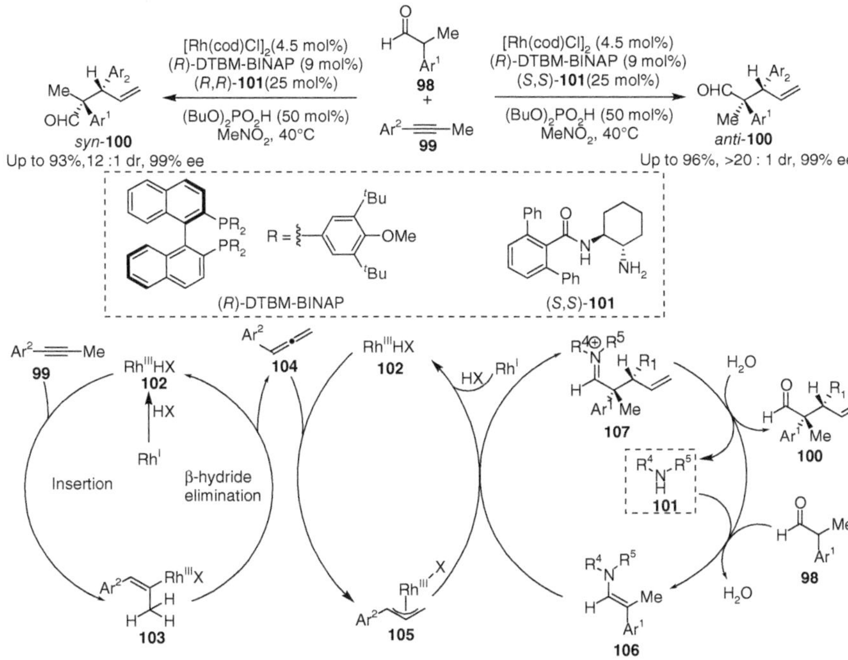

Scheme 3.19 Asymmetric α-allylation of aldehyde with alkyne by Rh/enamine cooperative catalysis.

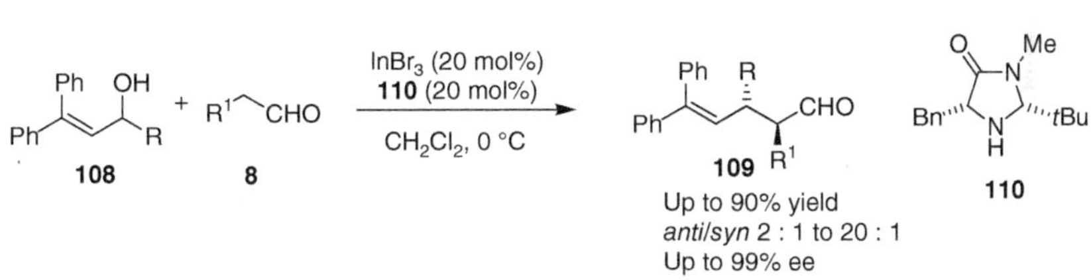

Scheme 3.20 Stereoselective substitution of allylic alcohols with aldehydes.

3.3 Catalytic Asymmetric Substitution

α-Alkylation of a carbonyl compound with alcohol derivatives represents a valuable method to install an alkyl substituent at the α-position (Scheme 3.22). Lewis acid catalysts can remove the hydroxyl or other groups from the substrates to generate a carbocation intermediate [27], which is then trapped by the enamine intermediate to give the substitution products.

Cozzi group reported an efficient enantioselective direct alkylation of aldehydes with benzyl alcohols [28a]. Later, Xiao developed a cooperative catalytic system consisting of diarylprolinol silyl ether and a Lewis acid for the highly enantioselective intermolecular α-alkylation of aldehydes **8** with an alcohol **116** [28b]. Either $InCl_3$ or CuCl appears to be an active Lewis acid catalyst, in concert with chiral amine **34**,

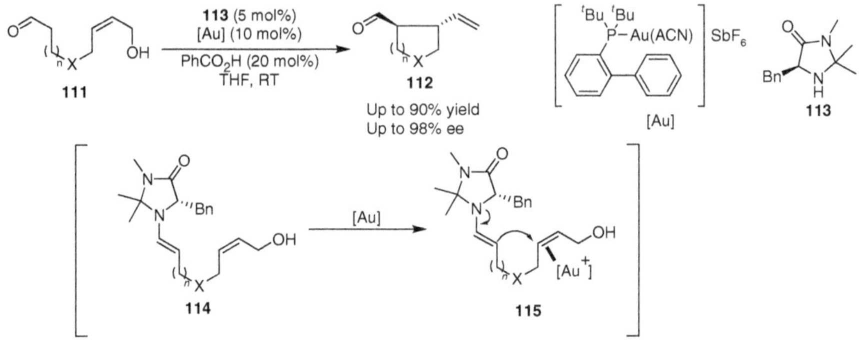

Scheme 3.21 Gold/enamine catalysis in the enantioselective α-allylic alkylation of aldehydes.

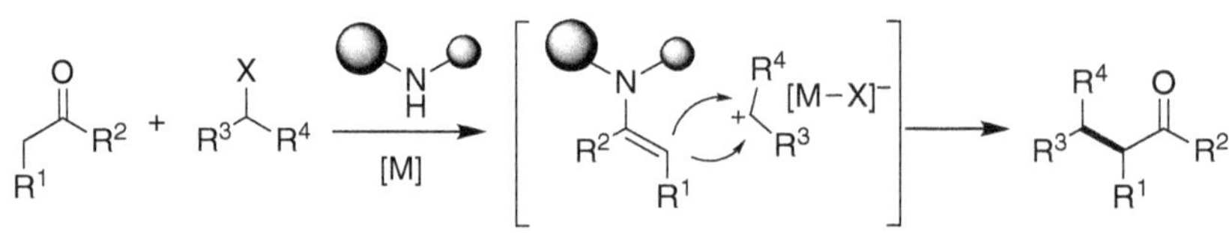

Scheme 3.22 Catalytic asymmetric substitution of carbonyl compound.

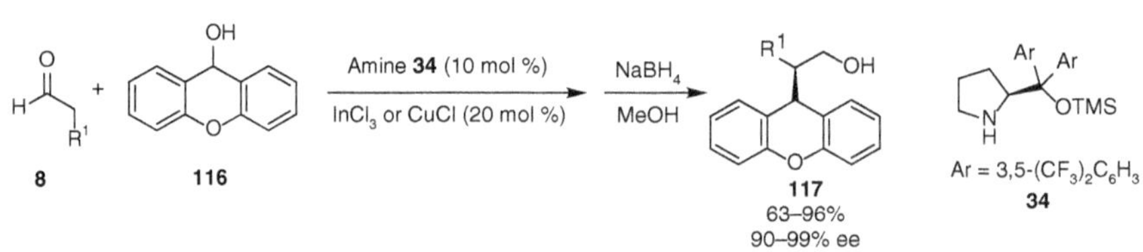

Scheme 3.23 Enantioselective direct alkylation of aldehydes with unfunctionalized alcohols.

to undergo the enantioselective S_N1-type reaction with aldehydes. Products **117** are obtained in high yields and with good diastereoselectivity and excellent enantioselectivity (Scheme 3.23).

Nishibayashi and coworkers created an asymmetric propargylation of aldehydes by Lewis acid/enamine cooperative catalysis [29]. In the presence of indium bromide and a MacMillan catalyst **113**, the asymmetric propargylation of aldehydes **8** with propargylic alcohols **118** proceeded smoothly, after reduction with $NaBH_4$, to give propargylic alkylated products **119** in high yields and high enantioselectivity, but with low diastereoselectivity. In this reaction, the $InBr_3$ undergoes dihydroxylation with the propargylic alcohol to generate a highly reactive ion pair **121**, which is subsequently attacked by the enamine **120** concurrently generated from aldehydes **8** and amine **113** to give propargylic alkylated products **123**, and followed by reduction, leading to the chiral alcohols **119** (Scheme 3.24).

Rueping et al. established an asymmetric substitution of aldehydes **8** with chromene acetals **124** enabled by ytterbium triflate and enamine cooperative catalysis [30]. The ytterbium triflate is a highly efficient catalyst to promote the

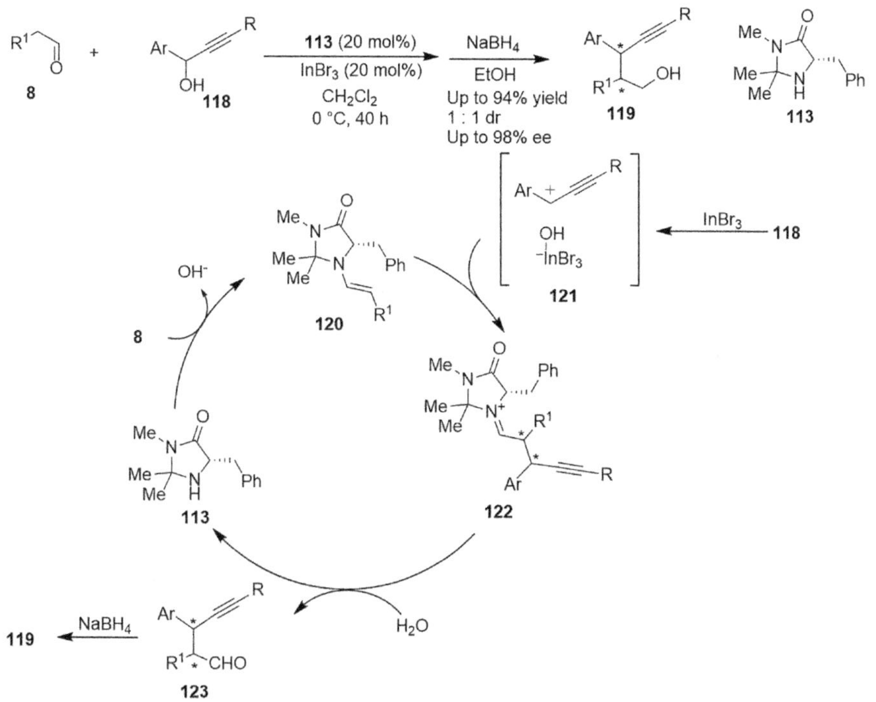

Scheme 3.24 Cooperative catalytic propargylic reaction of internal alkyne with aldehydes.

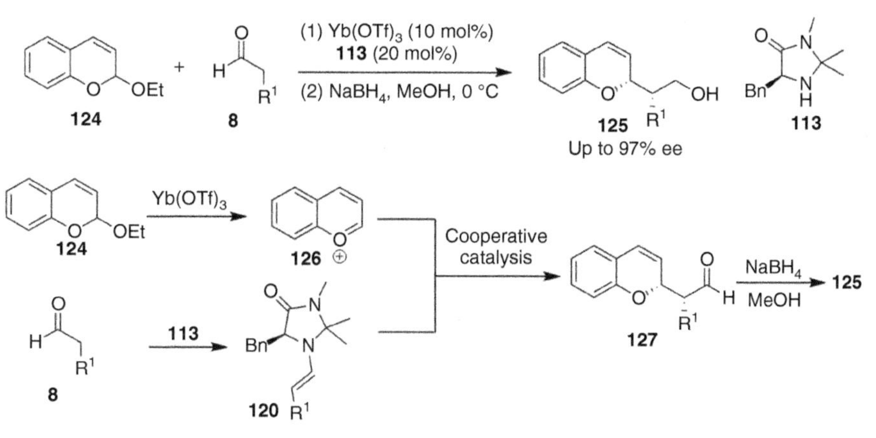

Scheme 3.25 Alkylation of cyclic acetals with aldehydes.

deoxygenation of chromene acetal **124** to generate a transient oxocarbenium ion **126**, which then undergoes an enantioselective addition reaction with the chiral enamine **120** *in situ* formed from the aldehydes **8** and a chiral amine catalyst **113** to give a variety of chiral 2*H*-chromenes **127**, after reduction with $NaBH_4$, leading to **125** with excellent enantioselectivities (Scheme 3.25).

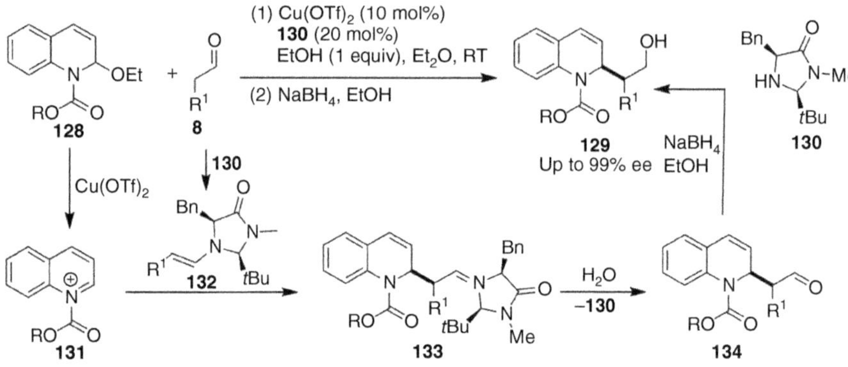

Scheme 3.26 Alkylation of cyclic *N*-acyl hemiaminals with aldehydes.

The cooperation of copper and amine catalysis allows an asymmetric formal allylic substitution of aldehydes **8** with 2-ethoxy-1-methoxycarbonyl-1,2-dihydroquinoline **128** [31]. $Cu(OTf)_2$ is the most efficient Lewis acid catalyst to promote the deoxygenation of the cyclic *N*-acyl hemiaminal substrate **128** into quinolinium ion **131**. The Mannich-type reaction between quinolinium ions **131** and the enamine **132** formed from the aldehyde **8** and a chiral imidazolidinone **130** proceeds well to yield the desired products with excellent enantioselectivity (Scheme 3.26).

Pineschi and coworkers accomplished an enantioselective S_N2'-type reaction between 2-phenylacetaldehyde **136** and 4-hydroxyl 1-benzyloxycarbonyl tetrahydropyridine **135** enabled by the cooperative catalysis of $Er(OTf)_3$ and the chiral imidazolidinone **130** [32]. The $Er(OTf)_3$ is considered to catalyze the dehydroxylation of **135** to generate 2,3-dihydropyridinium **138**. The dihydropyridinium **138** prefers to undergo an asymmetric Mannich-type reaction with the enamine **139** over Michael addition, leading to the formation of iminium **140**, which undergoes subsequent hydrolysis and followed by reduction with $NaBH_4$ to give **137** in 78% yield and with excellent enantioselectivity of 93% ee (Scheme 3.27).

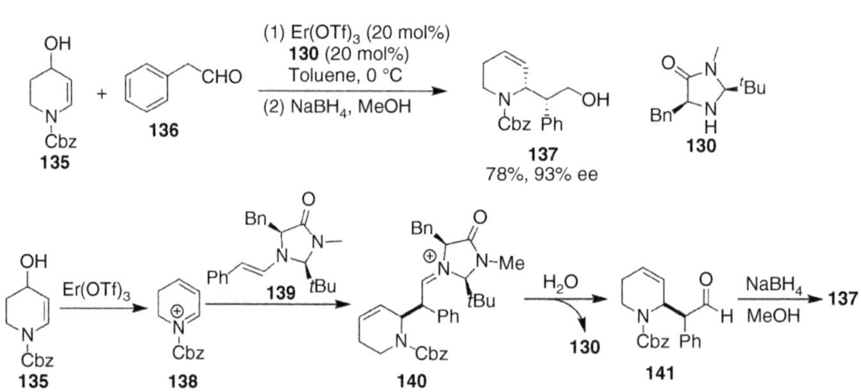

Scheme 3.27 $Er(OTf)_3$/enamine cooperatively catalyzed S_N2'-type reaction of aldehydes with quinolinium.

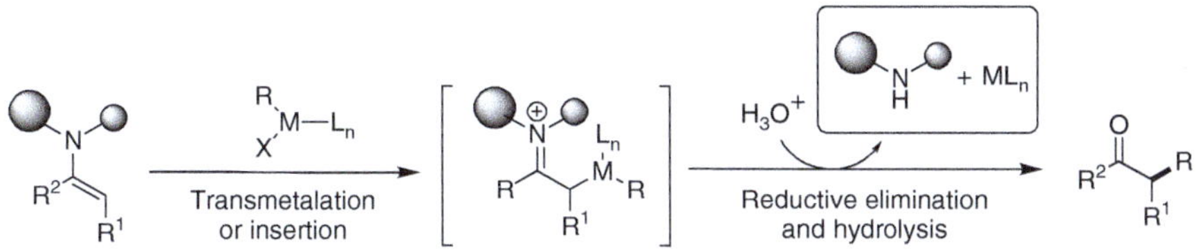

Scheme 3.28 General strategy for asymmetric α-functionalization of carbonyls with organometallics.

3.4 Catalytic Asymmetric α-Alkenylation, α-Arylation, and α-Trifluoromethylation of Carbonyl Compounds

Enamines would principally react with transient organometallic electrophiles through transmetalation of nucleophilic β-carbon or Heck-type insertion reaction of the electron-rich C—C double bond. Reductive elimination of the generated alkylmetallic intermediates enables new chemical bond formation. The combination of enamine activation and *in situ* generated organometallic species from a diverse range of electrophiles might become a general strategy for the creation of novel asymmetric transformations otherwise inaccessible (Scheme 3.28).

MacMillan and coworker reported an enantioselective α-arylation of aldehydes **8** with diaryliodonium salts **142** by synergistic catalysis of enamine and copper salt, providing a unique entry to optically active α-aryl aldehydes **143** [33]. Mechanistically, the copper catalyst undergoes the oxidative addition to the C—I bond of the diaryliodonium triflate **142** to generate an aryl–Cu(III) species **146**. The condensation of chiral imidazolidinone catalyst **144** with the aldehyde **8** furnishes a transient nucleophilic enamine **145**. The transmetalation proceeding via the complexation of the double bond of the enamine **145** with aryl–Cu(III) intermediate **146** leads to η^1-iminium organocopper species **147**. The subsequent reductive elimination of the η^1-iminium organocopper intermediate **147** forges the carbon–carbon bond and regenerates the Cu(I) catalyst (Scheme 3.29).

Later, MacMillan and coworker expanded such a combined catalyst to drive an enantioselective α-vinylation of carbonyls with alkenyl phenyliodonium triflates **148**, leading to optically active α-substituted-β,γ-unsaturated aldehydes **149** [34]. The combination of CuBr and imidazolidinone salt **150** is optimal for the reaction and gives the α-vinyl aldehyde adduct **149** in 97% yield and 97% ee (Scheme 3.30).

Catalytic introduction of a trifluoromethyl group to an organic molecule holds great potential applications in the pharmaceutical industry and material science [35]. The development of enantioselective trifluoromethylation turns out to be even more important but remains challenging [36]. MacMillan and coworker established a highly enantioselective α-trifluoromethylation of aldehydes **8** with Togni's reagent **151** rendered by the cooperative catalysis of CuCl and chiral imidazolidinone **113** [37]. Initially, Togni's reagent **113** undergoes a Lewis acid-catalyzed bond cleavage to generate a highly electrophilic iodonium salt **153**. Then, the chiral enamine **120** simultaneously formed from aldehyde **8** and amine **113** participates in an enantioselective S_N1-type substitution reaction to give an λ^3-iodane intermediate

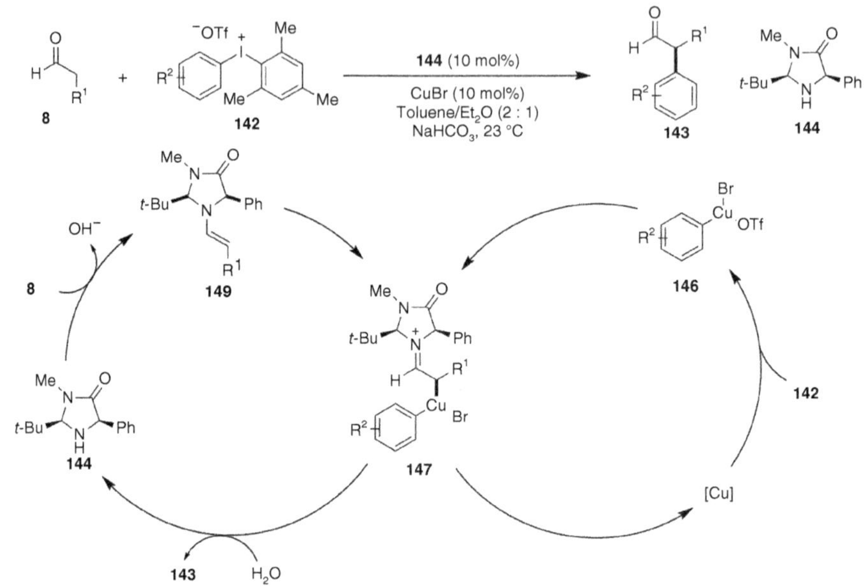

Scheme 3.29 Copper and amine catalyzed α-arylation of aldehydes.

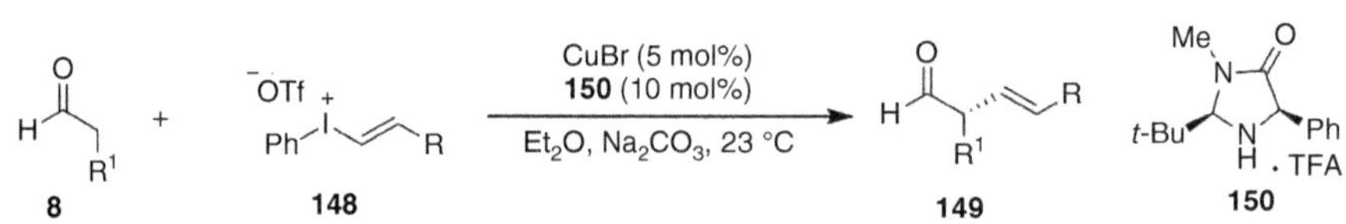

Scheme 3.30 Asymmetric α-alkenylation of the aldehyde with alkenyl phenyliodonium triflates.

154. The λ^3-iodane species **154** rapidly undergoes a reductive elimination with the alkyl transfer to forge the C—C bond, releasing the copper catalyst and byproduct 2-(2-iodophenyl)propan-2-ol. Subsequent hydrolysis affords the optically active α-formyl CF_3 product **152**. (Scheme 3.31).

The same group established an enantioselective oxidative α-alkenylation of aldehydes **8** with vinylboronic acids **155** by the synergistic catalysis of Cu(II) salt and chiral imidazolidinone **157** [38]. In this case, the transmetalation of copper acetate with a vinylboronic acid **155** generates an alkenylcopper(II) species, which is then oxidized to an alkenylcopper(III) intermediate **159**. Another transmetalation between the alkenylcopper(III) intermediate **159** and the enamine **158** occurs to result in an η^1-iminium alkenylcopper species **160**, which then undergoes reductive elimination to furnish α-alkenyl iminium and to release the copper(I) salt (Scheme 3.31). Regeneration of Cu(II) catalyst and amine catalysts is accomplished through hydrolysis of the iminium and single-electron oxidation of the copper(I) salt with oxygen, accompanying with the generation of enantioenriched α-alkenyl aldehyde products **156** (Scheme 3.32).

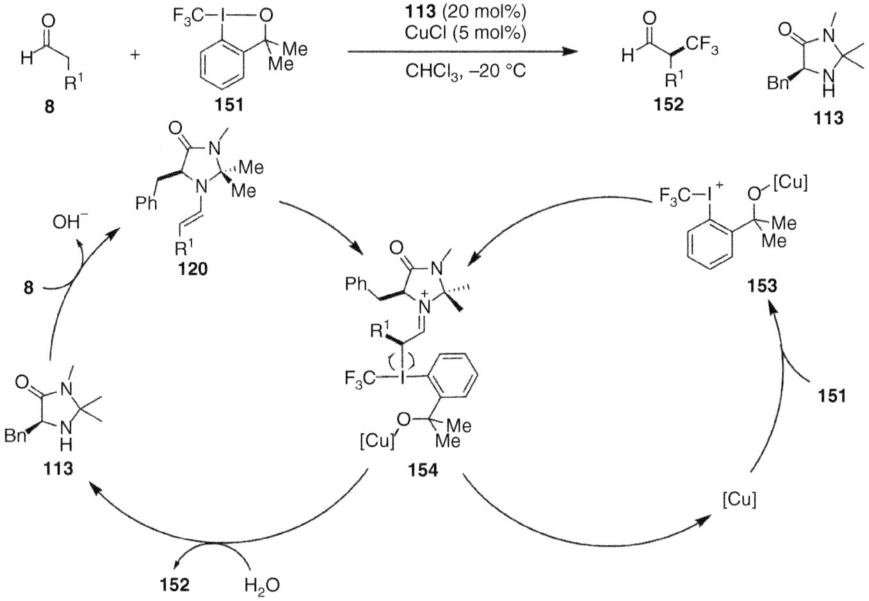

Scheme 3.31 Enantioselective α-trifluoromethylation of aldehydes.

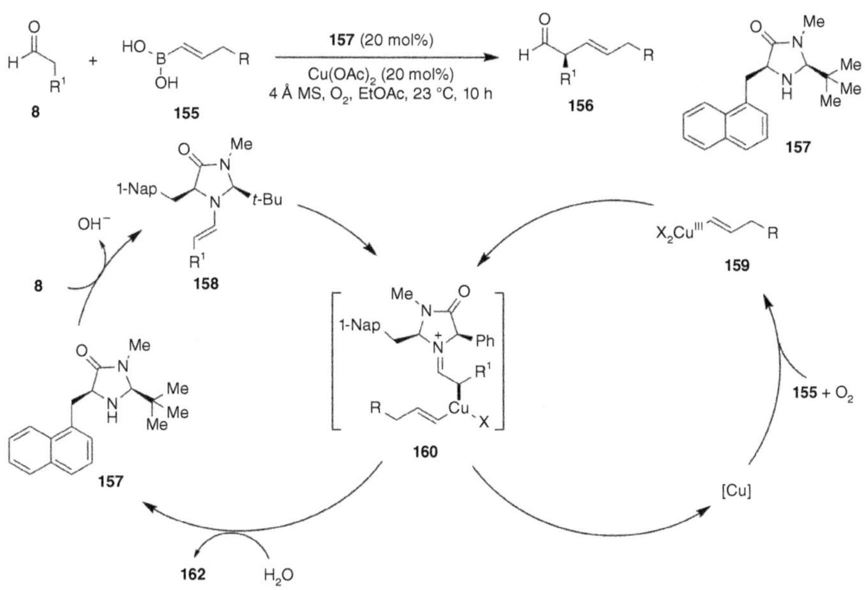

Scheme 3.32 Asymmetric oxidative α-alkenylation of aldehydes with alkenylboronic acids.

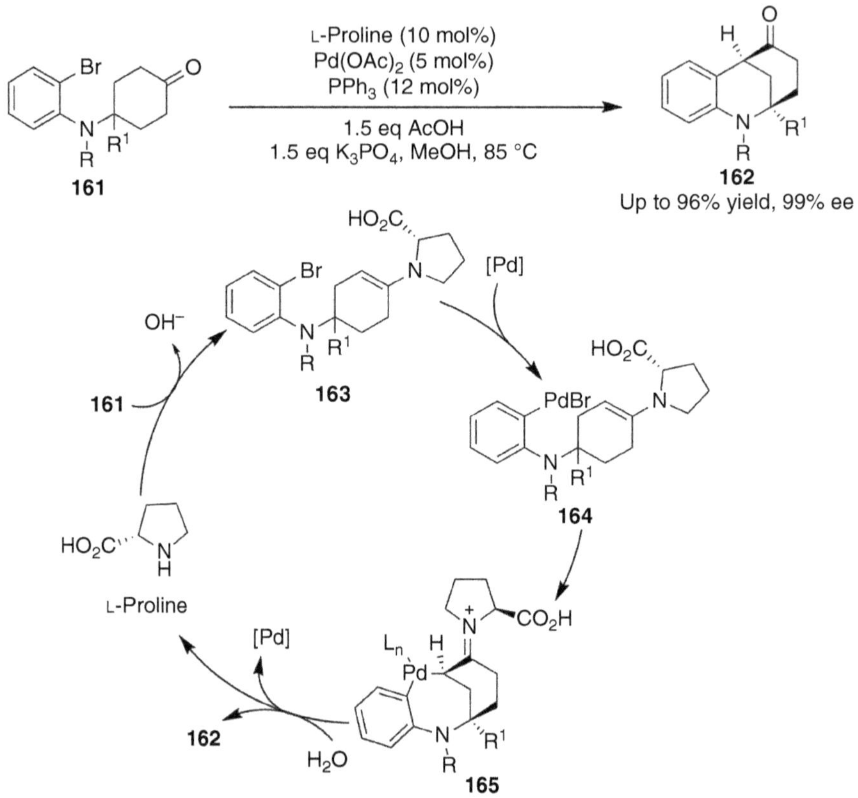

Scheme 3.33 Palladium/enamine catalysis for α-arylation of ketones.

Trapping of arylpalldium(II) species by chiral enamine intermediates through transmetalation or Heck-type insertion of π-system allows asymmetric α-arylation of carbonyls. Jia and coworkers reported a highly enantioselective desymmetric α-arylation of cyclohexanones by palladium and proline cooperative catalysis [39a]. The oxidative addition of aryl-bromide to the Pd(0) complex gives an arylpalladium complex **164** [39b]. An intramolecular transmetalation or a Heck-type insertion then occurs between the arylpalladium moiety and the enamine functionality concomitantly formed from the ketone functionality of the substrate **161** and proline, leading to an intermediate **165**, which after reductive elimination and hydrolysis is transformed into the final product **162**. A variety of optically active morphan derivatives are accessed in good yields and with excellent enantioselectivities (Scheme 3.33).

Lu and coworkers established an enantioselective intramolecular desymmetric α-arylation of cyclobutanones by cooperative catalysis of a chiral palladium complex and a chiral amino acid [40]. The reaction proceeds via a pathway similar to Jia's reaction [39a]. The matched combination of chiral bisphosphine ligand **168** and 2-indolinyl carboxylic acid **169** is required to induce excellent enantioselectivity for the intramolecular desymmetric arylation of *O*-tethered cyclobutanone **166**. Interestingly, enantioselective desymmetric α-arylation of *N*-tethered substrates **170**

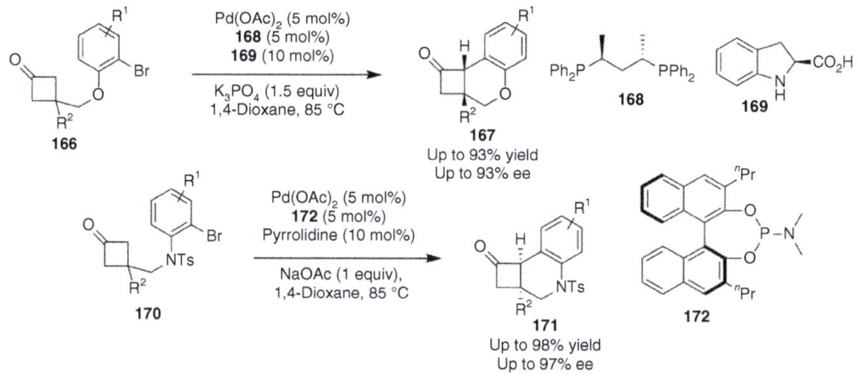

Scheme 3.34 Palladium/amine catalyzed intramolecular α-arylation of cyclobutanones.

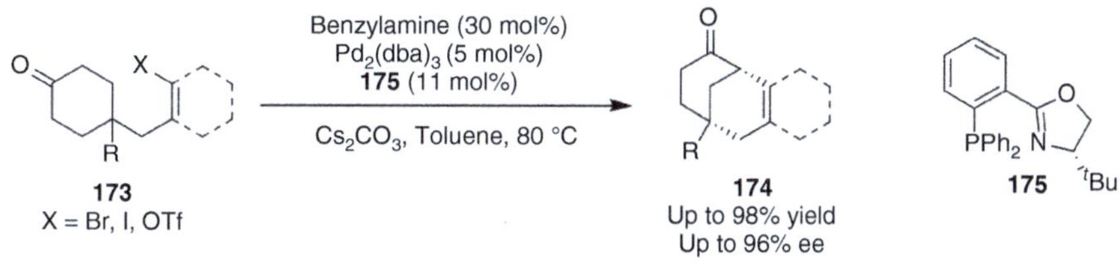

Scheme 3.35 Pd/amine-co-catalyzed asymmetric desymmetrization. Source: Modified from Wei et al. [41].

is achieved by achiral pyrrolidine and a chiral palladium complex *in situ* prepared from $Pd(OAc)_2$ and binol-derived phosphoramidite ligand **172** (Scheme 3.34).

Recently, Liu and coworkers developed an efficient desymmetrization of γ-quaternary carbon-containing cyclohexanones **173** via palladium/benzylamine cooperative catalysis (Scheme 3.35) [41]. The palladium complex adorned with phosphino-oxazoline ligand **175** enables the reaction to give high yields and moderate to good enantioselectivities, whereas the addition of benzylamine can substantially improve the stereoselectivity by enhancing the energetic difference between the transition states.

3.5 Asymmetric Addition to Alkynes by Cooperative Catalysis with π-Lewis Acids

π-Lewis acids, in particular soft Lewis acids, can activate unsaturated chemical bonds toward nucleophilic addition by π-coordination. Enamines are highly nucleophilic and are principally able to attack π-Lewis acid-activated alkenes or alkynes, leading to the formation of organometallic intermediates. After the subsequent protonation-demethylation and hydrolysis, the adducts would be generated (Scheme 3.36). Such enamine/π-Lewis acid cooperative catalysis can stereoselectively enable the direct coupling of alkenes or alkynes with the α-carbon of carbonyls to give value-added products.

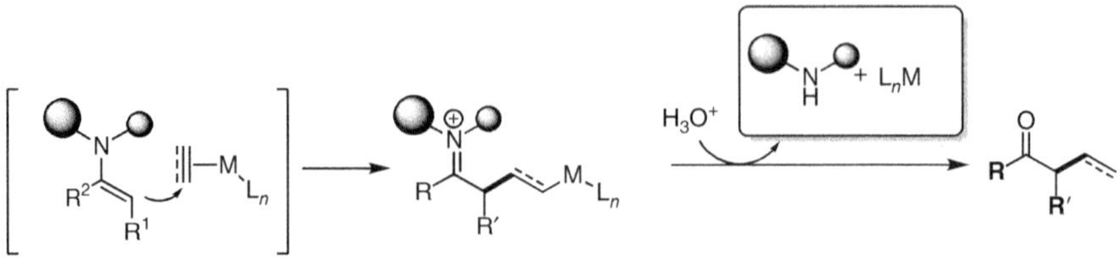

Scheme 3.36 Nucleophilic addition to inactivated unsaturated bonds by cooperative catalysis.

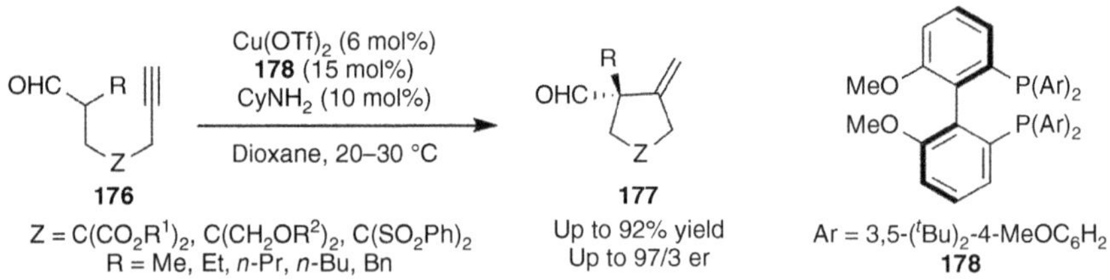

Scheme 3.37 Asymmetric carbocyclizaiton reactions by cooperative catalysis of chiral copper complex and amine.

In 2012, Ratovelomanana-Vidal and coworkers reported an intramolecular enantioselective α-alkenylation of aldehyde [42]. The formyl-alkyne substrate **176** undergoes an asymmetric carbocyclization reaction via an enantioselective addition of the *in situ* formed enamine to the alkynyl residue activated by the chiral copper complex adorned with bisphosphine **178**, leading to enantioenriched cyclopentane carbaldehydes **177** with moderate to excellent enantioselectivities (Scheme 3.37).

Dixon and coworkers described an enantioselective synthesis of 2-azabicyclo[3.3.1] nonanes via an intramolecular desymmetric cyclization of 4-propargylamino cyclohexanones **179** by cooperative catalysis of enamine and chiral silver complex [43]. The chiral silver complex serves as a π-Lewis acid to activate the triple bond of **179**, which then undergoes an enantioselective 6-*exo*-dig cyclization reaction with the enamine concomitantly formed with chiral amine **182** to give a bicyclic intermediate **184**. After hydrolysis, the intermediate **184** is transformed into 2-azabicyclo[3.3.1]nonanes **180** (Scheme 3.38). High levels of enantioselectivity can be achieved by combining the chiral pyrrolidine catalyst **182** and a cinchona alkaloid-derived aminophosphine ligand **181** for the silver co-catalyst, which exhibits a significant matched effect on the stereochemical control. Later, Dixon and coworkers expanded this enamine/π-Lewis acid cooperative catalysis to enantioselective desymmetrization of allene-linked cyclohexanones. Specifically, an L-prolinamide catalyst was used for the enamine activation while the CuOTf activated the allene moiety toward intramolecular nucleophilic addition, providing direct access to enantioenriched 4-vinyl-2-morphan and 4-vinyl-2-oxamorphan derivatives in high yields and enantioselectivity [44].

Direct aldol reaction of ynals with either aldehydes or ketones remains to be developed although it holds great synthetic potential. Chemoselectivity between Michael addition and aldol reaction, together with electrophilicity of the formyl of the ynal, might pose the challenge to access such an aldol reaction. The ynal–metal π-complexation can enhance the reactivity of the formyl aldol accepter and increases

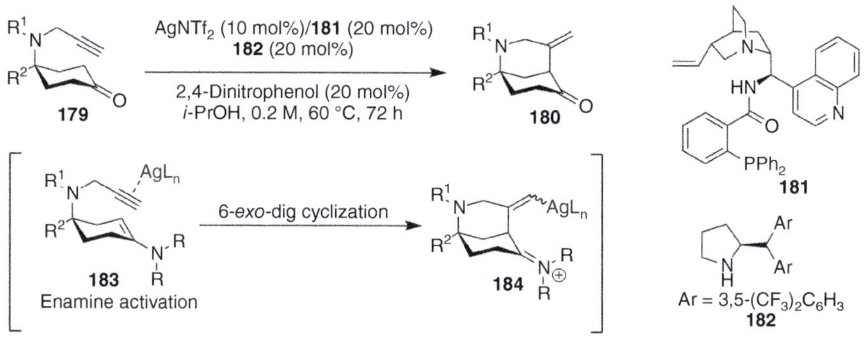

Scheme 3.38 Enantioselective desymmetric cyclization by silver/enamine cooperative catalysis.

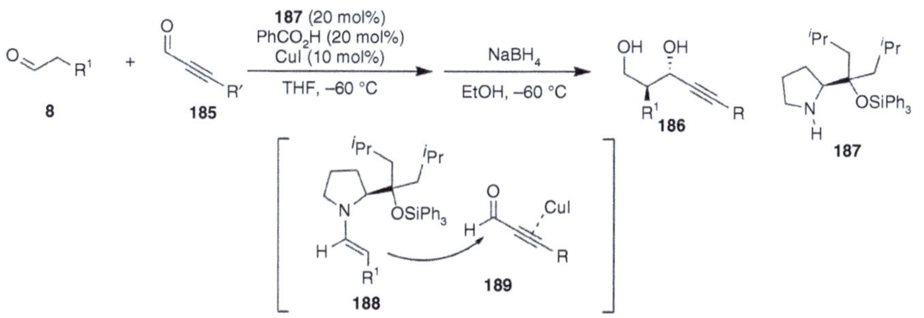

Scheme 3.39 Cross aldol reaction of ynals and aliphatic aldehydes.

the bulkiness of the triple bond to thereby tune the chemo- and stereoselectivities. Palomo established a highly stereoselective cross-aldol reaction between ynals **185** and enolizable aldehydes **8** by cooperative catalysis of copper and enamine [45]. The reversible complexation of copper and the carbon–carbon triple bond makes the cross aldol reaction of ynal much cleaner and more diastereoselective than that otherwise catalyzed by a chiral amine, individually. In particular, the combination of α,α-dialkylprolinol silyl ether, benzoic acid, and CuI offers high levels of diastereo- and enantioselectivity (anti/syn up to >20 : 1, ee up to >99%) (Scheme 3.39).

Gong and coworkers developed a highly enantioselective addition of ketones **191** to inactivated alkenes **190** cooperatively catalyzed by chiral amine and Pd(II). The combination of palladium acetate and chiral amine **193** successfully empowers the reaction to provide the γ-addition products **192** with efficient stereochemical control [46]. The $Pd(OAc)_2$ enables the inactivated alkenes to be reactive enough toward the chiral enamine **195** by coordinating with the double bond as shown in **194**, wherein the chiral amine controls the stereochemistry and offers high levels of stereochemical outcomes (Scheme 3.40).

3.6 Catalytic Asymmetric Propargylic Substitution Reaction of Carbonyl Compounds

Metal-allenylidene complexes generated from propargylic esters or their structural analogs are versatile intermediates capable of undergoing propargylic substitution

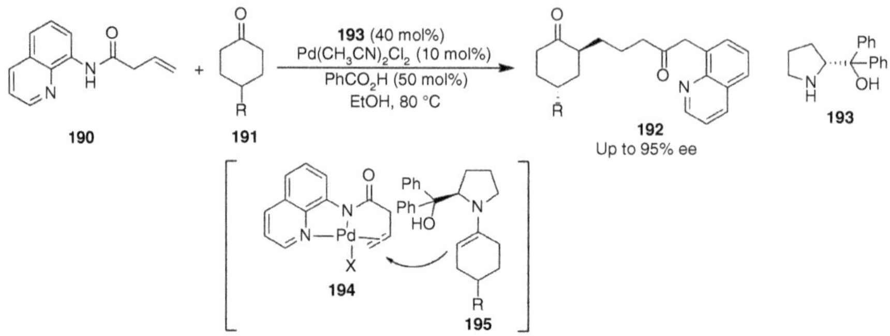

Scheme 3.40 Enantioselective hydroalkylation of ketones with inactivated alkenes.

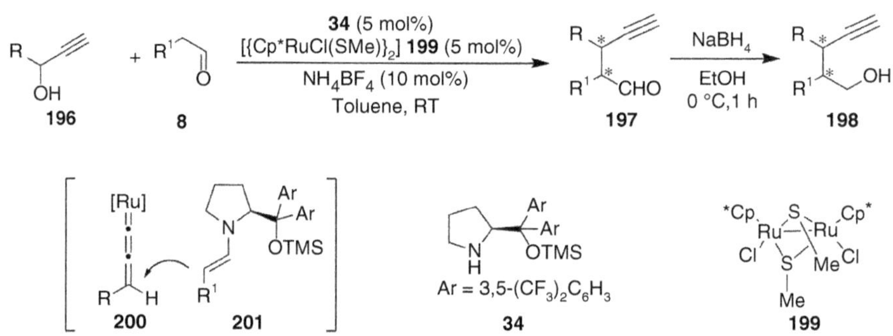

Scheme 3.41 α-Propargylation of aldehydes enabled by Ru/enamine cooperative catalysis.

reactions with a wide scope of nucleophiles, such as enamines [47]. The combination of transition metal and chiral amine would principally enable the asymmetric propargylic substitution of carbonyls.

Nishibayashi reported a ruthenium-catalyzed enantioselective propargylic alkylation of propargylic alcohols with aldehydes in concert with chiral amine catalysis [48]. The synergistic combination of the α,α-diarylprolinol silyl ether **34** with a ruthenium complex **199** efficiently promotes the enantioselective propargylic alkylation via the nucleophilic attack of the chiral enamine **201** to the ruthenium-allenylidene complex **200**. The protocol tolerates a broad scope of reaction components and gives the α-propargyl alcohols **198** in excellent yields and with high enantioselectivity after reduction (Scheme 3.41).

Later, the same group established an enantioselective α-propargylation of aldehydes **8** with propargylic esters **202** cooperatively catalyzed by a copper complex and chiral amine **34** [49]. The γ-carbon of a copper-allenylidene complex **204** formed from the reaction of a propargylic ester with a copper complex of racemic BINAP is attacked by the chiral enamine **205**, generated from aldehyde and the α,α-diarylprolinol silyl ether **34**, to result in the formation of propargylation product **203** in the excellent stereochemical outcome (Scheme 3.42).

An enantioselective γ-propargylation of α,β-unsaturated aldehydes with propargylic alcohols has been realized by cooperative catalysis of the thiolate-bridged diruthenium complex **199** and the chiral amine **34** [50]. The enamine **208**,

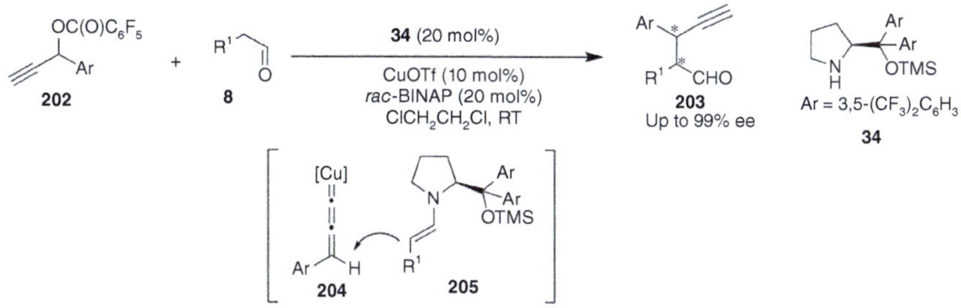

Scheme 3.42 Enantioselective α-propargylation of aldehydes with propargylic esters.

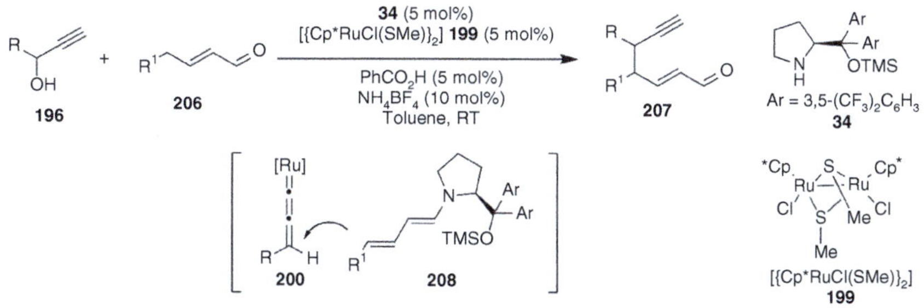

Scheme 3.43 γ-propargylation of enals enabled by Ru/enamine cooperative catalysis.

generated from the enal **206** and the chiral amine **34**, undergoes the propargylation reaction with ruthenium-allenylidene complex **200** to preferentially furnish γ-propargyl enal **207** in high yields as the mixture of two diastereoisomers. The use of 2-propanethiolate-bridged diruthenium complex [Cp*RuCl(μ2-SiPr)]$_2$ to replace [Cp*RuCl(μ2-SMe)]$_2$ leads to dramatically diminished enantioselectivity (Scheme 3.43).

3.7 Catalytic Asymmetric α-Oxidation of Aldehydes

Sibi group reported an asymmetric α-oxidation of aldehydes in the presence of the 2,2,6,6-tetramethylpiperidine-1-oxyl radical (TEMPO) under the cooperative catalysis of MacMillan imidazolidinone **113** and $FeCl_3$ [51]. Comprehensive mechanistic studies by MacMillan suggest that the nucleophilic addition of the transient enamine **120** to an electrophilic iron-TEMPO complex **212** leads to the enantioselective C—O bond formation and releases the $FeCl_2$, which is oxidized by TEMPO to generate catalytically active $FeCl_3$ [52] (Scheme 3.44).

An even more efficient and enantioselective α-oxidation can be established by cooperative catalysis of $CuCl_2$ and imidazolidinone **215** to give stable aldehyde products **210** (Scheme 3.45). The enantioselective C—O bond-forming event still occurs via the nucleophilic addition of the enamine **217** to the electrophilic Cu-TEMPO complex **216** [53].

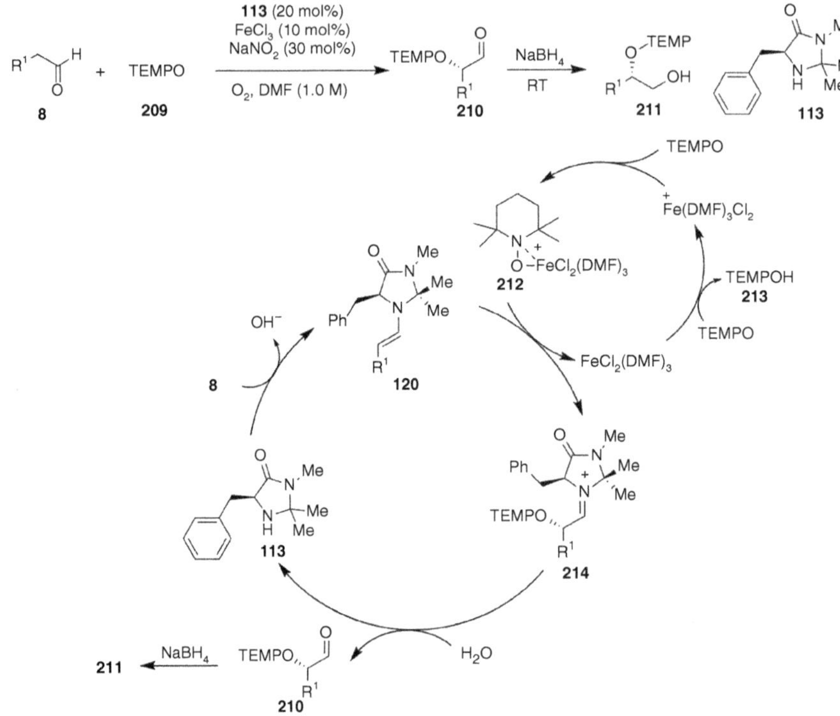

Scheme 3.44 Enantioselective α-oxyamination of aldehydes.

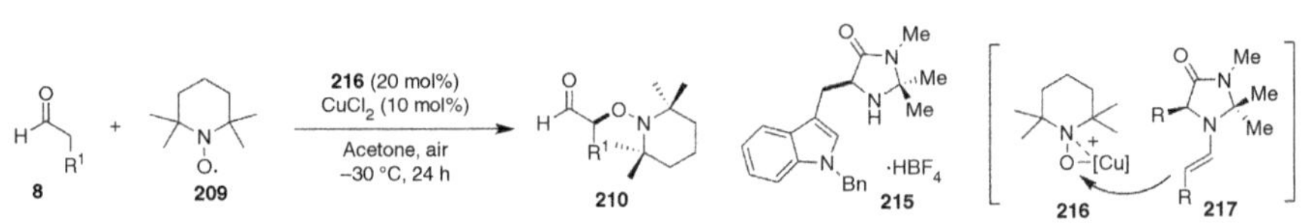

Scheme 3.45 Enantioselective α-oxidation of aldehydes via synergistic catalysis.

3.8 Relay Catalysis

3.8.1 Catalytic Asymmetric Cross Dehydrogenative Coupling

Cross dehydrogenative coupling (CDC) reaction is an efficient method to forge sp^3-chemical bonds from two C—H bonds. Dehydrogenation of a sp^3 C—H bond adjacent to either nitrogen or oxygen atom, usually mediated with a transition metal complex, produces a highly electrophilic iminium or oxonium intermediate. Either type of electrophiles, presumably activated by complexation with the same metal, are reactive toward the enamine. Thus, the combination of transition metal and chiral amine would be able to initiate an asymmetric cross-dehydrogenative coupling reaction (Scheme 3.46).

In 2009, Klussmann reported an oxidative coupling of cyclic tertiary amines **218** with acetone **71** enabled by vanadium and enamine combined catalysis [54].

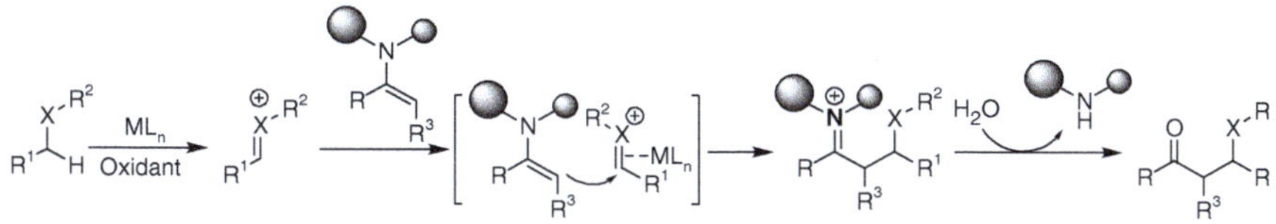

Scheme 3.46 Cross dehydrogenative coupling via cooperative catalysis.

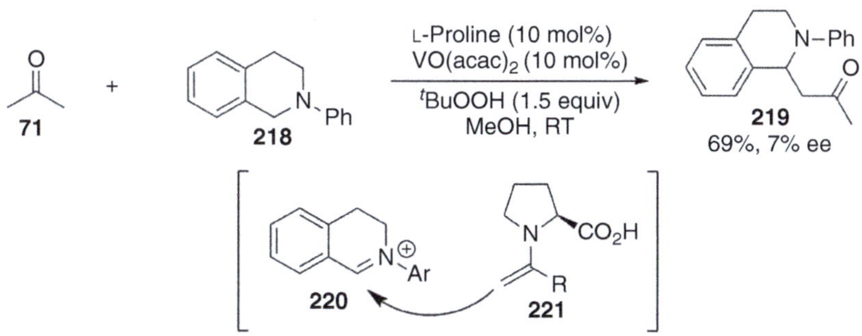

Scheme 3.47 Dehydrogenative coupling between amines and ketone by V(IV)/enamine cooperative catalysis.

An iminium **220**, formed from the oxidation of *N*-phenyl tetrahydroquinoline **218** catalyzed by vanadyl acetylacetonate, undergoes a Mannich-type reaction with the transient enamine **221**. Unfortunately, neither proline nor the prolinol derivative can induce a decent level of enantiocontrol (Scheme 3.47).

Huang and coworker developed a cross-dehydrogenative coupling reaction of amino esters **222** with acetone **71** by copper and enamine cooperative catalysis [55]. As indicated by Li and others [56], the $CuBr_2$ initiates oxidation of the amino ester **222** to an imino ester **226** that undergoes Mannich reaction with the ketone catalyzed by proline ester. The enantioselective version was attempted by using methyl L-prolinate as organocatalyst, however only resulted in moderate enantioselectivity of 15% ee (Scheme 3.48).

The highly enantioselective cross-dehydrgogenative coupling reaction of tetrahydroquinolines **218** with propionaldehyde **228** was established by synergistically merging the $CuBr_2$ and prolinol silyl ether catalysis **34** [57]. The use of **34** as the organocatalyst offers excellent enantioselectivity, but moderate diastereoselectivity (Scheme 3.49).

Luo reported an asymmetric oxidative α-amination of β-ketocarbonyls **230** with hydroxycarbamates **231** under aerobic conditions driven by enamine and copper cooperative catalysis [58]. The chiral primary amine tends to form an enamine **234** with β-ketocarbonyls stabilized by the intramolecular hydrogen-bonding interaction (Scheme 3.50). Meanwhile, the copper-catalyzed aerobic oxidation of benzyl hydroxycarbamate **231** proceeds to produce benzyl nitrosoformate **235** [59]. Then, the enantioselective nitroso aldol reaction of the enamine **234** and benzyl nitrosoformate **235** occurs via transition state **236** in good yields and with excellent enantioselectivities.

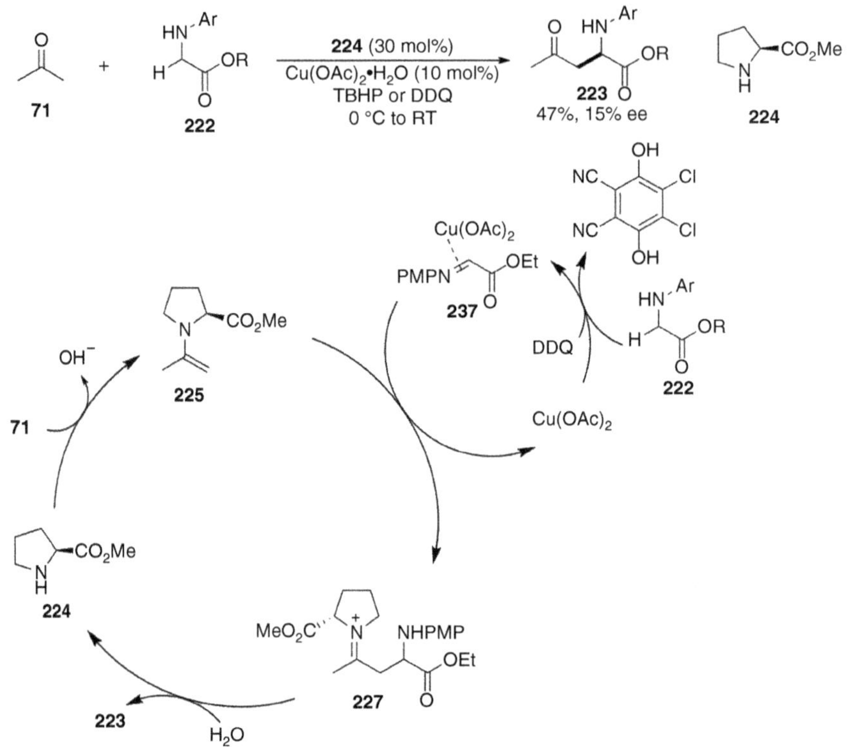

Scheme 3.48 Cross-dehydrogenative coupling of amino esters and acetone.

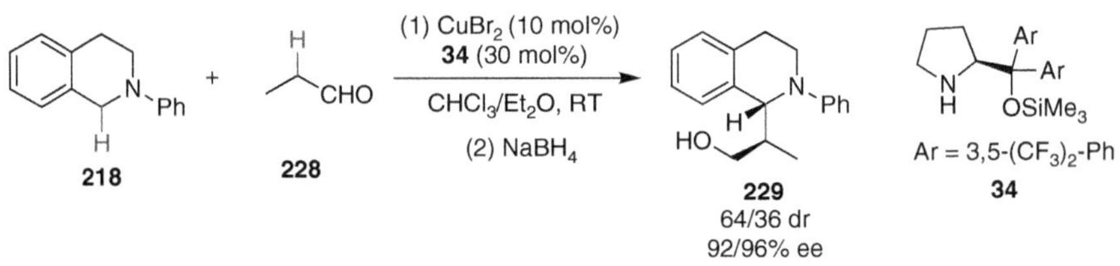

Scheme 3.49 Enantioselective dehydrogenative coupling of tetrahydroquinoline and aldehydes.

Later, the same group created an enantioselective N,O-ketalization of β-ketoesters **237** by using a similar catalytic system (Scheme 3.51) [60]. The process starts with the asymmetric hydroamination of β-ketoesters **237** enabled by enamine and copper relay catalysis to generate intermediates **241**. The presence of Brønsted acid facilitates the dehydration of the intermediate **241** to give diimine **242**, which subsequently undergoes an enantioselective addition reaction with benzyl hydroxycarbamate **231** to deliver hemiaminal-type product **238** in high yield and excellent enantioselectivity. Ketones are also reactive substrates and can participate in the reaction in the same way, but with much-diminished enantioselectivity.

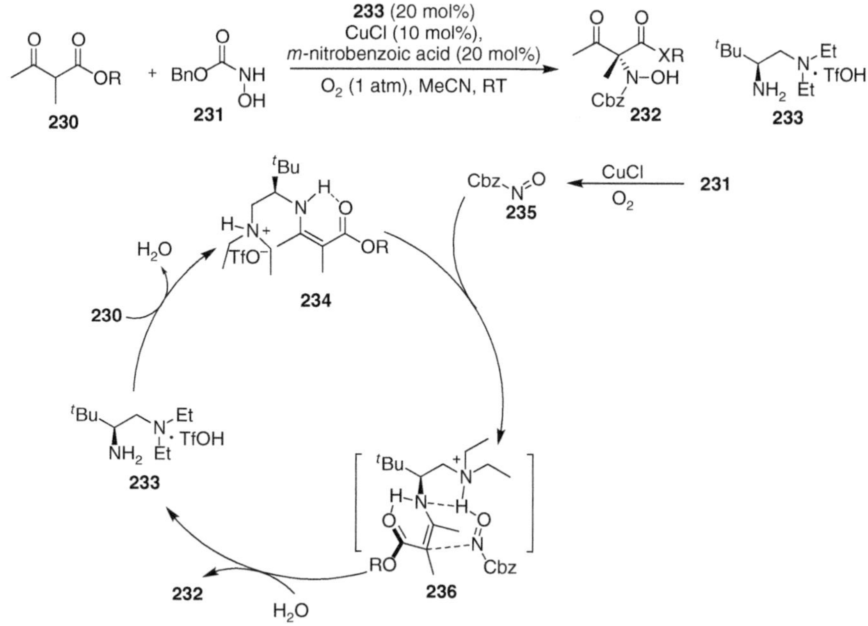

Scheme 3.50 Asymmetric α-amination of β-ketocarbonyls.

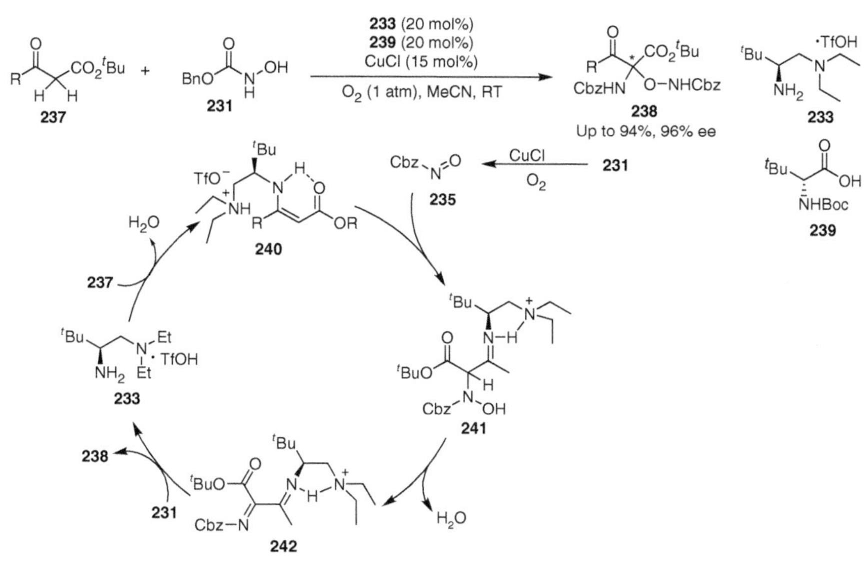

Scheme 3.51 α,α-Bis-functionalization of β-ketocarbonyls. Source: Modified from Xu et al. [60].

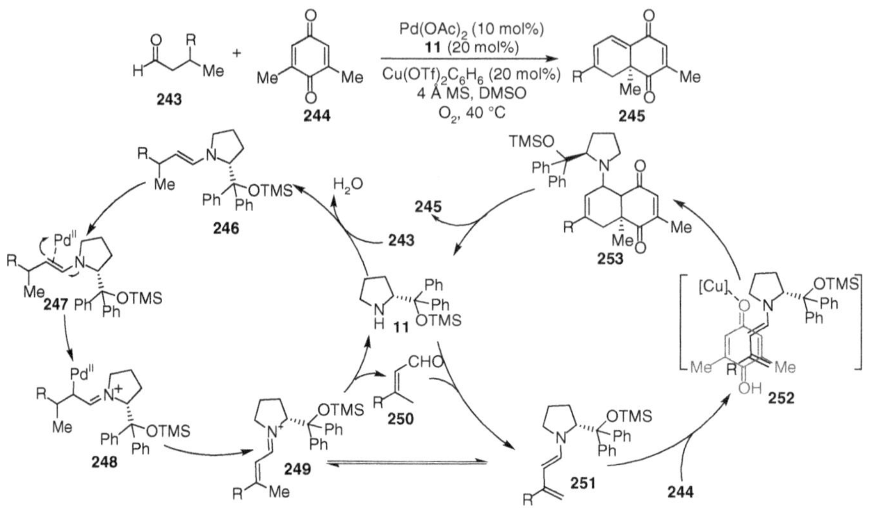

Scheme 3.52 Metal/organo cooperative and relay catalysis for the asymmetric dehydrogenative cycloaddition reaction. Source: Modified from Zhou et al. [61].

By merging catalytic activity of Pd(II) and chiral amine, Gong and coworkers established a stereoselective functionalization of inactive C—H bonds of aliphatic aldehydes **243**, enabling an asymmetric dehydrogenative [4+2] cyclization reaction with quinones to generate chiral bicyclic compounds **245** with excellent enantioselectivities (Scheme 3.52) [61]. This transformation occurs starting with Saegusa-type oxidation of enamine intermediate generated from chiral amine **11** and the aldehyde **243** to give an unsaturated iminium intermediate **249** [62], which then transforms to a dienamine **251**. Subsequently, an enantioselective [4+2] cycloaddition reaction between **251** and quinone **244** proceeds via transition state **252** to furnish the final adduct **245** after the elimination of the amine catalyst **11** from intermediate **253** [63]. Interestingly, the presence of the copper(II) triflate can enhance the reactivity of the quinone derivatives **244** while maintains the enantioselectivity.

3.8.2 Transformation of Olefins

Olefins can undergo transition metal-catalyzed hydroformylation and are thus the precursor of aldehydes. Based on this transformation, the groups of Eilbracht and Breit designed cascade hydroformylation and asymmetric aldol reaction enabled by relay catalysis of rhodium complex and proline, respectively. Eilbracht reported an enantioselective cascade reaction of styrenes **254** or cyclic olefins with acetone and syngas to assemble chiral β-hydroxy ketones **255** (Scheme 3.53a) [64]. In this process, α-branched aldehydes are initially produced from the hydroformylation of styrenes catalyzed by the rhodium complex *in situ* prepared from [Rh(acac)(CO)$_2$] and triphenyl phosphinite, and then undergo proline-catalyzed asymmetric aldol reaction to give the adducts **255** in high yields, but with low diastereoselectivity and moderate enantioselectivity. Even greater diastereo- and enantioselectivities are obtained

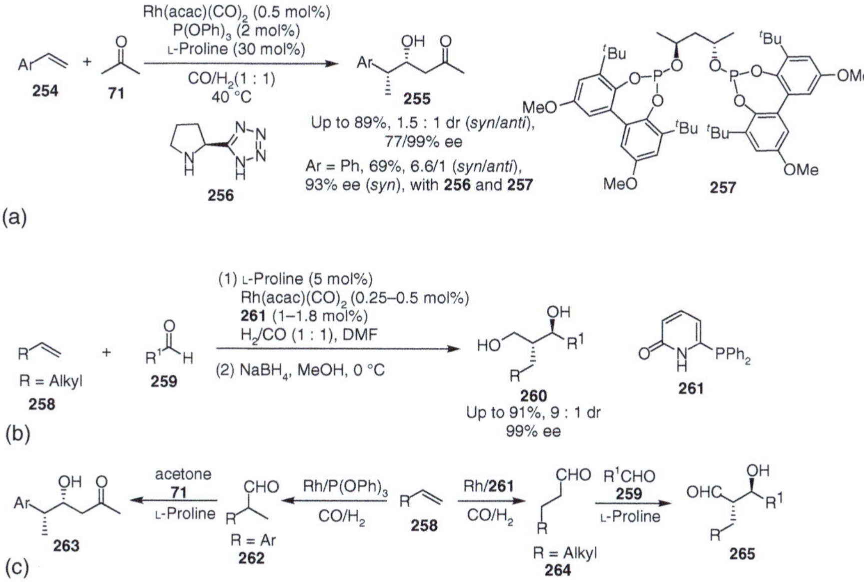

Scheme 3.53 Relay catalytic cascade hydroformylation and asymmetric aldol reaction. (a) Eilbracht's work. (b) Breit's work. (c) Reaction pathway. Source: (a) Chercheja et al. [64]. (b) Modified from Abillard and Breit [65].

by using chiral phosphinite ligand **257** and proline-tetrazole **256** (Scheme 3.53a). Breit and coworker investigated a similar cascade transformation of α-alkenes **258** with aldehydes **259** and syngas, leading to cross-aldol products **260** in excellent yield, high diastereoselectivity, and almost perfect enantioselectivity (Scheme 3.53b) [65]. The rhodium complex of the phosphine ligand **261** enables the hydroformylation of alkenes to give linear aldehydes that participate in the asymmetric cross aldol reaction promoted by proline (Scheme 3.53c).

Based on a similar strategy, Eilbracht established sequential hydroformylation and enantioselective Mannich reaction of aniline **267**, cyclopentene **266**, and acetone **71** under the atmosphere of syngas, giving rise to β-amino ketone **268** in moderate yield and enantiomeric excess (Scheme 3.54) [66].

In 2012, Behr and coworkers reported a cascade hydroformylation/α-alkylation of α-alkenes **258** with bis(4-(dimethylamino)phenyl)methanol **269** sequentially catalyzed by a ternary metal/organo hybrid catalyst system [67]. The presence of $Ru(C_5H_5)_2$ can enhance the rhodium complex-catalyzed hydroformylation and improve the conversion in the organocatalytic α-alkylation. The tandem sequence was initially optimized toward a high yield by the variation of substrate ratio, rhodium precursor, and ligands. High enantioselectivity can be realized by using diarylprolinol silyl ether **11** as a co-organocatalyst (Scheme 3.55).

Under the transition metal catalysis, allylic alcohols can smoothly be transformed into aldehydes by the double bond migration and thus, are used as the precursors of aldehydes. Alexakis and coworkers reported a sequential isomerization/enantioselective α-functionalization in the presence of a chiral iridium

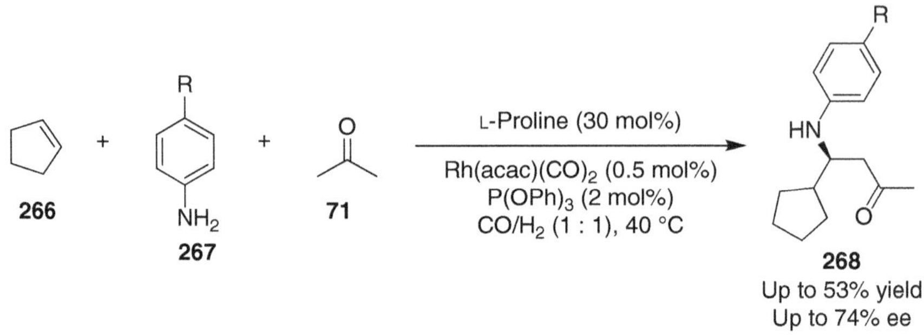

Scheme 3.54 Relay catalytic hydroformylation and Mannich reaction. Source: Modified from Chercheja et al. [66].

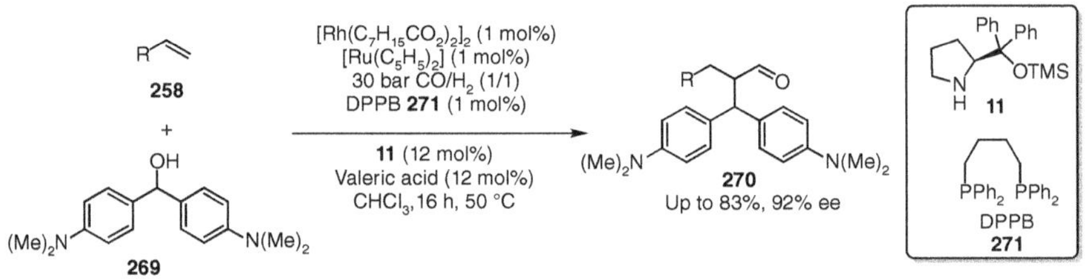

Scheme 3.55 Relay catalytic hydroformylation/α-alkylation.

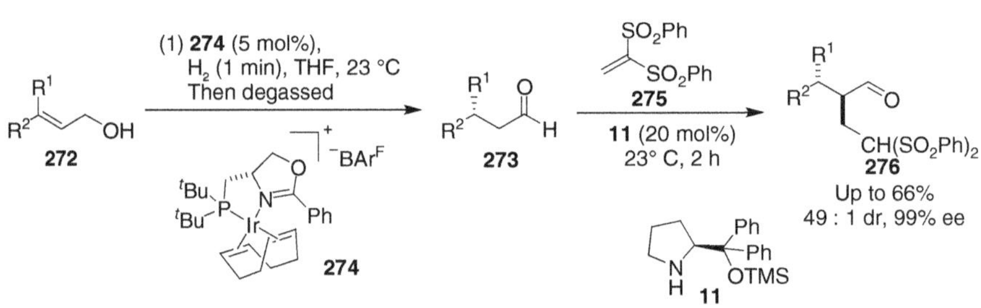

Scheme 3.56 Relay isomerization/enantioselective α-functionalization.

complex and a chiral amine [68]. The chiral iridium complex **274** efficiently catalyzes an enantioselective isomerization of 3,3-disubstituted primary allylic alcohols **272** to give β-chiral aldehydes **273**. Then, the chiral aldehydes participate in a diastereoselective Michael addition with vinyl sulfone **275** to afford highly α-functionalization of α,β-chiral aldehydes **276** with excellent diastereo- and enantioselectivities (Scheme 3.56).

3.9 Conclusion

The α-functionalization of carbonyl compounds represents one of the most important types of methods to assemble chemical bonds and holds widespread applications in synthetic chemistry. The combination of enamine activation and transition metal catalysis provides more opportunities to enable otherwise inaccessible reaction pathways for the enantioselective α-functionalization and transformation of aldehydes and ketones to construct complex chiral molecules. Indeed, such a concept has driven a wide scope of asymmetric transformations, including asymmetric alkylation, asymmetric arylation, asymmetric alkyne activation, and asymmetric propargylic substitution reaction. The salient features of these cooperative catalytic systems include the use of easily available commercial catalysts and substrates and efficient entry to carbonyl scaffolds with excellent levels of stereoselectivity under mild reaction conditions. However, most precedents of the combined enamine and transition metal catalysis require preactivated electrophiles and direct coupling of enamines with ubiquitous C—H bonds remains challenging. Besides, asymmetric α-heteroatom (e.g. boron, silicon, and chalcogens) functionalization of aldehydes or ketones via enamine and transition metal synergistic catalysis would be of great synthetic values. In general, there is still room for further development of new cooperatively catalytic protocols for the asymmetric synthesis of biologically relevant molecular structures.

References

1 (a) Stork, G., Brizzolara, A., Landesman, H. et al. (1963). *J. Am. Chem. Soc.* 85: 207–222. (b) Stork, G. and Saccomano, N.A. (1987). *Tetrahedron Lett.* 28: 2087–2090. (c) Rappoport, Z. (ed.) (1994). *The Chemistry of Enamines*. New York: Wiley.

2 (a) Hajos, Z.G. and Parrish, D.R. (1974). *J. Org. Chem.* 39: 1615–1621. (b) Eder, U., Sauer, R., and Wiechert, R. (1971). *Angew. Chem. Int. Ed.* 10: 496–497. (c) List, B., Lerner, R.A., and Barbas, C.F. III, (2000). *J. Am. Chem. Soc.* 122: 2395–2396. (c) List, B. (2000). *J. Am. Chem. Soc.* 122: 9336–9337. (d) Sakthivel, K., Notz, W., Bui, T., and Barbas, C.F. III, (2001). *J. Am. Chem. Soc.* 123: 5260–5267.

3 Mukherjee, S., Yang, J.W., Hoffmann, S., and List, B. (2007). *Chem. Rev.* 107: 5471–5569.

4 Melchiorre, P., Marigo, M., Carlone, A., and Bartoli, G. (2008). *Angew. Chem.* 120: 6232–6265; (2008). *Angew. Chem. Int. Ed.* 47: 6138–6171.

5 Afewerki, S. and Córdova, A. (2016). *Chem. Rev.* 116: 13512–13570.

6 (a) Hamilton, G.L., Kang, E.J., Mba, M., and Toste, F.D. (2007). *Science* 317: 496–499. (b) Du, Z.T. and Shao, Z.-H. (2012). *Chem. Soc. Rev.* 42: 1337–1378. (c) Chen, D.-F., Han, Z.-Y., Zhou, X.-L., and Gong, L.-Z. (2014). *Acc. Chem. Res.* 47: 2365–2377.

7 (a) Trost, B.M. and Vranken, D.L.V. (1996). *Chem. Rev.* 96: 395–422. (b) Trost, B.M. and Crawley, M.L. (2003). *Chem. Rev.* 103: 2921–2944.

8 Zhao, X., Liu, D., Xie, F., and Zhang, W. (2009). *Tetrahedron* 65: 512–517.

9 Ibrahem, I. and Córdova, A. (2006). *Angew. Chem.* 118: 1986–1990; (2006). *Angew. Chem. Int. Ed.* 45: 1952–1956.

10 Afewerki, S., Ibrahem, I., Rydfjord, J. et al. (2012). *Chem. Eur. J.* 18: 2972–2977.

11 (a) Yoshida, M., Terumine, T., Masaki, E., and Hara, S. (2013). *J. Org. Chem.* 78: 10853–10859. (b) Yoshida, M., Masaki, E., Terumine, T., and Hara, S. (2014). *Synthesis* 46: 1367–1373.

12 (a) Bihelovic, F., Matovic, R., Vulovic, B., and Saicic, R.N. (2007). *Org. Lett.* 9: 5063–5066. (b) Vulovic, B., Bihelovic, F., Matovic, R., and Saicic, R.N. (2009). *Tetrahedron* 65: 10485–10494.

13 Krautwald, S., Sarlah, D., Schafroth, M.A., and Carreira, E.M. (2013). *Science* 340: 1065–1068.

14 Bhaskararao, B. and Sunoj, R.B. (2015). *J. Am. Chem. Soc.* 137: 15712–15722.

15 (a) Krautwald, S., Schafroth, M.A., Sarlah, D., and Carreira, E.M. (2014). *J. Am. Chem. Soc.* 136: 3020–3023. (b) Sandmeier, T., Krautwald, S., Zipfel, H.F., and Carreira, E.M. (2015). *Angew. Chem.* 127: 14571–14575; (2015). *Angew. Chem. Int. Ed.* 54: 14363–14367. (c) Schafroth, M.A., Zuccarello, G., Krautwald, S. et al. (2014). *Angew. Chem.* 126: 14118–14121; (2014). *Angew. Chem. Int. Ed.* 53: 13898–13901.

16 Zhang, M.-M., Wang, Y.-N., Wang, B.-C. et al. (2019). *Nat. Commun.* 10: 2716.

17 Næsborg, L., Halskov, S.S., Tur, F. et al. (2015). *Angew. Chem.* 127: 10331–10335; (2015). *Angew. Chem. Int. Ed.* 54: 10193-10197.

18 (a) Skrzyńska, A., Przydacz, A., and Albrecht, Ł. (2015). *Org. Lett.* 17: 5682–5685. (b) Su, Y.-L., Han, Z.-Y., Li, Y.-H., and Gong, L.-Z. (2017). *ACS Catal.* 7: 7917–7922.

19 (a) Zhao, X.H., Liu, D.L., Guo, H. et al. (2011). *J. Am. Chem. Soc.* 133: 19354–19357. (b) Huo, X., Yang, G., Liu, D. et al. (2014). *Angew. Chem.* 126: 6894–6898; (2014). *Angew. Chem. Int. Ed.* 53: 6776–6780. (c) Huo, X., Quan, M., Yang, G. et al. (2014). *Org. Lett.* 16: 1570–1573.

20 (a) Zhou, H., Zhang, L., Xu, C., and Luo, S. (2015). *Angew. Chem.* 127: 12836–12839; (2015). *Angew. Chem. Int. Ed.* 54: 12645–12648.

21 (a) Trost, B.M. (1998). *Chem. Eur. J.* 4: 2405–2412. (b) Koschker, P. and Breit, B. (2016). *Acc. Chem. Res.* 49: 1524–1536.

22 Li, M., Datta, S., Barber, D.M., and Dixon, D.J. (2012). *Org. Lett.* 14: 6350–6353.

23 Zhou, H., Wang, Y., Zhang, L. et al. (2017). *J. Am. Chem. Soc.* 139: 3631–3634.

24 Cruz, F.A. and Dong, V.M. (2017). *J. Am. Chem. Soc.* 139: 1029–1032.

25 Capdevila, M.G., Benfatti, F., Zoli, L. et al. (2010). *Chem. Eur. J.* 16: 11237–11241.

26 Chiarucci, M., Lillo, M.d., Romaniello, A. et al. (2012). *Chem. Sci.* 3: 2859–2863.

27 Mayr, H., Kempf, B., and Ofial, A.R. (2003). *Acc. Chem. Res.* 36: 66–77.

28 (a) Cozzi, P.G., Benfatti, F., and Zoli, L. (2009). *Angew. Chem.* 121: 1339–1342; (2009). *Angew. Chem. Int. Ed.* 48: 1313. (b) Xiao, J. (2012). *Org. Lett.* 14: 1716–1719.

29 Motoyama, K., Ikeda, M., Miyake, Y., and Nishibayashi, Y. (2011). *Eur. J. Org. Chem.*: 2239–2246.

30 Rueping, M., Volla, C.M.R., and Atodiresei, I. (2012). *Org. Lett.* 14: 4642–4645.

31 Sun, S., Mao, Y., Lou, H., and Liu, L. (2015). *Chem. Commun.* 51: 10691–10694.

32 Berti, F., Malossi, F., Marchetti, F., and Pineschi, M. (2015). *Chem. Commun.* 51: 13694–13697.

33 Allen, A.E. and MacMillan, D.W.C. (2011). *J. Am. Chem. Soc.* 133: 4260–4263.

34 Skucas, E. and MacMillan, D.W.C. (2012). *J. Am. Chem. Soc.* 134: 9090–9093.

35 (a) Filler, R., Kobayashi, Y., and Yagupolskii, L.M. (1993). *Organofluorine Compounds in Medicinal Chemistry and Biomedical Applications*. New York: Elsevier. (b) Pursor, S., Moore, P.R., Swallow, S., and Gouverneur, V. (2008). *Chem. Soc. Rev.* 37: 320–330.

36 (a) Nagib, D.A., Scott, M.E., and MacMillan, D.W.C. (2009). *J. Am. Chem. Soc.* 131: 10875. (b) Noritake, S., Shibata, N., Nomura, Y. et al. (2009). *Org. Biomol. Chem.* 7: 3599–3604.

37 Allen, A.E. and MacMillan, D.W.C. (2010). *J. Am. Chem. Soc.* 132: 4986–4987.

38 Stevens, J.M. and MacMillan, D.W.C. (2013). *J. Am. Chem. Soc.* 135: 11756–11759.

39 (a) Liu, R.-R., Li, B.-L., Lu, J. et al. (2016). *J. Am. Chem. Soc.* 138: 5198–5201. (b) Solé, D., Vallverdú, L., Solans, X. et al. (2003). *J. Am. Chem. Soc.* 125: 1587–1594.

40 Wang, M., Chen, J., Chen, Z. et al. (2018). *Angew. Chem.* 130: 2737–2741; (2018). *Angew. Chem. Int. Ed.* 57: 2707–2711.

41 Wei, Q., Cai, J., Hu, X.-D. et al. (2020). *ACS Catal.* 10: 216–224.

42 Montaignac, B., Praveen, C., Vitale, M.R. et al. (2012). *Chem. Commun.* 48: 6559–6561.

43 Manzano, R., Datta, S., Paton, R.S., and Dixon, D.J. (2017). *Angew. Chem.* 129: 5928–5932; (2017). *Angew. Chem. Int. Ed.* 56: 5834–5838.

44 Zhang, L., Yamazaki, Y., Leitch, J.A. et al. (2020). *Chem. Sci.* 11: 7444–7450.

45 Gómez-Bengoa, E., García, J.M., Jiménez, S. et al. (2013). *Chem. Sci.* 4: 3198–3204.

46 Shen, H.-C., Zhang, L., Chen, S.-S. et al. (2019). *ACS Catal.* 9: 791–797.

47 Sakata, K. and Nishibayashi, Y. (2018). *Catal. Sci. Technol.* 8: 12–25.

48 Ikeda, M., Miyake, Y., and Nishibayashi, Y. (2010). *Angew. Chem.* 122: 7447–7451; (2010). *Angew. Chem. Int. Ed.* 49: 7289–7293.

49 Yoshida, A., Ikeda, M., Hattori, G. et al. (2011). *Org. Lett.* 13: 592–595.

50 Ikeda, M., Miyake, Y., and Nishibayashi, Y. (2012). *Organometallics* 31: 3810–3813.

51 Sibi, M.P. and Hasegawa, M. (2007). *J. Am. Chem. Soc.* 129: 4124–4125.

52 Humbeck, J.F.V., Simonovich, S.P., Knowles, R.R., and MacMillan, D.W.C. (2010). *J. Am. Chem. Soc.* 132: 10012–10014.

53 Simonovich, S.P., Humbeck, J.F.V., and MacMillan, D.W.C. (2012). *Chem. Sci.* 3: 58–61.

54 Sud, A., Sureshkumar, D., and Klussmann, M. (2009). *Chem. Commun.*: 3169–3171.

55 Xie, J. and Huang, Z.-Z. (2010). *Angew. Chem.* 122: 10379–10383; (2010). *Angew. Chem. Int. Ed.* 49: 10181–10185.

56 (a) Girard, S.A., Knauber, T., and Li, C.-J. (2014). *Angew. Chem.* 126: 76–103; (2014). *Angew. Chem. Int. Ed.* 53: 74–100. (b) Yeung, C.S. and Dong, V.M. (2011). *Chem. Rev.* 111: 1215–1292.

57 Zhang, J., Tiwari, B., Xing, C. et al. (2012). *Angew. Chem.* 124: 3709; (2012). *Angew. Chem. Int. Ed.* 51: 3649–3652.

58 Xu, C., Zhang, L., and Luo, S. (2014). *Angew. Chem.* 126: 4233–4237; (2014). *Angew. Chem. Int. Ed.* 53: 4149–4153.

59 Sandoval, D., Frazier, C.P., Bugarin, A., and Read deAlaniz, J. (2012). *J. Am. Chem. Soc.* 134: 18948–18951.

60 Xu, C., Zhang, L., and Luo, S. (2015). *Org. Lett.* 17: 4392–4395.

61 Zhou, X.-L., Wang, P.-S., Zhang, D.-W. et al. (2015). *Org. Lett.* 17: 5120–5123.

62 (a) Zhu, J., Liu, J., Ma, R. et al. (2009). *Adv. Synth. Catal.* 351: 1229–1232. (b) Liu, J., Zhu, J., Jiang, H. et al. (2009). *Chem. Asian J.* 4: 1712–1716.

63 Johansen, T.K., Gómez, C.V., Bak, J.R. et al. (2013). *Chem. Eur. J.* 19: 16518–16522.

64 (a) Chercheja, S. and Eibracht, P. (2007). *Adv. Synth. Catal.* 349: 1897–1905. (b) Chercheja, S., Nadakudity, S.K., and Eilbracht, P. (2010). *Adv. Synth. Catal.* 352: 637–643.

65 Abillard, O. and Breit, B. (2007). *Adv. Synth. Catal.* 349: 1891–1895.

66 Chercheja, S., Rothenbücher, T., and Eilbracht, P. (2009). *Adv. Synth. Catal.* 351: 339–344.

67 Stiller, J., Vorholt, A.J., Ostrowski, K.A. et al. (2012). *Chem. Eur. J.* 18: 9496–9499.

68 Quintard, A., Alexakis, A., and Mazet, C. (2011). *Angew. Chem.* 123: 2402–2406; (2011). *Angew. Chem. Int. Ed.* 50: 2354–2358.

4

Iminium and Metal Combined Catalysis

4.1 Introduction: Iminium Activation and Metal Combined Catalysis

The success of iminium activation of α,β-unsaturated aldehydes or ketones arises from the reversible condensation reaction of the carbonyl functionality and chiral amine catalysts. The *in situ* generated chiral iminium ions, often in equilibrium with electron-rich enamines, have become a general concept to bring to fruition a tremendous number of asymmetric transformations [1–3]. The LUMO (lowest unoccupied molecular orbital)-lowering activation inherent to the iminium catalysis increases the electrophilicity of β-carbon, facilitating the asymmetric β-functionalization of α,β-unsaturated carbonyl compounds through enantioselective conjugate addition reaction (left catalytic cycle, Scheme 4.1). The enamine intermediate, generated after the amine-catalyzed conjugate addition reaction, can undergo nucleophilic substitution, addition, or other related reactions to enable diverse α-functionalization with excellent stereo- and enantiocontrol (right catalytic cycle, Scheme 4.1) [4].

Catalytic organometallic reactions of many metal complexes and appropriate substrates provide versatile organometallic species (either nucleophilic or electrophilic) for further transformations of complex mechanisms. Theoretically, the organometallic nucleophiles can be trapped by α,β-unsaturated chiral iminium ions while electrophilic organometallic species or Lewis acid-activated electrophiles would react with the enamine intermediates [5]. Principally, the iminium activation and metal catalysis can merge in two fashions: cooperative catalysis and hybrid relay-cooperative catalysis. In cooperative catalysis, usually a transition metal complex catalyst is employed to activate an inert chemical bond of a substrate, affording a reactive nucleophilic organometallic intermediate to undergo addition reaction to the carbon–carbon double bond of α,β-unsaturated chiral iminium ion via a presumable intermediate **I** (Scheme 4.2). In hybrid relay-cooperative catalysis, an enamine intermediate is formed upon the conjugate addition of organic nucleophile (H–Nu) to the α,β-unsaturated chiral iminium ion. Then cooperative enamine activation and metal catalysis (usually late transition metal) enable the bond-forming event at α-position with corresponding organometallic electrophiles, presumably via an intermediate **II** (Scheme 4.2). The chiral conformation of product molecules can be induced by either chiral amines or chiral metal complexes.

Asymmetric Organo-Metal Catalysis: Concepts, Principles, and Applications, First Edition. Liu-Zhu Gong.
© 2022 WILEY-VCH GmbH. Published 2022 by WILEY-VCH GmbH.

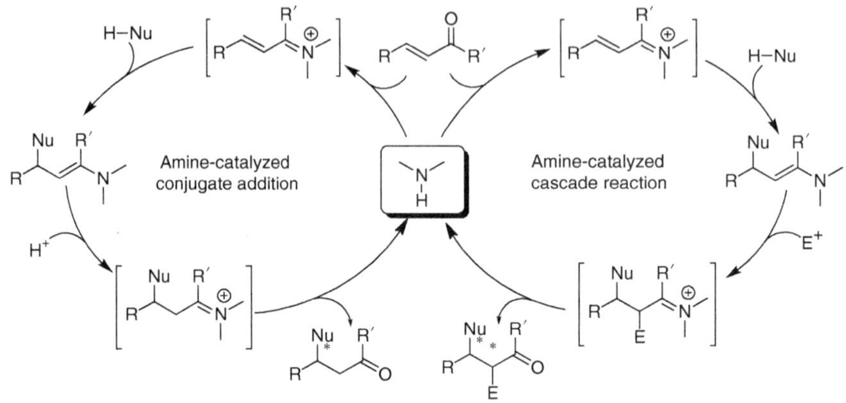

Scheme 4.1 Iminium catalysis. Source: Modified from List [4].

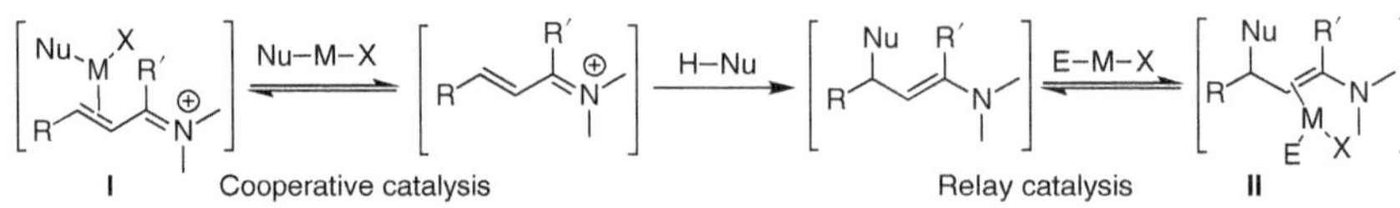

Scheme 4.2 General concepts in iminium and metal combined catalysis.

Excitingly, enantiodivergent synthesis may profit from the synergy of chiral imine activation and asymmetric metal catalysis, in which fine-tuning of each chiral element enables a high degree of stereo- and enantiocontrol. In this chapter, the seminal discovery of the iminium and metal combined catalysis, as well as recent applications of this unique catalytic system in asymmetric functionalization of α,β-unsaturated carbonyls will mainly be discussed.

4.2 Iminium Activation and Palladium Catalysis

4.2.1 Enantioselective Conjugate Addition Reaction

Transmetalation of organic nucleophiles to cationic palladium(II) complexes introduces conjugate addition reactivity to α,β-unsaturated carbonyl compounds [6]. Other than chiral palladium(II) catalysis, Córdova and coworkers disclosed $Pd(OAc)_2$ and a chiral prolinol derivative **4** co-catalyzed asymmetric β-arylation of α,β-unsaturated aldehydes **1** with aryl boronic acids **2** [7]. In this process, MeOH-accelerated transmetalation of **2** affords an aryl palladium acetate intermediate **6**, which can undergo nucleophilic addition to an *in situ* generated α,β-unsaturated iminium ion **5** from enal **1** and chiral amine catalyst **4**. The following protonolysis of the resultant alkyl palladium(II) intermediate **8**, yields iminium ion **9** and regenerates the active palladium(II) species for transmetalation [8]. Importantly, MeOH serves as the proton source to promote the cleavage of the palladium–carbon bond during protonolysis. The hydrolysis of **9**, a typical elementary step in iminium catalysis, occurs to give the product **3** and releases the chiral amine catalyst **4** (Scheme 4.3).

Rios group reported a highly enantioselective cyclopropanation of enals **1** with chloro-alkylbenzoxazoles **10** driven by cooperative catalysis of $Pd(OAc)_2$ and chiral amine **4** [9]. The nitrogen atom in benzoxazoles **10** coordinates to Lewis acidic $Pd(OAc)_2$ that promotes deprotonation of the chloromethyl group to give a Pd(II) enolate intermediate **12**. The stereoselective Michael addition of **12** to the chiral iminium ions **5**, followed by an intramolecular cyclization reaction and hydrolysis, leads to the formation of cyclopropane products **11**. The synergy of chiral iminium and Pd(II) Lewis acid catalysis delivers good yields, excellent enantioselectivity, and good diastereoselectivity (Scheme 4.4).

4.2.2 Asymmetric [3+2] Cycloaddition Via Ring-Opening Oxidative Addition

Oxidative addition of low valent palladium complexes to vinylaziridines, vinyl epoxides, and vinyl cyclopropanes generates palladium-stabilized 1,3-zwitterions, which engage in sequential conjugate addition to α,β-unsaturated iminium ions and intramolecular allylic alkylation reaction to furnish vinylcyclopentane derivatives (Scheme 4.5).

An asymmetric [3+2] cycloaddition of donor-acceptor vinyl cyclopropanes **14** with enals **1** driven by cooperative catalysis of Pd(0) and chiral amine was established by Michelet and coworkers to afford a broad range of multi-substituted cyclopentanes **15** in high yields and with excellent enantioselectivities (Scheme 4.6a) [10]. The palladium complex formed from $Pd_2(dba)_3$ and bis(1,2-diphenylphosphino)ethane (DPPE) turns out to be the most efficient metal catalyst. The ring-opening oxidative addition of vinyl cyclopropanes **14** to Pd(0) complex occurs to give a zwitterionic intermediate **16**, which then undergoes enantioselective conjugate addition to α,β-unsaturated iminium ions **5** to from an intermediate **17**, followed by an intramolecular allylic alkylation to generate iminium intermediate **18**. Subsequent hydrolysis and decomplexation of **18** yields the final product **15** and release the amine **4** and palladium catalyst (Scheme 4.6b). The Jørgensen group [11a] and Rios group [11b] found that $Pd(dba)_2$ and $Pd_2(dba)_3$ themselves could serve as excellent catalysts, in combination with chiral amine **4**, to promote highly enantioselective [3+2] cycloaddition of enals and vinyl cyclopropanes (Scheme 4.6a).

Vinylcyclopropane azlactones **19** are also capable of participating in the [3+2] cycloaddition with enals **1** enabled by the similar chiral amine/Pd(0) cooperative catalytic system to build up spirobicyclic molecules with structural complexity in excellent stereoselectivities (Scheme 4.7) [12]. Córdova and coworkers synthesized allyl acetates with tethered carbon nucleophiles, as surrogates of vinyl cyclopropanes, for chiral amine **4**/$Pd(OAc)_2$ co-catalyzed enantioselective dynamic Michael addition and intramolecular allylic alkylation cascade with enals [13].

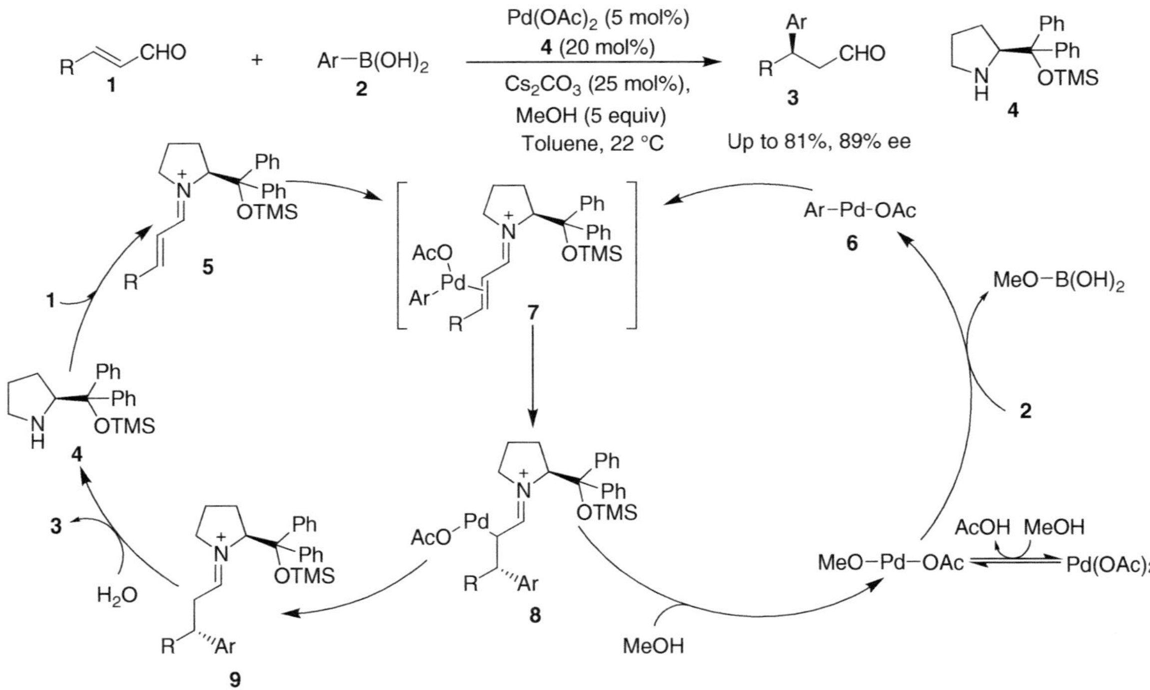

Scheme 4.3 Chiral amine/Pd(II) cooperatively catalyzed enantioselective β-arylation of enals.

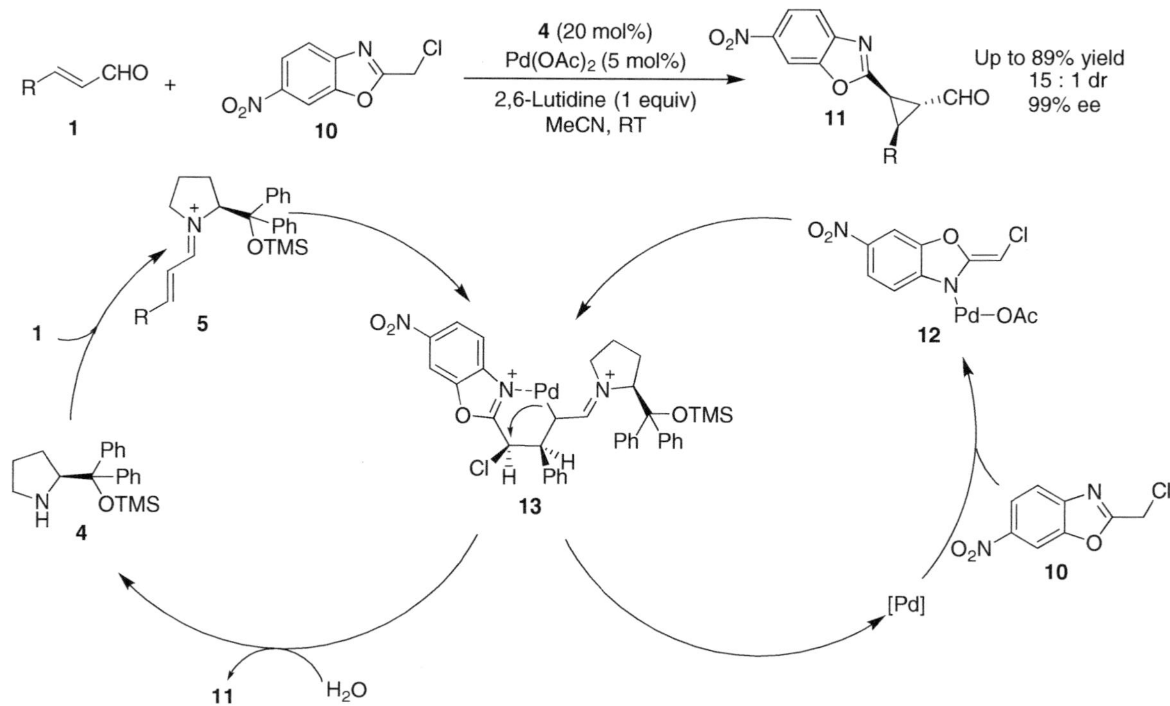

Scheme 4.4 Asymmetric cyclopropanation of enals and benzoxazoles.

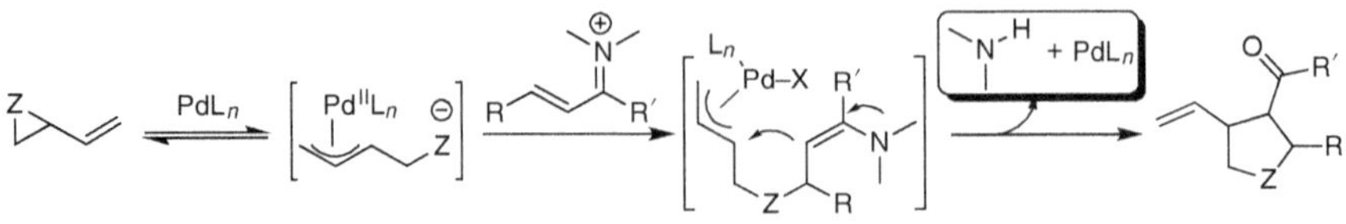

Scheme 4.5 A cooperative paradigm for asymmetric [3+2] cycloaddition.

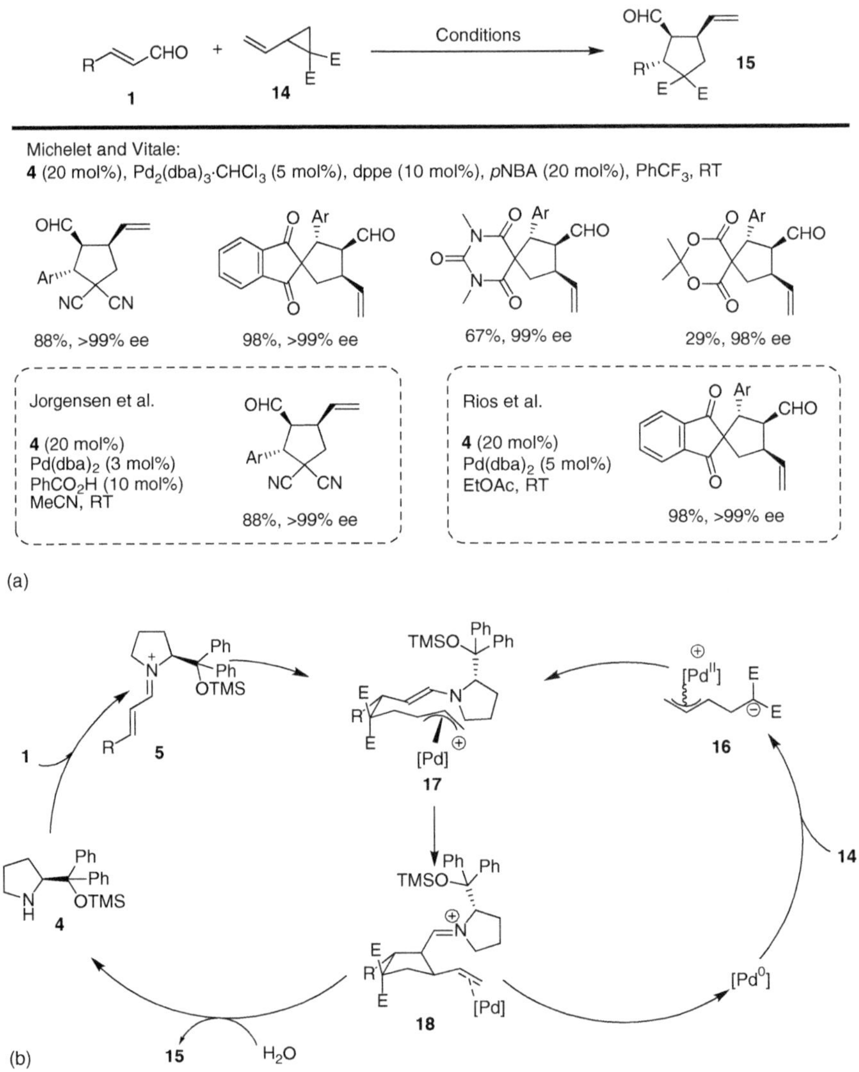

Scheme 4.6 [3+2] Annulation of vinyl cyclopropanes and enals by iminium/Pd cooperative catalysis. (a) [3+2] Cycloaddition of enals and vinyl cyclopropanes. (b) Proposed mechanism. Source: (a) Modified from Laugeois [10].

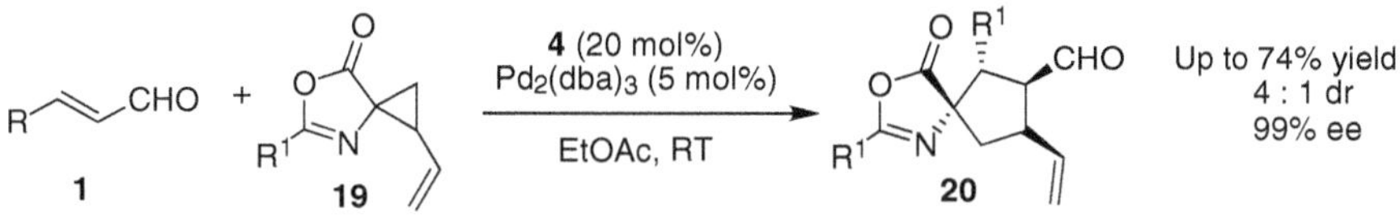

Scheme 4.7 [3+2] Annulation of enals with different vinyl cyclopropanes. Source: Modified from Kamlar et al. [12].

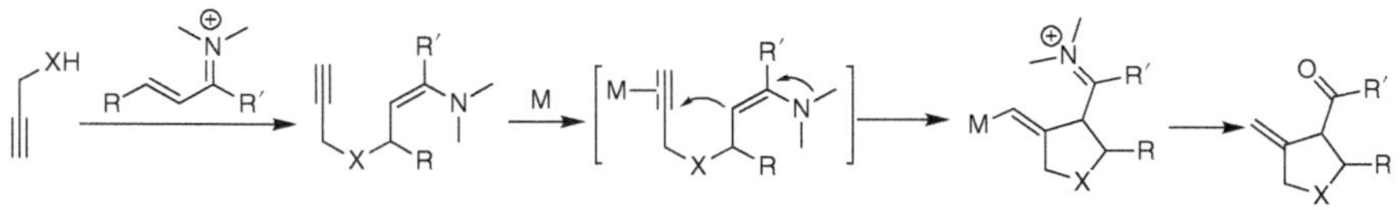

Scheme 4.8 General concept of asymmetric Michael addition and carbocyclization.

4.2.3 Asymmetric Michael Addition and Carbocyclization Cascade

The asymmetric Michael addition of propargyl nucleophiles to chiral unsaturated iminium ions generates alkynyl enamine intermediates, which subsequently undergo an intramolecular enamine addition to the carbon–carbon triple bond activated by a suitable transition metal π-Lewis acid catalyst. Privileged enantioselective 5-*exo*-dig cyclization provides unique access to (hetero)cyclopentanes with multiple substituents and an external or internal carbon–carbon double bond on the ring (Scheme 4.8).

In 2010, Wang and coworkers reported a highly enantioselective formal [3+2] cycloaddition consisting of asymmetric Michael addition and carbocyclization of enals **1** and propargyl malonate **21** enabled by the cooperative catalysis of chiral amine catalyst *ent*-**4** and $PdCl_2$ [14a]. The *in situ* generated chiral α,β-unsaturated iminium ions **23** are subjected to an asymmetric Michael addition to form an enamine intermediate **25**. π-Lewis acid activation of $PdCl_2$ toward the carbon–carbon triple bond promotes subsequent intramolecular cyclization of enamine **25**. Both $PdCl_2$ and chiral amine *ent*-**4** are essential for the success of this cascade reaction. In the absence of $PdCl_2$, only Michael addition reaction occurred with low yield (17%) and moderate enantioselectivity (67% ee); while no reactivity was observed without *ent*-**4** (Scheme 4.9). Córdova and coworkers independently developed the same type of cascade reaction by chiral amine/Pd(0) combined catalysis [14b].

Later, the Wang group expanded this chiral amine/$PdCl_2$ hybrid relay-cooperative catalysis to asymmetric Michael addition and carbocyclization cascade of enals **1** and 3-propargyl oxindoles **26**, affording spirocyclic oxindoles **27** in up to 95% yield, with >20 : 1 dr and >99% ee (Scheme 4.10) [15].

Córdova and coworkers disclosed a dynamic kinetic asymmetric oxa-Michael/carbocyclization by the combination of chiral amine and $PdCl_2$ catalysis [16a].

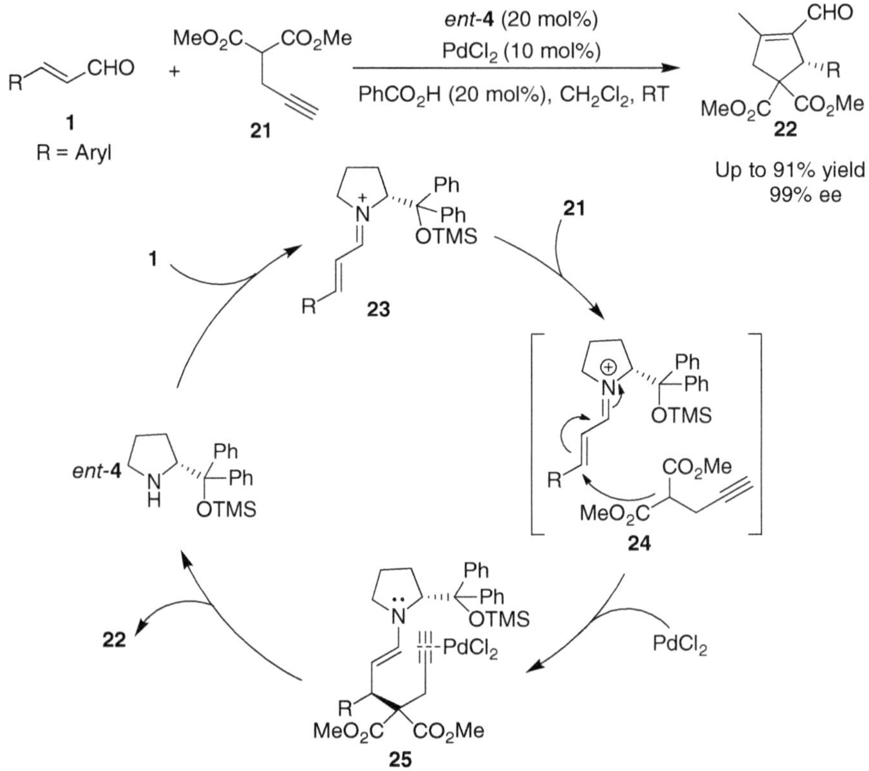

Scheme 4.9 Asymmetric Michael addition and carbocyclization of enals **1** and propargyl malonate.

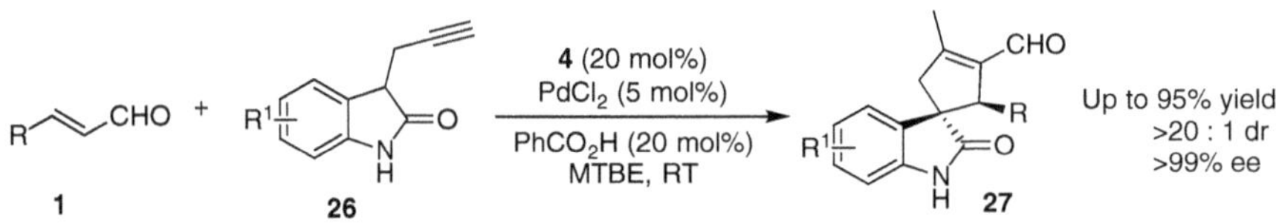

Scheme 4.10 Synthesis of spirocyclic oxindoles through chiral amine/$PdCl_2$ combined catalysis. Source: Modified from Sun et al. [15].

The equilibrium between starting materials (enals **1** and propargyl alcohol **28**) and oxa-Michael products during the chiral amine catalysis limits the transformation efficacy. Chiral amine **29** and $PdCl_2$ co-catalyzed carbocyclization would push the equilibrium toward the irreversible formation of final dihydrofuran products **30** (Scheme 4.11a). This cascade reaction with only chiral amine catalyst or $PdCl_2$ did not proceed. Other metals, such as Cu(I), Ag(I), and Zn(II), are less effective to drive the reaction (Scheme 4.11a). Later, Hong and coworkers developed an enantioselective aza-Michael/carbocyclization reaction of enals **1** and *N*-tosyl propargylamines **31** (Scheme 4.11b) [16b].

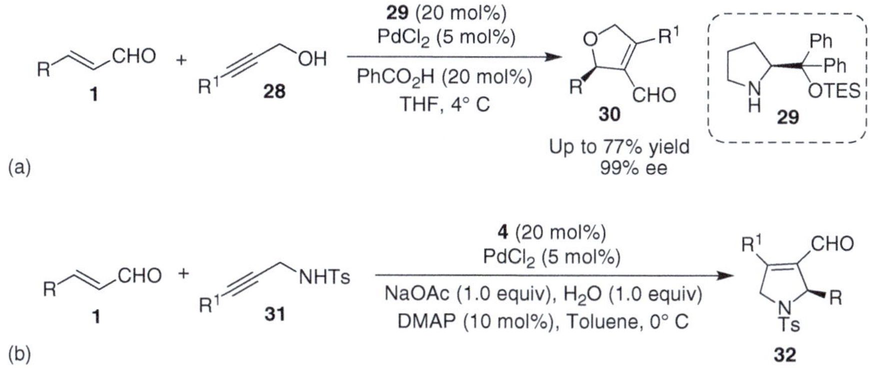

Scheme 4.11 Dynamic kinetic asymmetric hetero-Michael addition/cyclization. (a) Asymmetric oxa-Michael/cyclization. (b) Asymmetric aza-Michael/cyclization.

4.2.4 Asymmetric Oxidative Cascade Reaction

Wang and coworkers developed Saegusa-type oxidation of unmodified aldehydes into enals using molecular oxygen as the sole oxidant by enamine/Pd(II) cooperative catalysis [17]. Xu and coworkers merged Wang's protocol and iminium-catalyzed asymmetric Michael addition, affording enantiopure β-functionalized aldehydes [18]. Transmetallation of enamines **36** *in situ* generated from the condensation of aldehydes **33** and chiral amine **4** gives an alkyl Pd(II) species **38**, which then undergoes β-hydride elimination to furnish α,β-unsaturated iminium ions **39**. The asymmetric conjugate addition of malonates **34** to the iminium ions **39** leads to the formation of chiral aldehydes **35** in moderate to good yields and with excellent enantioselectivities (Scheme 4.12). Due to the concern about the storage stability of aldehydes under ambient conditions, Córdova and coworkers combined Pd(II)-catalyzed dehydrogenative oxidation of allylic alcohols into enals with chiral amine/Pd(II) co-catalyzed enantioselective Michael addition/carbocyclization in the one-pot manner [19].

4.3 Iminium Activation and Coinage Metal Catalysis

Group 11 elements, known as the coinage metals, consisting of copper (Cu), silver (Ag), and gold (Au), can serve as π-Lewis acid toward the activation of carbon–carbon triple bonds. In 2010, the Jørgensen group independently reported a chiral amine/coinage metal co-catalyzed enantioselective Michael addition and carbocyclization cascade reaction of propargyl malononitrile or cyanoacetates with enals [20a]. In this reaction, the CuOTf-PPh_3 complex serves as an excellent π-Lewis acid catalyst to promote the enamine-alkyne cyclization. [$Au(NTf_2)(PPh_3)$] was also found to be compatible with chiral amine **41**, albeit with a slower reaction rate than CuOTf-PPh_3. Notably, compared to Wang's work [14a], Jørgensen's chiral amine **41**/CuOTf-PPh_3 catalytic system substantially improved the scope from cinnamic aldehydes to β-alkyl-α,β-unsaturated aldehydes (Scheme 4.13a). Later, Jørgensen

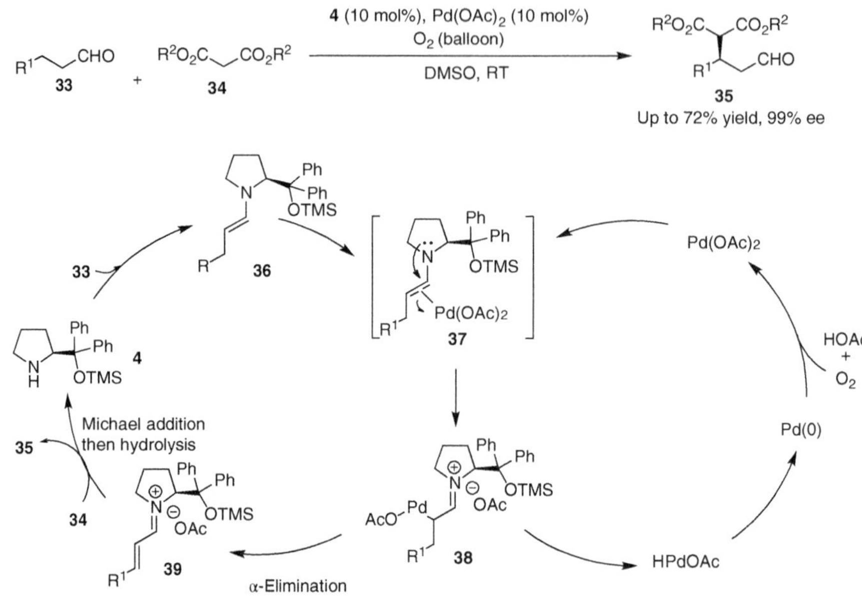

Scheme 4.12 Iminium and $Pd(OAc)_2$ co-catalyzed oxidative Michael addition reaction.

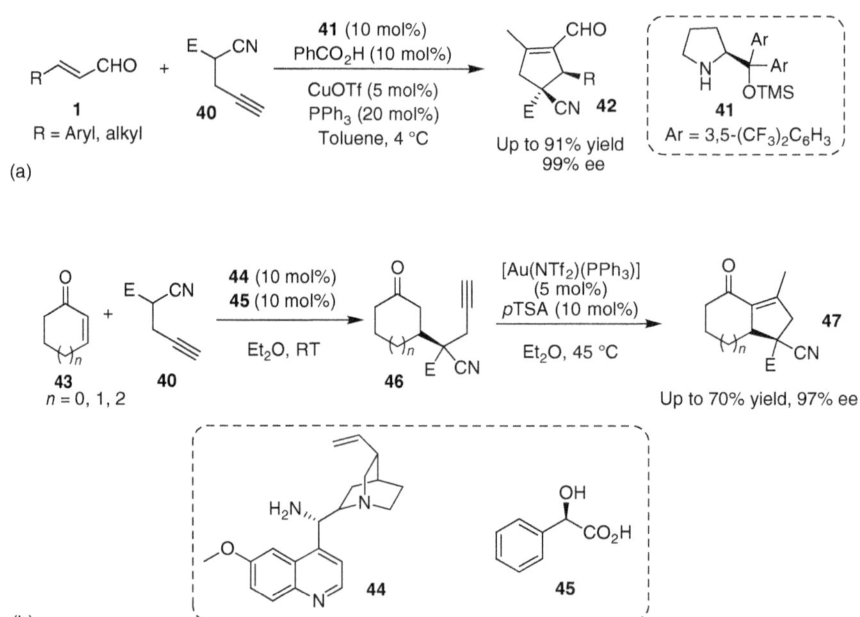

Scheme 4.13 Chiral amine/coinage metal co-catalyzed Michael addition and carbocyclization. (a) Michael addition/carbocyclization of enals. (b) Michael addition/carbocyclization of enones. Source: (b) Modified from Zweifel et al. [20b].

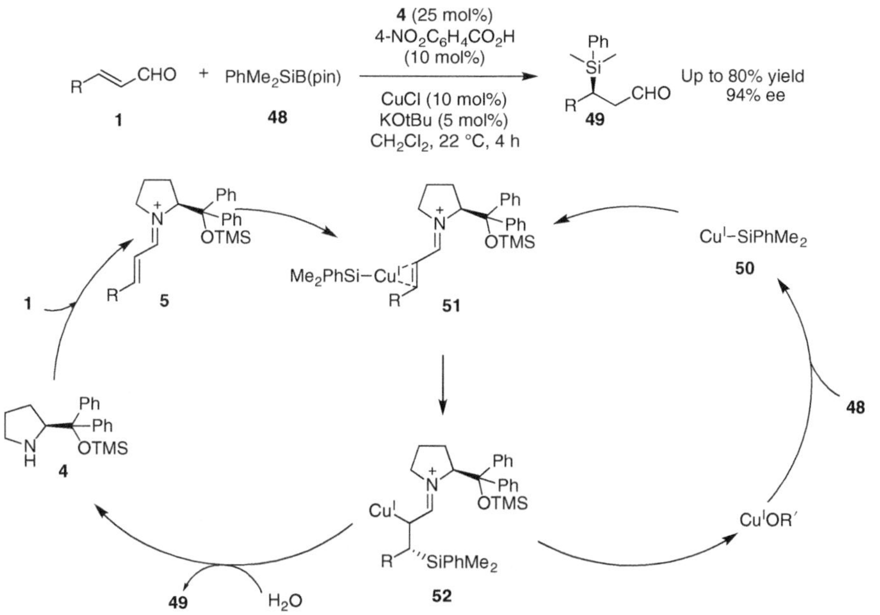

Scheme 4.14 Enantioselective β-silylation of enals by chiral amine/Cu(I) cooperative catalysis.

and coworkers developed a one-pot two-step Michael addition and carbocyclization reaction of enones and propargyl nucleophiles, in which the combination of a cinchona alkaloid-derived primary amine **44** and Brønsted acid **45** promotes the first Michael addition reaction, while [Au(NTf_2)(PPh_3)] and *para*-toluenesulfonic acid (*p*TSA) co-catalyze the cyclization reaction (Scheme 4.13b) [20b].

Córdova and coworkers developed an enantioselective β-silyl addition to enals **1** by virtue of iminium and copper cooperative catalysis [21]. The *in situ* generated Cu(I) alkoxide from CuCl and KOtBu would react with $PhMe_2SiB$(Pin) **48** to form a Cu(I)-$SiPhMe_2$ species **50**, which then adds to the β-carbon of unsaturated iminium **5** via a presumable transition state **51** to give an intermediate **52**. The protonolysis of the alkyl Cu(I) intermediate **52** and hydrolysis of the iminium moiety afford β-silyl aldehydes in moderate yields and with moderate to high enantioselectivities (Scheme 4.14).

In 2015, Quintard and Rodriguez demonstrated the utilization of keto diacid **53** as a 1,3-bis-nucleophile in an enantioselective cycloaddition reaction with enals **1** [22]. The addition of a co-catalyst such as a copper salt considerably increased both yield and enantioselectivity. The key to success is a synergistic activation of keto diacid **53** by copper catalysis and iminium activation of enals by amine catalyst **4**, providing direct access to valuable chiral cyclohexenones **54** through unprecedented di-decarboxylative Michael/aldol/dehydration cascade reaction (Scheme 4.15).

4.4 Iminium Activation and Other Metal Catalysis

Versatile iminium activation is also compatible with other transition metal complexes than palladium, copper, and gold. In 2016, Liu and coworkers developed

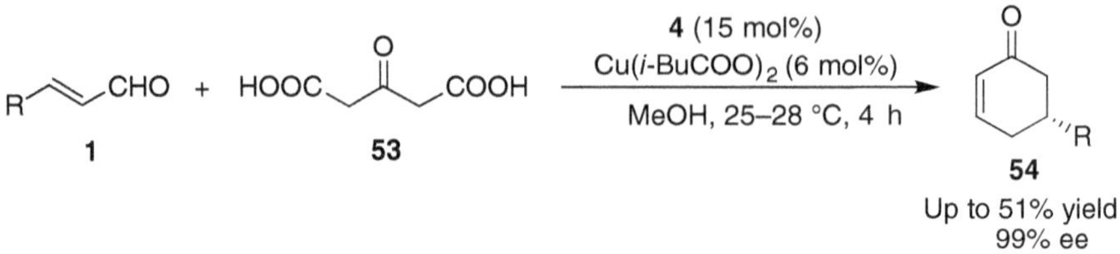

Scheme 4.15 Chiral amine/copper co-catalyzed decarboxylative cycloaddition reaction.

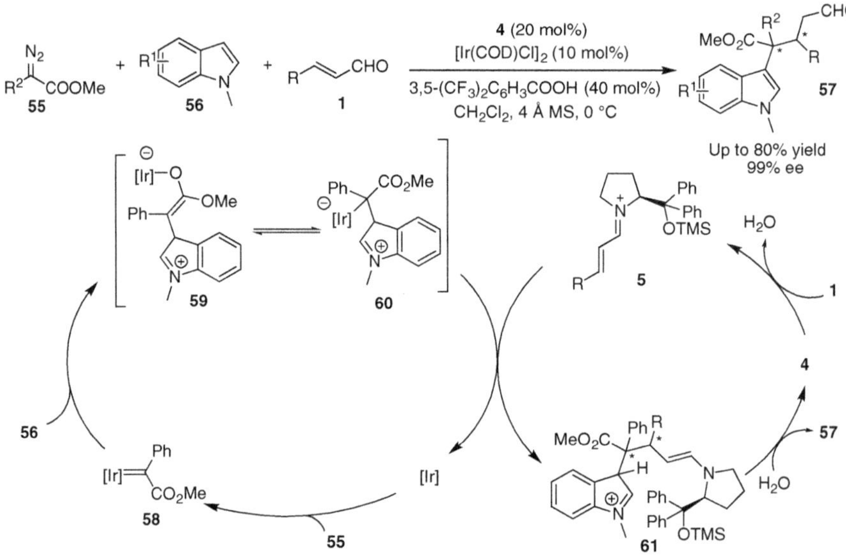

Scheme 4.16 Chiral amine and iridium(I) co-catalyzed enantioselective three-component reaction.

a chiral amine **4** and iridium(I) co-catalyzed enantioselective three-component coupling reaction of aryldiazoacetates **55**, indoles **56,** and enals **1** [23]. The key zwitterionic intermediates **59** is initially formed *in situ* from iridium carbenoids **58** and indole derivative **56**. The carboxylic acid additive accelerates the condensation reaction of enals **1** and chiral amine catalyst **4** to afford iminium ions **5**, which efficiently trap the zwitterionic intermediates **60** to produce enamine intermediates **61**. The following hydrolysis under the Brønsted acidic condition regenerates chiral amine catalyst **4** and yields indole derivatives **57** bearing aldehyde functionality and two consecutive stereocenters in good yields, with moderate diastereoselectivity and excellent enantioselectivity (Scheme 4.16). The matched reactivity of major intermediates and the comparable kinetics in both of the two catalytic cycles are believed to be critical for success.

Quintard et al. reported an enantioselective carbofunctionlization of allylic alcohols **62** by chiral amine and iron relay catalysis [24]. This cascade reaction makes a feature of three precise catalytic cycles: firstly, CO decoordination from the iron catalyst **64** with Me_3NO gives an active [Fe] species bearing a vacant site, which

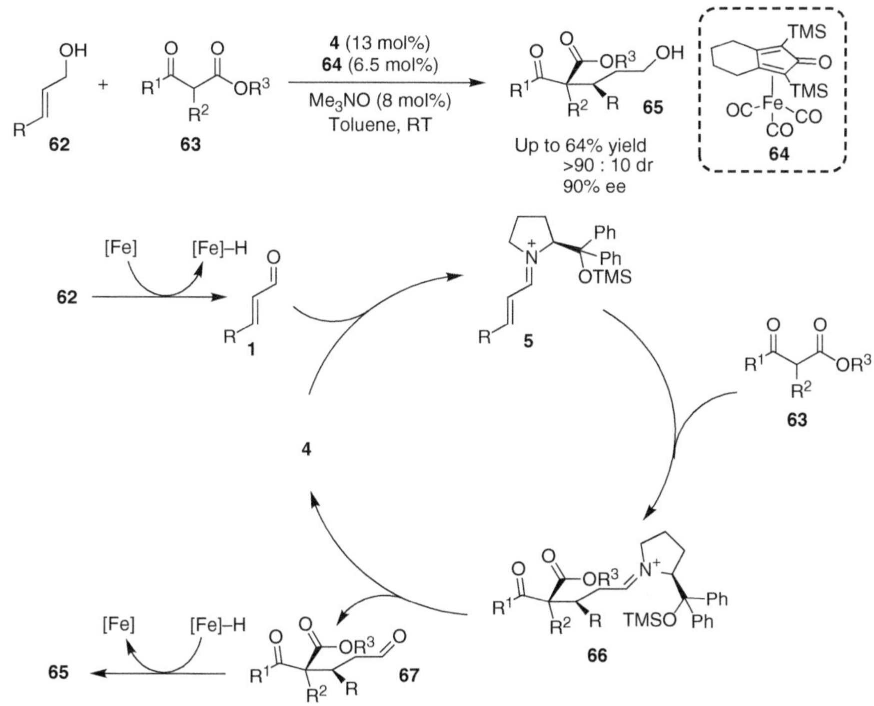

Scheme 4.17 Cascade reaction involving iminium intermediate and transfer hydrogenation.

can abstract hydrogen from allylic alcohols **62** to furnish enals **1** and a transient iron-hydrogen complex [Fe]–H; secondly, enantioselective Michael addition of enals **1** with β-ketoesters **63** via iminium activation from chiral amine catalyst **4** produces chiral aldehydes **67**; finally, chemoselective reduction of the aldehyde functionality of **67** by [Fe]–H delivers stable β-chiral alcohols **65** in moderate yields, with excellent enantioselectivities, and regenerates the active iron complex **[Fe]** (Scheme 4.17).

4.5 Conclusion

Dual activation from organocatalysts and metal complexes toward distinct substrates in a single operation have emerged as an enabling concept and powerful tool for the rapid buildup of molecular complexity otherwise inaccessible, among which the combination of iminium activation and diverse transition metal catalysis (such as palladium, copper, gold) has been applied in a series of enantioselective cascade bond-forming transformations, e.g. conjugate addition/cyclization, providing a conceptually new strategy to synthesize valuable heterocycles and other bioactive compounds. Despite those advances discussed in this chapter, the state-of-the-art reaction design through iminium and metal combined catalysis is mostly limited to conjugate addition that the remote (e.g. γ- and δ-position) functionalization of α,β-unsaturated carbonyls remain undeveloped. Moreover, merging iminium activation with the rich chemistry of the transition metal-catalyzed single-electron-transfer process would be promising.

References

1 (a) Erkkilä, A., Majander, I., and Pihko, P.M. (2007). *Chem. Rev.* 107: 5416–5470. (b) Melchiorre, P., Marigo, M., Carlone, A., and Bartoli, G. (2008). *Angew. Chem.* 120: 6232–6265; (2008). *Angew. Chem. Int. Ed.* 47: 6138–6171.

2 (a) Layer, R.W. (1963). *Chem. Rev.* 63: 489–510. (b) Langenbeck, W. and Sauerbier, R. (1937). *Chem. Ber.* 70: 1540–1541.

3 (a) Baum, J.S. and Viehe, H.G. (1976). *J. Org. Chem.* 41: 183–187. (b) Yamaguchi, M., Shiraishi, T., and Hirama, M. (1993). *Angew. Chem.* 105: 1243–1245; (1993). *Angew. Chem. Int. Ed.* 32: 1176–1178. (c) Ahrendt, K.A., Borths, C.J., and MacMillan, D.W.C. (2000). *J. Am. Chem. Soc.* 122: 4243–4244.

4 List, B. (2006). *Chem. Commun.*: 819–824.

5 (a) Jacobsen, E.N., Pfaltz, A., and Yamamoto, H. (eds.) (1999). *Comprehensive Asymmetric Catalysis*, vols. I–III, Suppl. I–II. New York: Springer. (b) Hamilton, G.L., Kang, E.J., Mba, M., and Toste, F.D. (2007). *Science* 317: 496–499. (c) Du, Z.T. and Shao, Z.-H. (2012). *Chem. Soc. Rev.* 42: 1337–1378. (d) Chen, D.-F., Han, Z.-Y., Zhou, X.-L., and Gong, L.-Z. (2014). *Acc. Chem. Res.* 47: 2365–2377. (e) Afewerki, S. and Córdova, A. (2016). *Chem. Rev.* 116: 13512–13570.

6 (a) Nishikata, T., Yamamoto, Y., and Miyaura, N. (2003). *Angew. Chem.* 115: 2874; (2003). *Angew. Chem. Int. Ed.* 42: 2768–2770. (b) Nishikata, T., Yamamoto, Y., and Miyaura, N. (2004). *Organometallics* 23: 4317–4324.

7 Ibrahem, I., Ma, G., Afewerki, S., and Córdova, A. (2013). *Angew. Chem.* 125: 912–916; (2013). *Angew. Chem. Int. Ed.* 52: 878–882.

8 (a) Nishikata, T., Yamamoto, Y., and Miyaura, N. (2003). *Angew. Chem.* 115: 2874–2876; (2003). *Angew. Chem. Int. Ed.* 42: 2768–2770. (b) Nishikata, T., Yamamoto, Y., and Miyaura, N. (2004). *Organometallics* 23: 4317–4324.

9 Meazza, M., Light, M.E., Mazzanti, A., and Rios, R. (2016). *Chem. Sci.* 7: 984–988.

10 Laugeois, M., Ponra, S., Ratovelomanana-Vidal, V. et al. (2016). *Chem. Commun.* 52: 5332–5335.

11 (a) Halskov, K.S., Næsborg, L., Tur, F., and Jørgensen, K.A. (2016). *Org. Lett.* 18: 2220–2223. (b) Meazza, M. and Rios, R. (2016). *Chem. Eur. J.* 22: 9923–9928.

12 Kamlar, M., Franc, M., Císařová, I. et al. (2019). *Chem. Commun.* 55: 3829–3832.

13 (a) Ma, G., Afewerki, S., Deiana, L. et al. (2013). *Angew. Chem.* 125: 6166–6170; (2013). *Angew. Chem. Int. Ed.* 52: 6050–6054. (b) Afewerki, S., Ma, G., Ibrahem, I. et al. (2015). *ACS Catal.* 5: 1266–1272.

14 (a) Yu, C., Zhang, Y., Zhang, S. et al. (2010). *Tetrahedron Lett.* 51: 1742–1744. (b) Zhao, G.-L., Ullah, F., Deiana, L. et al. (2010). *Chem. Eur. J.* 16: 1585–1591.

15 Sun, W., Zhu, G., Wu, C. et al. (2012). *Chem. Eur. J.* 18: 13959–13963.

16 (a) Lin, S., Zhao, G.-L., Deiana, L. et al. (2010). *Chem. Eur. J.* 16: 13930–13934. (b) Sun, W., Zhu, G., Hong, L., and Wang, R. (2011). *Chem. Eur. J.* 17: 13958–13962.

17 (a) Zhu, J., Liu, J., Ma, R. et al. (2009). *Adv. Synth. Catal.* 351: 1229–1232. (b) Liu, J., Zhu, J., Jiang, H. et al. (2009). *Chem. Asian J.* 4: 1712–1716.

18 Zhao, Y.-L., Wang, Y., Hu, X.-Q., and Xu, P.-F. (2013). *Chem. Commun.* 49: 7555–7557.

19 Santoro, S., Deiana, L., Zhao, G.-L. et al. (2014). *ACS Catal.* 4: 4474–4484.

20 (a) Jensen, K.L., Franke, P.T., Arróniz, C. et al. (2010). *Chem. Eur. J.* 16: 1750–1753. (b) Zweifel, T., Hollmann, D., Prüger, B. et al. (2010). *Tetrahedron: Asymmetry* 21: 1624–1629.

21 Ibrahem, I., Santoro, S., Himo, F., and Córdova, A. (2011). *Adv. Synth. Catal.* 353: 245–253.

22 Quintard, A. and Rodriguez, J. (2015). *Chem. Commun.* 51: 9523–9526.

23 Li, M., Guo, X., Jin, W. et al. (2016). *Chem. Commun.* 52: 2736–2739.

24 Quintard, A., Constantieux, T., and Rodriguez, J. (2013). *Angew. Chem.* 125: 13121–13125; (2013). *Angew. Chem. Int. Ed.* 52: 12883–12887.

5

Brønsted Acid and Transition Metal Cooperative Catalysis

5.1 Introduction

Brønsted acid catalysis, an important subfield of organocatalysis, has long been employed to catalyze the formation and cleavage of C—O bonds, such as hydrolysis, esterification, and acetal synthesis [1]. Commonly used achiral Brønsted acids range from weak acids, such as phenol, to super strong acids like trifluoromethanesulfonic acid. At the dawn of this century, chiral Brønsted acids have been recognized as powerful organocatalysts for a wide range of bond-forming reactions [2]. Chiral Brønsted acids mainly encompass weak or neutral Brønsted acids, such as thioureas, squaramides, α,α,α′,α′-tetraaryl-1,3-dioxolane-4,5-dimethanols (TADDOL), and 1,1′-Bi-2-naphthol (BINOL) derivatives (Scheme 5.1a) [3] and stronger Brønsted acids, exemplified by BINOL-derived phosphoric acids and phosphoramides (Scheme 5.1b) [4].

The weak or neutral Brønsted acids are also classified as hydrogen-bonding catalysts since the weak or neutral Brønsted acid-catalyzed reactions are usually driven by hydrogen-bonding interactions (Scheme 5.2) [5]. Stronger Brønsted acids, on the other hand, could activate carbonyls, imines, alkenes, alkynes, and hydroxyl groups through protonation to form oxonium, iminium salts, carbocations, and vinylic carbocations as well (Scheme 5.2) [6]. Up till now, chiral Brønsted acid catalysis has turned out to be a generally applicable platform for the development of asymmetric transformations [6, 7].

A wide range of reactive transition metal intermediates, including metal alkynylides, π-allyl-metals, metal hydrides, metal carbenes, metal allenylidenes, metal alkenyls, and metal alkenes (Scheme 5.3) have shown good compatibility with Brønsted acids and sufficient reactivity toward Brønsted acid-activated intermediates to provide a diverse range of nonclassical avenues for the development of asymmetric reactions. Over the past two decades, chiral Brønsted acid and transition metal cooperative catalysis [8] have indeed enabled the proliferation of a tremendous number of asymmetric transformations that are difficult or inefficient to be afforded by an individual catalyst. According to the role that chiral Brønsted acids and transition metal catalysts play, this chapter will summarize the proof of concept and evolution of chiral Brønsted acid/metal cooperative catalysis with the emphasis on the transition metal intermediates.

Asymmetric Organo-Metal Catalysis: Concepts, Principles, and Applications, First Edition. Liu-Zhu Gong.
© 2022 WILEY-VCH GmbH. Published 2022 by WILEY-VCH GmbH.

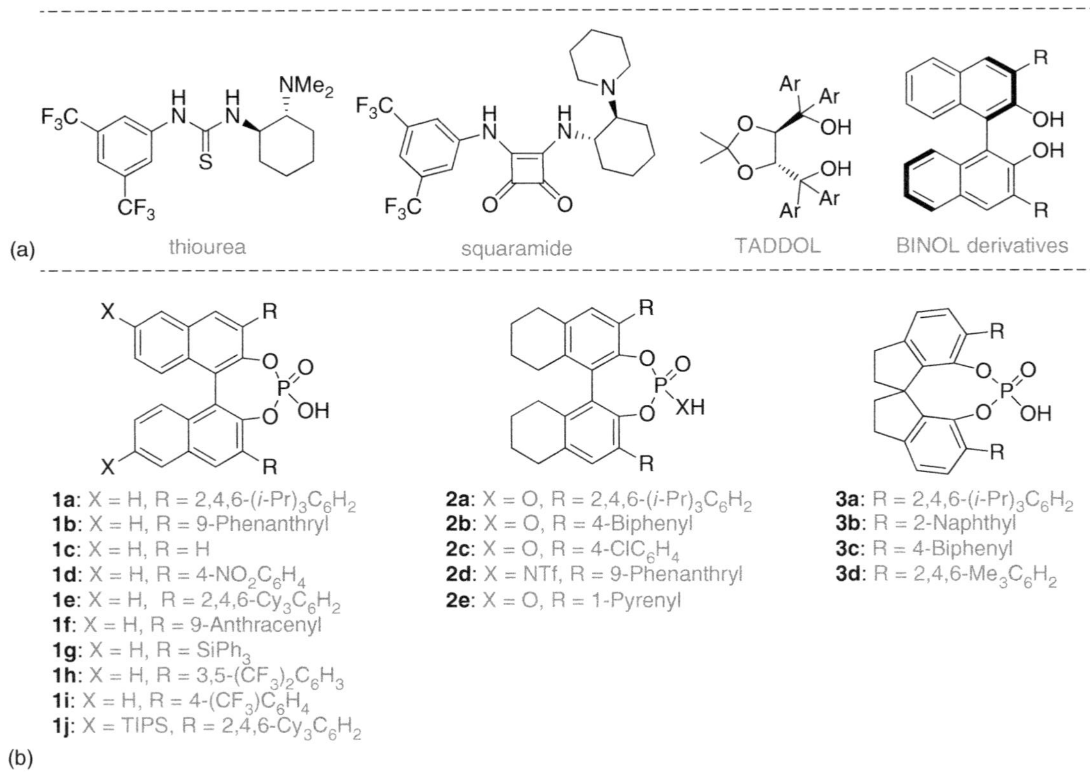

Scheme 5.1 Typical Brønsted acids commonly used for asymmetric catalysis. Source: (a) Modified from Doyle and Jacobsen [3]. (b) Modified from Parmar et al. [4].

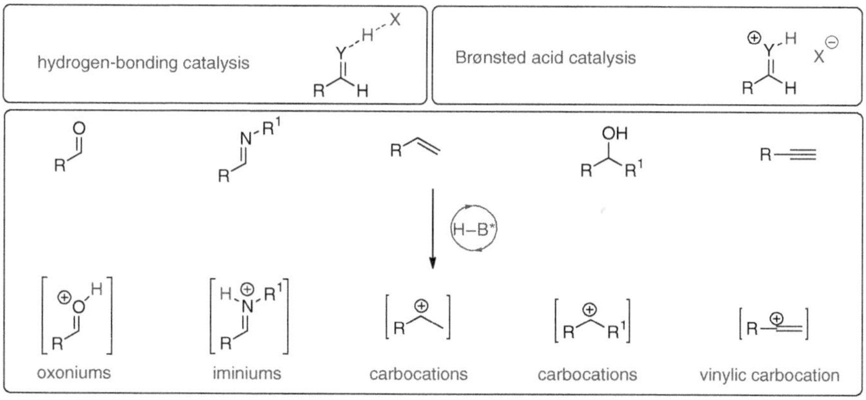

Scheme 5.2 Activation modes of chiral Brønsted acid catalysis.

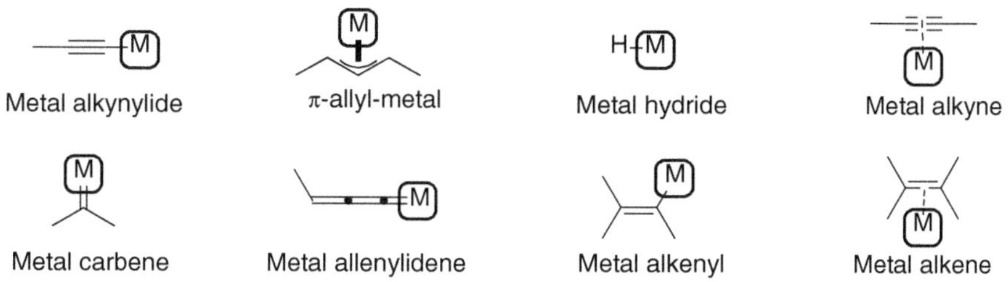

Scheme 5.3 Representative examples of reactive transition metal intermediates.

5.2 Early Stage of Metal/Brønsted Acid Cooperative Catalysis

The first example of transition metal/chiral phosphoric acid cooperative catalysis was reported by Krische and coworker in 2006 [9]. They established a chiral ion pair-mediated enantioselective reductive coupling of 1,3-enyne **5** to picolinaldehyde **4** (Scheme 5.4). The combination of an achiral rhodium catalyst and a chiral phosphoric acid (*R*)-**1a** allows the reaction to give high levels of enantioselectivity. The chiral phosphoric acid is supposed to protonate 2-pyridinecarboxaldehyde **4** to give an ion pair **7**, which then undergoes the Rh(I)-mediated asymmetric reductive coupling with a conjugated enyne **5** to form a metallacycle intermediate **8**. Subsequently, intermediate **8** undergoes hydrogenolysis to release chiral phosphoric acid and to generate a cationic Rh(III) intermediate **9**. Finally, the coupling product **6** is delivered through reductive elimination and with the regeneration of cationic Rh(I) catalyst.

5.3 Metal Alkynylide-Mediated Transformations

The addition of alkynes to imines represents one of the most straightforward and reliable methods to access propargylamines [10], which can work as versatile

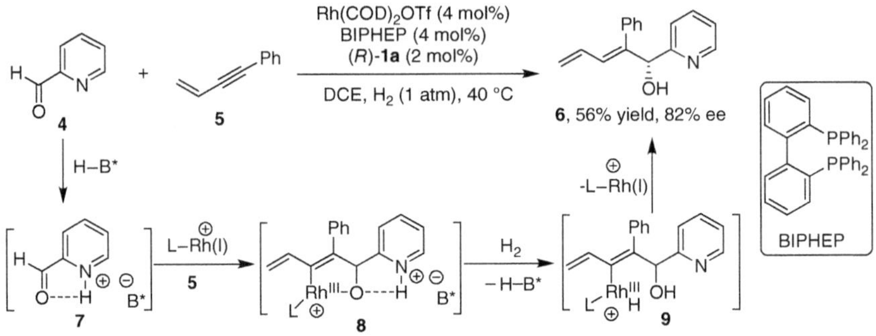

Scheme 5.4 Early example of the Rh complex/chiral phosphoric acid cooperative catalysis.

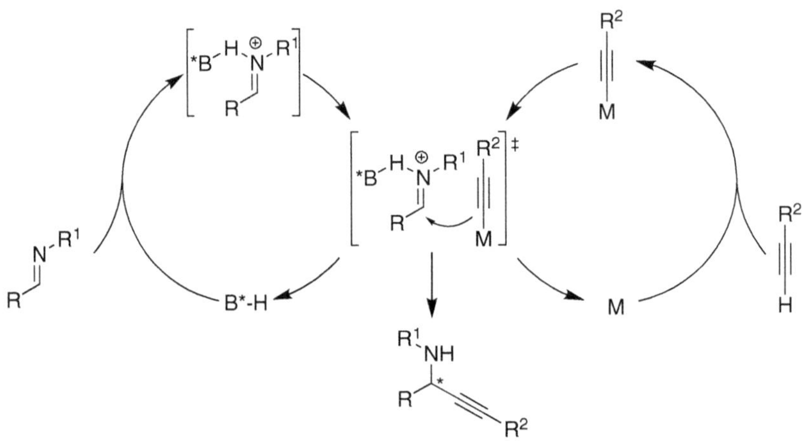

Scheme 5.5 General concept for chiral Brønsted/metal cooperatively catalyzed alkynylation of imines.

building blocks in organic synthesis [11]. Classical asymmetric addition of alkynes to imines for the synthesis of chiral propargylic amines relies on the introduction of chiral ligands to form chiral alkynyl-metal reagents, especially the use of chiral copper catalysts [10, 12]. In these transformations, catalytically generated chiral metal alkynylide serves as an active nucleophile to attack the imine functionality. In contrast to this traditional concept of using chiral ligands, chiral Brønsted acid/metal cooperative catalysis can simultaneously activate both imine by the chiral acid and alkyne by transition metal catalyst to synergistically drive the asymmetric addition event (Scheme 5.5), basically excluding the use of chiral ligand to control the stereochemistry.

In 2007, Rueping et al. reported an asymmetric alkynylation of imines **10** synergistically catalyzed by a chiral phosphoric acid and silver acetate (Scheme 5.6a) [13], providing methyl 2-amino-3-butynoates **12** in good yields and with high levels of enantioselectivity. The addition of an alkynyl silver intermediate to a chiral phosphoric acid activated imine (**TS-1**) turns out to be the enantiodetermining step. Interestingly, such a concept is also viable to asymmetric alkynylation of cyclic imines

13 to access optically active 11-substituted-10,11-dihydrodibenzo[*b,f*][1,4]oxazepine derivatives **14** with excellent enantioselectivities (Scheme 5.6b) [14].

In addition to silver catalysts, cheaper copper complexes are also robust catalysts to activate terminal alkynes [15], and the combination of chiral Brønsted acids with copper salts is also amenable to asymmetric alkynylation of imines [12]. For example, Arndtsen and coworkers found that a copper complex combined with *N*-Boc proline was able to efficiently drive asymmetric addition of acetylene derivatives **11** to imines **15** (**TS-2**), allowing the rapid access to chiral propargyl amines in high yields and with excellent levels of enantioselectivity (Scheme 5.7) [16].

Afterward, Hashimoto and Maruoka reported a copper/chiral carboxylic acid synergistically catalyzed asymmetric alkynylation of C,N-cyclic azomethine imines **17** and alkynes (Scheme 5.8) [17]. In this transformation, chiral Brønsted acid **19** is proposed to protonate azomethine imine and induces the stereoselectivity of nucleophilic addition of copper acetylide (**TS-3**) to give chiral tetrahydroisoquinoline derivatives **20**.

Although these combined catalyst systems show notable catalytic performance, they are principally not capable of controlling the stereochemistry of a three-component variant of aldehydes **21**, secondary amines **22**, and acetylenes **11**, as the resulting iminium intermediate **25** has no lone pair required for hydrogen-bonding interaction, protonation, or metal coordination. In 2015, Seidel and coworker found that chiral Brønsted acid **23** bearing both a carboxylic acid and a thiourea could serve as an effective chiral co-catalyst for a Cu(I)-catalyzed enantioselective A^3 reactions involving secondary amines (Scheme 5.9) [18]. In this case, a neutral copper acetylide **26** presumably reacts with an iminium ion pair **25** *in situ* generated from the condensation of aldehyde **21**, pyrrolidine **22**, and chiral Brønsted acid **23**.

5.4 π-Allyl-Metal-Mediated Transformation

The transition metal-catalyzed asymmetric allylic substitution reaction represents one of the most important transformations in organic synthesis [19], enabling rapid access to alkene-bearing substances for building up molecular complexity with versatile late-stage functionalizations [20]. The commonly used strategies to establish asymmetric allylic substitution mainly rely on the use of chiral ligands to induce chirality [21]. In contrast, chiral Brønsted acid/metal cooperative catalysis provides a fresh perspective to the control of enantioselection. Mechanistic analysis suggests that in the presence of a chiral Brønsted acid, oxidative addition of a transition metal to an allyl substrate would generate a π-allylmetal species bearing a chiral counterion as an internal base, which is principally capable of controlling the stereochemistry through a hydrogen-bonding interaction with soft nucleophiles (Scheme 5.10).

Murahashi et al. found that a Pd-catalyzed α-allylation of aldehydes with allyl amines only proceeded in the presence of an achiral acid as cocatalyst [22]. Based on this reaction, List and coworker first described the proof of concept of the Pd/chiral

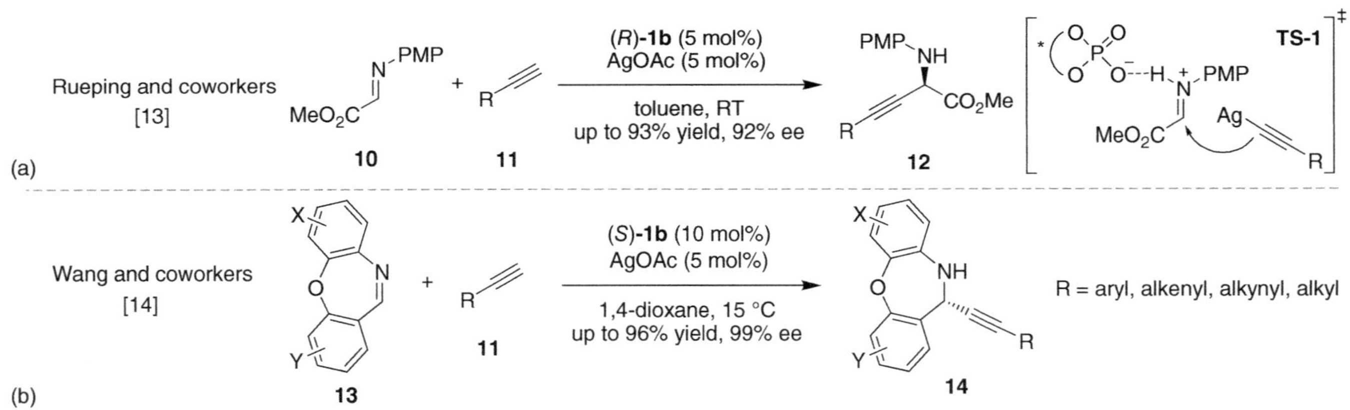

Scheme 5.6 Alkynylation of imines via chiral phosphoric acid and silver combined catalysis. Source: (a) Modified from Rueping et al. [13]. (b) Modified from Ren et al. [14].

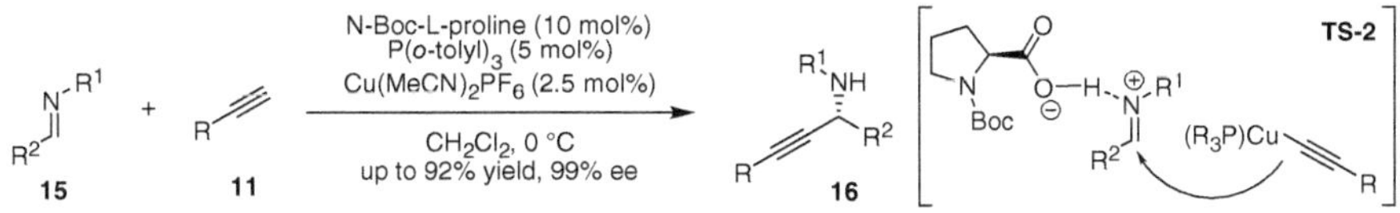

Scheme 5.7 Asymmetric alkynylation of imines by chiral carboxylic acid and copper cooperative catalysis. Source: Modified from Lu et al. [16].

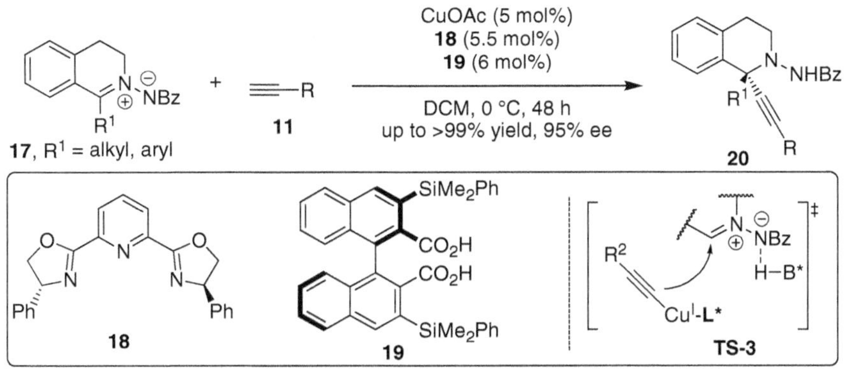

Scheme 5.8 Asymmetric alkynylation of C,N-cyclic azomethine imines by copper/chiral Brønsted acid cooperative catalysis. Source: Modified from Hashimoto et al. [17].

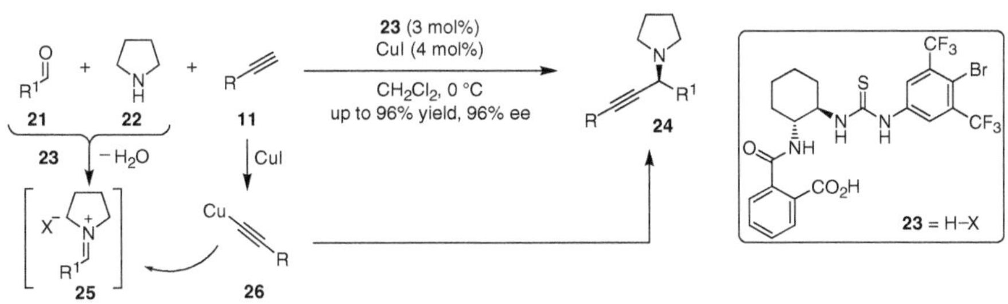

Scheme 5.9 Asymmetric alkynylation of imines enabled by chiral thiourea/copper cooperative catalysis. Source: Modified from Zhao et al. [18].

phosphoric acid cooperative catalysis for the development of asymmetric allylic substitution reactions [23]. A highly enantioselective Tsuji–Trost type α-allylation of aldehydes **27** with allylamines **28** was established by a combined catalyst system consisting of a chiral phosphoric acid ***(R)*-1a** and a $Pd(PPh_3)_4$ (Scheme 5.11). In this case, the phosphoric acid plays a dual role: (i) protonating the allylic amine to activate the allylic C—N bond and facilitating the oxidative addition to generate the π-allyl-Pd species; (ii) serving as a chiral counterion of the cationic π-allyl-Pd species to activate the enamine by hydrogen-bonding interaction and to control stereoselectivity of the C—C bond-forming step (**TS-4**). Afterward, the same group expanded such a concept to asymmetric α-allylation of aldehydes **27** with allylic alcohols **30** [24]. In comparison with the former case, an external primary amine is required for the formation of enamine intermediate. This ternary catalyst system allows allylic

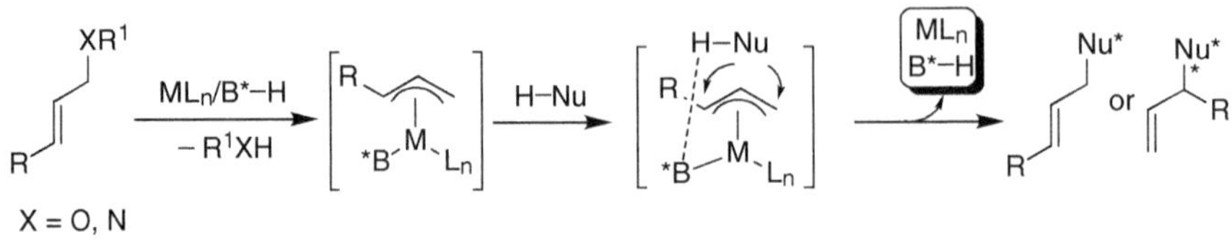

Scheme 5.10 General strategy for the metal/Brønsted acid cooperatively catalyzed asymmetric allylic substitution.

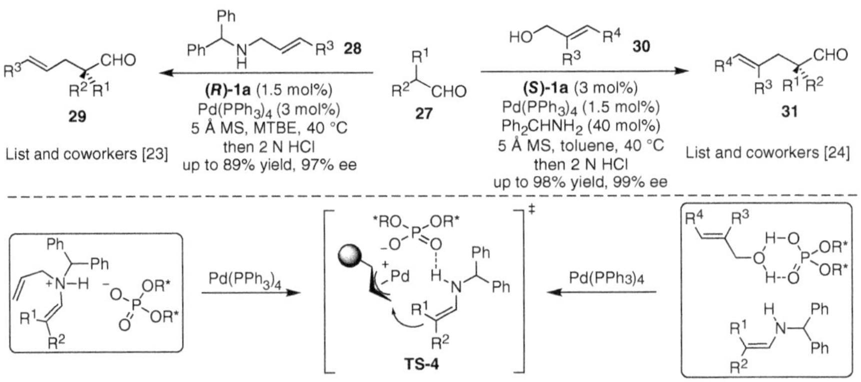

Scheme 5.11 α-Allylation of aldehydes by chiral phosphoric acid and palladium cooperative catalysis.

alkylation to give high levels of enantioselectivity. Again, chiral phosphoric acid assists the palladium to cleave the C—O bond of the allylic alcohols **30** and to control the stereochemistry of the bond-forming step (**TS-4**) by hydrogen bonding interaction (Scheme 5.11).

The direct use of readily accessible olefins as allylating reagents essentially minimizes the preoxidation and functionality manipulation steps required for the preparation of the substrates, as such the allylic C–H functionalization has been considered an even more efficient alternative [25]. Inspired by List's findings [23, 24], Gong and coworkers found that a ternary catalyst system consisting of a primary amine, a chiral phosphoric acid, and a palladium complex could smoothly promote an enantioselective α-allylation of aldehydes **27** with allylarenes **32** (Scheme 5.12) [26]. Mechanistically, in the presence of $Pd(PPh_3)_4$, 2,6-DMBQ (dimethylbenzo-1,4-quinone) and chiral phosphoric acid (*R*)-**1a**, terminal olefin **32** undergoes an allylic C–H cleavage to generate a π-allyl palladium phosphate intermediate **34**, which then couples with an enamine **35** *in situ* generated from a primary amine and an enolizable aldehyde via a similar transition state **TS-5** proposed by List and coworkers [23, 24], leading to the formation of allylation products **33** in high yields and with good to excellent levels of enantioselectivity. Notably, diphenylmethanamine, previously used to activate aldehydes [24], can be easily oxidized into an imine and subsequently decomposes to benzophenone under oxidizing conditions. As a consequence, a more stable achiral primary amine, such as cumylamine, is vital to ensure the success of this catalytic process. Recently, such

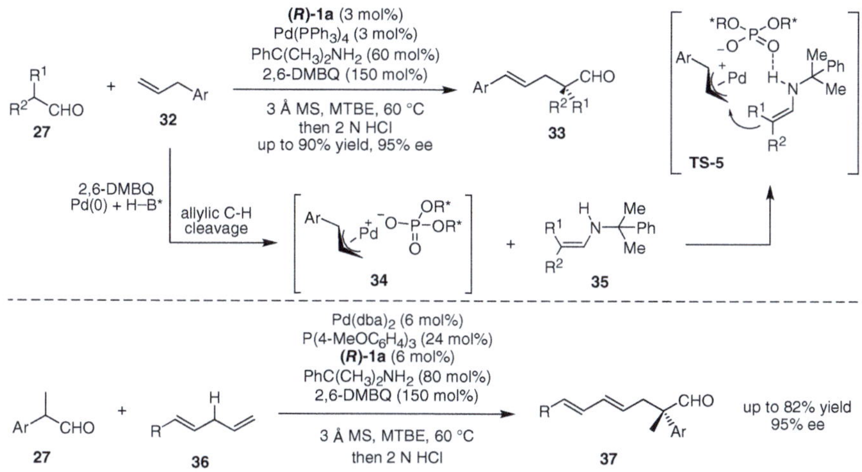

Scheme 5.12 α-Allylation of aldehydes with terminal alkenes.

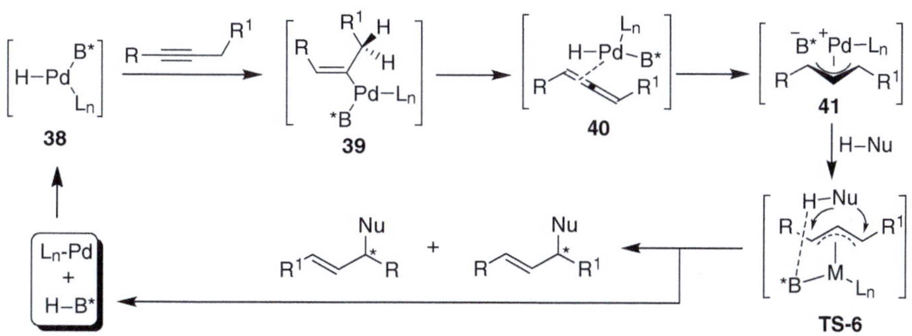

Scheme 5.13 Palladium-catalyzed allylic substitution using alkynes as allylating reagents.

a strategy was successfully applied to the oxidative α-allylation of aldehydes with 1,4-dienes **36** (Scheme 5.12), providing even higher stereoselectivity for a broader scope of substrates [27].

The oxidative addition of Pd(0) complex to a chiral Brønsted acid basically gives palladium hydride complex **38** that can undergo an insertion reaction with the triple bond of an alkyne to give an intermediate **39**. After sequential β-hydride elimination and another insertion of the double bond, the intermediate **39** can be transformed into a π–allylpalladium complex **41** bearing a chiral counterion, which can undergo the allylic substitution with soft nucleophiles. Again, the chiral counterion can activate the nucleophiles by hydrogen-bonding interaction (**TS-6**) and to control the stereochemistry (Scheme 5.13). In comparison with olefins as the allylating reagents, the allylation with alkynes is relatively superior with a 100% atom economy [28].

Based on this general concept, Gong and coworkers accomplished an asymmetric α-allylation of aldehydes **27** with alkynes **42** [29]. The combined catalyst system consisting of an achiral palladium complex, an achiral primary amine, and a chiral phosphoric acid turns out to be preeminent for this transformation (Scheme 5.14a).

The chiral π-allylpalladium phosphate species of type **41**, generated via the reaction pathway shown in Scheme 5.13, undergoes an enantioselective substitution with the enamine via transition state **TS-7** to give the desired products **43**. Interestingly, skipped enynes **44** also undergo a similar process, as shown in Scheme 5.13, to generate vinyl π-allylpalladium intermediates **47** and subsequently isomerizes to intermediates **48**, which then couple with the enamine via the transition state **TS-8** to give allylic alkylation products **45** in excellent stereoselectivity (Scheme 5.14b).

In addition to the enamine, enol is also a reactive nucleophile to undergo palladium-catalyzed allylic substitution reaction. In 2016, List and coworkers reported an asymmetric Tsuji–Trost type α-allylation of branched cycloketones **49** with allyl methyl carbonate **50** under the cooperative catalysis of chiral phosphoric acid and palladium complex (Scheme 5.15) [30]. In the presence of chiral phosphoric acid **(*S*)-2a**, the oxidative addition of Pd(0) to allyl methyl carbonate **50** gives a π-allyl palladium phosphate complex. The enol tautomerized from a cyclohexanone undergoes the asymmetric allylic substitution with the π-allyl palladium phosphate complex assisted by the hydrogen-bonding interaction between the phosphoryl oxygen and enol hydroxyl as shown in the transition state **TS-9**. Interestingly, allylic alcohol **51** is also able to undergo a similar procedure under the CO_2 atmosphere. CO_2 activates the allylic alcohol by the *in situ* formation of a more reactive allyl carbonate **53**, which can react with the palladium(0) complex to give a π-allyl palladium phosphate intermediate. Asymmetric allylic alkylation of the π-allyl palladium phosphate intermediate with enol form of the cycloketone proceeds via the same transition state **TS-9** to generate **52**. Notably, this protocol has been applied to a concise enantioselective formal total synthesis of (+)-crinane through a chiral building block **54** obtained from this reaction. Recently, Trauner and coworkers adopted this protocol to prepare a key chiral building block **55** for asymmetric total synthesis of stephadiamine [31].

Pyrazol-5-ones are an important class of heterocyclic nucleophiles capable of participating in a wide range of asymmetric transformations [32]. In 2013, Gong and coworkers established a highly efficient asymmetric allylic alkylation of pyrazol-5-ones **57** with allylic alcohols **56** under the cooperative catalysis of a chiral phosphoramidite–palladium complex and a chiral phosphoric acid, affording various functionalized chiral pyrazol-5-ones **59** in high yields and with excellent enantioselectivities (Scheme 5.16) [33]. By the hydrogen-bonding activation with chiral phosphoric acid **(*R*)-1c**, the allylic alcohol **56** becomes reactive enough to undergo oxidative addition with chiral phosphoramidite–palladium complex to give a π-allylpalladium phosphate **61**, which then undergoes asymmetric allylic substitution with enol form of the pyrazol-5-one **57**, wherein the counterion acts as a base to activate the nucleophile by means of hydrogen-bonding interaction (**TS-10**) and to control the stereochemistry in collaboration with chiral ligand **58** (Scheme 5.16).

As demonstrated previously [26], allylarenes can undergo allylic C–H cleavage upon exposing to a mixture of the palladium complex, chiral phosphoric acid, and a suitable oxidant to generate a π-allyl palladium phosphate intermediate, which can principally undergo the similar asymmetric allylic substitution with the pyrazol-5-ones [33]. Indeed, the combination of a chiral palladium complex adorned with chiral phosphoramidite **63** and the chiral phosphoric acid **(*R*)-2b**

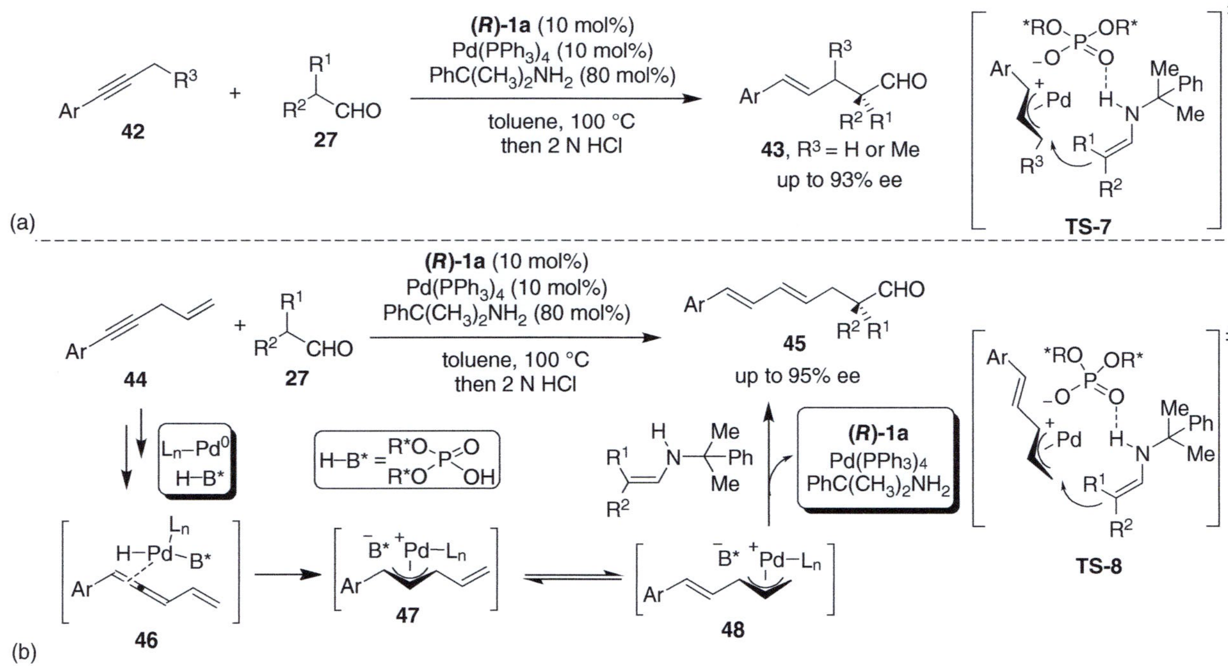

Scheme 5.14 α-Allylation of aldehydes with internal alkynes.

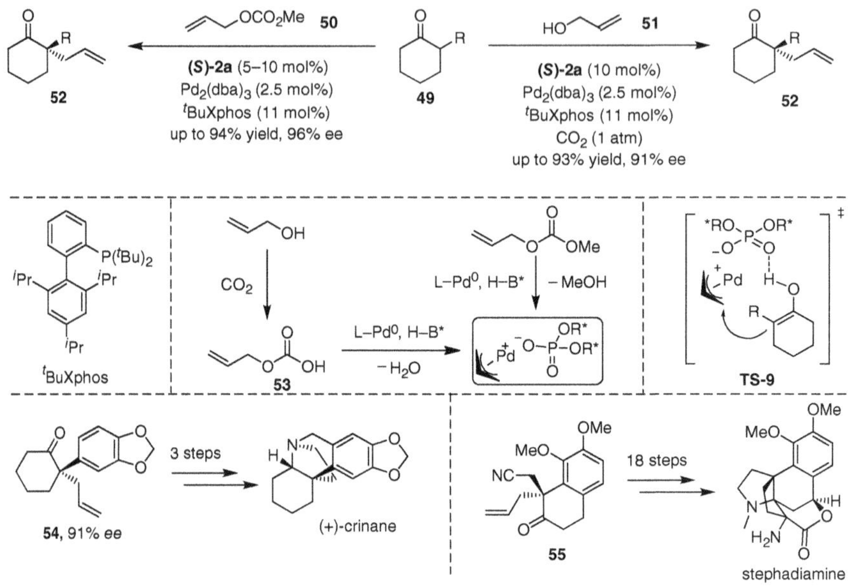

Scheme 5.15 α-Allylation of branched ketones by chiral phosphoric acid and palladium cooperative catalysis. Source: Modified from Pupo et al. [30].

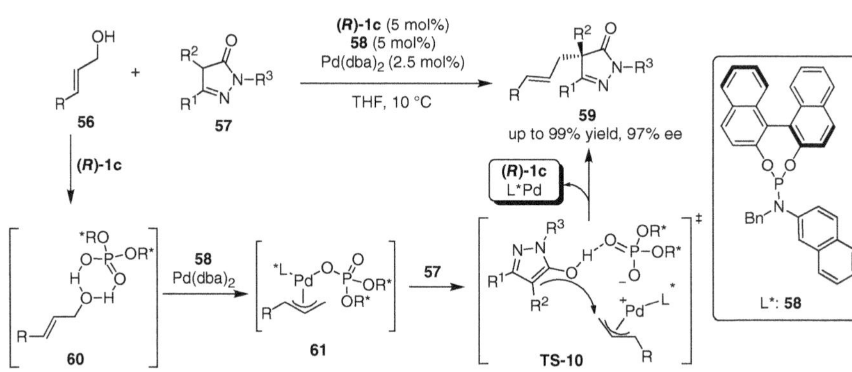

Scheme 5.16 Asymmetric allylic alkylation of pyrazol-5-ones with allylic alcohols.

allows the asymmetric allylic C–H alkylation of allylarenes with pyrazol-5-ones **57** to proceed successfully and to give the desired products **59** in excellent yields and enantioselectivities by using 2,5-DMBQ as external oxidant (Scheme 5.16) [34]. Notably, inactivated terminal alkenes are also tolerable to provide the corresponding alkylation products in good results by using chiral ligand **64** and chiral phosphoric acid ***(R)*-2c** at a higher temperature (Scheme 5.17) [35].

Recently, Shi and coworkers reported a catalytic asymmetric α-arylation of pyrazol-5-ones **57** with 2-indolylmethanols **65** as electrophilic arylation reagents under the cooperative catalysis of chiral phosphoric acid and achiral palladium complex (Scheme 5.18) [36]. The chiral phosphoric acid not only protonates the

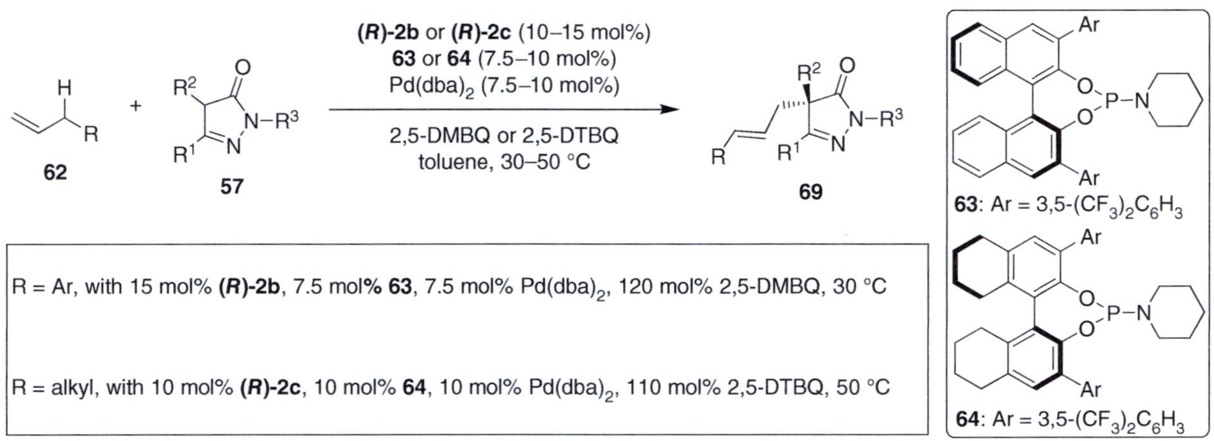

Scheme 5.17 Asymmetric allylic C–H alkylation of pyrazol-5-ones. Source: Modified from Fan et al. [35].

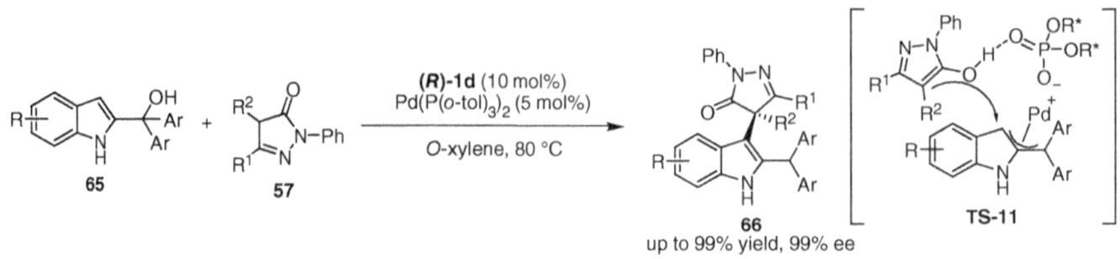

Scheme 5.18 Asymmetric α-arylation of pyrazol-5-ones. Source: Modified from Zhu et al. [36].

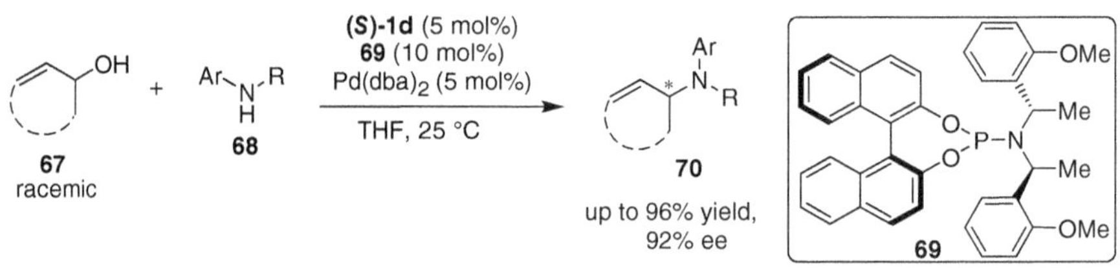

Scheme 5.19 Asymmetric allylic amination of racemic allylic alcohols. Source: Modified from Banerjee et al. [37].

2-indolylmethanols **65** to promote the formation of a π-allyl-Pd species, but also facilitates the enantioselective nucleophilic substitution of the pyrazol-5-ones with the π-allyl-Pd species via transition state **TS-11**.

In addition to the stereoselective carbon–carbon bond formation, chiral phosphoric acid and palladium cooperative catalysis is amenable to carbon–heteroatom bond-forming reactions. In 2014, Beller et al. expanded the cooperative catalysis of chiral phosphoric acid and chiral phosphoramidite–palladium complex [33] to an enantioselective allylic amination of racemic allylic alcohols **67** (Scheme 5.19) [37]. The presence of chiral phosphoramidite **69** and phosphoric acid (*S*)-**1d** enables the palladium-catalyzed allylic amination of both cyclic and acyclic allylic alcohols with anilines to result in a variety of chiral allylic amines **70** in excellent yields and with excellent enantioselectivities.

Chiral hydrogen-bonding organocatalysts can serve as neutral Brønsted acids and are compatible with some transition metal catalysts [8g]. In 2016, Lu and Xiao combined chiral bis thiourea-based hydrogen-bonding catalyst **73** and palladium complex to catalyze an asymmetric allylic alkylation of 3-aryloxindoles **71**, affording 3,3′-disubstituted oxindoles **74** in high yields and with excellent enantioselectivities (Scheme 5.20) [38]. The oxidative addition of palladium complex to **72** gave a π-allylpalladium species and *tert*-butoxide anion, which functions as a base to deprotonate the 3-aryl oxindole **71**. The resulting oxindole anion **75** is stabilized by the thiourea **73** through the hydrogen bonding interaction and attacks the π-allylpalladium species to afford the desired chiral products **74** (Scheme 5.20).

Zhang and coworkers established an asymmetric decarboxylative cycloaddition of vinyl ethylene carbonates **76** with β-nitroolefins **77** enabled by cooperative catalysis of palladium complex and chiral squaramide **79** to access the chiral multisubstituted tetrahydrofurans (Scheme 5.21) [39]. By using a palladium complex generated *in situ*

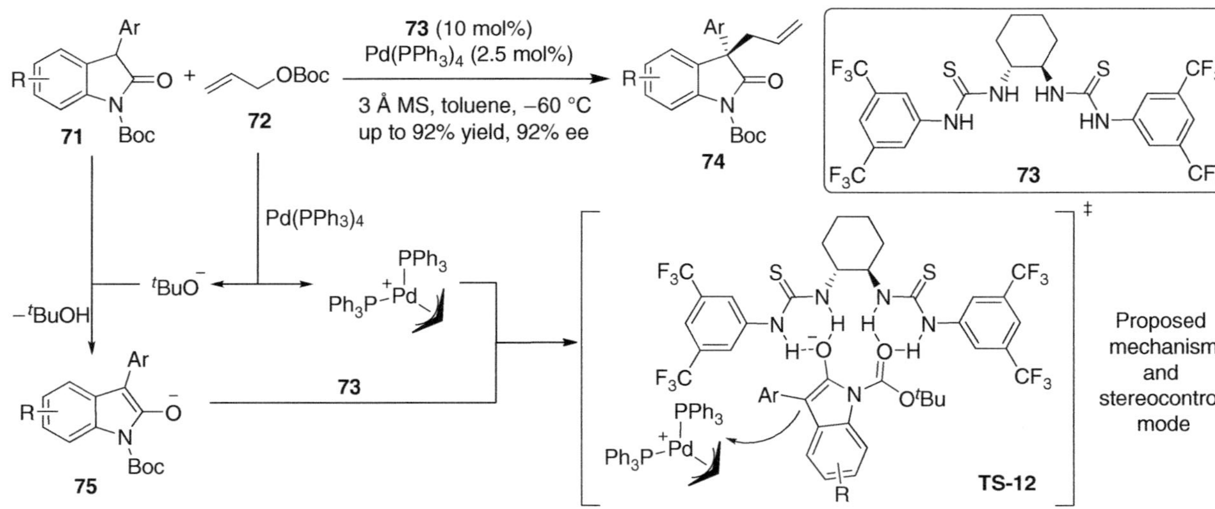

Scheme 5.20 Asymmetric allylic alkylation of 3-aryloxindoles by chiral thiourea/Pd cooperative catalysis.

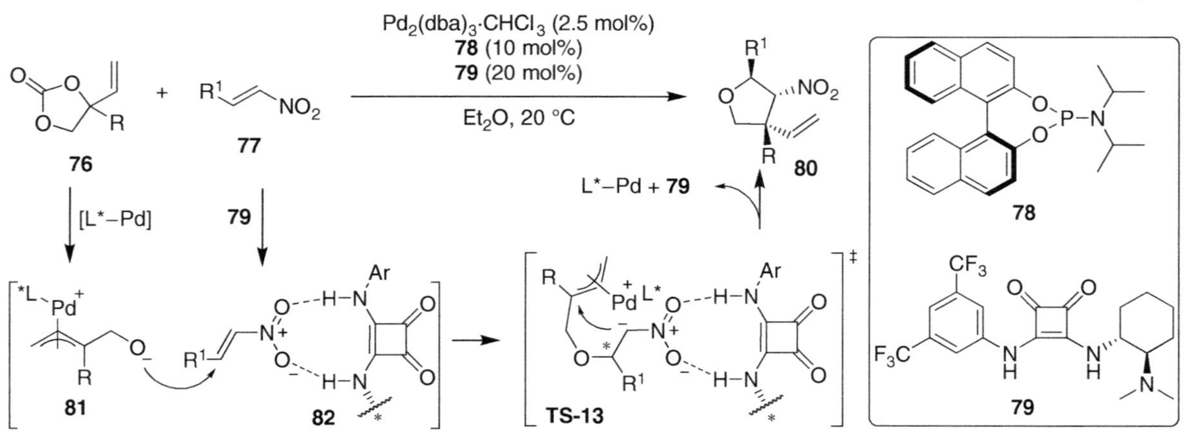

Scheme 5.21 Asymmetric cycloaddition of vinylethylene carbonates with β-nitroolefins. Source: Modified from Liu et al. [39].

from $Pd_2(dba)_3{\cdot}CHCl_3$ and chiral phosphoramidite **78**, a zwitterionic allylpalladium intermediate **81** is generated from the oxidative addition with vinyl ethylene carbonates **76**. Then, the π-allylpalladium species **81** attacks the activated unsaturated electrophiles **82** *in situ* generated from β-nitroolefins **77** and chiral squaramide **79**, followed by cycloaddition to afford the desired products **80** in good to high yields and with high enantio- and diastereoselectivities.

5.5 Asymmetric Hydrogenation of C—N Double Bond

Transition metal-catalyzed asymmetric hydrogenation is a scalable method for the manufacture of chiral molecules [40]. The vast majority of this process relies on the employment of chiral ligand to induce chirality [41]. The combination of chiral Brønsted acid and transition metal catalyst has been considered as an important alternative to upgrading this process cooperatively. In 2008, Xiao and coworkers reported that a binary catalyst made up of a chiral phosphoric acid and an iridium complex **84** adorned with Noyori–diamine ligand could afford a highly efficient asymmetric hydrogenation of acyclic imines **83** (Scheme 5.22) [42]. The iridium complex **84** incorporating a chiral phosphate as the counteranion can undergo hydrogenolysis to generate an iridium hydride complex and to release chiral phosphoric acid, which then protonates the acyclic imines **83** to facilitate the transfer hydrogenation with the iridium hydride complex via **TS-14**. The matched chirality between the chiral diamine ligand of the iridium complex **84** and chiral phosphoric acid **(*R*)-1a** is required to deliver high stereoselectivity. Although iridium complex **84** can promote this hydrogenation reaction alone, the presence of chiral phosphoric acid obviously enhances the catalytic reaction performance. The extension of the hybrid catalyst system to asymmetric reductive amination of ketones **85** with anilines **86** is also successful (Scheme 5.22) [43]. Experimental and computational investigations suggest that the noncovalent interaction of iridium complex and chiral phosphoric acid is responsible for catalytic activity and stereoselectivity [44].

The combination of iridium complex **90** and chiral Brønsted acid (*S*)-**2d** is also able to render asymmetric hydrogenation of quinolones **89**. Notably, even the racemic iridium complex **90**, in collaboration with chiral *N*-triflyl phosphoramide (*S*)-**2d**, still enables the reaction to give good enantioselectivity (Scheme 5.23) [45].

In addition to asymmetric hydrogenation, chiral phosphoric acid and iridium cooperative catalysis is amenable to borrow hydrogen methods. In 2014, Zhao and coworkers reported an enantioselective amination of alcohols **92** synergistically catalyzed by a chiral iridium complex **93** and chiral phosphoric acid (*R*)-**1a** (Scheme 5.24) [46]. The iridium complex **93** initially catalyzes the dehydrogenation of the alcohol **92** to a ketone intermediate and itself becomes a reactive iridium hydride complex. The imine *in situ* formed from the ketone and aniline is protonated by the chiral phosphoric acid (*R*)-**1a** and then undergoes the asymmetric transfer hydrogenation with the iridium hydride complex to furnish chiral amine **94**. Afterward, the same group established a dynamic kinetic asymmetric amination

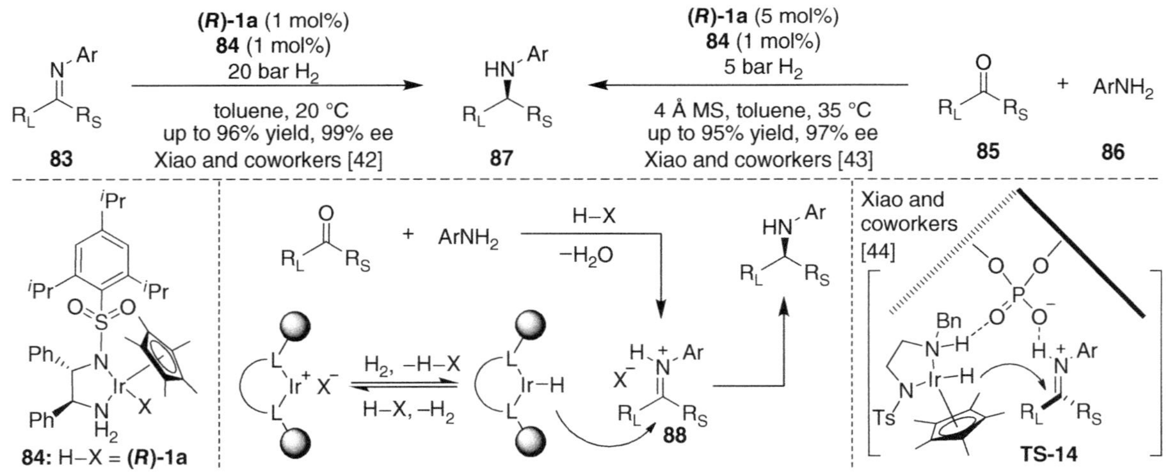

Scheme 5.22 Asymmetric hydrogenation of *N*-aryl imines. Source: Modified from Li et al. [42].

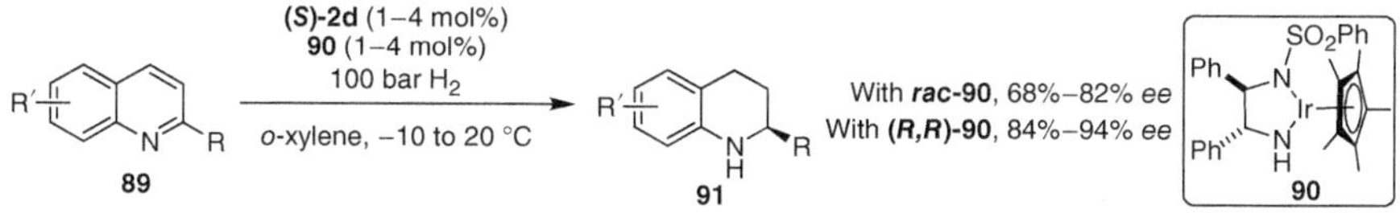

Scheme 5.23 Asymmetric hydrogenation of quinolines. Source: Modified from Rueping and Koenigs [45].

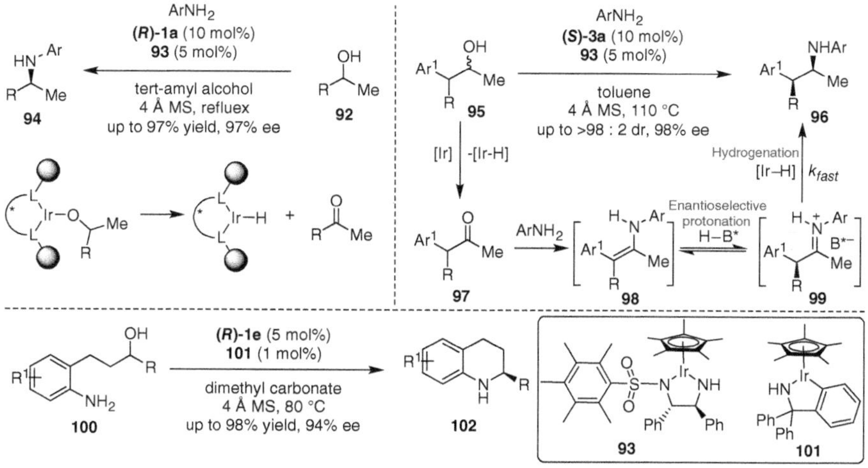

Scheme 5.24 Asymmetric amination of alcohols by chiral phosphoric acid and iridium cooperative catalysis. Source: Modified from Zhang et al. [46].

of alcohols **95** promoted by a similar combined catalyst system (Scheme 5.24) [47]. Racemic α-branched alcohols **95** are converted to highly enantioenriched chiral amines **96** by a redox dynamic kinetic resolution process proceeding via sequential dehydrogenation of alcohols **95** to ketones **97**, dynamic imine **99** formation, and asymmetric transfer hydrogenation. Density functional theory (DFT) calculations suggest that the enantioselectivity is induced at the early stage of enantioselective protonation before the diastereo-determining hydrogenation step [48]. Moreover, this borrowing hydrogen concept is also applicable to an intramolecular variant for the synthesis of chiral 2-substituted tetrahydroquinolines **102** (Scheme 5.24) [49].

Aside from late transition metals, nonprecious metals can also collaborate with chiral phosphoric acids to catalyze asymmetric hydrogenation. Beller and coworkers found that the combination of chiral phosphoric acid **(*S*)-1a** and the Knölker iron complex **103** allocated high enantioselectivity to the asymmetric hydrogenation of imines [50] (Scheme 5.25). Soon after this event, a similar combined catalyst system was applied to the asymmetric hydrogenation of quinoxalines **104** and benzoxazines **105** (Scheme 5.25) [51].

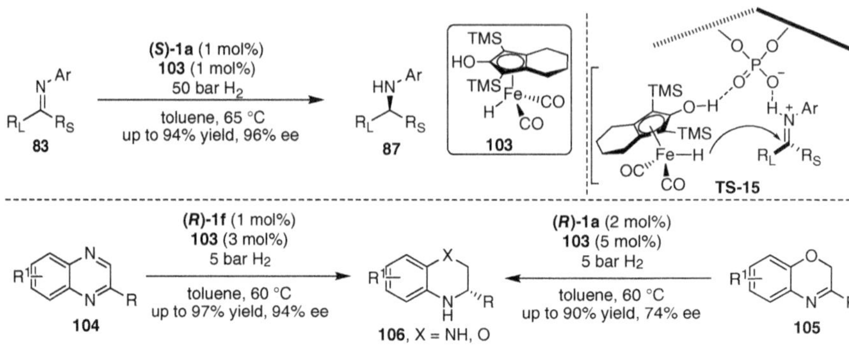

Scheme 5.25 Asymmetric hydrogenation via chiral phosphoric acid and iron cooperative catalysis. Source: Modified from Fleischer et al. [51].

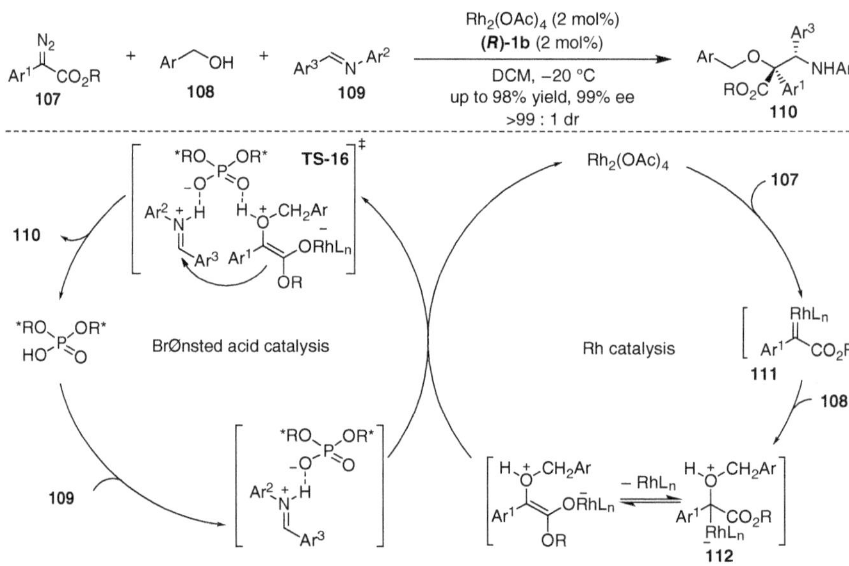

Scheme 5.26 Three-component coupling reaction of diazo-carbonyls, benzyl alcohols, and imines. Source: Modified from Hu et al. [54].

5.6 Metal Carbene-Mediated Transformations

Metal carbenes, generally *in situ* generated from diazo compounds, are highly important and active synthetic intermediates capable of undergoing many valuable transformations [52]. In particular, α-diazo carbonyl compounds have been suitable substrates for the generation of highly reactive ylides or zwitterionic intermediates, which can be trapped by appropriate electrophiles [53]. In 2008, Hu and Gong reported a three-component coupling reaction of diazo-carbonyls **107**, benzyl alcohols **108**, and imines **109** synergistically catalyzed by $Rh_2(OAc)_4$ and chiral phosphoric acid **(*R*)-1b**, leading to a wide range of β-amino-α-hydroxyl acid derivatives **110** with up to 98% yield, >99 : 1 dr and >99% ee (Scheme 5.26) [54]. In this

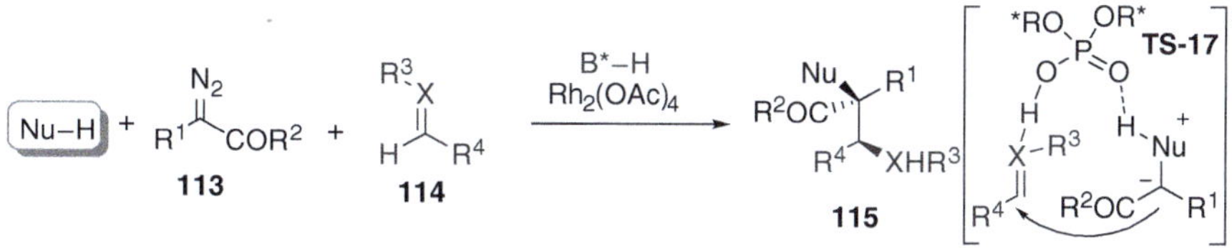

Scheme 5.27 General paradigm for the creation of multicomponent reactions involving diazo-carbonyls enabled by Rh/Brønsted acid cooperative catalysis. Source: Modified from Hu et al. [54].

reaction, the $Rh_2(OAc)_4$ mediates the decomposition of the diazo compound **107** to generate a rhodium carbene **111**, which undergoes the OH insertion reaction with the alcohol **108** to form an oxonium ylide intermediate **112**. Meanwhile, the bifunctional phosphoric acid not only activates the imine **109** by protonation but also interacts with the oxonium ylide intermediate **112** through a hydrogen bond (**TS-16**) to enable a highly enantioselective Mannich-type reaction. Moreover, either diazo cyclopropanes [55] or *in situ* generated imines from aldehydes and amines [56] are well tolerated in this protocol, providing a diverse range of structural frameworks.

Since many nucleophiles can undergo carbene insertion to generate a wide spectrum of zwitterionic intermediates, which can be trapped by abundant electrophiles activated by chiral Brønsted acids via a proposed model (**TS-17**), a large number of asymmetric multicomponent reactions can be created by following the general concept implicit in this work (Scheme 5.27) [54].

Indeed, the binary catalyst systems of such type are able to orchestrate dozens of asymmetric multicomponent reactions involving a wide scope of nucleophilic reagents, including carbamates [57], amines [58], alcohols [59], water [60], and thiols [61], and electrophiles, such as imines [57a, 60b], aldehydes [58b], ketones [61b] and Michael acceptors [60c], affording a huge number of chiral molecules in structural and functional diversity (Scheme 5.28). By rational design of substrates and reaction sequence, chiral cyclic compounds **116–127** have been accessed by using a similar catalytic strategy [62].

Moreover, the rhodium/chiral phosphoric acid cooperative catalysis offers a good avenue to access chiral benzoxazines **129** from imines **109** and 2-amidophenyl diazoesters **128** (Scheme 5.29) [63]. The key zwitterionic intermediate **131** is generated from the intramolecular oxygen addition of the amide functionality to the rhodium carbene. Chiral phosphoric acid co-catalyst orchestrates the enantioselective trapping process of the zwitterionic intermediate **131** with imines **109**.

Electronically rich arenes are also reactive nucleophiles capable of undergoing C–H insertion to metal carbenoids. Hu and coworkers have revealed that the rhodium carbene formed from *N*-aryl diazoamide **132** can generate a reactive zwitterionic intermediate **135** via intramolecular aryl C–H insertion, and chiral phosphoric acid-activated imine is capable of tapping this intermediate to deliver chiral oxindole **133** in high yields and with excellent levels of stereoselectivity [64] (Scheme 5.30). In addition, intermolecular C–H insertion of indoles **137** to rhodium carbenes formed from **136** and **141** smoothly occurs to give zwitterionic intermediates **140** and **144**, respectively. In the presence of chiral phosphoric acid

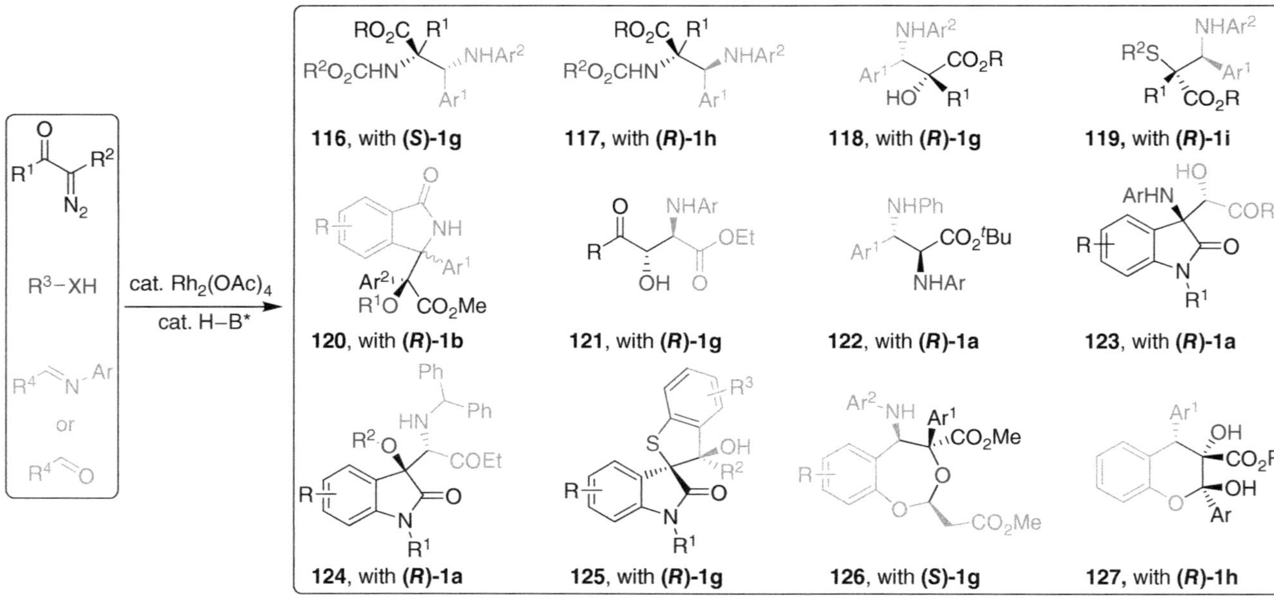

Scheme 5.28 Electrophilic trapping of oxonium ylides via chiral phosphoric acid and rhodium combined catalysis.

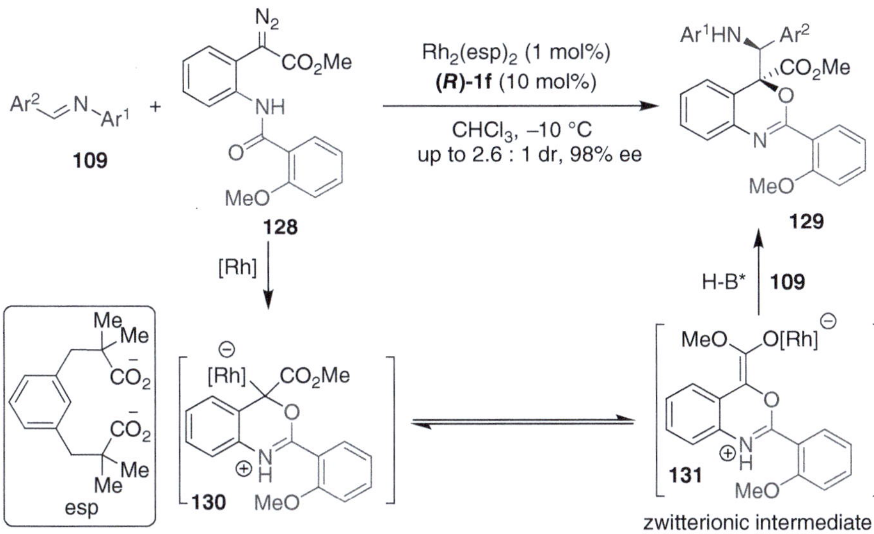

Scheme 5.29 Asymmetric synthesis of chiral benzoxazines. Source: Modified from Jia et al. [63].

(*S*)-**1g**, these intermediates readily undergo a Mannich-type reaction with either imine **109** or *in situ* generated imine from aniline **86** and aldehyde **142** to furnish chiral indoles derivatives **138** and **143** in excellent levels of enantioselectivity (Scheme 5.30) [65].

Under the cooperative catalysis of rhodium acetate and chiral phosphoric acid, a three-component reaction of *N,N*-disubstituted anilines **145**, diazoesters **107**, and imines **109** provides α,α-diaryl β-amino esters **146** in extremely high enantioselectivities [66] (Scheme 5.31). Similar to the precedent reactions [64], this transformation also proceeds from the aromatic C–H insertion to the rhodium carbene and generates a zwitterionic intermediate **147**, which is stereoselectively trapped by chiral phosphoric acid-activated imines **109**. In addition, alkynyldiazoacetates **148** have been convinced to be nice substrates to participate in the three-component reaction with the imines **109** and a broad scope of nucleophiles, including alcohols, indoles, and *N,N*-disubstituted anilines, to yield **149** (Scheme 5.31) [67].

The zwitterionic intermediates generated from the X—H bond insertion to the transition metal carbene can be trapped either by electrophiles beyond imines [68] or protonated via proton transfer [69]. The control of stereoselectivity of X—H bond insertion reaction has been a longstanding challenge and very few chiral transition metal complexes can offer synthetically useful stereochemical outcomes [70]. In 2011, Zhu, Zhou and coworkers first introduced the metal/chiral Brønsted acid cooperative catalysis to enantioselective N–H insertion reaction of amide **150** to α-diazophenylacetates **113** (Scheme 5.32) [71]. Chiral spiro phosphoric acids (SPAs) are identified as superior co-catalysts of the rhodium-catalyzed asymmetric N–H insertion reaction, in which chiral SPAs actually act as chiral proton-transfer shuttle (CPTS) catalysts, that is, providing a proton and accepting another proton

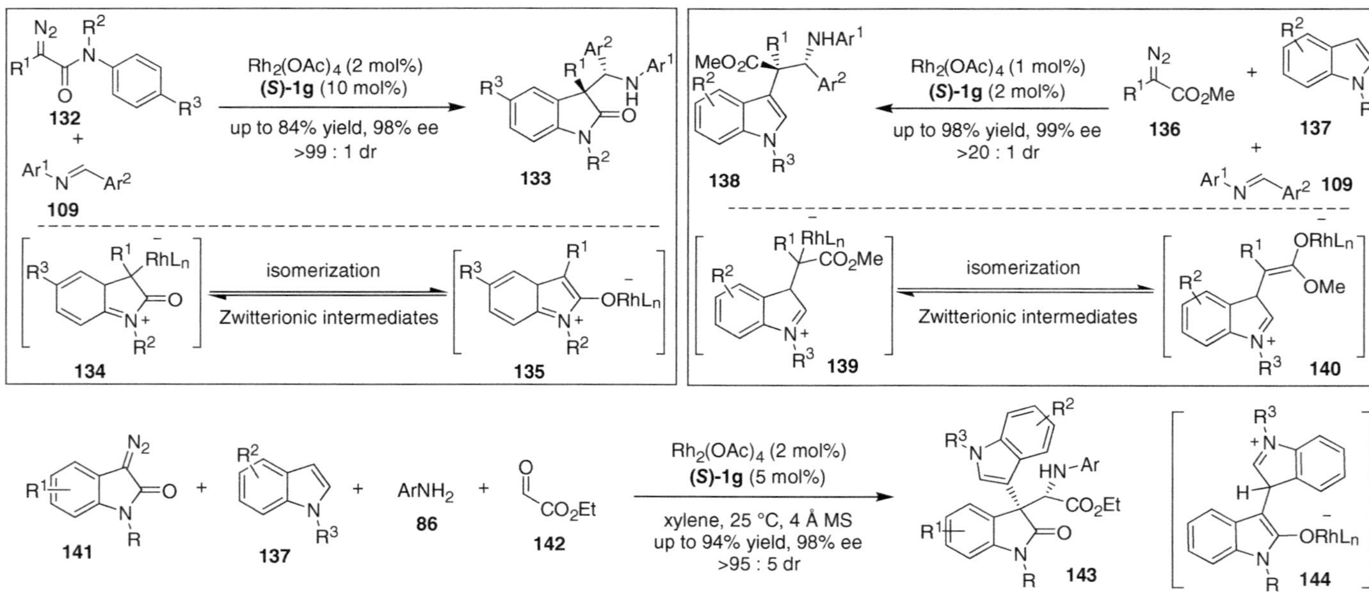

Scheme 5.30 Asymmetric reactions based on aryl C–H insertion to metal carbenoid. Source: Modified from Jing et al. [65].

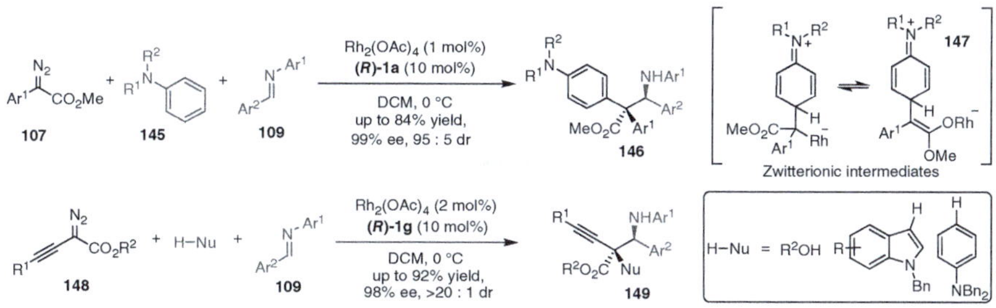

Scheme 5.31 Three-component reaction of electron-rich arenes, diazoesters, and imines. Source: Modified from Yu et al. [67].

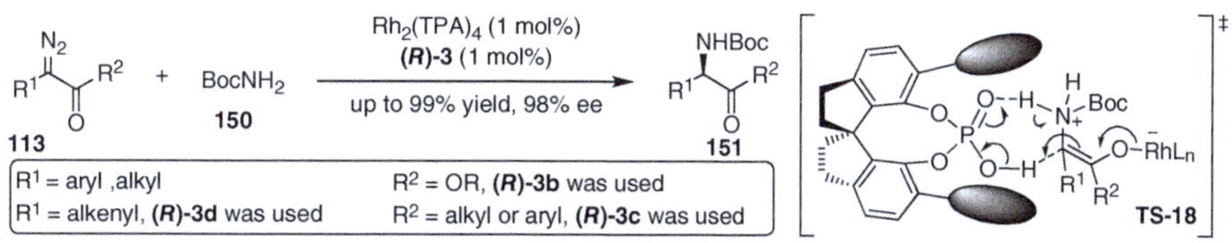

Scheme 5.32 Asymmetric N–H insertion reaction by chiral spiro phosphoric acid/rhodium cooperative catalysis. Source: Modified from Xu et al. [71].

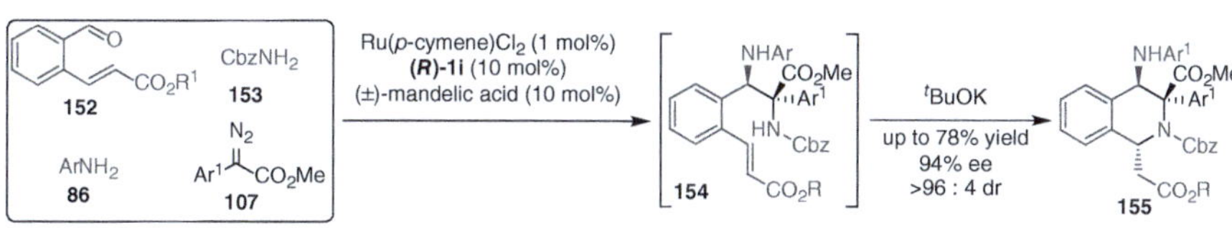

Scheme 5.33 Asymmetric four-component reaction to assemble tetrahydroisoquinolines.

synchronously from a rhodium-enolate intermediate through a cyclic transition state **TS-18** (Scheme 5.32). Notably, such a concept, now known as CPTS, has been a generally applicable strategy to address the stereoselection issue in asymmetric N–H insertion reaction of diazocarbonyls [72, 73].

In addition to rhodium complexes, other transition metal complexes have also been found compatible with chiral phosphoric acids to cooperatively catalyze metal-carbene-involved reactions. In 2014, Hu and coworkers developed a cascade Mannich/aza-Michael reaction enabled by chiral Brønsted acid and ruthenium complex combined catalysis to assemble densely functionalized tetrahydroisoquinolines **155** from four reaction components [74], including 2-formylphenylacrylates **152**, amide , anilines **86**, and diazoesters **107** (Scheme 5.33).

The combination of π-allylpalladium chloride complex with chiral phosphoric acid has been identified as a proper binary catalyst system to render a highly stereoselective three-component reaction of diazo esters **107**, pyrroles **156**, and imines **109** (Scheme 5.34) [75]. Both *syn*- and *anti*-products can be accessed by simply varying chiral Brønsted acid co-catalysts. The use of L-tartaric acid and **(*R*)-1g** favors the generation of *anti*-products **157**, while chiral phosphoric acid **(*R*)-1a** alone prefers to afford *syn*-diastereoisomers **158**. The palladium carbene phosphate **159** is believed to undergo the C–H insertion reaction to form a zwitterionic intermediate **160** or

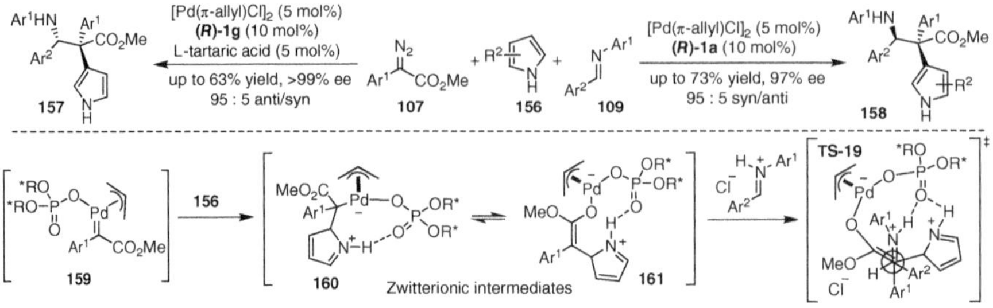

Scheme 5.34 Three-component reaction enabled by Pd/Brønsted acid cooperative catalysis. Source: Modified from Zhang et al. [75].

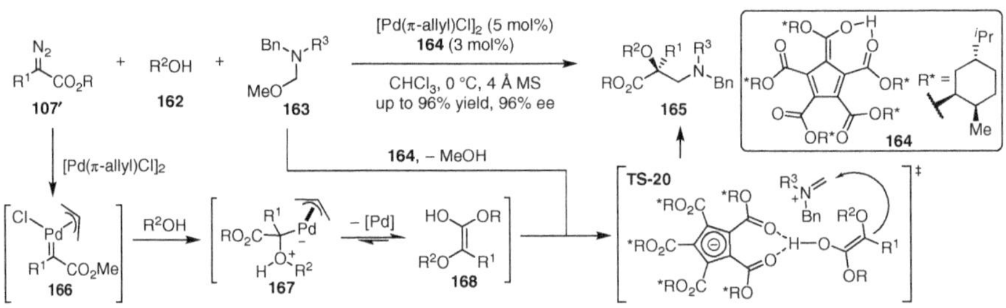

Scheme 5.35 Asymmetric aminomethylation enabled by chiral PCCP acid and Pd(II) cooperative catalysis. Source: Modified from Kang et al. [77].

161. Subsequently, asymmetric Mannich-type reaction proceeds via **TS-19** assisted by hydrogen-bonding interaction of the phosphoryl oxygen with pyrrole and the protonated imine, respectively.

Chiral pentacarboxycyclopentadiene (PCCP) acid has emerged as another type of promising chiral Brønsted acid catalysts for asymmetric synthesis [76]. Recently, Xu and Hu established an enantioselective aminomethylation reaction of diazo compounds **107**, alcohols **162**, and α-aminomethyl ethers **163** enabled by the combined catalysis of π-allylpalladium chloride complex and chiral PCCP acid **164** (Scheme 5.35) [77]. The PCCP acid initially reacts with α-aminomethyl ether **163** to give chiral methylene iminium ion pair, which subsequently undergoes an asymmetric Mannich reaction with an enol intermediate **168** generated from the diazoester **107** and alcohol **162** promoted by the palladium catalyst. The chiral PCCP anion controls the enantioselectivity by the electrostatic and H-bonding interactions with iminium ion and enol **168**, as shown in **TS-20**.

5.7 π-Lewis Acid Mediated Transformations

A wide range of transition metals can serve as π-Lewis acids to activate unsaturated carbon–carbon bonds, and in particular, palladium and gold can incorporate with chiral Brønsted acids for asymmetric transformations of alkenes and alkynes [8f, 78]. In 2012, Chai and Rainey reported an asymmetric migratory ring expansion

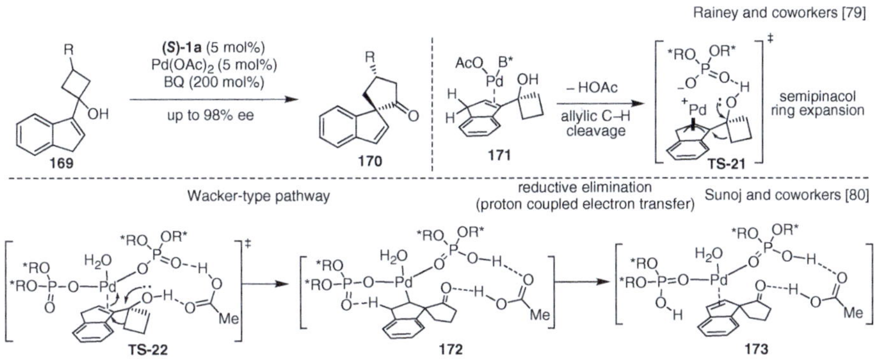

Scheme 5.36 Asymmetric migratory ring expansion via chiral phosphoric acid/$Pd(OAc)_2$ combined catalysis. Source: Modified from Chai et al. [79].

of cyclobutanols **169** for the synthesis of chiral spiro-ketones **170** catalyzed by a chiral phosphoric acid and palladium acetate (Scheme 5.36) [79]. This transformation was initially proposed to proceed via an allylic C–H cleavage and subsequent π-allylpalladium species-mediated semipinacol rearrangement of spirocyclic alcohols (**TS-21**). However, the detailed DFT computations by Sunoj and coworker suggest that a Wacker-type pathway [80], consisting of a Pd-*bis*-phosphate species-mediated semipinacol ring expansion of the indenyl cyclobutanol via transition state **TS-22** and reductive elimination of intermediate **172** (proton-coupled electron transfer), is energetically preferred in comparison with the allylic C–H cleavage pathway.

Using chiral phosphoric acid and palladium cooperative catalysis, Yao and coworkers accomplished an asymmetric cascade annulation between 2-hydroxystyrenes **175** and 2-alkynylbenaldehyes or 1-(2-alkynylphenyl)ketones **174**, affording variable bridged ring compounds **176** bearing multiple chiral centers (Scheme 5.37) [81]. This reaction proceeds from a $Pd(OAc)_2$-mediated cycloisomerization of **174** to generate a Pd(II)-isochromenylium intermediate **177**. Then **177** undergoes anion exchange with (*S*)-**1a** and a subsequent asymmetric [4+2] cycloaddition with 2-hydroxystyrenes **175** via **TS-23** to generate a carbocation intermediate **178**, in which the carbocation is captured by the internal phenol hydroxyl and followed by protonation with HOAc to furnish the products and to regenerate $Pd(OAc)_2$ and chiral phosphoric acid, successively.

Recently, Liu and Engle described an enantioselective α-alkylation of azlactones **181** with terminal alkenes **180** based on the palladium and chiral phosphoric acid cooperative catalysis (Scheme 5.38) [82]. DFT calculations suggest that the bidentate directing group orientates the regioselectivity in the Wacker-type nucleopalladation of intermediate **183** and stabilizes the resultant alkylpalladium(II) intermediate **184**. Simultaneously, the chiral phosphoric acid (*R*)-**1j** not only serves as a chiral ligand to control the stereoselection and to activate the nucleophile by hydrogen-bonding interaction in the nucleopalladation step but also a Brønsted acid to undergo the subsequent protodepalladation step via intermediate **184** for the regeneration of Pd(II) catalyst after decomplexation of **185**.

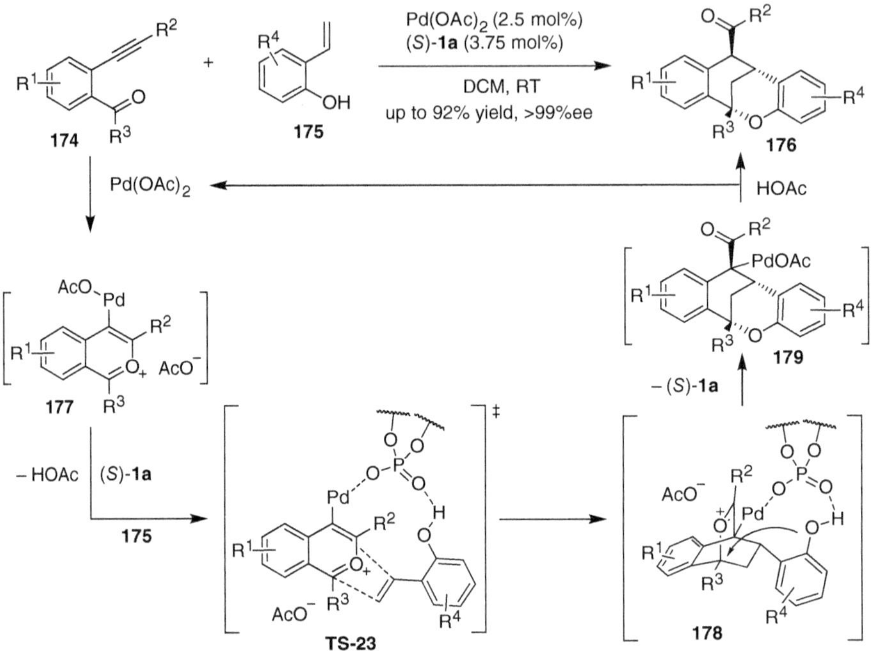

Scheme 5.37 Asymmetric annulation of 2-hydroxystyrenes and 1-(2-alkynylphenyl)carbonyls. Source: Modified from Yu et al. [81].

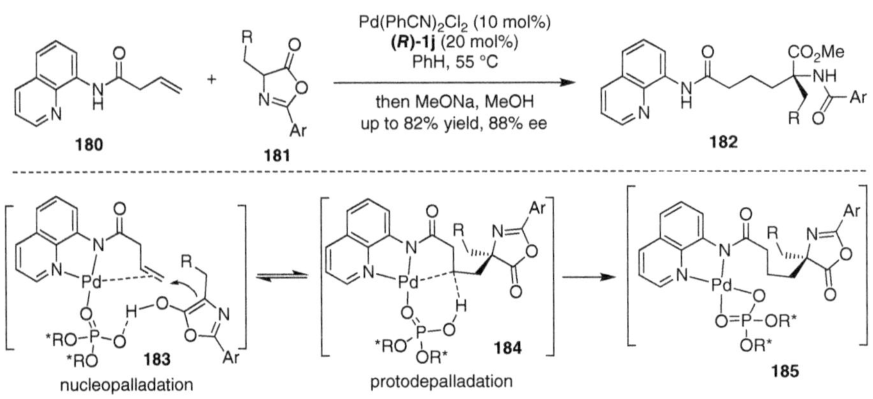

Scheme 5.38 Enantioselective α-alkylation of azlactones with terminal alkenes. Source: Modified from Nimmagadda et al. [82].

The combination of chiral Pd(II) complex and chiral Brønsted acid is also amenable to the asymmetric functionalization of alkenes. In 2014, Gong and coworkers reported an enantioselective oxidative cascade cyclization reaction of *N*-(2,2-disubstituted hex5-en-1-yl)acrylamides **186** by using a combined catalyst of a Pd(II) complex of a chiral Bu-QUOX (quinoline-oxazoline) ligand **187** and a chiral phosphoric acid (*S*)-**1a** (Scheme 5.39) [83]. Mechanistic results show that chiral phosphoric acid (*S*)-**1a** and $Pd(CF_3COO)_2$ bearing chiral Bu-QUOX ligand

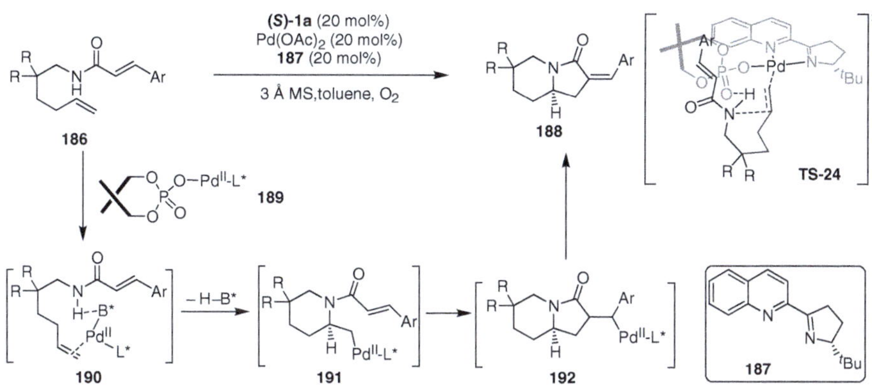

Scheme 5.39 Enantioselective oxidative tandem cyclization reaction. Source: Modified from He et al. [83].

187 are capable of undergoing anion exchange to generate a catalytically active Pd(II) complex **189**. Then a tandem Wacker-type oxidation/cyclization reaction of **186** proceeds via intermediates **190-192** under the catalysis of the Pd(II) complex **189** to enable the synthesis of chiral 6,5-bicyclic aza-heterocycles **188** in moderate yields and with excellent levels of enantioselectivity. With alkene coordination to Pd(II) and a hydrogen-bonding interaction with phosphate, the bond-forming step favors a *Re*-face transition state **TS-24**, highlighting the synergistic effect between chiral ligand and chiral counterion in the palladium catalysis.

Gold complexes are typical π-Lewis acids and have widespread applications in the transformation of alkynes [84]. The chiral phosphoric acid/gold binary catalysts have enabled a broad scope of asymmetric reactions [85]. Recently, Xu and coworkers accomplished an asymmetric cascade oxidative cyclization and Mannich reaction of 3-butynol **193** and nitrones **194** catalyzed by a gold complex and chiral phosphoric acid (Scheme 5.40) [86]. The protocol commences with gold-catalyzed intermolecular alkyne oxidation of **193** with the nitrone **194** to deliver an α-oxo gold carbene intermediate **196** and imines **109** [87]. The gold carbene intermediate **196** is attacked by a tethered hydroxy group to form an intermediate **197** and then isomerizes to an enolate species **198**, which undergoes an asymmetric Mannich-type reaction with the protonated imine intermediate **199** to yield the corresponding products **195** with high levels of enantioselectivity.

5.8 Summary and Outlook

Chiral Brønsted acid and transition metal cooperative catalysis has provided a wide range of unprecedented activation modes and unique transformations of chemical bonds by taking the advantage of the dual catalyst systems, in which the catalyst evaluation procedure can be largely simplified by applying available modular chiral pools of both transition-metal catalysts and chiral Brønsted acids. Despite these advances, this field is still in the primary stage. The catalyst compatibility and

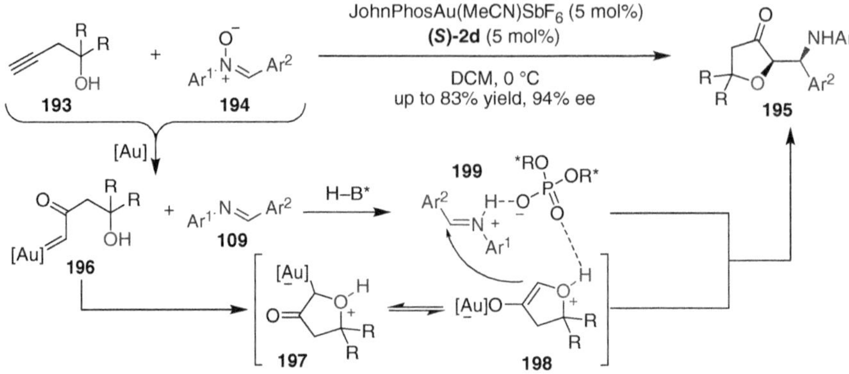

Scheme 5.40 Oxidative cyclization/Mannich reaction by chiral phosphoric acid/gold cooperative catalysis. Source: Modified from Wei et al. [86].

interoperability between the kinetics of generating reactive intermediates for the stereoselective bond formation remain big issues yet to be addressed in future studies. Inspired by the functions of many enzymes in nature, the use of the hybrid catalysts involving both transition-metal catalysts and chiral Brønsted acids with a linker might be a new solution to these challenges. Considering the outstanding ability of metal catalysis to activate plenty of chemical bonds, the introduction of versatile reactive transition metal intermediates in this cooperative catalysis will continue to be highly desired for the future development of new chemistry. Another orientation might be the stereodivergent construction of continuous stereogenic centers through a simultaneously controllable activation of both nucleophiles and electrophiles. Although cooperative catalysis is convenient and fruitful, more mechanistic and theoretical insights are also required to fully understand the role of each catalyst and the related active species. The growth of fundamental knowledge in catalytic activation modes and the origin of cooperativity would greatly assist chemists to design new transformations that could efficiently provide valuable building blocks in previously unknown processes. With the rapid growth in the development and understanding of reactions and catalytic systems, this field will continue to grow and new chemistry beyond imagination can be anticipated.

References

1 Akiyama, T. (2007). *Chem. Rev.* 107: 5744–5758.

2 (a) Akiyama, T., Itoh, J., Yokota, K. et al. (2004). *Angew. Chem. Int. Ed.* 43: 1566–1568. (b) Uraguchi, D. and Terada, M. (2004). *J. Am. Chem. Soc.* 126: 5356–5357.

3 Doyle, A.G. and Jacobsen, E.N. (2007). *Chem. Rev.* 107: 5713–5743.

4 Parmar, D., Sugiono, E., Raja, S., and Rueping, M. (2014). *Chem. Rev.* 114: 9047–9153.

5 Taylor, M.S. and Jacobsen, E.N. (2006). *Angew. Chem. Int. Ed.* 45: 1520–1543.

6 Terada, M. (2010). *Synthesis* 12: 1929–1982.

7 (a) Yu, J., Shi, F., and Gong, L.Z. (2011). *Acc. Chem. Res.* 44: 1156–1171. (b) Wu, H., He, Y.P., and Shi, F. (2015). *Synthesis* 47: 1990–2016.

8 (a) Park, Y.J., Park, J.W., and Jun, C.H. (2008). *Acc. Chem. Res.* 41: 222–234. (b) Shao, Z. and Zhang, H. (2009). *Chem. Soc. Rev.* 38: 2745–2755. (c) Rueping, M., Koenigs, R.M., and Atodiresei, I. (2010). *Chem. Eur. J.* 16: 9350–9365. (d) Zhong, C. and Shi, X. (2010). *Eur. J. Org. Chem.* 2010: 2999–3025. (e) Du, Z. and Shao, Z. (2013). *Chem. Soc. Rev.* 42: 1337–1378. (f) Wu, X., Li, M.L., and Gong, L.Z. (2013). *Acta Chim. Sinica* 71: 1091–1100. (g) Chen, D.F., Han, Z.Y., Zhou, X.L. et al. (2014). *Acc. Chem. Res.* 47: 2365–2377.

9 Komanduri, V. and Krische, M.J. (2006). *J. Am. Chem. Soc.* 128: 16448–16449.

10 Lauder, K., Toscani, A., Scalacci, N., and Castagnolo, D. *Chem. Rev.* 117: 14091–14200.

11 (a) Vessally, E., Edjlali, L., Hosseinian, A. et al. (2016). *RSC Adv.* 6: 49730–49746. (b) Vessally, E., Hosseinian, A., Edjlali, L. et al. (2016). *RSC Adv.* 6: 71662–71675.

12 Zani, L. and Bolm, C. (2006). *Chem. Commun.*: 4263–4275.

13 Rueping, M., Antonchick, A.P., and Brinkmann, C. (2007). *Angew. Chem. Int. Ed.* 46: 6903–6906.

14 Ren, Y.Y., Wang, Y.Q., and Liu, S. (2014). *J. Org. Chem.* 79: 11759–11767.

15 (a) Thomas, A.M., Sujatha, A., and Anilkumar, G. (2014). *RSC Adv.* 4: 21688–21698. (b) Dong, X.Y., Zhang, Y.F., Ma, C.L. et al. (2019). *Nat. Chem.* 11: 1158–1166.

16 Lu, Y., Johnstone, T.C., and Arndtsen, B.A. (2009). *J. Am. Chem. Soc.* 131: 11284–11285.

17 Hashimoto, T.O. and Maruoka, K. (2011). *Angew. Chem. Int. Ed.* 50: 8952–8955.

18 Zhao, C. and Seidel, D. (2015). *J. Am. Chem. Soc.* 137: 4650–4653.

19 (a) Trost, B.M. and Van Vranken, D.L. (1996). *Chem. Rev.* 96: 395–422. (b) Trost, B.M. (2004). *J. Org. Chem.* 69: 5813–5837. (c) Lu, Z. and Ma, S. (2008). *Angew. Chem. Int. Ed.* 47: 258–297.

20 Trost, B.M. and Crawley, M.L. (2003). *Chem. Rev.* 103: 2921–2944.

21 Trost, B.M., Machacek, M.R., and Aponick, A. (2006). *Acc. Chem. Res.* 39: 747–760.

22 Murahashi, S., Makabe, Y., and Kunita, K. (1988). *J. Org. Chem.* 53: 4489–4495.

23 Mukherjee, S. and List, B. (2007). *J. Am. Chem. Soc.* 129: 11336–11337.

24 Jiang, G. and List, B. (2011). *Angew. Chem. Int. Ed.* 50: 9471–9474.

25 (a) Liu, G. and Wu, Y. (2009). *Top. Curr. Chem.* 292: 195–209. (b) Liron, F., Oble, J., Lorion, M.M. et al. (2014). *Eur. J. Org. Chem.*: 5863–5883. (c) Tang, H., Huo, X., Meng, Q. et al. (2016). *Acta Chim. Sinica* 74: 219–233. (d) Wang, R., Luan, Y., and Ye, M. (2019). *Chin. J. Chem.* 37: 720–743.

26 Wang, P.S., Lin, H.C., Zhai, Y.J. et al. (2014). *Angew. Chem. Int. Ed.* 53: 12218–12221.

27 Zhou, X.L., Su, Y.L., Wang, P.S. et al. (2018). *Acta Chim. Sinica* 76.

28 (a) Kadota, I., Shibuya, A., Gyoung, Y.S. et al. (1998). *J. Am. Chem. Soc.* 120: 10262–10263. (b) Kadota, I., Shibuya, A., Lutete, L.M., and Yamamoto, Y. (1999). *J. Org. Chem.* 64: 4570–4571.

29 Su, Y.L., Li, L.L., Zhou, X.L. et al. (2018). *Org. Lett.* 20: 2403–2406.

30 Pupo, G., Properzi, R., and List, B. (2016). *Angew. Chem. Int. Ed.* 55: 6099–6102.

31 Hartrampf, N., Winter, N., Pupo, G. et al. (2018). *J. Am. Chem. Soc.* 140: 8675–8680.

32 (a) Ebner, S., Wallfisch, B., Andraos, J. et al. (2003). *Org. Biomol. Chem.* 1: 2550–2555. (b) Kimata, A., Nakagawa, H., Ohyama, R. et al. (2007). *J. Med. Chem.* 50: 5053–5056. (c) Mariappan, G., Korim, R., Joshi, N. et al. (2010). *J. Adv. Pharm. Technol. Res.* 1: 396–400.

33 Tao, Z.L., Zhang, W.Q., Chen, D.F. et al. (2013). *J. Am. Chem. Soc.* 135: 9255–9258.

34 Lin, H.C., Wang, P.S., Tao, Z.L. et al. (2016). *J. Am. Chem. Soc.* 138: 14354–14361.

35 Fan, L.F., Wang, P.S., and Gong, L.Z. (2019). *Org. Lett.* 21: 6720–6725.

36 Zhu, Z.Q., Shen, Y., Liu, J.X. et al. (2017). *Org. Lett.* 19: 1542–1545.

37 Banerjee, D., Junge, K., and Beller, M. (2014). *Angew. Chem. Int. Ed.* 53: 13049–13053.

38 Boucherif, A., Duan, S.W., Yuan, Z.G. et al. (2016). *Adv. Synth. Catal.* 358: 2594–2598.

39 Liu, K., Khan, I., Cheng, J. et al. (2018). *ACS Catal.* 8: 11600–11604.

40 Xie, J. and Zhou, Q. (2012). *Acta Chim. Sinica* 70: 1427.

41 (a) Minnaard, A.J., Feringa, B.L., Lefort, L. et al. (2007). *Acc. Chem. Res.* 40: 1267–1277. (b) Zhang, W., Chi, Y., and Zhang, X. (2007). *Acc. Chem. Res.* 40: 1278–1290.

42 Li, C., Wang, C., Villa-Marcos, B. et al. (2008). *J. Am. Chem. Soc.* 130: 14450–14451.

43 Li, C., Villa-Marcos, B., and Xiao, J. (2009). *J. Am. Chem. Soc.* 131: 6967–6969.

44 Tang, W., Johnston, S., Iggo, J.A. et al. (2013). *Angew. Chem. Int. Ed.* 52: 1668–1672.

45 Rueping, M. and Koenigs, R.M. (2011). *Chem. Commun.* 47: 304–306.

46 Zhang, Y., Lim, C.S., Sim, D.S. et al. (2014). *Angew. Chem. Int. Ed.* 53: 1399–1403.

47 Rong, Z.Q., Zhang, Y., Chua, R.H. et al. (2015). *J. Am. Chem. Soc.* 137: 4944–4947.

48 Tribedi, S., Hadad, C.M., and Sunoj, R.B. (2018). *Chem. Sci.* 9: 6126–6133.

49 Lim, C.S., Quach, T.T., and Zhao, Y. (2017). *Angew. Chem. Int. Ed.* 56: 7176–7180.

50 Zhou, S., Fleischer, S., Junge, K. et al. (2011). *Angew. Chem. Int. Ed.* 50: 5120–5124.

51 Fleischer, S., Zhou, S., Werkmeister, S. et al. (2013). *Chem. Eur. J.* 19: 4997–5003.

52 (a) Doyle, M.P. (1986). *Chem. Rev.* 86: 919–939. (b) Doyle, M.P. and Forbes, D.C. (1998). *Chem. Rev.* 98: 911–936.

53 Ye, T. and McKervey, M.A. (1994). *Chem. Rev.* 94: 1091–1160.

54 Hu, W., Xu, X., Zhou, J. et al. (2008). *J. Am. Chem. Soc.* 130: 7782–7783.

55 Li, M.F., Zheng, Q., Jin, W.F. et al. (2016). *Tetrahedron* 72: 2929–2934.

56 Xu, X., Zhou, J., Yang, L. et al. (2008). *Chem. Commun.*: 6564–6566.

57 (a) Jiang, J., Xu, H.D., Xi, J.B. et al. (2011). *J. Am. Chem. Soc.* 133: 8428–8431. (b) Kisan, H.K. and Sunoj, R.B. (2016). *Org. Lett.* 18: 3746–3749.

58 (a) Hu, W., Jiang, L., Zhang, D. et al. (2013). *Synthesis* 45: 452–458. (b) Ren, L., Lian, X.L., and Gong, L.Z. (2013). *Chem. Eur. J.* 19: 3315–3318.

59 Jia, S.K., Lei, Y.B., Song, L.L. et al. (2017). *Chin. Chem. Lett.* 28: 213–217.

60 (a) Qian, Y., Jing, C., Shi, T. et al. (2011). *ChemCatChem* 3: 653–656. (b) Qian, Y., Jing, C., Liu, S. et al. (2013). *Chem. Commun.* 49: 2700–2702. (c) Alamsetti, S.K., Spanka, M., and Schneider, C. (2016). *Angew. Chem. Int. Ed.* 55: 2392–2396. (d) Qiu, L., Guo, X., Qian, Y. et al. (2016). *Chem. Commun.* 52: 11831–11833. (e) Kang, Z., Zhang, D., Shou, J. et al. (2018). *Org. Lett.* 20: 983–986.

61 (a) Xiao, G., Ma, C., Xing, D. et al. (2016). *Org. Lett.* 18: 6086–6089. (b) Xiao, G., Chen, T., Ma, C. et al. (2018). *Org. Lett.* 20: 4531–4535.

62 Qiu, L., Guo, X., Ma, C. et al. (2014). *Chem. Commun.* 50: 2196–2198.

63 Jia, S., Yang, X., Dong, G. et al. (2019). *Org. Lett.* 21: 4014–4018.

64 Qiu, H., Li, M., Jiang, L.Q. et al. (2012). *Nat. Chem.* 4: 733–738.

65 Jing, C., Xing, D., and Hu, W. (2015). *Org. Lett.* 17: 4336–4339.

66 Jia, S., Xing, D., Zhang, D. et al. (2014). *Angew. Chem. Int. Ed.* 53: 13098–13101.

67 Yu, S., Hua, R., Fu, X. et al. (2019). *Org. Lett.* 21: 5737–5741.

68 Zhang, D. and Hu, W. (2017). *Chem. Rec.* 17: 739–753.

69 Ren, Y.Y., Zhu, S.F., and Zhou, Q.L. (2018). *Org. Biomol. Chem.* 16: 3087–3094.

70 (a) Chen, C., Zhu, S.F., Liu, B. et al. (2007). *J. Am. Chem. Soc.* 129: 12616–12617. (b) Liu, B., Zhu, S.F., Zhang, W. et al. (2007). *J. Am. Chem. Soc.* 129: 5834–5835. (c) Zhang, Y.Z., Zhu, S.F., Wang, L.X., and Zhou, Q.L. (2008). *Angew. Chem.* 120: 8624–8626. (d) Zhu, S.F., Chen, C., Cai, Y. et al. (2008). *Angew. Chem. Int. Ed.* 47: 932–934. (e) Zhu, S.F., Cai, Y., Mao, H.X. et al. (2010). Zhou. *Nat. Chem.* 2: 546–551.

71 Xu, B., Zhu, S.F., Xie, X.L. et al. (2011). *Angew. Chem. Int. Ed.* 50: 11483–11486.

72 Guo, J.X., Zhou, T., Xu, B. et al. (2016). *Chem. Sci.* 7: 1104–1108.

73 Xu, B., Zhu, S.F., Zuo, X.D. et al. (2014). *Angew. Chem. Int. Ed.* 53: 3913–3916.

74 Jiang, J., Ma, X., Ji, C. et al. (2014). *Chem. Eur. J.* 20: 1505–1509.

75 Zhang, D., Qiu, H., Jiang, L. et al. (2013). *Angew. Chem. Int. Ed.* 52: 13356–13360.

76 Gheewala, C.D., Collins, B.E., and Lambert, T.H. (2016). *Science* 351: 961–965.

77 Kang, Z., Wang, Y., Zhang, D. et al. (2019). *J. Am. Chem. Soc.* 141: 1473–1478.

78 Han, Z.Y. and Gong, L.Z. (2018). *Prog. Chem.* 30: 505–512.

79 Chai, Z. and Rainey, T.J. (2012). *J. Am. Chem. Soc.* 134: 3615–3618.

80 Jindal, G. and Sunoj, R.B. (2014). *J. Am. Chem. Soc.* 136: 15998–16008.

81 Yu, S.Y., Zhang, H., Gao, Y. et al. (2013). *J. Am. Chem. Soc.* 135: 11402–11407.

82 Nimmagadda, S.K., Liu, M., Karunananda, M.K. et al. (2019). *Angew. Chem. Int. Ed.* 58: 3923–3927.

83 He, Y.P., Wu, H., Xu, L. et al. (2014). *Org. Chem. Front.* 1: 473–476.

84 (a) Hashmi, A.S. and Hutchings, G.J. (2006). *Angew. Chem. Int. Ed.* 45: 7896–7936. (b) Hashmi, A.S. and Rudolph, M. (2008). *Chem. Soc. Rev.* 37: 1766–1775. (c) Sengupta, S. and Shi, X. (2010). *ChemCatChem* 2: 609–619.

85 (a) Inamdar, S.M., Konala, A., and Patil, N.T. (2014). *Chem. Commun.* 50: 15124–15135. (b) Wang, P.S., Chen, D.F., and Gong, L.Z. (2019). *Top. Curr. Chem.* 378: 9.

86 Wei, H., Bao, M., Dong, K. et al. (2018). *Angew. Chem. Int. Ed.* 57: 17200–17204.

87 (a) Yeom, H.S. and Shin, S. (2014). *Acc. Chem. Res.* 47: 966–977. (b) Zhang, L. *Acc. Chem. Res.* 47: 877–888.

6

Metal-Brønsted Acid Relay Catalysis

6.1 Introduction

The extraordinary compatibility of Brønsted acid catalysts with a broad scope of metal complexes opens a novel window for dual catalysis to enable asymmetric transformations synergistically or in a relay manner [1, 2]. Since the asymmetric metal/Brønsted acid cooperative catalysis has been highlighted in Chapter 5, we will in this chapter focus on asymmetric cascade transformations accessed by the relay catalysis of metal complexes and chiral Brønsted acids (Scheme 6.1). The premier example of metal/chiral Brønsted acid relay catalysis is the transition metal-catalyzed *in situ* generation of reactive enamine (enol ether) intermediates, chiral Brønsted acid-catalyzed enamine-iminium (oxonium) tautomerization, and subsequent enantioselective trapping of either enamine (enol ether) or iminium (oxonium) ions (Scheme 6.2a). In addition, metal hydride-mediated alkene isomerization, π-Lewis acid-catalyzed addition of alcohols or amines to alkynes, and transition metal-catalyzed dehydrogenative oxidation of amines or ethers all tolerate chiral Brønsted acid catalysis for relay catalytic enantioselective bond-forming transformations (Scheme 6.2a). Moreover, the insertion to carbenoids is also able to generate nucleophilic species that can be trapped by Brønsted acid-activated electrophiles, leading to new asymmetric processes (Scheme 6.2b).

6.2 π-Lewis Acid-Chiral Brønsted Acid Relay Catalysis

π-Lewis acids are able to activate alkynes toward nucleophilic attack. The addition of oxygen- or nitrogen-containing nucleophiles to π-Lewis acid-activated alkynes, under the relay catalysis of chiral Brønsted acids, often introduces cascade enol ether-oxonium or enamine-iminium reactivity to enable enantioselective conversion of sp-hybridized carbon into sp^3-carbon stereocenters (Scheme 6.3) [3].

Gold and platinum complexes are typical π-Lewis acid catalysts for diverse transformations of C—C triple bonds, such as hydroamination, hydroxylation,

Asymmetric Organo-Metal Catalysis: Concepts, Principles, and Applications, First Edition. Liu-Zhu Gong.
© 2022 WILEY-VCH GmbH. Published 2022 by WILEY-VCH GmbH.

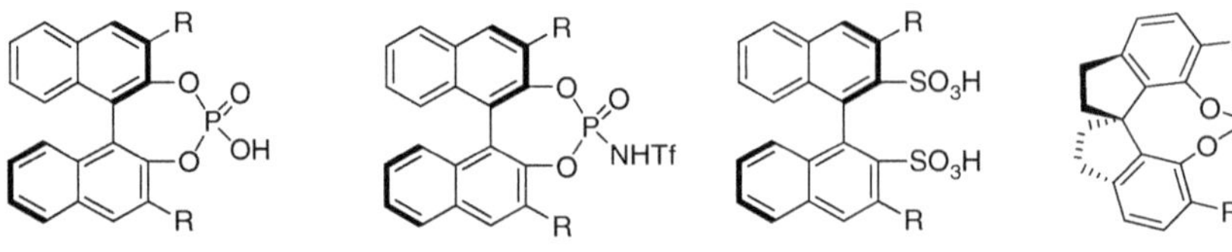

1 2 3 4

Scheme 6.1 Representative chiral phosphoric acids, phosphoramides, and disulfonic acids.

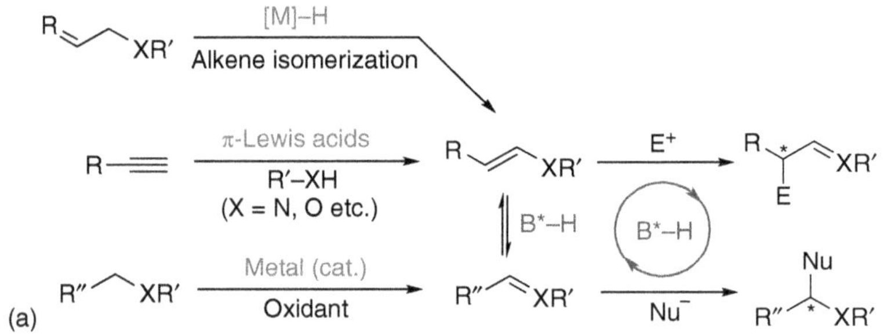

Scheme 6.2 General concepts of metal-chiral Brønsted acid relay catalysis. (a) Catalytic generation of reactive nucleophiles and electrophiles. (b) Carbene insertion to generate nucleophiles.

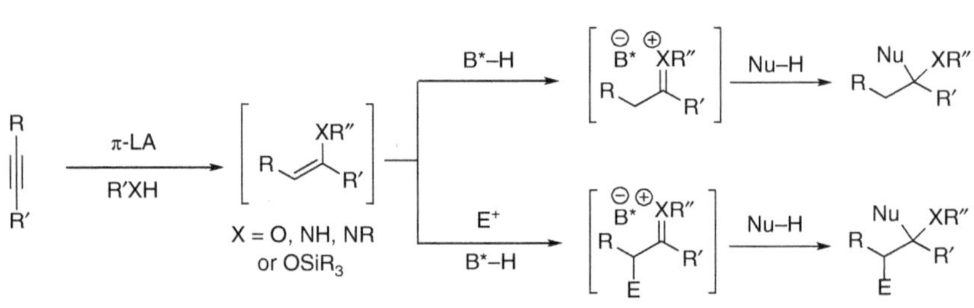

Scheme 6.3 General reaction scheme by π-Lewis acid/chiral Brønsted acid relay catalysis. Source: Chen et al. [3].

and enyne cycloisomerization [4–6]. In 2006, Krause and coworker reported a gold/TsOH relay catalytic 5-*endo*-dig cyclization of homopropargylic alcohols **5** and hydroalkoxylation cascade reaction to give tetrahydrofuran ethers **6** (Scheme 6.4a) [7]. By combining Pt(II) and achiral Brønsted acid (HBF_4) catalysis, Barluenga et al. developed a three-component cycloaddition reaction of alkynols **7**, aldehydes **8**, and aromatic amines **9** to access spiroquinoline derivatives **10** in good yields (Scheme 6.4b) [8].

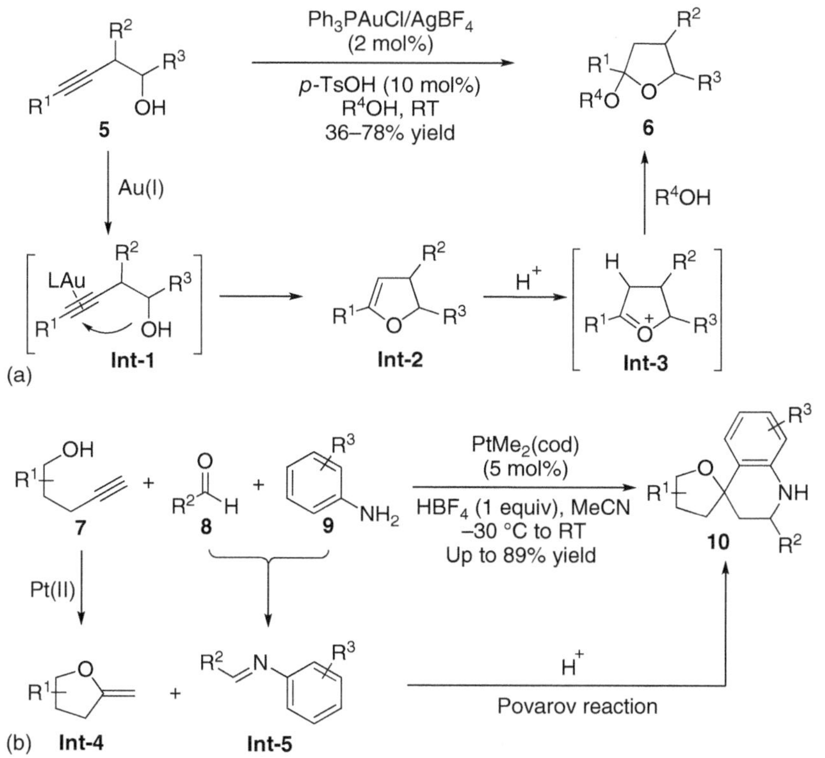

Scheme 6.4 Examples of π-Lewis acid-achiral Brønsted acid relay catalysis. Source: (a) Modified from Belting and Kraus [7], (b) Modified from Barluenga et al. [8].

6.2.1 Hydroamination-Initiated Cascade Reaction

Based on the enamine synthesis by π-Lewis-catalyzed hydroamination of alkynes, Gong and coworkers established a gold complex/chiral Brønsted acid relay catalytic transformation to directly convert the propargylic anilines **11** to a number of chiral tetrahydroquinolines **13** in excellent yields and with high levels of enantioselectivity (Scheme 6.5) [9]. This cascade reaction proceeds via an Au(I)-catalyzed intramolecular hydroamination of the C—C triple bond and the chiral phosphoric acid-catalyzed enantioselective transfer hydrogenation [10, 11]. Control experiments indicate that the *in situ* formed Au(I) phosphate from Ph_3PAuMe and **1a** is a much less effective catalyst than **1a** itself for the asymmetric transfer hydrogenation reaction. In the presence of an achiral ligand, chiral gold phosphate can also efficiently catalyze the cascade reaction at relatively elevated temperature [12]. Independently, Liu and Che reported a cascade intermolecular hydroamination/transfer hydrogenation reaction of alkynes **14** and anilines **9** to furnish chiral secondary amine products **15** in excellent yields and enantioselectivities [13].

Later, Che and coworkers developed a gold(I)/chiral phosphoric acid sequential catalysis for cascade hydroamination/quinoline formation and asymmetric transfer hydrogenation reaction of 2-aminobenzaldehydes **16** or 2-aminophenones and

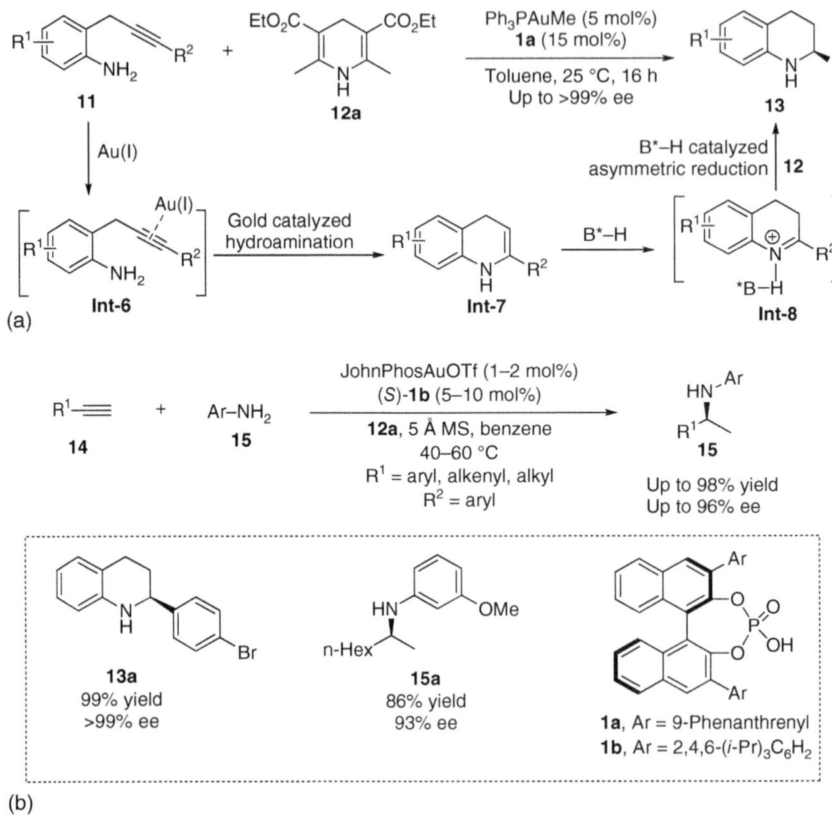

Scheme 6.5 Au(I)/chiral Brønsted acid relay catalytic cascade hydroamination and asymmetric transfer hydrogenation. Source: Modified from Han et al. [9].

simple alkynes **17** (Scheme 6.6) [14]. This method features the efficient synthesis of 2-substituted tetrahydroquinolines **Int-9** with excellent selectivity, allowing access to 2,3- or 2,4-disubstituted tetrahydroquinolines **18** bearing continuous stereogenic centers with high regio-, diastereo- and enantioselectivities.

Imines have proven to be excellent hydride acceptors. As depicted in Scheme 6.7, a rational designed iminium ion **Int-11**, generated from the Au(I)-catalyzed hydroamination of 1-(2-ethynylphenyl)pyrrolidine **20** with aniline **9a** and chiral phosphoric acid-catalyzed tautomerization, undergoes a 1,5-H shift and then a cyclic aminalization reaction, furnishing a cyclic aminal **21** in a high yield [15]. The combination of chiral phosphoric acid **(*R*)-1b** and the highly cationic triphenylphosphine-gold complex leads to an optically active **21**. However, an excess amount of **(*R*)-1b** is required to deliver up to 99% ee (Scheme 6.7).

The condensation of aniline with enol ether generated from the gold-catalyzed intramolecular hydroalkoxylation of homopropargylic alcohols enables the generation of imine, which is able to undergo the addition reaction with nucleophiles. On the basis of this event, Patil and coworkers established a gold(I)/chiral phosphoric acid relay catalytic enantioselective hydroamination/hydroarylation reaction of

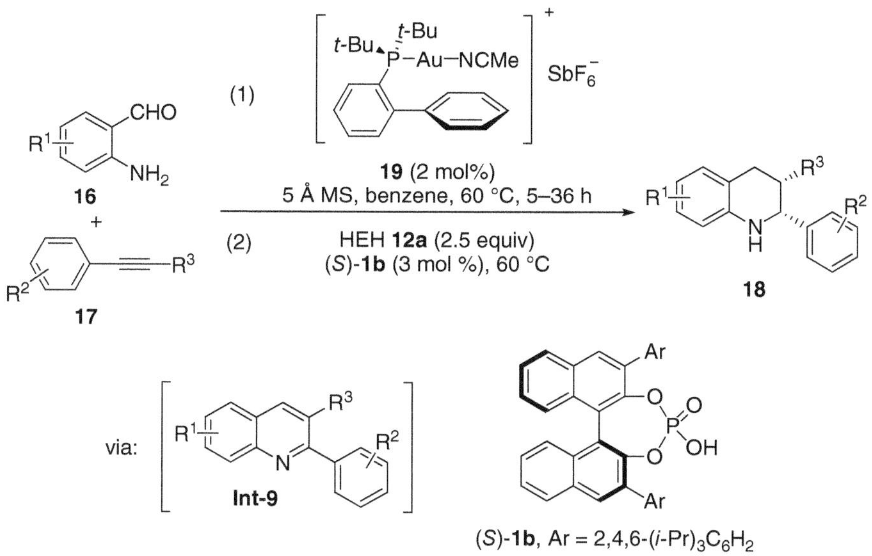

Scheme 6.6 Au(I)/phosphoric acid sequentially catalyzed cascade reaction to access tetrahydroquinolines. Source: Modified from Liu et al. [14].

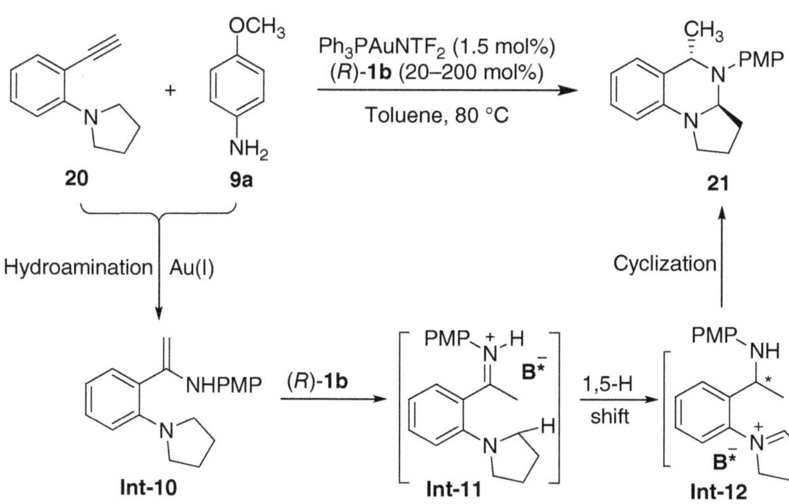

With (*R*)-**1b** (20 mol%): 78% yield, 63 : 37 *anti*/*syn*, 19%/25% ee
With (*R*)-**1b** (200 mol%): 61% yield, 63 : 37 *anti*/*syn*, 97%/99% ee

Scheme 6.7 Au(I)/chiral phosphoric acid relay catalytic cascade hydroamination /1,5-H shift/cyclization reaction.

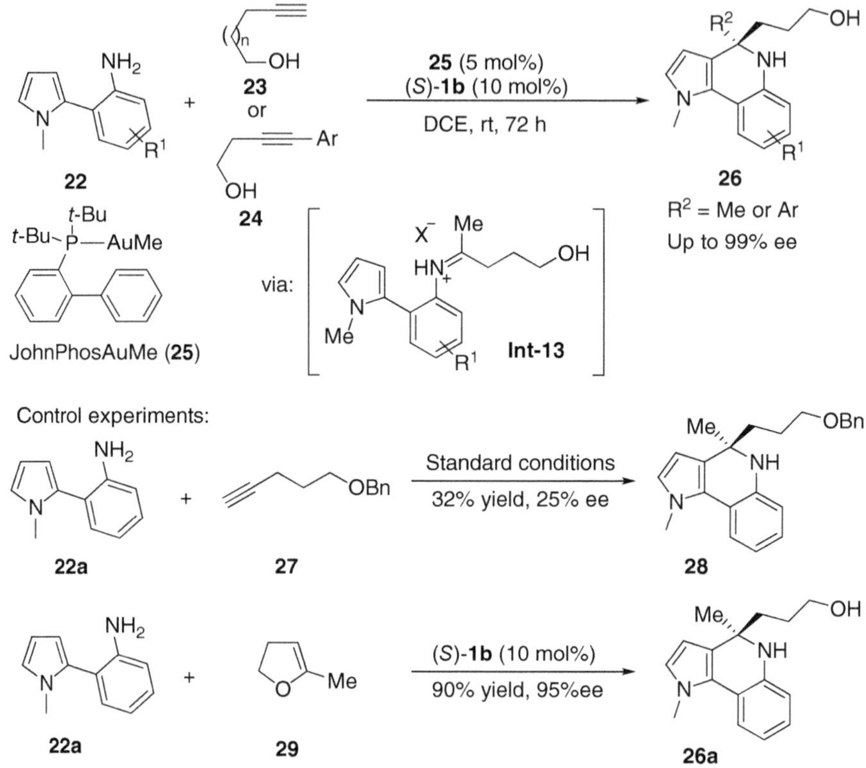

Scheme 6.8 Au(I)/chiral phosphoric acid relay catalytic enantioselective hydroamination/hydroarylation reaction of pyrrole-based aromatic amines and alkynols. Source: Modified from Shinde et al. [16].

pyrrole-based aromatic amines **22** with alkynols **23** and **24** (Scheme 6.8) [16]. The presence of tethered hydroxyl groups in **23** or **24** turns out to be essential to the stereocontrol, and leads to much higher yield and enantioselectivity. In contrast, simple alkynes furnish the desired products in low yield and enantioselectivity. Density functional theory (DFT) studies suggest a transition state involving the H-bonding interaction between the transient imino alcohol and the phosphoric acid. A commercially available 2-methylenetetrahydrofuran **29** can also react with **22a** to give the desired product with excellent enantioselectivity under the catalysis of chiral phosphoric acid **1c**, suggesting that the gold-catalyzed intramolecular hydroalkoxylation provides the enol ether intermediate to access imine by condensation with the aniline **22**.

Under Au(I) catalysis, intramolecular hydroamination of densely functionalized molecules **30** gives indole intermediates **Int-14**, which then undergoes a chiral phosphoric acid-catalyzed asymmetric intramolecular Friedel–Crafts reaction via **TS-1** to afford chiral tetrahydrocarbazoles **31** bearing two adjacent stereogenic centers in excellent yields and with high enantioselectivities (Scheme 6.9) [17].

Amides are prevalent nucleophiles that undergo the gold-catalyzed hydroamination of acetylenes. In 2013, Dixon designed a cascade hydroamination/

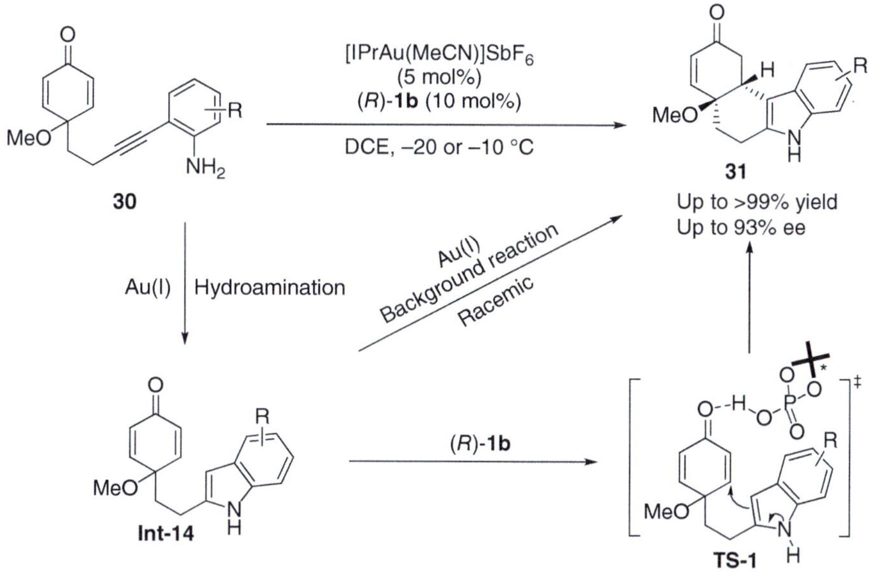

Scheme 6.9 Enantioselective construction of functionalized tetrahydrocarbazoles by asymmetric Au(I)/chiral phosphoric acid relay catalysis. Source: Modified from Zhao et al. [17].

Pictet–Spengler type cyclization reaction consecutively catalyzed by a gold complex and chiral phosphoric acid **(*R*)-1b** to afford polycyclic products **33** in good to high yields and with high levels of enantioselectivity (Scheme 6.10). Interestingly, analogous amide substrates are also amenable to this relay catalytic cascade cyclization reaction. Mechanistic studies disclosed a Au(I)-catalyzed non-enantioselective background reaction, which interfered with the chiral phosphoric acid-catalyzed highly enantioselective *N*-sulfonyliminium cyclization. A judiciously selected combination of the Au(I) complex and chiral phosphoric acid is crucial to achieving high levels of stereocontrol [18].

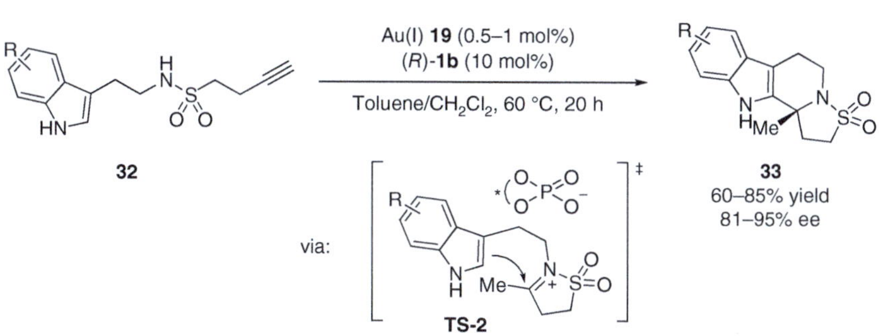

Scheme 6.10 Au(I)/chiral phosphoric acid relay catalytic cascade hydroamination/ *N*-sulfonyliminium cyclization reaction.

2-(2-Propynyl)anilines **11** are able to undergo a Brønsted acid-catalyzed three-component Povarov reaction with aldehydes **8** and enamides **34**, yielding intermediates **Int-16**, which then undergo a subsequent hydroamination reaction catalyzed by a gold complex, leading to polycyclic nitrogenous heterocycles. The chiral phosphoric acid **1c** and the *in situ* generated gold phosphate from JohnPhosAuMe were identified to be the combined catalyst for this relay catalytic cascade reaction and delivered excellent enantioselectivities (Scheme 6.11) [19]. Since products **35** are unstable for isolation, a one-pot reduction with $NaBH(OAc)_3$ was conducted to generate julolidine derivatives **36** in high yields and with high enantioselectivities. Kinetic experiments indicate that the chiral phosphoric acid not only catalyzes the asymmetric Povarov reaction but also facilitates the gold-catalyzed hydroamination reaction.

Under Brønsted acid catalysis, an asymmetric aminalization between 2-alkynylbenzaldehydes **37** and 2-aminobenzamides **38** was reported to give chiral aminals **Int-18**, in which the amine moiety can participate in an intramolecular 6-*endo*-dig cyclohydroamination to give 1,2-dihydroisoquinolines **39** under the gold catalysis (Scheme 6.12) [20]. The binary catalytic system consisted of chiral phosphoric acid **(*R*)-1b**, and an *in situ* generated gold phosphate from Ph_3PAuMe enabled the relay catalytic cascade reaction to offer excellent yields and stereoselectivity. Although the cationic gold complex is a highly active catalyst for hydroamination, it can also catalyze the aminalization, leading to racemic products, thus is excluded as the co-catalyst.

6.2.2 Hydroalkoxylation Mediated Relay Catalysis

The gold-catalyzed intramolecular hydroalkoxylation of 3-butylnoic acids **40** principally generates furan-2(3*H*)-one intermediates **41**. In the presence of chiral phosphoric acid, the condensation between **41** and tryptamine **42** affords *N*-acyliminium phosphate **Int-19**, which then undergoes an enantioselective *N*-acyliminium cyclization to generate **43** (Scheme 6.13). The combination of triphenylphosphine gold triflate complex and chiral phosphoric acid **1d** allows access to optically active heterocycles **43** from readily available substrates in good yield and high ee (Scheme 6.13) [21].

A 5-*exo*-dig cyclization of the alkynol **7** driven by an *in situ* generated gold phosphate from **45** and chiral phosphoric acid (*S*)-**1b** furnishes an enol ether **Int-20**. The subsequent protonation of **Int-20** with the chiral phosphoric acid (*S*)-**1c** forms 3,4-dihydro-1-furanium phosphate, which then undergoes an asymmetric aldol-type reaction with the azlactones **44**, allowing for the generation of conformationally restricted amino acid precursors **46** bearing vicinal quaternary stereogenic centers with high levels of enantioselectivity (Scheme 6.14) [22]. Mechanistic studies reveal that both *in situ* generated cationic gold phosphate and the excess phosphoric acid are able to catalyze the asymmetric aldol-type reaction of azlactones with the enol ether intermediate, whereas the latter leads to significantly higher reaction rate and slightly better stereocontrol.

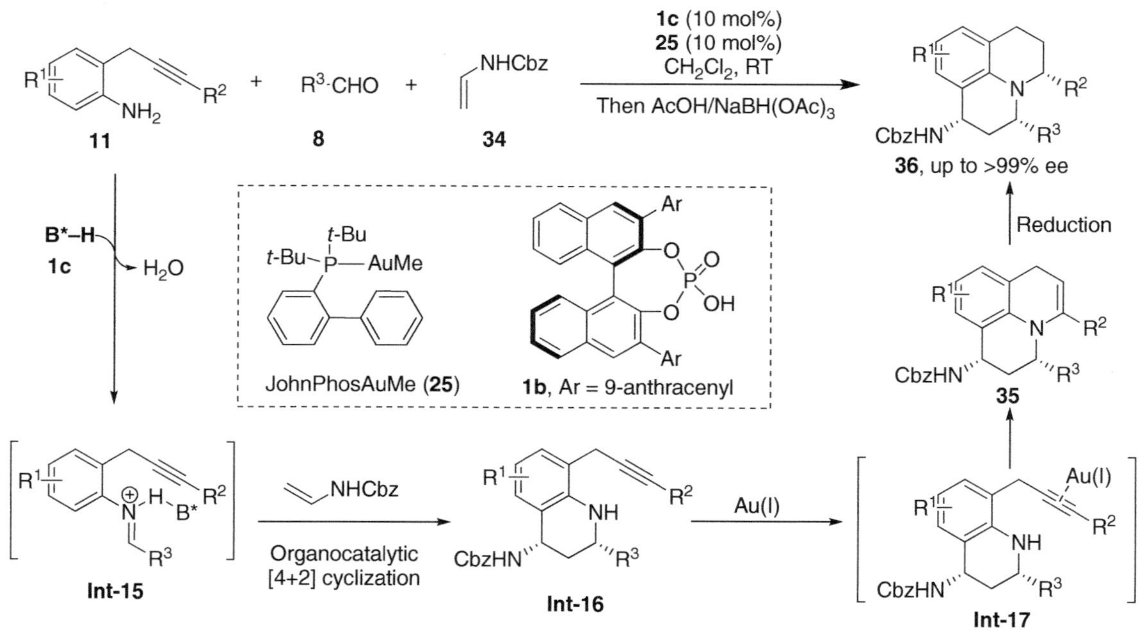

Scheme 6.11 Au(I)/chiral phosphoric acid relay catalytic three-component reaction of 2-(2-propynyl)anilines, aldehydes, and enamides. Source: Modified from Wang et al. [19].

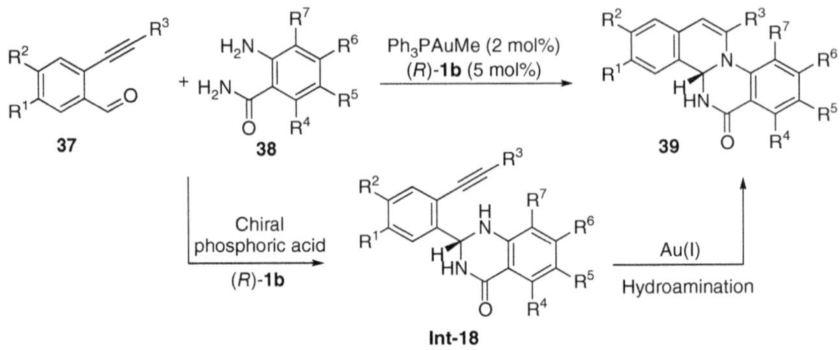

Scheme 6.12 Cascade asymmetric aminalization/hydroamination of 2-alkynylbenzaldehydes with 2-aminobenzamides. Source: Modified from Patil et al. [20].

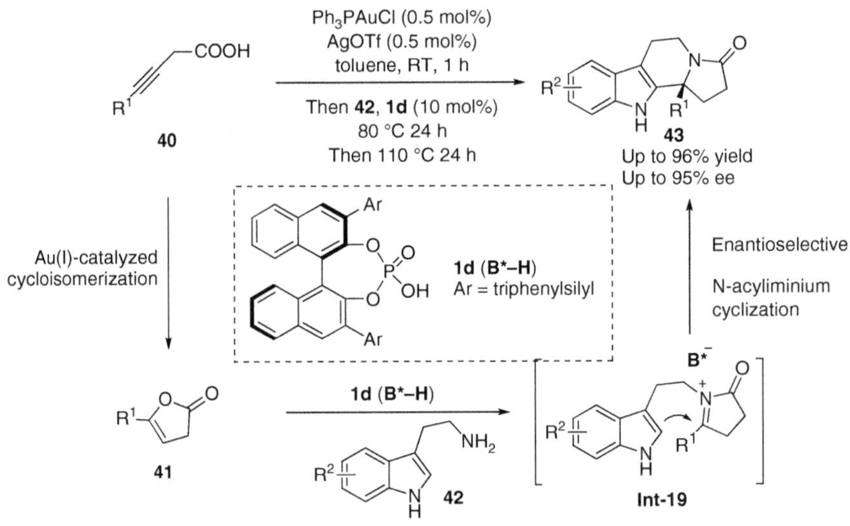

Scheme 6.13 Cycloisomerization/enantioselective *N*-acyliminium cyclization reaction. Source: Modified from Yang et al. [21].

Similarly, in the presence of a Au(I) complex, 2-ethynyl benzyl alcohols **47** can undergo a 5-*exo*-dig cyclization to form an enol ether **Int-22** for subsequent chiral Brønsted acid-catalyzed asymmetric Mannich-type reaction/acetalization with *in situ* generated salicylaldehyde imines **Int-22** to give chiral spiroacetals **49**. The use of **(*R*)-1b** in concert with *in situ* generated gold phosphate enables the asymmetric three-component cascade reaction of 2-ethynyl benzyl alcohols **47**, salicylaldehydes **48**, and anilines **9** to access enantioenriched spiroacetals **49** in excellent yields and stereoselectivities (Scheme 6.15) [23].

A three-component reaction of alkynols **7**, anilines **9**, and glyoxylic acid **50** for the enantioselective synthesis of spiroacetals **51** bearing an α-amino acid motif was established by Fañanás and coworkers by virtue of the similar relay catalytic strategy (Scheme 6.16) [24]. In contrast to the previous report using excess amounts of chiral

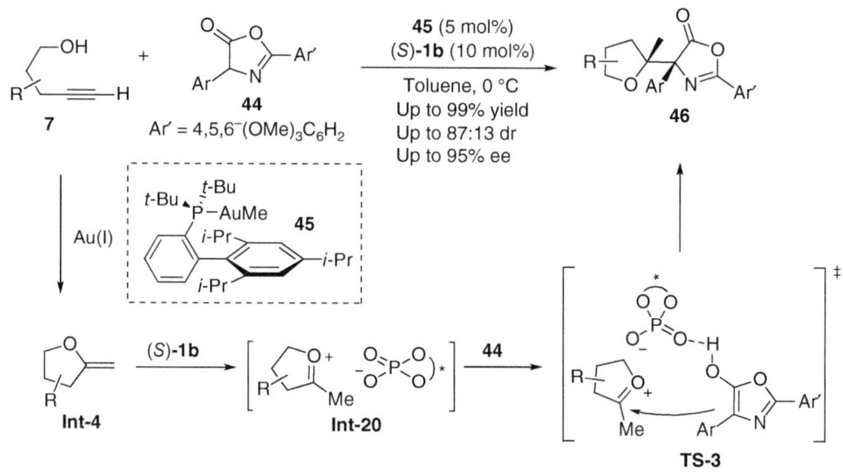

Scheme 6.14 Cascade alkynols cyclization and asymmetric addition of azlactones. Source: Modified from Han et al. [22].

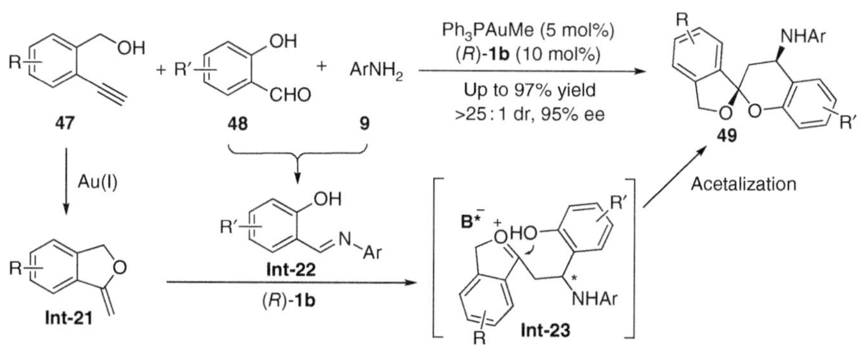

Scheme 6.15 Three-component reaction to access enantioenriched spiroacetals. Source: Modified from Wu et al. [23].

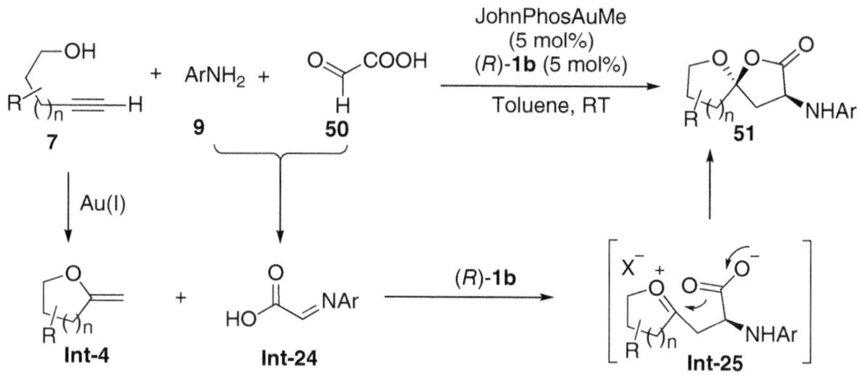

Scheme 6.16 Asymmetric three-component reaction of alkynols, anilines, and glyoxylic acid. Source: Modified from Cala et al. [24].

phosphoric acid to enhance the stereocontrol [23], stoichiometric JohnPhosAuMe and chiral phosphoric acid **(*R*)-1b** are sufficient to promote this cascade reaction. It is known that JohnPhosAuMe reacts easily with the chiral phosphoric acid to generate a chiral gold phosphate [19], which is presumably the active catalyst in this report.

The gold-catalyzed intramolecular hydroalkoxylation of homopropargyl alcohols **5** principally generates dihydrofuran derivatives **Int-2**, typical dienophiles amenable to hetero-Diels–Alder reaction. Fañanás and coworkers found that such dihydrofuran intermediates can be trapped by the *N*-arylaldimine generated *in situ* from aldehydes and anilines to undergo chiral phosphoric acid-catalyzed Povarov reaction (Scheme 6.17) [25]. Chiral phosphoric acid **(*R*)-1b**, combined with its gold phosphate, efficiently drives the three-component reaction and delivers chiral hexahydrofuro[3,2-*c*]quinolines **52** with up to 98% ee. DFT studies suggest that the Povarov reaction proceeds more like a sequential Mannich/intramolecular Friedel–Crafts process rather than a concerted [4+2] cycloaddition.

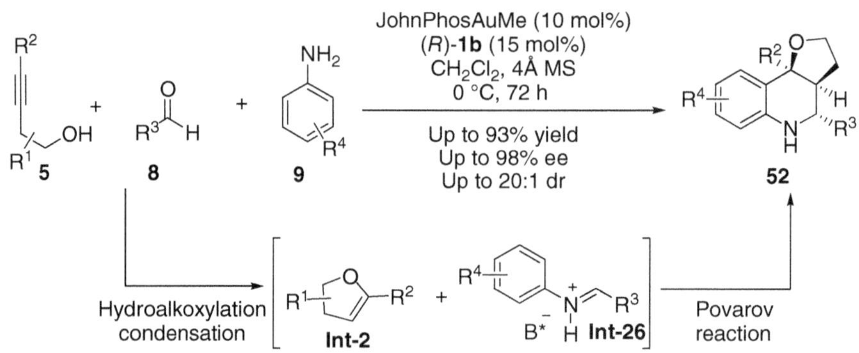

Scheme 6.17 Cascade hydroalkoxylation/Povarov reaction of alkynols, aldehydes, and anilines. Source: Modified from Calleja et al. [25].

Chiral gold(I) complexes can catalyze a cascade of intramolecular hydroalkoxylation/asymmetric acetalization of alkynyl glycols to give optically active benzannulated spiroacetals, but with moderate stereoselectivities [26]. By employing gold complex and chiral Brønsted acid relay catalysis, a highly enantioselective intramolecular cascade reaction of alkynyl glycols **53** was achieved to afford spiroacetals **54** in high yields and with excellent enantioselectivities (Scheme 6.18) [27].

6.2.3 Hydrosiloxylation Mediated Relay Catalysis

While the addition of water or alcohols to alkynes bond has long been established by transition metal catalysis, the use of silanols as nucleophiles to produce synthetically useful silyl enol ethers has been rarely described [28]. In 2012, Gong and coworkers reported a Au(I)/chiral Brønsted acid relay catalytic cascade intramolecular hydrosiloxylation/asymmetric Diels–Alder reaction (Scheme 6.19a) [29]. The homocoupling of the silanols, leading to silicon ethers, was identified to be the major side reaction competing with the hydrosiloxylation process. The

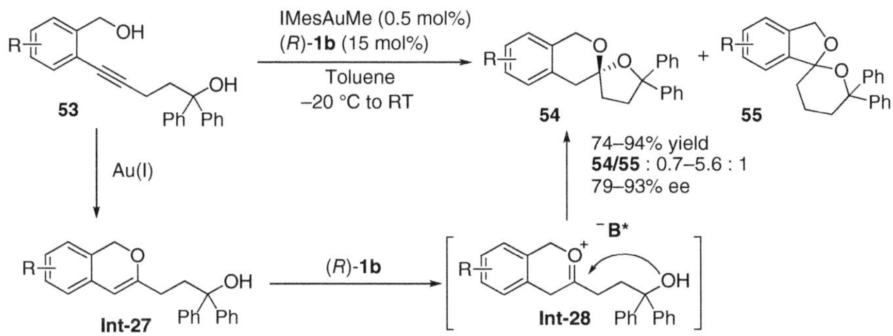

Scheme 6.18 Synthesis of chiral spiroacetals from alkynyl glycols by Au(I)/chiral Phosphoric acid relay catalysis. Source: Modified from Rexit and Mailikezati [27].

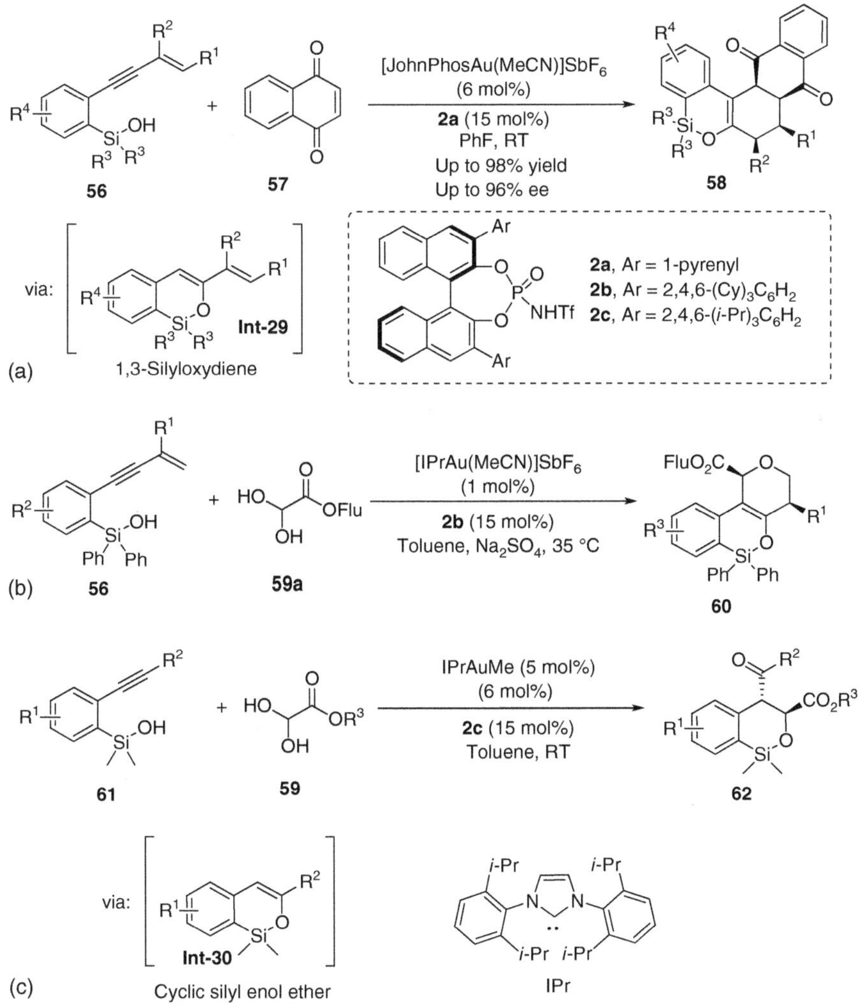

Scheme 6.19 Relay catalytic cascade reactions initiated by alkyne hydrosiloxylation. Source: (a) Modified from Han et al. [29], (b) Modified from Li et al. [30], (c) Modified from Wang et al. [31].

use of Au(I) complexes bearing bulky ligands, as well as aromatic solvents, could suppress the homocoupling and facilitate the hydrosiloxylation process. BINOL (1,1′-bi-2-naphthol)-derived phosphoramide **2a** efficiently catalyzes the enantioselective Diels–Alder reaction of the 1,3-silyloxydiene intermediates **Int-29** and quinones [27, 32]. Notably, after a one-pot reduction with diisobutylaluminium hydride, multicyclic diols bearing five adjacent stereocenters were obtained in good yields with excellent diastereo- and enantioselectivities. Under the relay catalysis of [IPrAu(MeCN)]SbF_6 and chiral phosphoramide **2b**, an intramolecular hydrosiloxylation/asymmetric hetero-Diels–Alder reaction also works well (Scheme 6.19b) [30]. Expanding this dual catalytic strategy to a cascade hydrosiloxylation/asymmetric Mukaiyama aldol reaction turns out to be successful, where the siloxy group is not only a precursor to form the enol silyl ether **Int-30** but also an active functionality that is able to undergo a variety of transformations for additional molecular modification (Scheme 6.19c) [31].

6.2.4 Relay Catalysis Involving the Addition of Nitrone or Nitro Group to Alkynes

Besides the classic addition of alcohols or amines to alkynes, another important activation mode in gold catalysis is intermolecular or intramolecular oxidation of alkynes generating reactive α-oxo or α-imino gold carbenes, which serve as versatile intermediates [33–35]. Zhang and coworker demonstrated that α-oxo gold carbenes could participate in chiral Brønsted acid-catalyzed transformations. Under the Au(I)/chiral Brønsted acid relay catalysis, an enantioselective cascade reaction was reported to convert tertiary propargylic alcohol **63** bearing a nitrone moiety into spirocyclic diketones **64** (Scheme 6.20) [36], which was initially reported as a racemic version by Shin and coworkers [37]. The cascade reaction involves the initial formation of α-oxo gold carbene intermediate **Int-31** from Au(I)-catalyzed internal

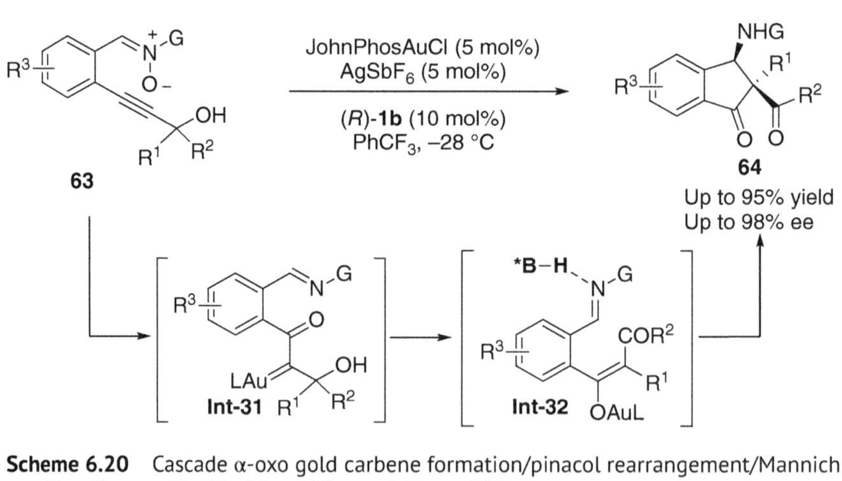

Scheme 6.20 Cascade α-oxo gold carbene formation/pinacol rearrangement/Mannich reaction. Source: Modified from Qian and Zhang [36].

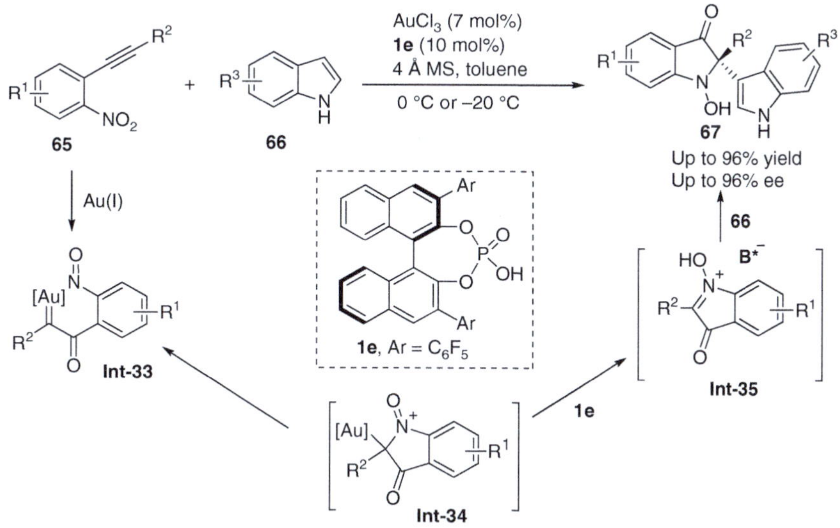

Scheme 6.21 Redox annulation of nitroalkynes and enantioselective addition with indoles.. Source: Modified from Liu et al. [38].

oxygen-transfer of **63**, subsequent pinacol rearrangement to generate gold enolate **Int-32**, and a chiral phosphoric acid-catalyzed asymmetric Mannich-type reaction.

Jia and coworkers established an enantioselective redox annulation of nitroalkynes **65** with indoles **66** enabled by the relay catalysis of $AuCl_3$ and chiral phosphoric acid **1e**, allowing for rapid synthesis of chiral indolin-3-ones **67** bearing C2 quaternary stereocenters in high yields and with excellent enantioselectivities (Scheme 6.21) [38]. This cascade reaction starts with a redox annulation between **65** and $AuCl_3$ to generate a gold carbene **Int-33**, which is captured by an internal nitroso group to give an intermediate **Int-34**. Subsequent protonolysis of **Int-34** by chiral phosphoric acid **1e** furnishes a chiral ion pair **Int-35** that undergoes an enantioselective addition with indoles to give the final product.

6.2.5 Relay Catalysis Involving the Addition of Carbon Nucleophiles to Alkynes

The asymmetric assembly of the structurally complex polycyclic skeleton from easily accessible 2-alkynyl indoles **69** and indolylmethanols **68** has readily been realized by the multiple relay catalysis of $Ph_3PAuOTf$ and chiral phosphoric acid **1a** (Scheme 6.22) [39]. Initially, the chiral phosphoric acid-catalyzed asymmetric substitution reaction of indolymethanol **68** with indole **69** [40] furnishes an enantioenriched compound, which then undergoes a 6-*endo*-dig cyclization between the C3 of one indole substituent and the alkyne moiety catalyzed by the gold complex, leading to spirocyclic intermediate **Int-37**. A subsequent 1,2-shift process [41] enables a ring-expanding event to transform the intermediate **Int-37** into **Int-39**, which after protonolysis furnishes the product **70** and releases the gold catalyst.

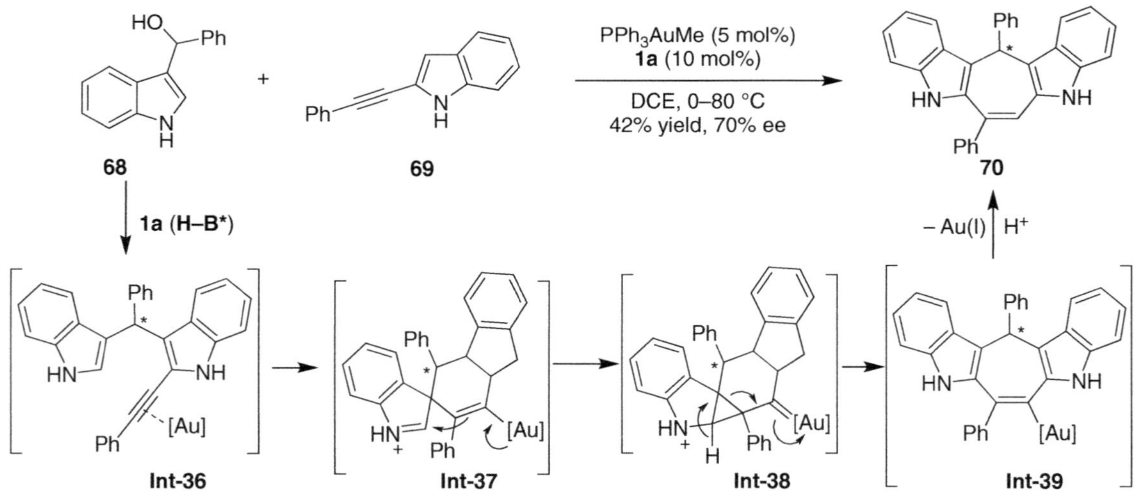

Scheme 6.22 Au(I)/chiral phosphoric acid relay catalytic enantioselective synthesis of polycyclic indole derivatives. Source: Modified from Inamdar et al. [39].

6.3 Metal/Brønsted Acid Relay Catalysis Involving Alkene Metathesis

In 2009, Xiao and coworkers reported a Ru-catalyzed tandem cross-metathesis/intramolecular Friedel–Crafts alkylation of indoles [42]. The asymmetric variant of the cascade cross-metathesis/Friedel–Crafts alkylation was accessed by the relay catalysis of a ruthenium complex and chiral Brønsted acid (Scheme 6.23a) [43]. In the presence of chiral phosphoric acid (*S*)-**1a** and Ru catalyst **73**, allyl ether or amide **71** and vinyl ketone **72** can smoothly undergo the cross-metathesis/asymmetric intramolecular Friedel–Crafts-type Michael addition cascade to give tetrahydropyrano[3,4-*b*]indoles (THPI) and tetrahydro-β-carbolines

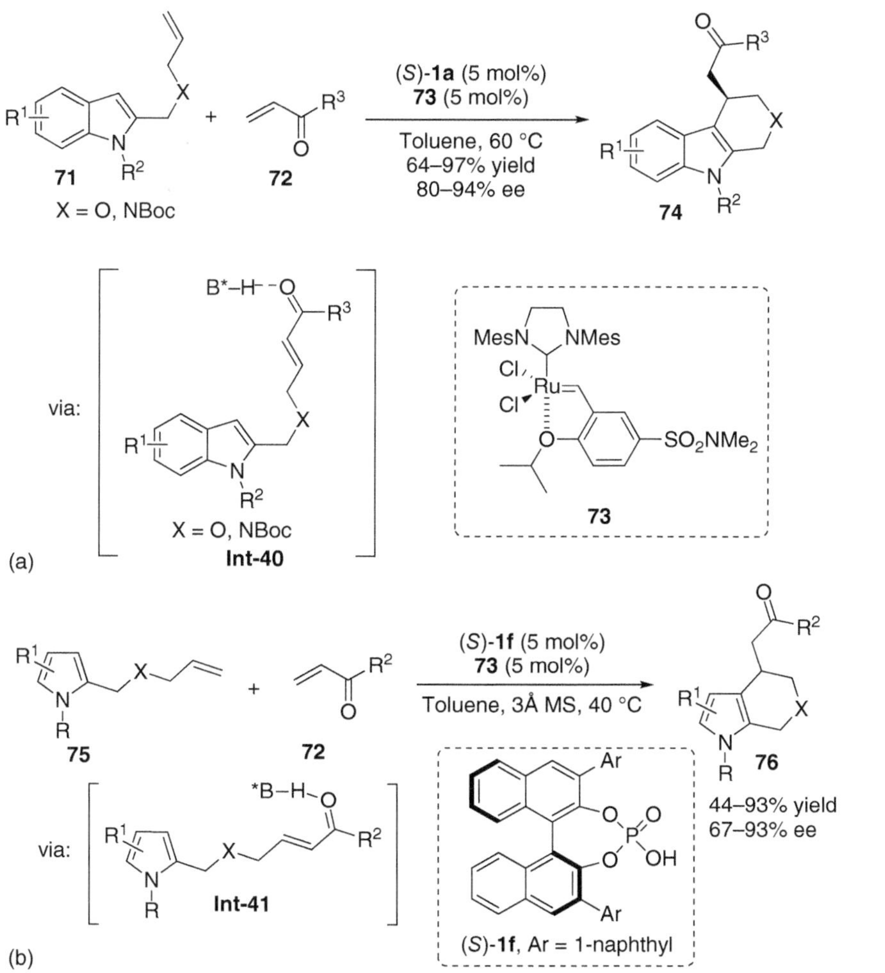

Scheme 6.23 Ruthenium/chiral phosphoric acid relay catalytic cascade cross-metathesis/asymmetric Friedel–Crafts-type reaction. Source: (a) Modified from Cai et al. [43], (b) Modified from Zhang et al. [44].

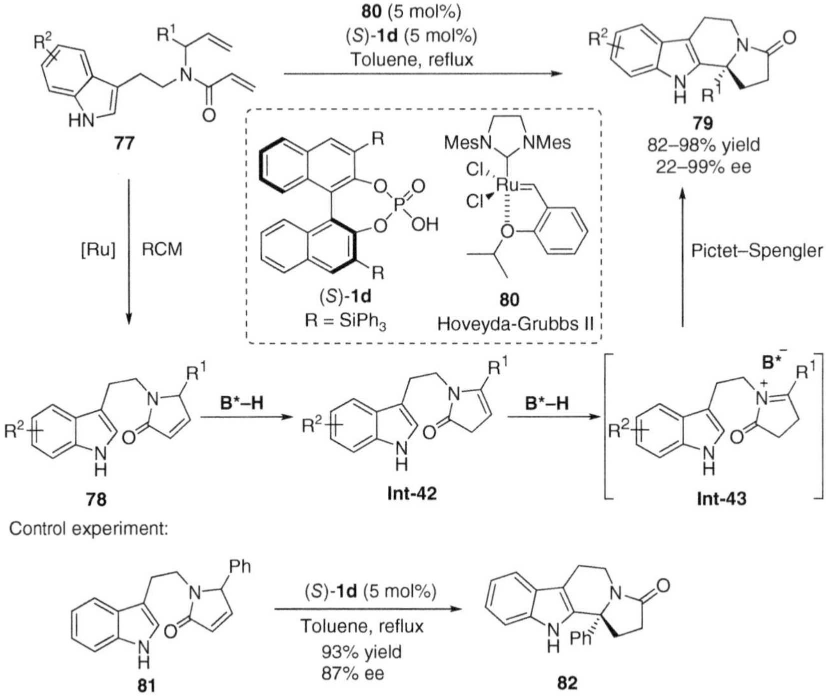

Scheme 6.24 Enantioselective cascade RCM/isomerization/Pictet–Spengler reaction. Source: Modified from Cai et al. [45].

(THBCs) in excellent yields and with excellent enantioselectivity. Extension of this phosphoric acid **(*S*)-1f** and ruthenium complex **73** binary catalyst system to pyrrole-based substrate **75** successfully provides 4,5,6,7-tetrahydroindoles **76** with up to 93% ee (Scheme 6.23b) [44].

By relay catalysis of Hoveyda-Grubbs II catalyst **80** and phosphoric acid **1d**, You and coworkers developed a cascade ring-closing metathesis (RCM)/isomerization/Pictet–Spengler reaction of readily available **77** to provide tetrahydro-β-carbolines with up to 98% yield and 99% ee (Scheme 6.24) [45]. Treating pre-synthesized **78a** with 5 mol% of phosphoric acid **1d** in refluxing toluene leads to the desired product in 95% yield and 87% ee, which is similar to those of the relay catalytic process. Further control experiments suggest that the alkene isomerization and the *N*-acyl iminium cyclization are dominantly catalyzed by the phosphoric acid **1f**.

Expansion of the similar catalyst system to a cascade RCM/isomerization/Mannich reaction allows efficient conversion of *N*-allyl-*N*-benzylacrylamide **83** and 2,3-substituted indoles **84** to *N*-alkyl indole products **85** (Scheme 6.25a) [46]. Interestingly, the relay catalytic process provides a higher yield and enantioselectivity than the stepwise procedure. Later, Patil and coworkers reported an intramolecular variant of the cascade RCM/isomerization/Mannich-type reaction of **86**, furnishing pyrrole-embedded aza-heterocyclic scaffolds **87** in high yields [47]. Unfortunately,

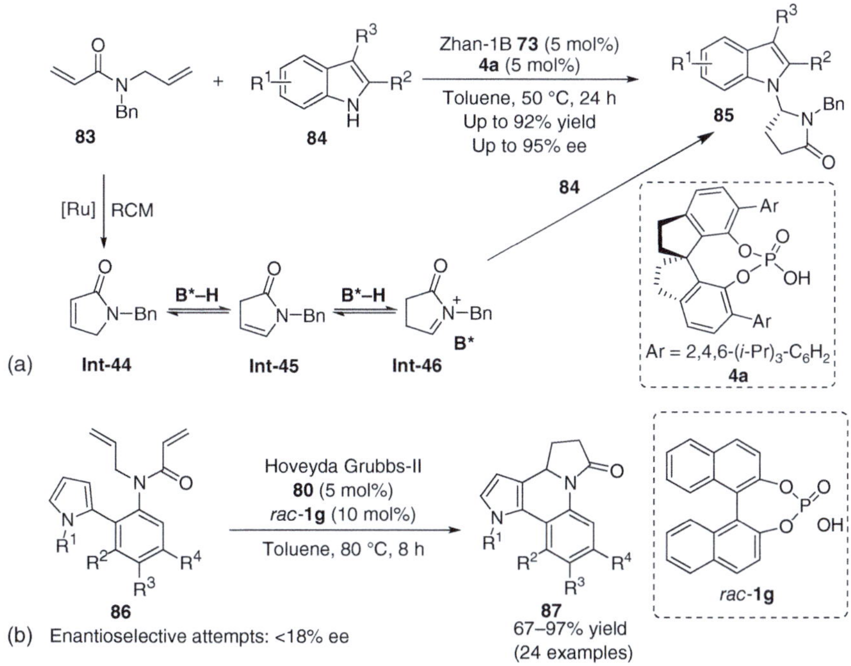

Scheme 6.25 Cascade RCM/isomerization/ Mannich-type reaction. Source: (a) Modified from Shi et al. [46].

chiral phosphoric acids failed to provide excellent stereocontrol that only 18% ee was achieved (Scheme 6.25b).

Chiral 2,3-dihydrobenzofurans and 3,4-dihydro-2*H*-1,4-benzoxazines are common structural motifs in numerous pharmaceuticals and biologically active compounds. You and coworkers created a cascade cross-metathesis/asymmetric intramolecular oxo-Michael reaction to access these structures enabled by ruthenium/Brønsted acid relay catalysis. The combination of a chiral phosphoric acid (*S*)-**1d** and Hoveyda-Grubbs II **80** allows direct transformation of **88** and **89** to 2,3-dihydrobenzofurans **90** and 3,4-dihydro-2*H*-1,4-benzoxazines **91**, respectively (Scheme 6.26) [48].

The asymmetric allylboration of aldehydes **92** with allyl borates **93** catalyzed by a chiral phosphoric acid gives chiral alcohol intermediates **Int-49**, which are able to undergo the RCM to furnish six- and seven-membered benzo-fused cyclic homoallylic alcohols **95**. Thus, the combination of chiral phosphoric acid **(*R*)-1b** and the Grubbs II catalyst **94** renders a one-pot cascade process to assemble **95** in high yields and with excellent enantioselectivities (Scheme 6.27) [49]. The high enantioselectivity of the allylboration is generally induced at −30°C, while the subsequent RCM undergoes at room temperature. Notably, one of the products, **95a,** is a key intermediate in Lautens' synthesis of the antidepressant sertraline [50].

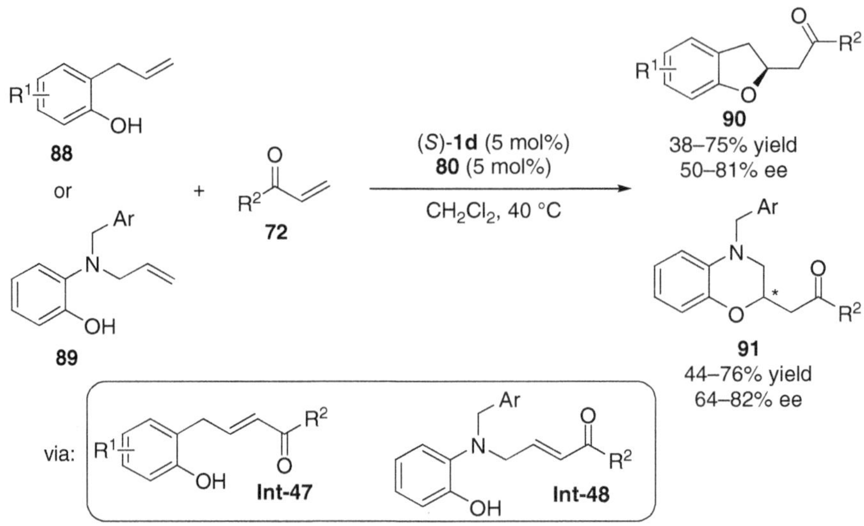

Scheme 6.26 Cascade cross-metathesis/asymmetric intramolecular oxo-Michael reaction. Source: Modified from Zhang et al. [48].

6.4 Metal/Brønsted Acid Relay Catalysis Involving Alkene Isomerization

Remote C—C double bonds in many functionalized alkenes can undergo multi-positional migration mediated with transition metal hydride [51], leading to more reactive alkenes that could be trapped by a broad range of reagents. Principally, the constant evolution of transition metal complex/chiral Brønsted acid relay catalysis would enable a cascade reaction consisting of olefin migration and enantioselective trapping of the resulted reactive alkenes, which is termed remote C—C double bond functionalization (Scheme 6.28).

The feasibility of relay catalytic cascade reaction based on metal hydride-mediated double migration was first showcased by Terada and coworker [52]. A rapid double bond migration of allylamine **97** catalyzed by a ruthenium hydride complex proceeds to generate an enamine intermediate **Int-50**. A phosphoric acid-promoted tautomerization of enamine **Int-50** to iminium **Int-51** facilitates subsequent Friedel–Crafts reaction with an electronically rich arene **98** (Scheme 6.29) [52]. The development of an enantioselective cascade alkene isomerization/Friedel–Crafts reaction remains challenging. In 2013, the same group reported cascade C—C double bond isomerizations and enantioselective Pictet–Spengler-type cyclization of **101**, using $RuClH(CO)(PPh_3)_3$ and chiral phosphoric acid **1c** as catalysts, to afford tetrahydroisoquinoline derivatives in moderate to good yields and with moderate enantioselectivities (Scheme 6.30a) [57]. Higher stereoselectivity was observed in a similar cascade process of *N*-allyl tryptamine derivatives **103** to furnish 1,2,3,4-tetrahydro-β-carbolines under the promotion of Hoveyda–Grubbs II catalyst and chiral spiro-phosphoric acid **4b** (Scheme 6.30b) [54]. Interestingly,

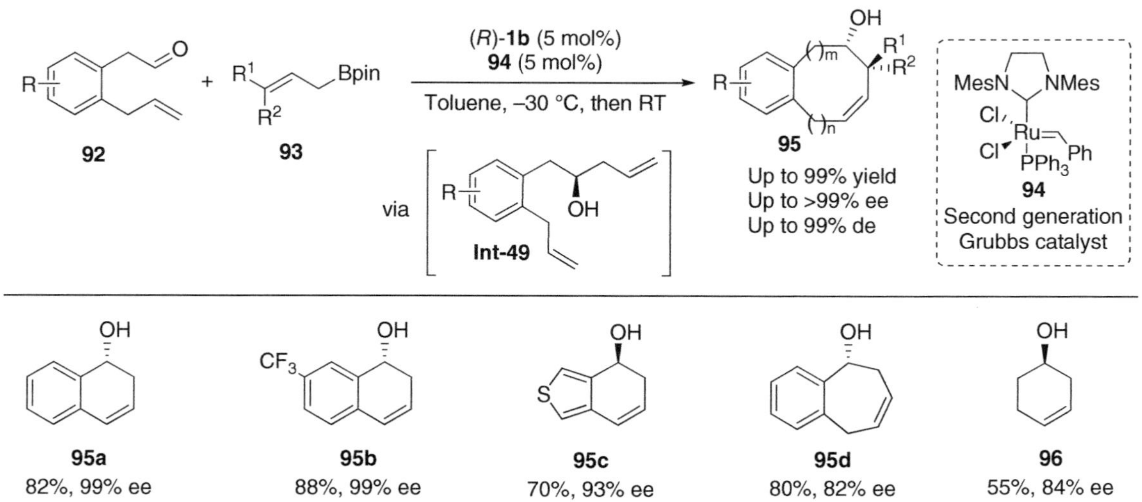

Scheme 6.27 Asymmetric one-pot allylboration/ring closing metathesis reaction. Source: Modified from Fustero et al. [49].

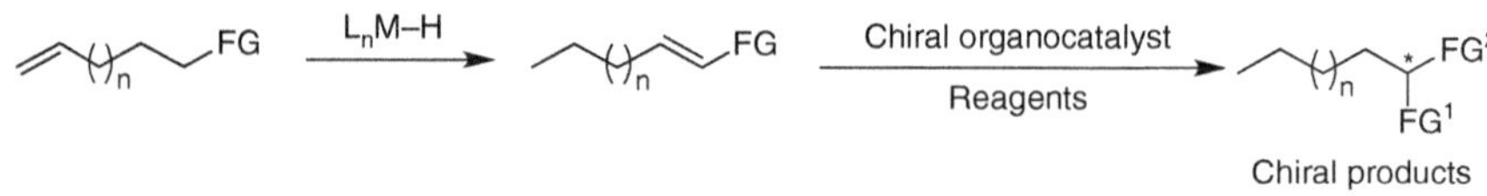

Scheme 6.28 Metal/chiral Brønsted acid relay catalysis for remote C—C double bond functionalization.

the use of *N*-allyl tryptamine derivatives **105** as substrates that build up quaternary stereogenic centers results in diminished enantioselectivity (Scheme 6.30c) [55, 56].

A cascade long-distance chain-walking/hemiaminalization reaction of 2-hydroxy(alkyl) substituted *N*-alkenyl aniline, and benzylamine derivatives can be accessed by Ru/chiral phosphoric acid relay catalysis. Commercially available $RuH_2(CO)(PPh_3)_3$ efficiently catalyzes the double-bond migration to give an enamine intermediate **108**. Both TsOH·H_2O and chiral phosphoric acid enable the subsequent hemiaminalization to furnish 1,3-oxazaheterocycles **109** in moderate to excellent yields (Scheme 6.31) [58]. Especially, chiral phosphoric acid **1d** allows the production of **109** in 88% yield and with 62% ee.

Frauenrath and coworkers found that $LiBHEt_3$-activated $NidppbCl_2$ turned out to be a highly efficient catalyst for a *Z*-selective double-bond migration of acyclic allyl acetals and allyl ethers [59]. Based on this process, Terada established a nickel/chiral phosphoric acid sequentially catalyzed C—C double bond migration/asymmetric aza-Petasis–Ferrier rearrangement reaction, furnishing anti-β-amino aldehydes (Scheme 6.32) [53]. Upon activation from $LiBHEt_3$, the NiI_2 complex with either dppf or *p*-tolbiphep gave highly *Z*-selective enol ether intermediate **111**. The chiral phosphoric acid **1h** efficiently catalyzes the asymmetric aza-Petasis–Ferrier rearrangement of **111** to give high yields and excellent enantioselectivities, presumably through transition state **TS-4**. Interestingly, the racemic hemiaminal intermediate **111** could be efficiently be resolved by **1h** in the aza-Petasis–Ferrier rearrangement process.

The $NiCl_2$ complex of triphos activated by zinc can also enable the double bond migration of *N*-allylcarbamates to give enamides, which can undergo the subsequent asymmetric addition reaction as anamphiphile [60]. Thus, the combination of nickel complex with chiral phosphoric acid **(*R*)-1b** can enable the cascade isomerization/Friedel–Crafts reaction to give α-indolyl carbamates **114** in good yields and uniformly excellent enantioselectivities (Scheme 6.33a). Moreover, the isomerization/enantioselective Povarov process of N-allylcarbamates **115** and imines **116** has been accessed by Ni(II)/chiral phosphoric acid **1i** relay catalysis. Interestingly, the addition of solid $NaHCO_3$ after the completion of the isomerization step is required to obtain the tetrahydroquinoline products **117** in high yields and stereoselectivity (Scheme 6.33b) [60].

Cationic iridium(I) complexes of phosphine Ligands activated with hydrogen have emerged as efficient catalysts for C—C double bond migration at room temperature [61, 62]. Murakami and coworkers found that an iridium tetrafluoroborate complex with tricyclohexylphosphine (PCy_3), upon the activation with hydrogen, was able to catalyze efficient olefin transposition of (*E*)-1-alkenylboronates **118** to generate (*E*)-2-alkenylboronate intermediate **93** [63], which then underwent a

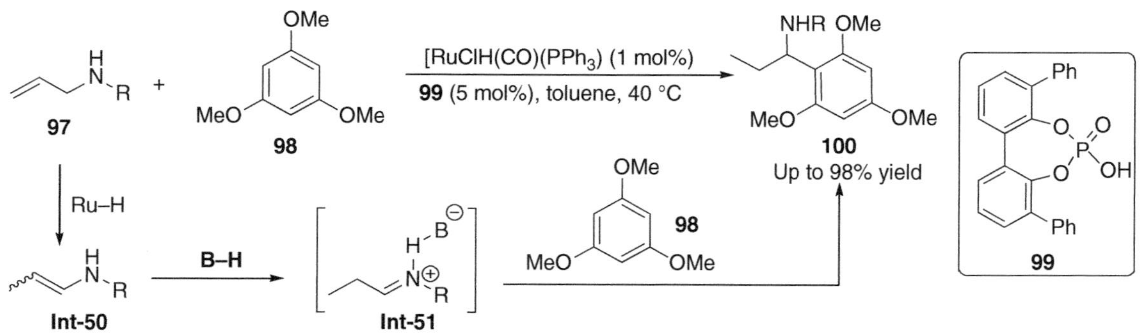

Scheme 6.29 Relay catalytic cascade allylamine isomerization/Friedel–Crafts reaction. Source: Modified from Sorimachi and Terada [52].

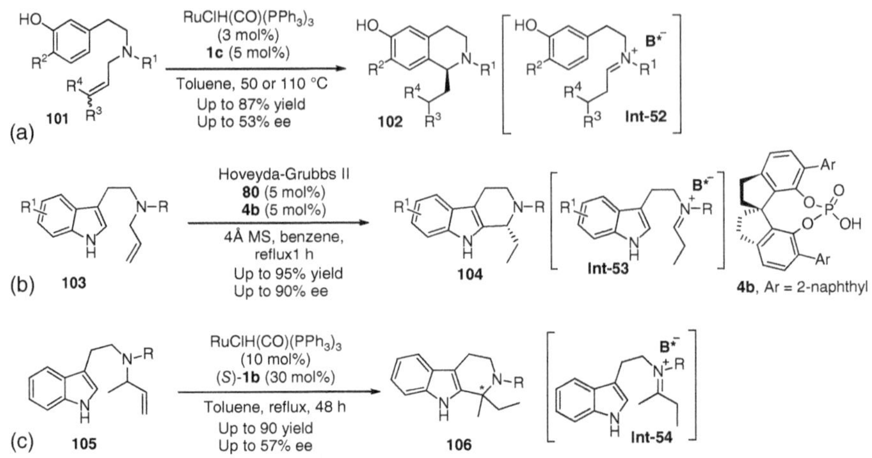

Scheme 6.30 Relay catalytic cascade double bond migration/enantioselective Pictet–Spengler-type cyclization reaction. Source: (a) Modified from Terada and Toda [53], (b) Modified from Cai et al. [54], (c) Hansen et al. [55], Ascic et al. [56].

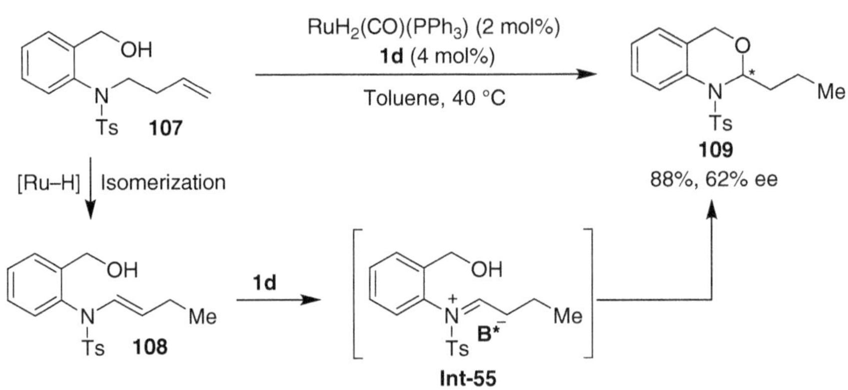

Scheme 6.31 Cascade double bond migration/enantioselective hemiaminalization reaction. Modified from Rexit and Mailikezati [27].

chiral phosphoric acid-catalyzed asymmetric allylation of aldehydes [64, 65]. The combination of the achiral cationic iridium complex and chiral phosphoric acid **(*R*)-1b** enables the direct conversion of (*E*)-1-alkenylboronates and aldehydes into chiral *anti*-homoallylic alcohols **119** in high yields and with high levels of diastereo- and enantioselectivity (Scheme 6.34a) [63]. Control experiments suggest that the generation of kinetically favored allylboron (*E*)-**93** dominates in the olefin transposition process at low conversion. The *E*/*Z* ratio of the allylboronate approaches c. 4 : 1 in a prolonged reaction. As such, the rapid trap of the allylboronate by an aldehyde via a six-membered chair-like transition state is crucial for a successful

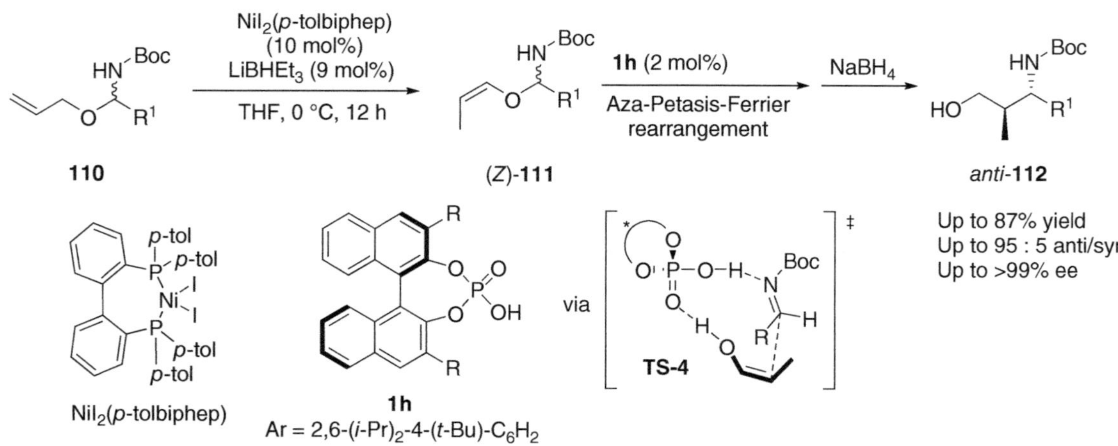

Scheme 6.32 Ni(II)/chiral phosphoric acid catalyzed sequential double bond isomerization/asymmetric aza-Petasis–Ferrier rearrangement reaction. Source: Modified from Terada and Toda [53].

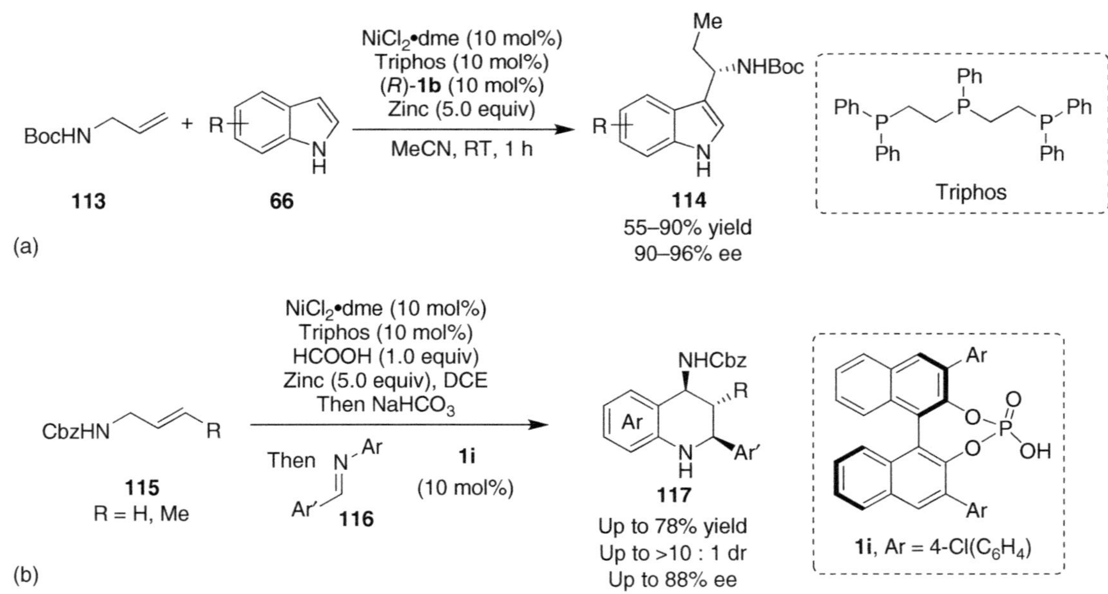

Scheme 6.33 Isomerization and divergent functionalization of *N*-allylcarbamates. Source: (b) Modified from Richmond et al. [60].

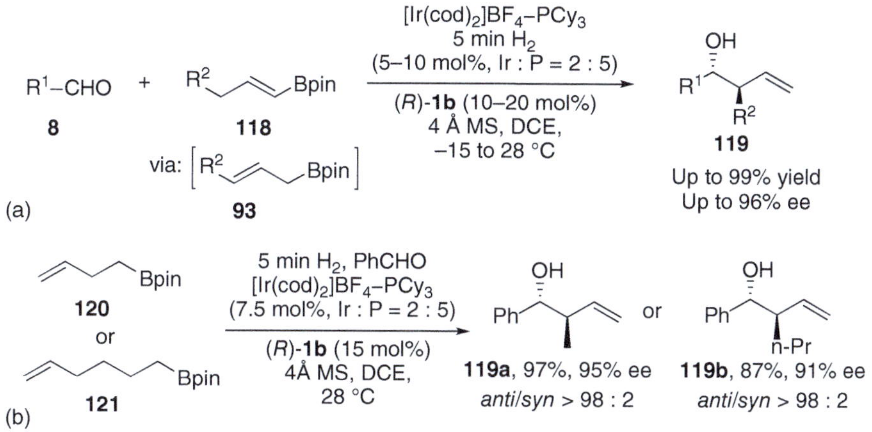

Scheme 6.34 Iridium complex/chiral phosphoric acid relay catalytic olefin migration/asymmetric allylation reaction. Source: (a) Modified from Miura et al. [63].

relay catalytic process [65]. Notably, both 3- and 5-alkenylboronates **120** and **121** undergo the iridium(I)-catalyzed olefin migration, providing allylboronates for subsequent asymmetric allylation (Scheme 6.34b).

The same group developed a Pd(I)/chiral phosphoric acid relay catalytic double bond migration and asymmetric allylation cascade between 1,1-di(boryl)alk-3-enes **122** and aldehydes, giving (*E*)-δ-boryl-substituted *anti*-homoallylic alcohols **125** with high diastereo- and enantioselectivity (Scheme 6.35a) [66]. The cationic Pd(I) complex exhibits high efficiency to catalyze the olefin transposition of the 1,1-di(boryl)alk-3-enes **122** to 1,1-di(boryl)alk-2-enes **123** that participates in a chiral phosphoric acid-catalyzed allylation with aldehydes via a chair-like transition state **TS-5** to give *Z*-selective products **124**. Pd(I)-mediated geometrical isomerization from *Z*- to *E*-isomer proceeds during the reaction, leading to the (*E*)-anti-homoallylic alcohols **125**. A cationic Ru(II) complex **126** can catalyze C—C double bond migration as well but prevents the geometrical isomerization process, allowing the production of cyclic 1,2-oxaborinan-3-enes **127** with high enantioselectivities (Scheme 6.35b) [67].

6.5 Metal/Brønsted Acid Relay Catalysis Involving Hydrogenation

Asymmetric hydrogenation represents an ideal atom-economic transformation to manufacture chiral molecules and holds a historical impact on organic synthesis. The stereochemical control in asymmetric hydrogenation basically relies on the chiral ligand bonded to the transition metal that activates both hydrogen gas and unsaturated chemical bonds [68]. On the other hand, organocatalytic asymmetric transfer hydrogenation has been well established that excess highly reactive organic hydride donors, such as Hantzsch esters and benzothiazolines [10, 69–72], are

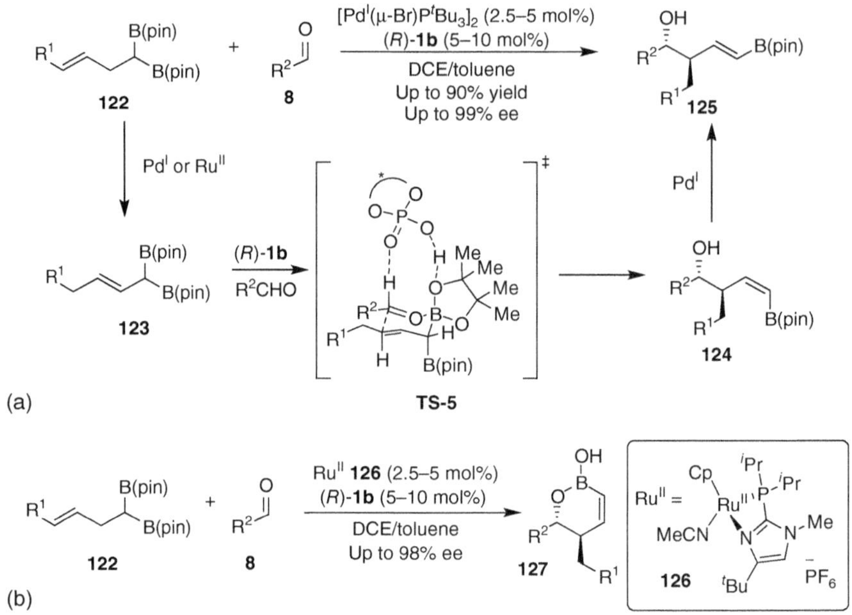

Scheme 6.35 Cationic Pd(I) or Ru(II) complex/chiral phosphoric acid relay catalytic olefin migration/asymmetric allylation reaction. Source: (a) Modified from Miura et al. [66], (b) Modified from Miura et al. [67].

generally required for full conversion. The relay catalysis of transition metals and chiral organocatalysts has allowed the use of hydrogen gas as the ultimate reductant, which is a principally new strategy from the classical chiral-ligand-controlled asymmetric hydrogenation [73].

In 2011, Zhou and coworkers developed highly enantioselective hydrogenation of quinoxalines **128** enabled by achiral ruthenium complex and chiral Brønsted acid relay catalysis (Scheme 6.36) [74]. This relay catalytic hydrogenation reaction proceeds from the Ru(II)-catalyzed hydrogenation of quinoxaline **128**, giving rise to dihydroquinoxaline **129**. Two pathways, Ru(II)-catalyzed non-enantioselective hydrogenation and chiral phosphoric acid (*S*)-**1c**-catalyzed asymmetric disproportionation through a self-transfer hydrogenation model **TS-6**, were proposed to access tetrahydroquinoxaline **130** from **129**. Control experiments reveal that the chiral phosphoric acid-catalyzed self-transfer hydrogenation is much faster than the Ru-catalyzed process, ensuring the high enantioselectivity of the entire relay catalytic hydrogenation reaction.

Hantzsch esters **12** can be generated from Ru(II)-catalyzed hydrogenation of corresponding pyridines, allowing a practical Hantzsch ester shuttle catalysis for asymmetric hydrogenation of heterocycles (Scheme 6.37). In 2011, Zhou and coworkers disclosed a triple Ru(II)/chiral Brønsted acid/Hantzsch ester catalysis to enable a highly enantioselective relay catalytic hydrogenation of benzoxazinones **131** [75].

Dihydrophenanthridine (DHPD, **135**) is an even more active hydrogen transfer reagent that can be regenerated from transition metal-catalyzed hydrogenation

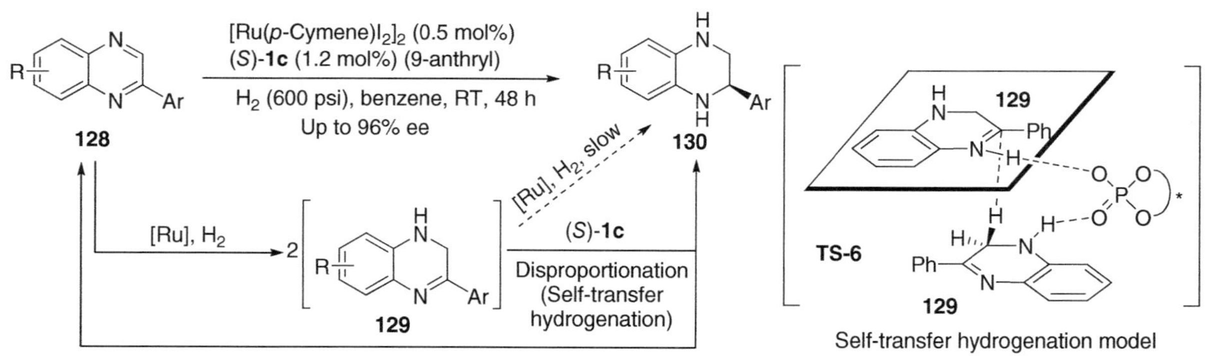

Scheme 6.36 Ru(II)/chiral phosphoric acid relay catalytic asymmetric hydrogenation of quinoxalines. Source: Modified from Chen et al. [74].

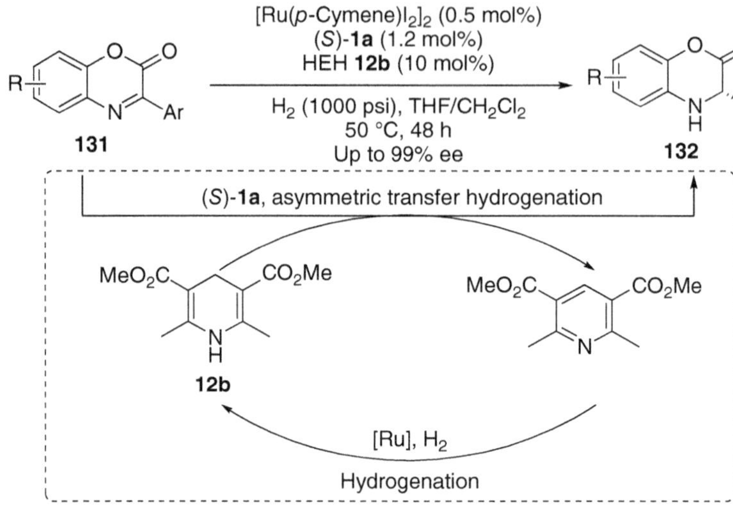

Scheme 6.37 Asymmetric hydrogenation of benzoxazinones enabled by multiple hybrid catalysts.

under milder conditions. In the presence of the catalytic amount of phenanthridine, the combination of Ru(II) and chiral Brønsted acid facilitates highly enantioselective hydrogenation of a wide scope of heterocycles, including benzoxazinones, benzoxazines, quinoxalines, and quinolines (Scheme 6.38a) [76]. Notably, the DHPD-mediated reduction features a 1,2-hydride transfer pathway other than the common 1,4-hydride transfer of Hantzsch esters and other NAD(P)H models. Later, the same group demonstrated another tunable and regenerable biomimetic hydrogen-transfer reagent, 4,5-dihydropyrrolo[1,2-*a*]quinoxalines **138**, for the Ru(II)/chiral phosphoric acid relay catalytic asymmetric hydrogenation of 3-aryl-2*H*-benzo[*b*][1,4]oxazines and 1-alkyl-3-aryl-quinoxalin-2(1*H*)-ones (Scheme 6.38b) [77].

Beller and coworkers developed a Fe/chiral Brønsted acid relay catalytic asymmetric hydrogenation of benzoxazinones using a catalytic amount of phenanthridine to generate the hydrogen transfer reagent (Scheme 6.39) [78]. Advantageous to previous reports, the current method allows for the highly enantioselective reduction of 3-alkyl-substituted benzoxazinones with high yields [75, 76].

By using Pd(II)/Brønsted acid relay catalysis, Zhou and coworkers developed an asymmetric hydrogenation of 2-hydroxypyrimidines **139**, enabling facile access to chiral cyclic ureas **140** (Scheme 6.40) [79]. It is proposed that one of the two tautomeric oxo forms of 2-hydroxypyrimidine **139**, **Int-56** and **Int-57**, undergoes partial hydrogenation to give intermediate **Int-59**, which then tautomerizes to its imine form **Int-60** in the presence of the Brønsted acid, followed by the Pd-catalyzed asymmetric hydrogenation of the imine. Notably, the hydrogenation of tri-substituted 2-hydroxypyrimidines provides partially hydrogenated 2-hydroxypyrimidines, and the configuration of the chiral phosphoric acid does not affect the stereoselectivity of the reaction.

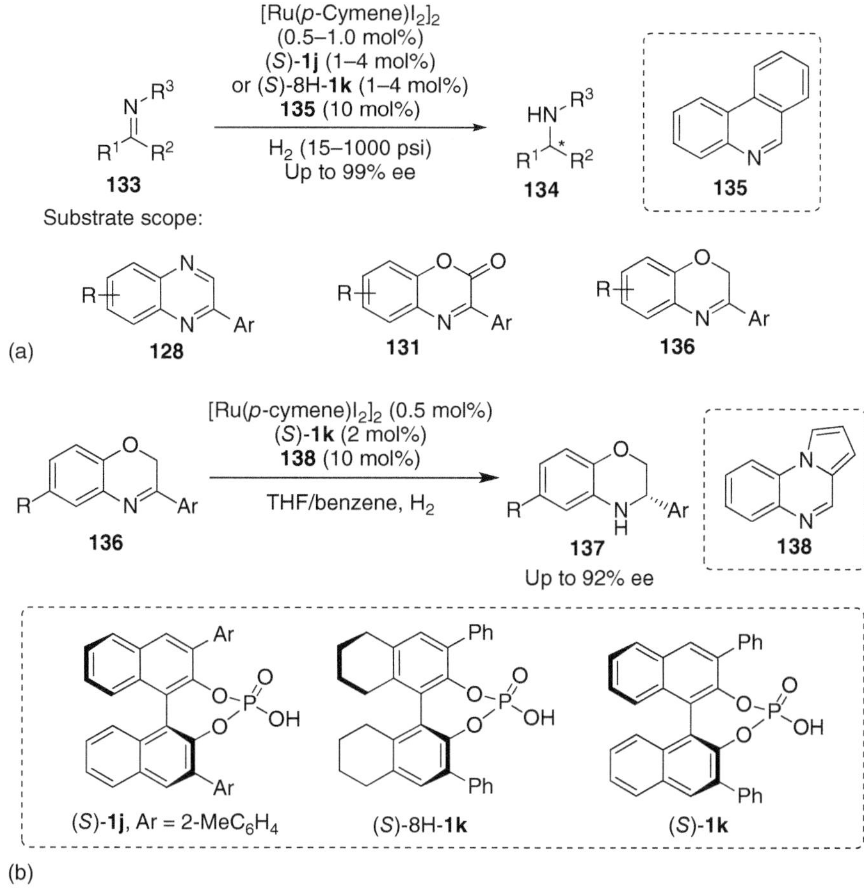

Scheme 6.38 Relay catalytic asymmetric hydrogenation of heterocycles. Source: (a) Modified from Chen et al. [76], (b) Modified from Chen et al. [77].

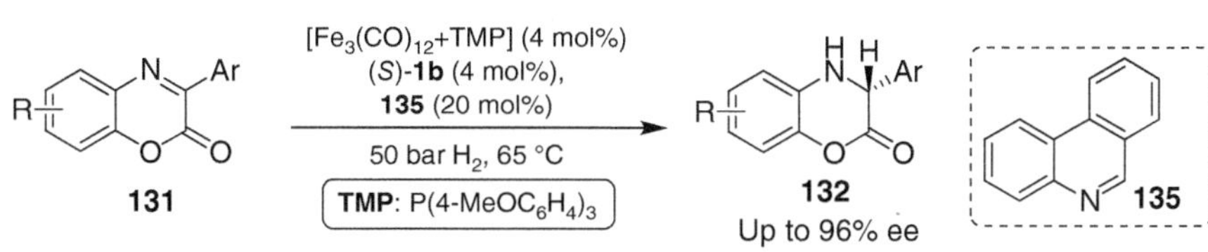

Scheme 6.39 Fe/chiral Brønsted acid relay catalytic asymmetric reduction of benzoxazinones. Source: Modified from Lu et al. [78].

6.6 Palladium/Brønsted Acid Relay Catalytic Asymmetric Allylation of Carbonyls

Allylborons are common and versatile synthons in organic synthesis. A mild and practical approach to access allylborons involves the allylic reaction of allylic

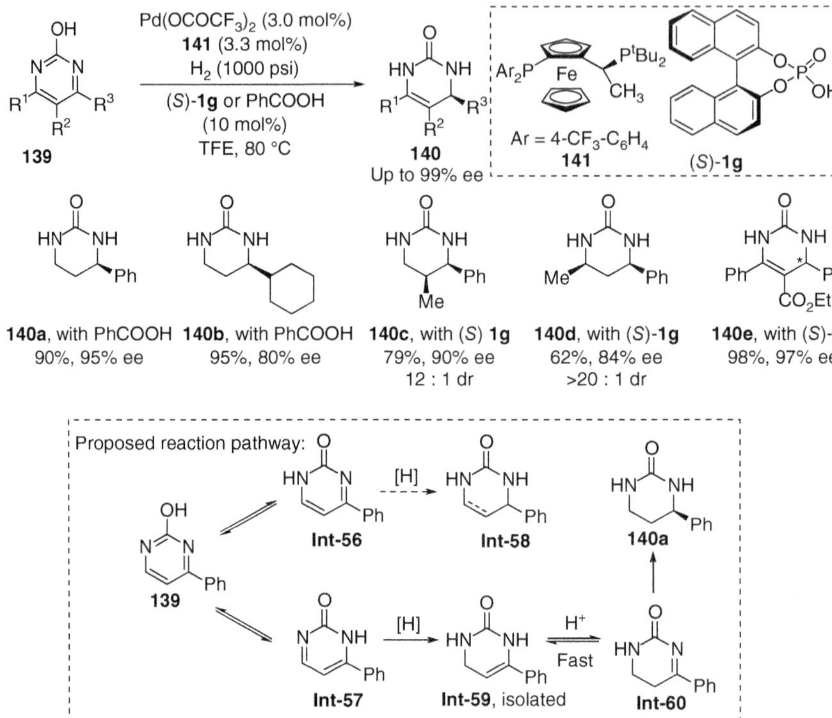

Scheme 6.40 Pd/Brønsted acid relay catalytic hydrogenation of 2-hydroxypyrimidines. Source: Modified from Feng et al. [79].

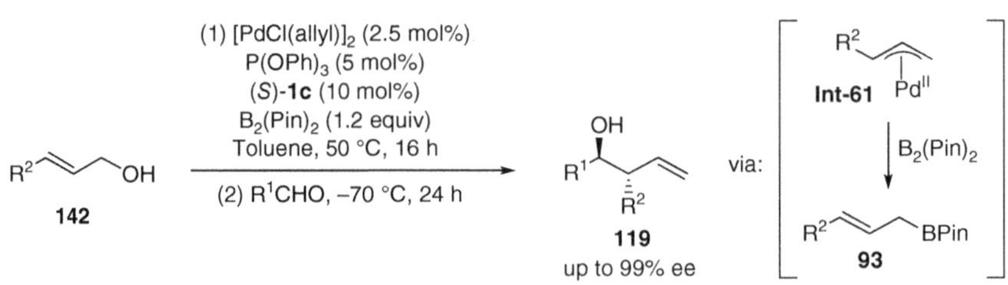

Scheme 6.41 Asymmetric allylation of aldehydes with allylic alcohols and pinacodiborate. Source: Modified from Zhang et al. [83].

alcohols and other π-allylpalladium precursors with boron-based pro-nucleophiles [80–82]. Antilla and others found that chiral Brønsted acids can efficiently catalyze asymmetric allylboration reactions of aldehydes [63, 64, 80]. Based on these concepts, Gong and coworkers developed a one-pot asymmetric carbonyl allylation of aldehydes, where a Pd/chiral phosphoric acid co-catalyzed allylic borylation of allylic alcohols proceeded at 50 °C, and a subsequent chiral phosphoric acid-catalyzed enantioselective allylation of aldehydes was performed at low temperature (Scheme 6.41) [83].

Some of the α-alkenes can undergo allylic activation with Pd(II) complex to give π-allylpalladium complexes, which are principally able to undergo allylic

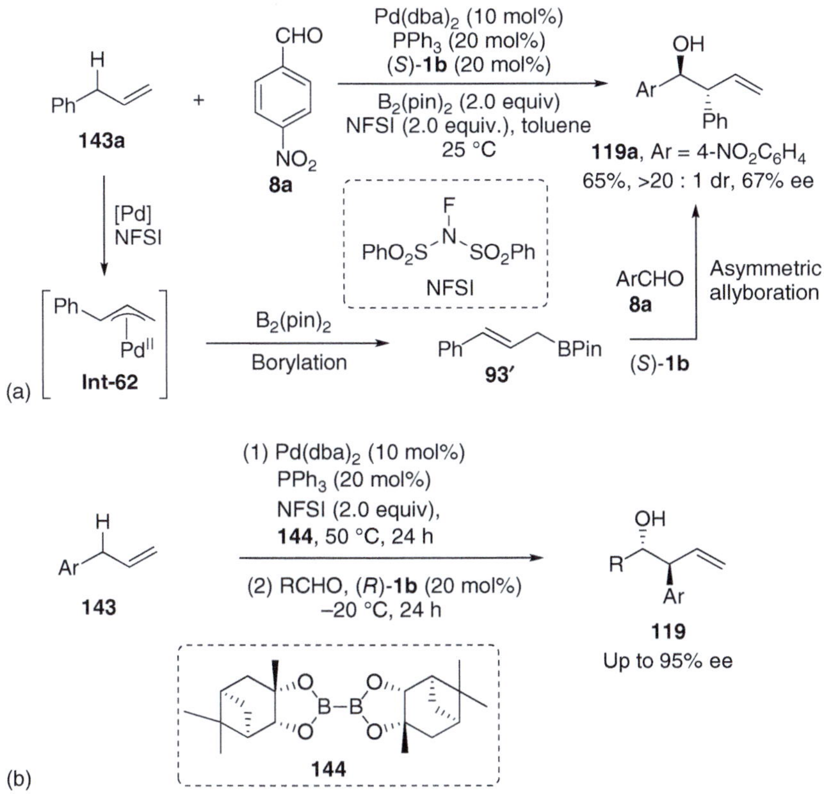

Scheme 6.42 Sequential Pd-catalyzed allylic C–H borylation and asymmetric allylboration of aldehydes. Source: Modified from Li et al. [85].

borylation to give allylborates [80–82]. Szabó and coworkers [82] and Gong and coworkers [83–85] successively reported Pd-catalyzed allylic C–H borylation reaction. Interestingly, the combination of chiral phosphoric acid with Pd(II) complexes allows a relay catalytic asymmetric carbonyl allylation of aldehydes using allylbenzene as the allylating reagent, leading to homoallylic alcohol with excellent diastereoselectivity and moderate 67% ee (Scheme 6.42a). The use of chiral bis(pinanediolato)diboron **144** and co-catalyst chiral Brønsted acid **(*R*)-1b** enables the Pd-catalyzed asymmetric carbonyl allylation of aldehydes with a significantly improved level of stereocontrol (Scheme 6.42b) [85].

6.7 Metal/Brønsted Acid Relay Catalysis Involving Hydroformylation

Metal-catalyzed hydroformylation of alkenes is one of the most practical homogeneous chemical processes in the world [87–89]. Hydroformylation could also integrate with other transformations, such as reduction, nucleophilic addition, or aldol condensation, in tandem or domino reactions [90–93]. In 2015, Xiao and coworker demonstrated a Rh/chiral Brønsted acid relay catalytic asymmetric hydroaminomethylation of styrenes with Hantzsch ester and syngas (Scheme 6.43a) [86]. The cascade reaction involves Rh-catalyzed hydroformylation, imine formation, and subsequent chiral phosphoric acid-catalyzed dynamic kinetic transfer hydrogenation with the Hantzsch ester to give chiral amine products. Han and coworkers refined this process, which features milder conditions (1 bar of syngas, room temperature in most cases), broader substrate scope, higher yields (up to 99%), and high enantioselectivities (up to >99 ee) [95]. Notably, acrylamides are also well tolerated, furnishing the desired chiral amines with high yields and enantioselectivities (Scheme 6.43b). Aliphatic alkenes and vinyl esters are also applicable, albeit lower yields and enantioselectivities are obtained.

The Rh-catalyzed hydroformylation of styrenes generates α-branched aldehyde **Int-63′**. Under the cooperative catalysis of chiral phosphoric acid, primary amine, and palladium complex, the α-branched aldehyde **Int-63′** undergoes the asymmetric α-allylation with allylic alcohols, as indicated by List and coworker [96]. Thus, a multiple relay catalytic cascade hydroformylation/asymmetric allylation was developed (Scheme 6.44) [94].

The combination of rhodium complex of biphosphine, $Pd(PPh_3)_4$, chiral phosphoric acid **(*R*)-1b**, and primary amine **152** efficiently drives the cascade reaction to directly transform styrenes, allylic alcohols, and syngas to α-quaternary chiral aldehydes with up to 97% yield and with up to 99% ee. As reported by List [96] and Gong [98], the enamine intermediate **Int-65** generated from amine **152** and the α-branched aldehyde **Int-63′** after the hydroformylation process reacts with the π-allyl Pd(II) complex **Int-66** or **Int-67**, which is generated from the allyl alcohol in the presence of palladium and Brønsted acid, giving products **153** or **154** and regenerating the achiral amine **152**. Since carbon monoxide could deactivate the palladium catalyst via coordination to slow down the allylation process, it is crucial to use low-pressure syngas. Furthermore, the bidentate phosphine ligand for the rhodium catalyst also inhibits the palladium-catalyzed allylation process that a higher loading of Rh catalyst (2/2.4 mol% Rh precursor and ligand) results in a much slower reaction.

The chiral homoallylic alcohols **Int-68** generated from the chiral phosphoric acid-catalyzed asymmetric allylation of aldehydes **8** with allylboronates **93** can participate in the rhodium-catalyzed hydroformylation to give linear-selective chiral aldehydes **Int-69**, which then undergo hemiacetalization to give highly enantioenriched tetrahydropyran intermediates **155** (Scheme 6.45) [97]. The hemiacetal intermediates **155** are isolatable and can be oxidized to tetrahydropyranones **156** or reduced to tetrahydropyrans **157**. The method is amenable to the concise synthesis of key chiral intermediates to access natural products herboxidiene (Scheme 6.46a) [99] and leucascandrolide A (Scheme 6.46b) [100].

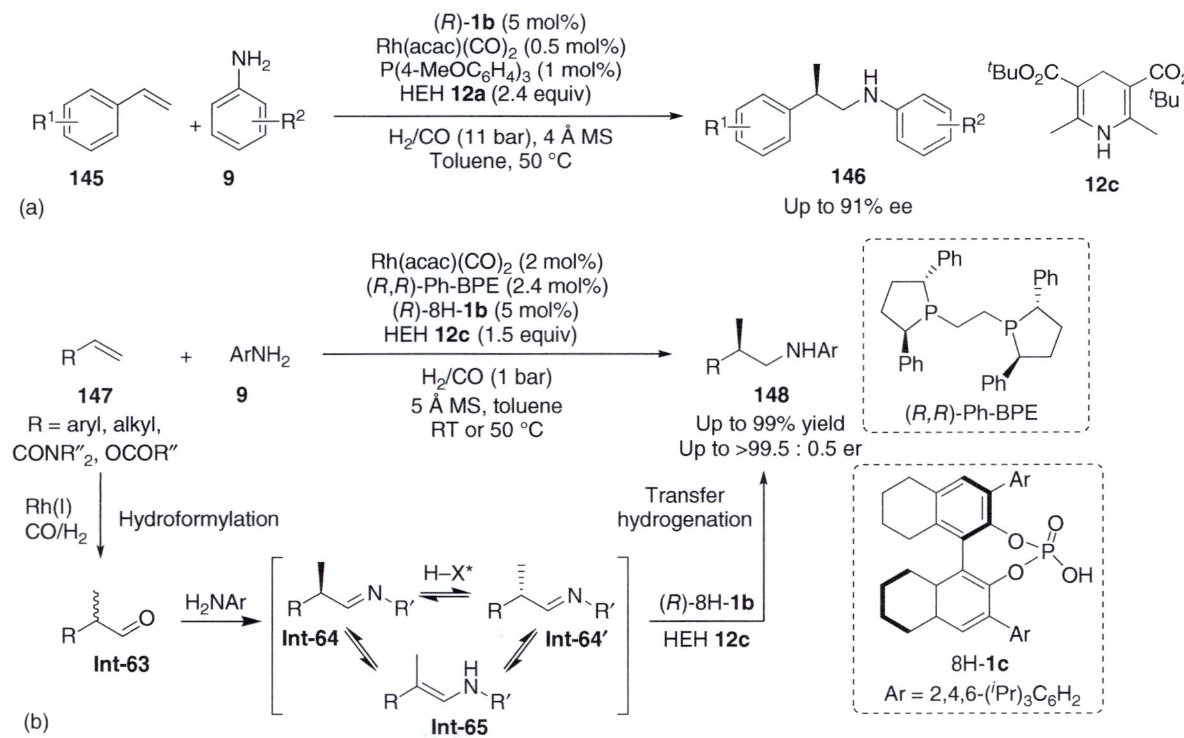

Scheme 6.43 Rh/chiral Brønsted acid relay catalytic asymmetric hydroaminomethylation of alkenes. Source: Modified from Villa-Marcos and Xiao [86].

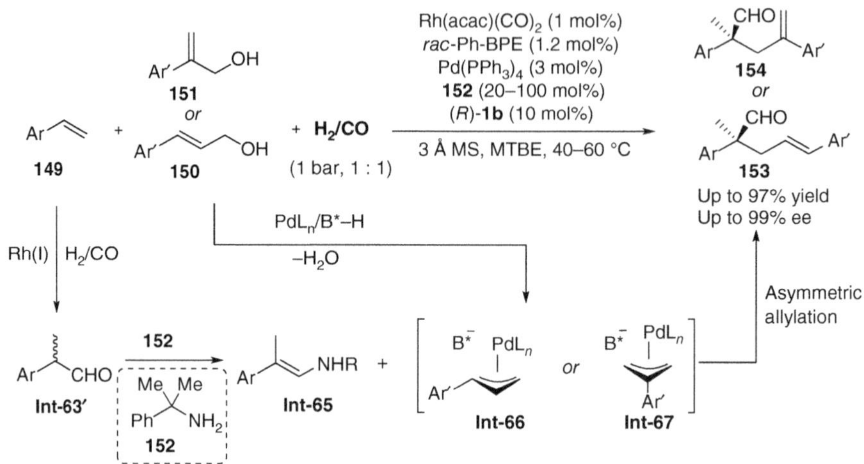

Scheme 6.44 Multiple relay catalytic cascade hydroformylation/asymmetric allylation reaction. Source: Modified from Meng et al. [94].

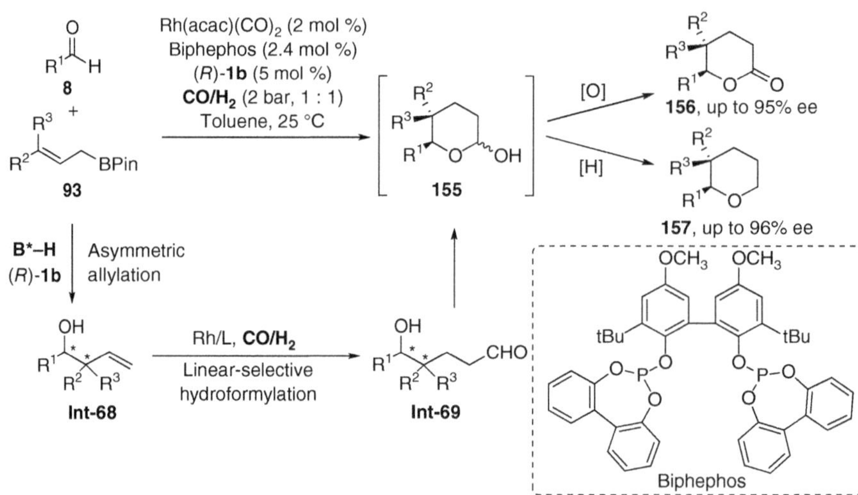

Scheme 6.45 Rh(I)/chiral Brønsted acid relay catalytic asymmetric carbonyl allylation/hydroformylation. Source: Modified from Li et al. [97]

6.8 Metal/Brønsted Acid Relay Catalysis Involving Metal Carbene Formation

6.8.1 Cascade Metal Carbene Formation and Asymmetric Protonation

Chiral Brønsted acid-catalyzed asymmetric protonation of prochiral enol derivatives has been recognized as a straightforward approach to access enantioenriched α-carbonyl compounds [101–103]. Combined with transition metal catalysts, chiral phosphoric acids could work efficiently as chiral proton-transfer shuttle catalysts

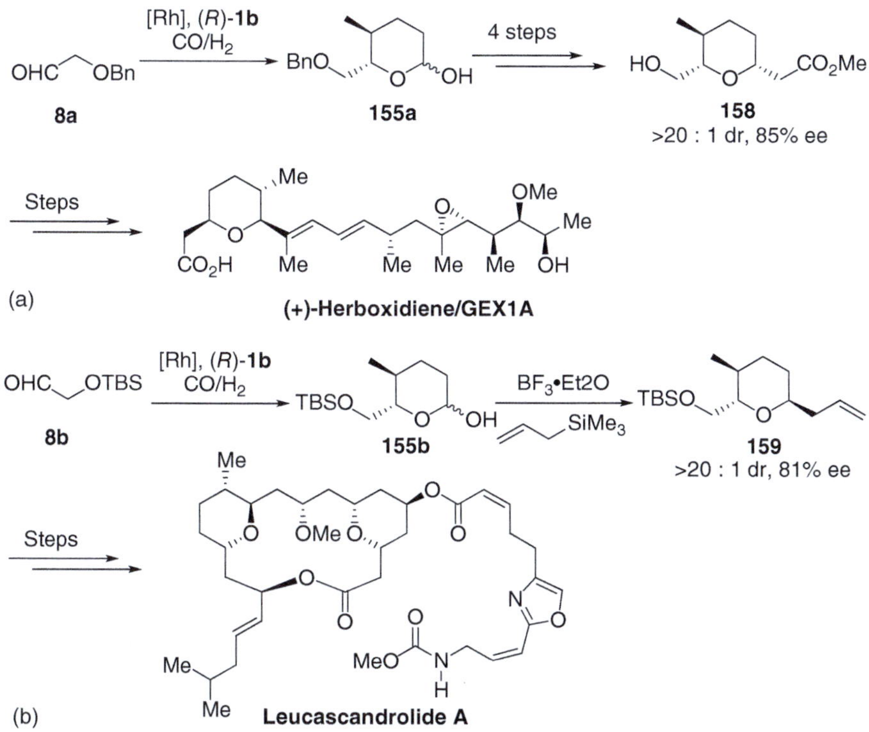

Scheme 6.46 Formal synthesis of (+)-herboxidiene and leucascandrolide A. Source: (a) Modified from Ghosh and Li [99], (b) Modified from Wipf and Reeves [100].

in carbene insertion reactions [104–106]. As shown in Scheme 6.47, this type of reactions initially proceed through the formation of a highly electrophilic metal carbene intermediate **Int-70**, which can be attacked by the lone-pair electrons of a heteroatom to generate a metal-associated ylide **Int-71** or enolate **Int-72**. These intermediates can undergo enantioselective proton transfer to give chiral products **161** (path a). However, the metal dissociation from these intermediates often occurs to form a metal-free intermediate, either ylide **Int-73** or enol **Int-74**, which then undergoes non-enantioselective proton transfer (path b). The proton transfer process of either the metal-associated or metal-free ylides and enols can be accelerated by a chiral Brønsted acid shuttle catalysis. The former case falls into the category of metal/Brønsted acid cooperative catalysis (Chapter 5), as the metal catalyst could affect the enantioselectivity-determining proton transfer process [106, 110]. In the latter case, the enantioselective proton-transfer process is solely controlled by the chiral Brønsted acid catalyst, wherein the proton transfer occurs synchronously between chiral phosphoric acid and the ylide or enol through a cyclic transition state **TS-7** (path d). Generally, chiral metal complexes have little effect on the enantioselectivity that metal/Brønsted acid relay catalysis is likely dominating the asymmetric carbene insertion reaction.

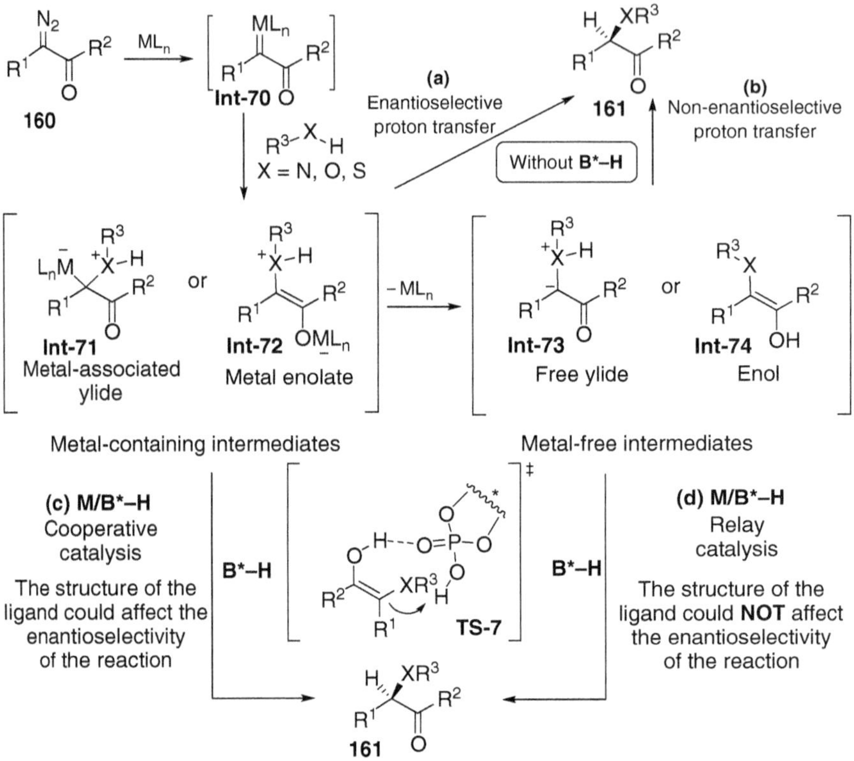

Scheme 6.47 Asymmetric carbene insertion into the X—H bonds via transition metal catalysis (path a and path b [non-enantioselective]), metal/Brønsted acid cooperative catalysis (path c), and metal/Brønsted acid relay catalysis (path d).

Zhou, Zhu and coworkers developed an enantioselective N–H insertion of α-diazoketones with $BocNH_2$ by the combined catalysis of $Rh_2(TFA)_4$ and chiral phosphoric acid **4b** (Scheme 6.48a) [107]. However, the use of chiral rhodium(II) complexes, [Rh_2(*R*-DOSP)$_4$], [Rh_2(*S*-PTAD)$_4$], or a chiral copper complex with a spiro bis(oxazoline) ligand is unable to induce enantioselectivity. Therefore, the rhodium complex dissociates from the metal-bonded ylide, and the chirality-determining proton transfer catalyzed by the chiral phosphoric acid **4b** is believed to proceed via a transition state **TS-8** involving metal-free ylide or enol intermediate, which has been verified by DFT calculations [90]. Moreover, the binary catalytic system of $Rh_2(TPA)_4$ [tetrakis(triphenylacetato)dirhodium(II)] and chiral phosphoric enables S—H and O—H bonds insertion to deliver high levels of enantioselectivity (Scheme 6.48b,c) [108, 109].

Other types of chiral Brønsted acid catalysts, such as ureas and protonated guanidines, are also compatible with transition metal catalysts in relay catalysis. Liu and coworkers developed an enantioselective N–H insertion of α-diazoesters employing a multi-catalyst system consisting of an achiral dirhodium(II) acetate, chiral sulfonamide urea, and an achiral sulfonic acid (Scheme 6.49) [111]. As Jacobsen and

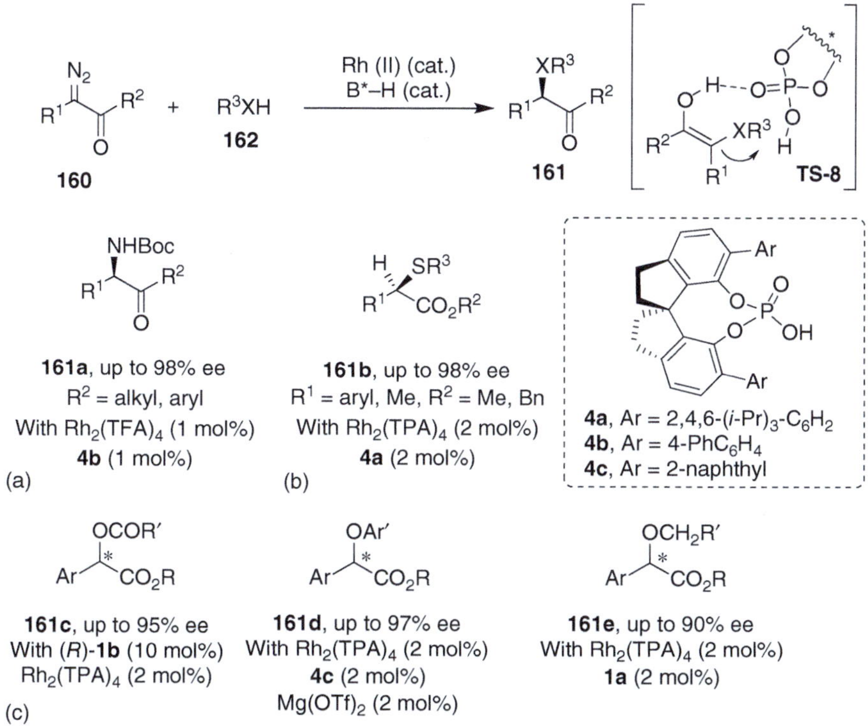

Scheme 6.48 Asymmetric carbene insertion into the X–H bonds by Rh(II)/chiral Brønsted acid relay catalysis. Source: (a) Modified from Xu et al. [107], (b) Modified from Xu et al. [108], (c) Modified from Zhang et al. [109].

coworkers have disclosed, the noncovalent interactions between chiral sulfonamide urea **163** and a sulfonic acid **164** would lead to an assembled chiral Brønsted acid [113], which turns out to be a chiral proton shuttle assisting the asymmetric proton transfer process through **TS-9**.

The relatively strong basicity and coordinating ability of aliphatic amines pose additional constraints on the development of highly enantioselective insertion of metal carbenes into the N—H bonds of alkyl amines, which represents a long-standing challenge in traditional chiral transition metal catalysis. Very recently, Zhou and coworkers identified that the orthogonal combination of an achiral copper complex (Tp*Cu) [114] and a chiral amino-thiourea [115] rendered a highly enantioselective carbene insertion into N—H bonds of aliphatic amines **165** (Scheme 6.50) [112]. Tp*Cu shows high catalytic activity for the carbene insertion into aliphatic amines. A copper(I)/thiourea complex (Tp*Cu·**166**) actually promotes the enantioselective proton transfer process of the free enol intermediate **Int-77** through a transition state **TS-10**.

In addition to typical Brønsted acids, guanidines are also compatible with various transition metal catalysts. The combination of $Rh_2(OAc)_4$ and chiral guanidine **G1** can successfully drive a highly enantioselective insertion reaction of α-diazoesters and α-diazoketones into the O—H bonds of carboxylic acids (Scheme 6.51) [116].

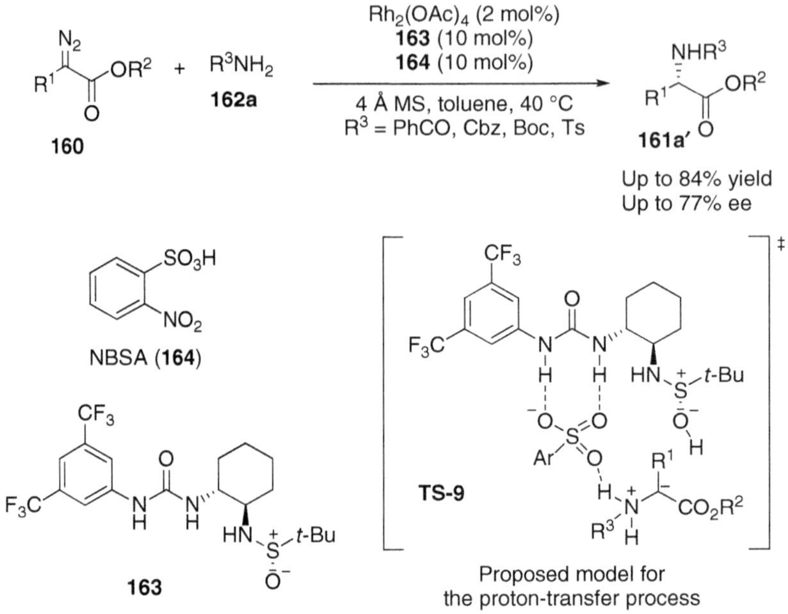

Scheme 6.49 Enantioselective insertion of α-diazoesters into amide N—H bonds enabled by combined catalysis of Rh(II) and assembled chiral Brønsted acid. Source: Modified from Ni et al. [111].

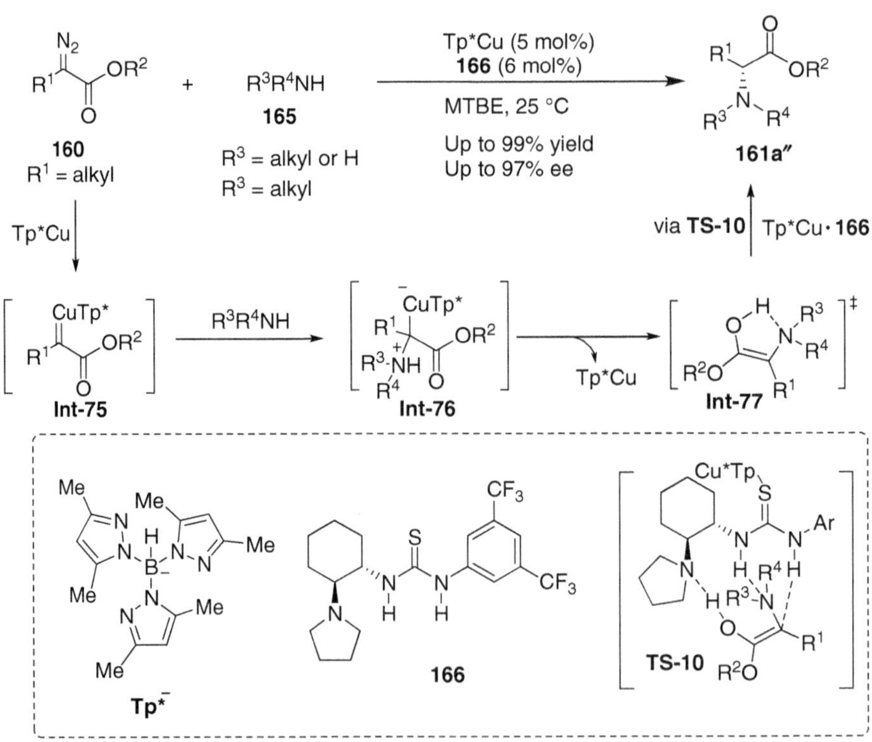

Scheme 6.50 Cu(I)/chiral amino thiourea relay catalytic enantioselective insertion of α-diazoesters into N—H bonds of aliphatic amines. Source: Modified from Li et al. [112].

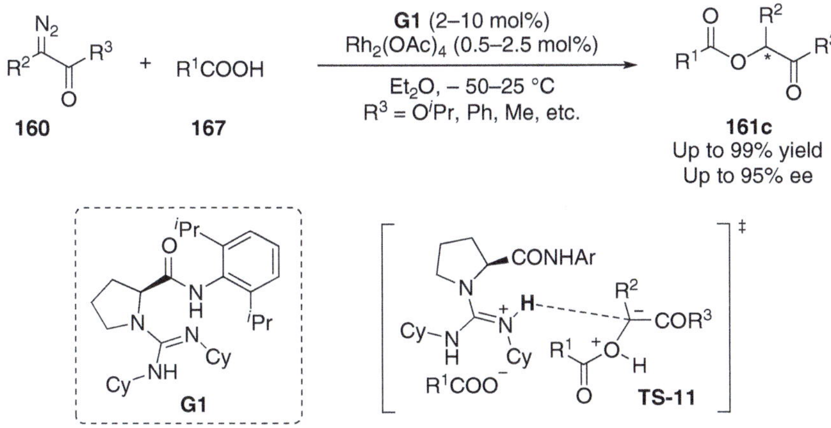

Scheme 6.51 Enantioselective O–H insertion by Rh(II) and chiral guanidine relay catalysis. Source: Modified from Tan et al. [116].

In situ generated guanidinium salt from carboxylic acids and the guanidine catalyst actually act as chiral proton-transfer catalyst for the enantioselective protonation of either metal-free ylide or enol intermediate.

The insertion of rhodium carbene into aniline C—H bonds generates a zwitterion with or without the rhodium, as shown in **Int-79** or **Int-80**, which undergoes an asymmetric 1,2-proton shift mediated with chiral phosphoric acid via **TS-12** to give chiral arylation products **169**. Zhu and Zhou found that the combination of $Rh_2(TFA)_4$ and chiral phosphoric acid (*R*)-**4d** allows the relay catalytic arylation reaction to give **169** with excellent enantioselectivity (Scheme 6.52) [117].

The gold-catalyzed decomposition of diazoesters **160** generates a gold carbene complex **Int-81** with the major cation resonance of type **Int-82**, which is highly carbophilic toward the nucleophilic addition of enamide **170** to generate an intermediate **Int-83**. DFT calculation suggests that the intermediate **Int-83** undergoes a proton transfer to generate a more thermodynamically stable enol species **Int-85** via gold enolate species **Int-84**. Chiral phosphoric acid-catalyzed asymmetric protonation of **Int-85** via **TS-13** furnishes the chiral imine product. Ten mole percentage of XPhos-gold(I) complex, in combination with 0.1 mol% of chiral phosphoric acid **11**, enables this aza-ene-type reaction of diazo esters **160** with enamides **170** to deliver γ-keto esters **171** in up to 99% yield and with 97% ee (Scheme 6.53) [118].

6.8.2 Multiple Cascade Reaction Initiated with Metal Carbene

Terada and coworker accomplished a dirhodium(II) tetracarboxylate/chiral phosphoric acid relay catalytic cascade carbonyl ylide formation/enantioselective reduction reaction (Scheme 6.54) [119]. The decomposition of α-diazocarbonyl compounds **172** by dirhodium(II) tetracarboxylate generates the rhodium carbene complex **Int-86**. The subsequent intramolecular cyclization of the rhodium carbene complex **Int-86** affords a carbonyl ylide **Int-87** or oxidopyrylium **Int-87′** via

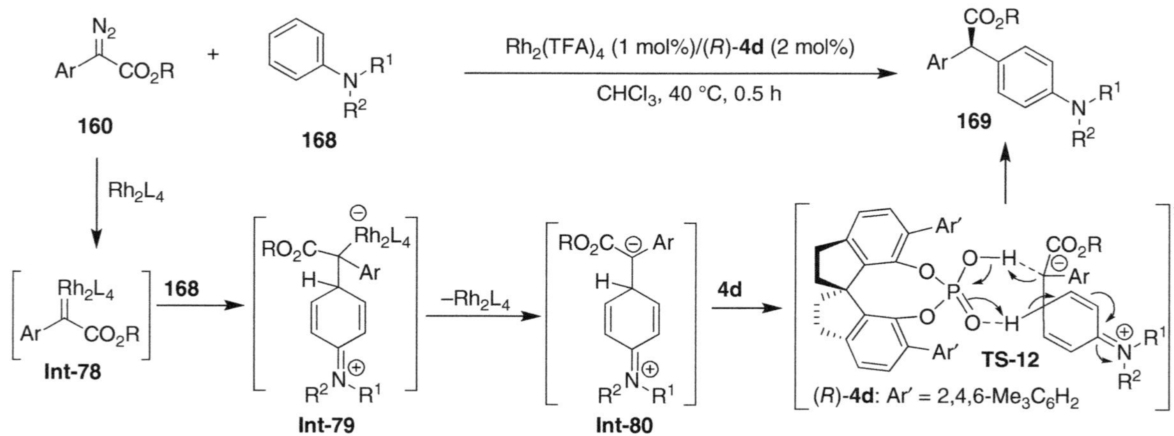

Scheme 6.52 Asymmetric arylation of diazoesters with aniline derivatives. Source: Modified from Xu et al. [117].

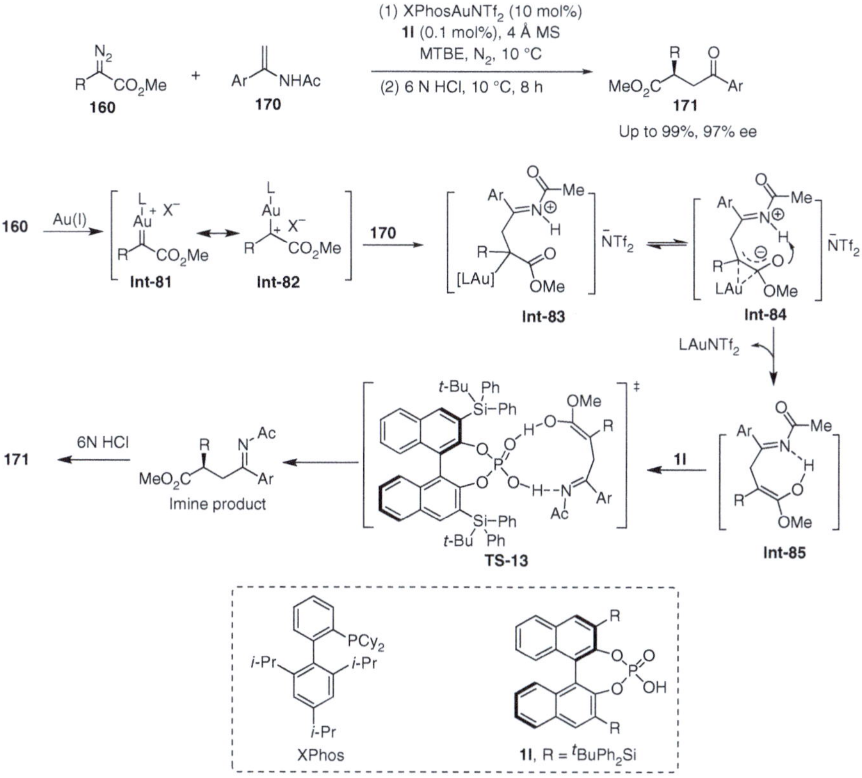

Scheme 6.53 Enantioselective aza-carbene ene-type reaction of α-diazo esters with enamides. Source: Modified from Zhao et al. [118].

tautomerization, which then undergoes protonation with chiral phosphoric acid **1a** to form an isobenzopyrylium ion pair **Int-88**. The enantioselective transfer hydrogenation of the isobenzopyrylium ion **Int-88** with Hantzsch ester occurs to give **173**, after being treated with benzoyl chloride, to afford the benzoyloxy isochromene derivative **174** [119]. Chiral rhodium complex barely induces enantioselectivity in this reaction, indicating that the rhodium-free intermediate **Int-88** plays a crucial role in the enantioselective transfer hydrogenation step.

6.9 Lewis Acid/Chiral Brønsted Acid Relay Catalysis

Friedländer condensation is an operative method to access quinolones [121]. A catalytic amount of either Lewis acids or Brønsted acids is able to promote the formation of quinolones. On the other hand, chiral phosphoric acids are able to efficiently catalyze the enantioselective transfer hydrogenation of quinolones. The relay catalysis by a Lewis acid and a chiral phosphoric acid would principally enable the enantioselective conversion of anilines **175** and β-keto esters to optically active tetrahydroquinolines. A cascade Friedländer condensation and transfer

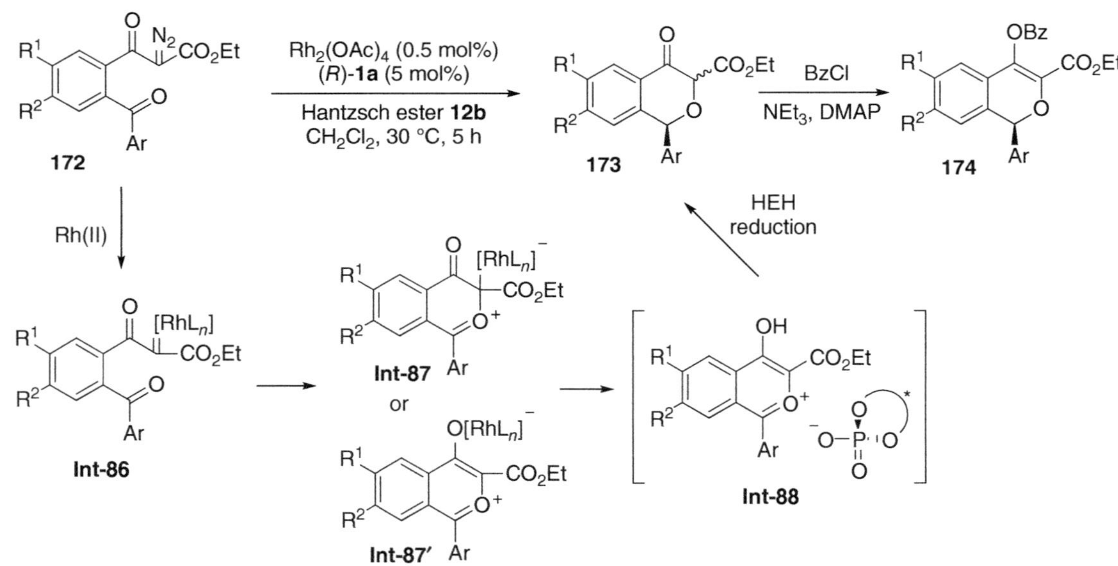

Scheme 6.54 Catalytic cascade carbene formation/asymmetric reduction. Source: Modified from Terada and Toda [119].

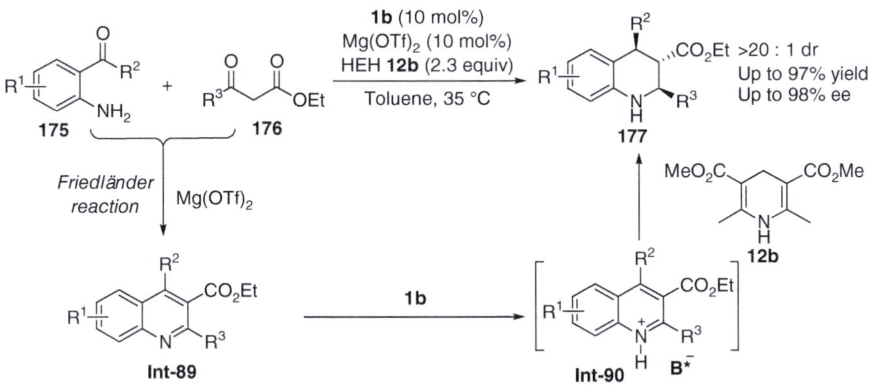

Scheme 6.55 Mg(II)/chiral phosphoric acid relay catalytic Friedländer condensation/transfer hydrogenation. Source: Modified from Ren et al. [120].

hydrogenation reaction proceeds efficiently under the relay catalysis of $Mg(OTf)_2$ and chiral phosphoric acid **1b** to furnish highly substituted tetrahydroquinoline derivatives **177** in high yield and with excellent levels of diastereo- and enantioselectivity (Scheme 6.55) [120]. Despite both $Mg(OTf)_2$ and the chiral phosphoric acid are effective in accelerating the condensation process, the combined catalysis considerably improves the conversion, indicating the synergistic effect of $Mg(OTf)_2$ and chiral phosphoric acid. In contrast, the asymmetric transfer hydrogenation is solely catalyzed by the chiral phosphoric acid as indicated by the kinetic studies.

Zhou and coworkers established an enantioselective Nazarov cyclization of indolyl enones **178** enabled by $ZnCl_2$/chiral phosphoric acid relay catalysis (Scheme 6.56) [122]. The cyclization of the dienone **178** is catalyzed by cooperative Lewis acid and the phosphoric acid catalysis, while the enantioselectivity of the subsequent asymmetric proton transfer of the enol intermediate is controlled by the chiral Brønsted acid (*R*)-**4d** via transition state **TS-14**.

6.10 Miscellaneous

The development of highly enantioelective radical-based reactions has long been a formidable challenge in organic synthesis. New concepts and catalyst systems are highly demanded to address the issues associated with stereochemical control. Liu and coworkers successfully established an enantioselective formal C—H bond alkoxylation, triggered by radical trifluoromethylation of alkenes by copper cyanide and chiral phosphoric acid relay catalysis (Scheme 6.57) [123]. The single-electron reduction of the Togni reagent by Cu(I) generates Cu(II) species and a trifluoromethyl radical, which then adds to terminal carbon–carbon double bonds, leading to a transient radical **Int-92**. Thermodynamically favorable 1,5-hydrogen atom transfer of the intermediate **Int-92** proceeds to form a radical species **Int-93**, which is then oxidized by Cu(II) to give an imine intermediate. Chiral phosphoric

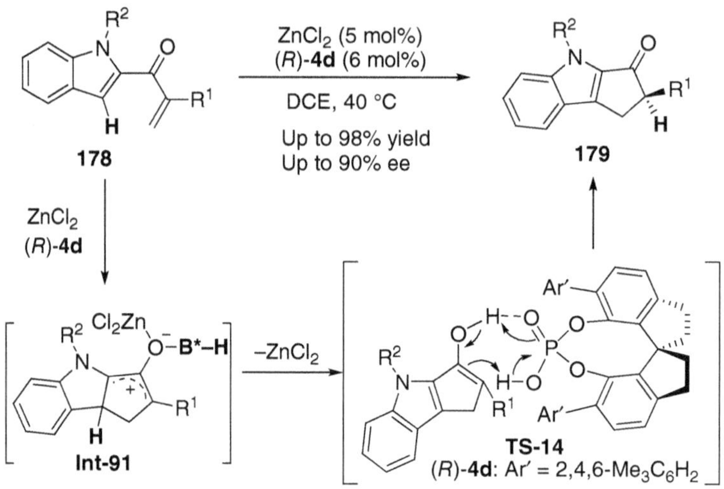

Scheme 6.56 $ZnCl_2$/chiral spirophosphoric acid relay catalytic enantioselective Nazarov cyclization of indolyl enones. Source: Modified from Wang et al. [122].

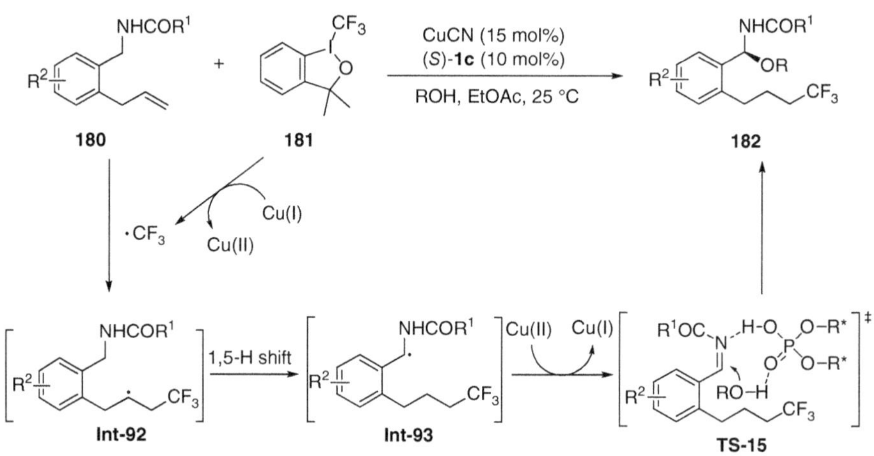

Scheme 6.57 Enantioselective formal C—H bond functionalization triggered by radical trifluoromethylation of inactivated alkene. Source: Modified from Yu et al. [123].

acid-catalyzed addition of alcohols to imines yields chiral N,O-hemiaminal **182** via transition state **TS-15**.

A single electron transfer (SET) reaction between a polyfluoroalkyl sulfuric chloride **184** and CuI generates a Cu(II) species and an oxidative radical that undergoes addition to the 1,1-diarylalkene **183** and gives a transient alkyl radical **Int-94**. The subsequent single-electron oxidation of radical **Int-94** by Cu(II) leads to a benzylic carbocation **Int-95**, which undergoes a chiral phosphoric acid-catalyzed asymmetric substitution reaction with indole via transition state **TS-16**. The combination of CuI and chiral phosphoric acid **4e** allows this relay catalytic three-component reaction to proceed with high conversion and excellent enantioselectivity (Scheme 6.58) [124].

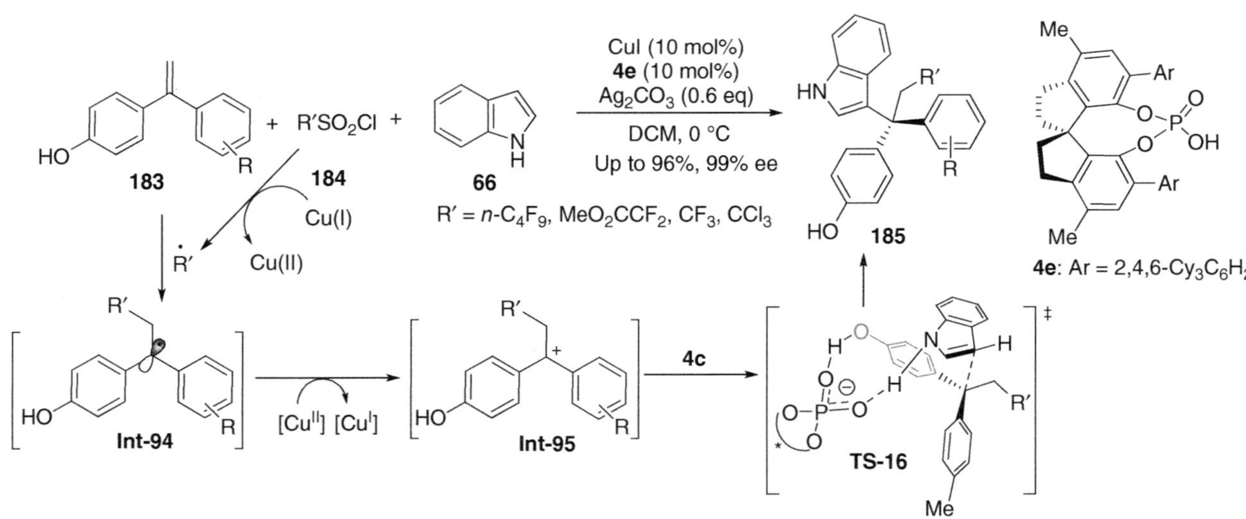

Scheme 6.58 Copper/chiral phosphoric acid relay catalytic radical-initiated dicarbofunctionalization of 1,1-diarylalkenes. Source: Modified from Lin et al. [124].

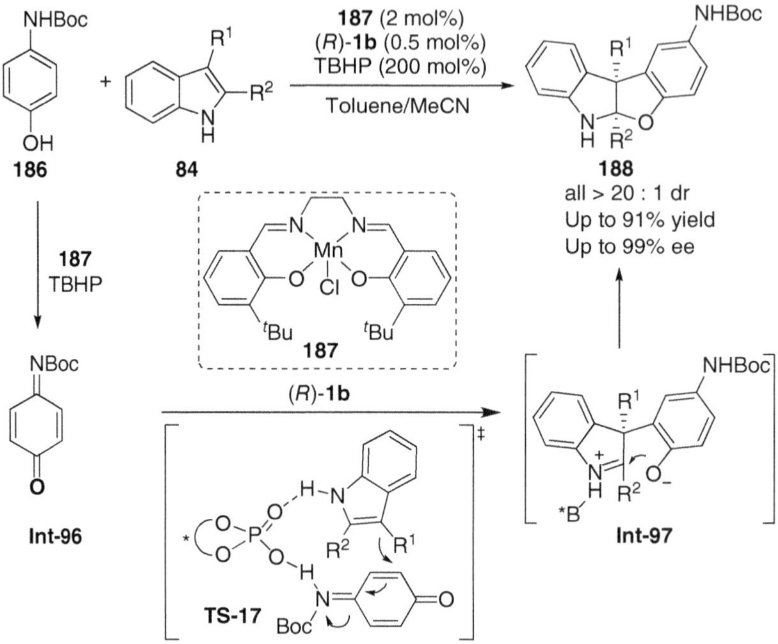

Scheme 6.59 Mn(III)/chiral phosphoric acid relay catalytic asymmetric oxidative phenol-indole [3+2] coupling. Source: Modified from Yu et al. [125].

Oxidation of phenols mediated with redox active transition-metal salts provides electrophilic quinones or quinone monoimines capable of undergoing coupling reaction with nucleophiles. Wu and coworkers developed a Mn(III) complex/chiral phosphoric acid relay catalytic direct oxidative phenol indole [3+2] annulation reaction. The combination of Mn(III) complex **187** and the chiral phosphoric acid **(*R*)-1b** enabled the direct conversion of readily available *p*-aminophenol derivatives **186** and indoles **84** into a structurally diverse range of chiral benzofuroindolines **188** proceeding through asymmetric Michael addition via **TS-17** and cyclization via intermediate **Int-97** (Scheme 6.59) [125].

6.11 Summary and Outlook

Metal and chiral Brønsted acid relay catalysis has gone through its proof-of-concept stage and has been recognized as a versatile strategy for the development of non-classical cascade reactions. Nevertheless, vast space still remains for the future development of this area. As highlighted in this chapter, most examples of metal/Brønsted acid asymmetric relay catalysis employed BINOL-derived chiral phosphoric acids, while other structurally distinct chiral Brønsted acids, such as chiral dicarboxylic acids, chiral disulfonic acids, and chiral sulfonimides, remain undeveloped in the relay catalysis. Moreover, the utilization of metal/chiral Brønsted acid relay catalysis in cost-efficient synthetic chemistry, i.e. asymmetric photocatalysis and electrochemical transformations, represents a promising research field. In conclusion, along with the development of transition-metal catalysis and chiral Brønsted acid catalysis, the combination of both in asymmetric relay catalysis will undoubtedly be creative and of high efficiency in pursuit of novel and unique cascade reactivity.

References

1 (a) Uraguchi, D. and Terada, M. (2004). *J. Am. Chem. Soc.* 126: 5356–5357. (b) Akiyama, T., Itoh, J., Yokota, K., and Fuchibe, K. (2004). *Angew. Chem. Int. Ed.* 43: 1566–1568.

2 (a) Akiyama, T. and Mori, K. (2015). *Chem. Rev.* 115: 9277–9306. (b) Parmar, D., Sugiono, E., Raja, S., and Rueping, M. (2014). *Chem. Rev.* 114: 9047–9153.

3 Chen, D.F., Han, Z.Y., Zhou, X.L., and Gong, L.Z. (2014). *Acc. Chem. Res.* 47: 2365–2377.

4 (a) Widenhoefer, R.A. and Han, X. (2006). *Eur. J. Org. Chem.* 2006: 4555–4563. (b) Zhang, L., Sun, J., and Kozmin, S.A. (2006). *Adv. Synth. Catal.* 348: 2271–2296. (c) Fürstner, A. and Davies, P.W. (2007). *Angew. Chem. Int. Ed.* 46: 3410–3449. (d) Hashmi, A.S.K. (2007). *Chem. Rev.* 107: 3180–3211.

5 Jiménez-Núñez, E.s. and Echavarren, A.M. (2008). *Chem. Rev.* 108: 3326–3350.

6 Müller, T.E., Hultzsch, K.C., Yus, M. et al. (2008). *Chem. Rev.* 108: 3795–3892.

7 Belting, V. and Krause, N. (2006). *Org. Lett.* 8: 4489–4492.

8 Barluenga, J., Mendoza, A., Rodríguez, F., and Fañanás, F.J. (2008). *Angew. Chem. Int. Ed.* 47: 7044–7047.

9 Han, Z.-Y., Xiao, H., Chen, X.-H., and Gong, L.-Z. (2009). *J. Am. Chem. Soc.* 131: 9182–9183.

10 List, B., Hoffmann, S., and Nicoletti, M. (2006). *J. Am. Chem. Soc.* 128: 13074–13075.

11 Rueping, M. and Antonchick, A.P. (2007). *Angew. Chem. Int. Ed.* 46: 4562–4565.

12 Du, Y.-L., Hu, Y., Zhu, Y.-F. et al. (2015). *J. Org. Chem.* 80: 4754–4759.

13 Liu, X.-Y. and Che, C.-M. (2009). *Org. Lett.* 11: 4204–4207.

14 Liu, X.-Y., Xiao, Y.-P., Siu, F.-M. et al. (2012). *Org. Biomol. Chem.* 10: 7208–7219.

15 He, Y.-P., Wu, H., Chen, D.-F. et al. (2013). *Chem. Eur. J.* 19: 5232–5237.

16 Shinde, V.S., Mane, M.V., Vanka, K. et al. (2015). *Chem. Eur. J.* 21: 975–979.

17 Zhao, F., Li, N., Zhu, Y.-F., and Han, Z.-Y. (2016). *Org. Lett.* 18: 1506–1509.

18 Gregory, A.W., Jakuec, P., Turner, P., and Dixon, D.J. (2013). *Org. Lett.* 15: 4330–4333.

19 Wang, C., Han, Z.-Y., Luo, H.-W., and Gong, L.-Z. (2010). *Org. Lett.* 12: 2266–2269.

20 Patil, N.T., Mutyala, A.K., Konala, A., and Tella, R.B. (2012). *Chem. Commun.* 48: 3094–3096.

21 Yang, T., Campbell, L., and Dixon, D.J. (2007). *J. Am. Chem. Soc.* 129: 12070–12071.

22 Han, Z.-Y., Guo, R., Wang, P.-S. et al. (2011). *Tetrahedron Lett.* 52: 5963–5967.

23 Wu, H., He, Y.-P., and Gong, L.-Z. (2013). *Org. Lett.* 15: 460–463.

24 Cala, L., Mendoza, A., Fananas, F.J., and Rodriguez, F. (2013). *Chem. Commun.* 49: 2715–2717.
25 Calleja, J., Gonzalez-Perez, A.B., de Lera, A.R. et al. (2014). *Chem. Sci.* 5: 996–1007.
26 Quach, R., Furkert, D.P., and Brimble, M.A. (2013). *Tetrahedron Lett.* 54: 5865–5868.
27 Rexit, A.A. and Mailikezati, M. (2015). *Tetrahedron Lett.* 56: 2651–2655.
28 Grimster, N.P., Wilton, D.A.A., Chan, L.K.M. et al. (2010). *Tetrahedron* 66: 6429–6436.
29 Han, Z.-Y., Chen, D.-F., Wang, Y.Y. et al. (2012). *J. Am. Chem. Soc.* 134: 6532–6535.
30 Li, N., Chen, D.-F., Wang, P.-S. et al. (2014). *Synthesis* 46: 1355–1361.
31 Wang, P.-S., Li, K.N., Zhou, X.-L. et al. (2013). *Chem. Eur. J.* 19: 6234–6238.
32 Nakashima, D. and Yamamoto, H. (2006). *J. Am. Chem. Soc.* 128: 9626–9627.
33 Shapiro, N.D. and Toste, F.D. (2007). *J. Am. Chem. Soc.* 129: 4160–4161.
34 Li, G. and Zhang, L. (2007). *Angew. Chem. Int. Ed.* 46: 5156–5159.
35 Zhang, L. (2014). *Acc. Chem. Res.* 47: 877–888.
36 Qian, D. and Zhang, J. (2013). *Chem. Eur. J.* 19: 6984–6988.
37 Yeom, H.S., Lee, Y., Jeong, J. et al. (2010). *Angew. Chem. Int. Ed.* 49: 1611–1614.
38 Liu, R.-R., Ye, S.-C., Lu, C.-J. et al. (2015). *Angew. Chem. Int. Ed.* 54: 11205–11208.
39 Inamdar, S.M., Gonnade, R.G., and Patil, N.T. (2017). *Org. Biomol. Chem.* 15: 863–869.
40 Mei, G.J. and Shi, F. (2017). *J. Org. Chem.* 82: 7695–7707.
41 Ferrer, C. and Echavarren, A.M. (2006). *Angew. Chem. Int. Ed.* 45: 1105–1109.
42 Chen, J.-R., Li, C.-F., An, X.-L. et al. (2008). *Angew. Chem. Int. Ed.* 47: 2489–2492.
43 Cai, Q., Zhao, Z.-A., and You, S.-L. (2009). *Angew. Chem. Int. Ed.* 48: 7428–7431.
44 Zhang, J.-W., Liu, X.-W., Gu, Q. et al. (2015). *Org. Chem. Front.* 2: 476–480.
45 Cai, Q., Liang, X.-W., Wang, S.-G. et al. (2012). *Org. Lett.* 14: 5022–5025.
46 Shi, Y.-C., Wang, S.-G., Yin, Q., and You, S.-L. (2014). *Org. Chem. Front.* 1: 39–43.
47 Inamdar, S.M., Chakrabarty, I., and Patil, N.T. (2016). *RSC Adv.* 6: 34428–34433.
48 Zhang, J.-W., Cai, Q., Gu, Q. et al. (2013). *Chem. Commun.* 49: 7750–7752.
49 Fustero, S., Rodriguez, E., Lazaro, R. et al. (2013). *Adv. Synth. Catal.* 355: 1058–1064.
50 Lautens, M. and Rovis, T. (1997). *J. Org. Chem.* 62: 5246–5247.
51 Vasseur, A., Bruffaerts, J., and Marek, I. (2016). *Nat. Chem.* 8: 209–219.
52 Sorimachi, K. and Terada, M. (2008). *J. Am. Chem. Soc.* 130: 14452–14453.
53 Terada, M. and Toda, Y. (2009). *J. Am. Chem. Soc.* 131: 6354–6355.
54 Cai, Q., Liang, X.-W., Wang, S.-G., and You, S.-L. (2013). *Org. Biomol. Chem.* 11: 1602–1605.
55 Hansen, C.L., Clausen, J.W., Ohm, R.G. et al. (2013). *J. Org. Chem.* 78: 12545–12565.

56 Ascic, E., Hansen, C.L., Le Quement, S.T., and Nielsen, T.E. (2012). *Chem. Commun.* 48: 3345–3347.
57 Toda, Y. and Terada, M. (2013). *Synlett* 24: 752–756.
58 Bernardez, R., Suarez, J., Fananas-Mastral, M. et al. (2016). *Org. Lett.* 18: 642–645.
59 Wille, A., Tomm, S., and Frauenrath, H. (1998). *Synthesis* 1998: 305–308.
60 Richmond, E., Khan, I.U., and Moran, J. (2016). *Chem. Eur. J.* 22: 12274–12277.
61 Mantilli, L., Gérard, D., Torche, S. et al. (2009). *Angew. Chem. Int. Ed.* 48: 5143–5147.
62 Li, J.-Q., Peters, B., and Andersson, P.G. (2011). *Chem. Eur. J.* 17: 11143–11145.
63 Miura, T., Nishida, Y., Morimoto, M., and Murakami, M. (2013). *J. Am. Chem. Soc.* 135: 11497–11500.
64 Jain, P. and Antilla, J.C. (2010). *J. Am. Chem. Soc.* 132: 11884–11886.
65 Grayson, M.N., Pellegrinet, S.C., and Goodman, J.M. (2012). *J. Am. Chem. Soc.* 134: 2716–2722.
66 Miura, T., Nakahashi, J., and Murakami, M. (2017). *Angew. Chem. Int. Ed.* 56: 6989–6993.
67 Miura, T., Nakahashi, J., Zhou, W. et al. (2017). *J. Am. Chem. Soc.* 139: 10903–10908.
68 Zhang, W., Chi, Y., and Zhang, X. (2007). *Acc. Chem. Res.* 40: 1278–1290.
69 Akiyama, T. and Zhu, C. (2010). *Adv. Synth. Catal.* 352: 1846–1850.
70 Akiyama, T., Henseler, A., Kato, M., and Mori, K. (2011). *Angew. Chem. Int. Ed.* 50: 8180–8183.
71 You, S.-L. (2007). *Chem. Asian J.* 2: 820–827.
72 Rueping, M., Azap, C., Sugiono, E., and Theissmann, T. (2005). *Synlett*: 2367–2369.
73 Shi, F. and Gong, L.-Z. (2012). *Angew. Chem. Int. Ed.* 51: 11423–11425.
74 Chen, Q.A., Wang, D.S., Zhou, Y.G. et al. (2011). *J. Am. Chem. Soc.* 133: 6126–6129.
75 Chen, Q.-A., Chen, M.-W., Yu, C.-B. et al. (2011). *J. Am. Chem. Soc.* 133: 16432–16435.
76 Chen, Q.-A., Gao, K., Duan, Y. et al. (2012). *J. Am. Chem. Soc.* 134: 2442–2448.
77 Chen, Z.-P., Chen, M.-W., Guo, R.-N., and Zhou, Y.-G. (2014). *Org. Lett.* 16: 1406–1409.
78 Lu, L.Q., Li, Y.H., Junge, K., and Beller, M. (2015). *J. Am. Chem. Soc.* 137: 2763–2768.
79 Feng, G.-S., Chen, M.-W., Shi, L., and Zhou, Y.-G. (2018). *Angew. Chem. Int. Ed.* 57: 5853–5857.
80 Diner, C. and Szabo, K.J. (2017). *J. Am. Chem. Soc.* 139: 2–14.
81 Ishiyama, T., Ahiko, T.-a., and Miyaura, N. (1996). *Tetrahedron Lett.* 37: 6889–6892.
82 Selander, N., Sebelius, S., Estay, C., and Szabó, K.J. (2006). *Eur. J. Org. Chem.* 2006: 4085–4087.
83 Zhang, Z.-J., Tao, Z.-L., Adele, A., and Gong, L.-Z. (2017). *Acta Chim. Sin.* 75: 1196–1201.

84 Tao, Z.-L., Li, X.-H., Han, Z.-Y., and Gong, L.-Z. (2015). *J. Am. Chem. Soc.* 137: 4054–4057.

85 Li, L.-L., Tao, Z.-L., Han, Z.-Y., and Gong, L.-Z. (2017). *Org. Lett.* 19: 102–105.

86 Villa-Marcos, B. and Xiao, J. (2015). *Chin. J. Catal.* 36: 106–112.

87 Agbossou, F., Carpentier, J.-F., and Mortreux, A. (1995). *Chem. Rev.* 95: 2485–2506.

88 Hebrard, F. and Kalck, P. (2009). *Chem. Rev.* 109: 4272–4282.

89 Franke, R., Selent, D., and Börner, A. (2012). *Chem. Rev.* 112: 5675–5732.

90 Eilbracht, P., Bärfacker, L., Buss, C. et al. (1999). *Chem. Rev.* 99: 3329–3366.

91 Abillard, O. and Breit, B. (2007). *Adv. Synth. Catal.* 349: 1891–1895.

92 Ahmed, M., Buch, C., Routaboul, L. et al. (2007). *Chem. Eur. J.* 13: 1594–1601.

93 Chercheja, S. and Eilbracht, P. (2007). *Adv. Synth. Catal.* 349: 1897–1905.

94 Meng, J., Fan, L.-F., Han, Z.-Y., and Gong, L.-Z. (2018). *Chem* 4: 1047–1058.

95 Meng, J., Li, X.-H., and Han, Z.-Y. (2017). *Org. Lett.* 19: 1076–1079.

96 Jiang, G.X. and List, B. (2011). *Angew. Chem. Int. Ed.* 50: 9471–9474.

97 Li, L.-L., Su, Y.-L., Han, Z.-Y., and Gong, L.-Z. (2018). *Chem. Eur. J.* 24: 7626–7630.

98 Wang, P.-S., Lin, H.-C., Zhai, Y.-J. et al. (2014). *Angew. Chem. Int. Ed.* 53: 12218–12221.

99 Ghosh, A.K. and Li, J. (2011). *Org. Lett.* 13: 66–69.

100 Wipf, P. and Reeves, J.T. (2002). *Chem. Commun.*: 2066–2067.

101 Duhamel, L., Duhamel, P., and Plaquevent, J.-C. (2004). *Tetrahedron: Asymmetry* 15: 3653–3691.

102 Cheon, C.H. and Yamamoto, H. (2008). *J. Am. Chem. Soc.* 130: 9246–9247.

103 Guin, J., Varseev, G., and List, B. (2013). *J. Am. Chem. Soc.* 135: 2100–2103.

104 Ren, Y.-Y., Zhu, S.-F., and Zhou, Q.-L. (2018). *Org. Biomol. Chem.* 16: 3087–3094.

105 Liang, Y., Zhou, H., and Yu, Z.-X. (2009). *J. Am. Chem. Soc.* 131: 17783–17785.

106 Xu, B., Zhu, S.-F., Xie, X.-L. et al. (2011). *Angew. Chem. Int. Ed.* 50: 11483–11486.

107 Xu, B., Zhu, S.-F., Zuo, X.-D. et al. (2014). *Angew. Chem. Int. Ed.* 53: 3913–3916.

108 Xu, B., Zhu, S.-F., Zhang, Z.-C. et al. (2014). *Chem. Sci.* 5: 1442–1448.

109 Zhang, Y., Yao, Y., He, L. et al. (2017). *Adv. Synth. Catal.* 359: 2754–2761.

110 Guo, J.-X., Zhou, T., Xu, B. et al. (2016). *Chem. Sci.* 7: 1104–1108.

111 Ni, Y., Guo, X., Hu, W., and Liu, S. (2014). *Chin. J. Org. Chem.* 34: 107–111.

112 Li, M.-L., Yu, J.-H., Li, Y.-H. et al. (2019). *Science* 366: 990–994.

113 Xu, H., Zuend, S.J., Woll, M.G. et al. (2010). *Science* 327: 986–990.

114 Bromberg, S.E. (1997). *Science* 278: 260–263.

115 Sigman, M.S. and Jacobsen, E.N. (1998). *J. Am. Chem. Soc.* 120: 4901–4902.

116 Tan, F., Liu, X., Hao, X. et al. (2016). *ACS Catal.* 6: 6930–6934.

117 Xu, B., Li, M.-L., Zuo, X.-D. et al. (2015). *J. Am. Chem. Soc.* 137: 8700–8703.

118 Zhao, F., Li, N., Zhang, T. et al. (2017). *Angew. Chem. Int. Ed.* 56: 3247–3251.

119 Terada, M. and Toda, Y. (2012). *Angew. Chem. Int. Ed.* 51: 2093–2097.

120 Ren, L., Lei, T., Ye, J.-X., and Gong, L.-Z. (2012). *Angew. Chem. Int. Ed.* 51: 771–774.

121 Marco-Contelles, J., Pérez-Mayoral, E., Samadi, A. et al. (2009). *Chem. Rev.* 109: 2652–2671.

122 Wang, G.-P., Chen, M.-Q., Zhu, S.-F., and Zhou, Q.-L. (2017). *Chem. Sci.* 8: 7197–7202.

123 Yu, P., Lin, J.-S., Li, L. et al. (2014). *Angew. Chem. Int. Ed.* 53: 11890–11894.

124 Lin, J.-S., Li, T.-T., Liu, J.-R. et al. (2019). *J. Am. Chem. Soc.* 141: 1074–1083.

125 Yu, Q., Fu, Y., Huang, J. et al. (2019). *ACS Catal.* 9: 7285–7291.

7

Lewis Base–Lewis Acid Cooperative Catalysis

7.1 Introduction: Combined Lewis Base and Lewis Acid Activations

Lewis base (LB) is defined as a compound or ionic species, which can donate an electron pair to an acceptor compound [1]. Lewis base catalysis represents the process of accelerating a reaction by interacting an electron-pair donor of the Lewis base catalyst with an acceptor atom in one of the reagents or substrates [2]. The interaction between the nonbonding electron pair of a Lewis base catalyst and the anti-bonding π orbitals (π^*) of an acceptor presented in carbonyls, electronically deficient olefins, alkynes, and many other unsaturated functional groups, represent the most commonly recognized form of Lewis base catalysis (Scheme 7.1) [2, 3]. This n–π^* interaction can generate reactive intermediates with either enhanced nucleophilicity by raising the HOMO (the highest occupied molecular orbital) (Scheme 7.1a), or electrophilicity by lowering LUMO (the lowest occupied molecular orbital) (Scheme 7.1b) to allow catalyzing synthetic transformations. The Lewis base catalysis via the n–π^* interaction concept dates back to the discovery of the accelerating effect of pyridine on the acylation of alcohols one hundred years ago [4], and the first chiral Lewis base catalyzed asymmetric acylation was reported by Wegler in 1932 [5]. Since then, the rich chemistry that occurs with both nucleophiles and electrophiles covalently bonding to the chiral Lewis bases opens up a new field of asymmetric organocatalysis.

Chiral nucleophilic amines [6] and N-heterocyclic carbenes (NHCs) [7] are among the most common Lewis base catalysts frequently applied to drive asymmetric reactions (Figure 7.1a). The earliest example of nucleophilic chiral amine catalysis was demonstrated by Pracejus and coworkers in the 1960s [8a, b]. Since then, cinchona alkaloids have gradually been the versatile concept in asymmetric catalysis and, in particular, found widespread applications for the creation of new transformations in recent years [8]. The "planar-chiral" 4-dimethylaminopyridine (DMAP) Lewis bases introduced by Fu and coworkers have enabled a diverse array of enantioselective processes [9]. Chiral isothioureas (ITUs), another important family of Lewis base organocatalysts, have become visible since the seminal work by Birman and coworkers in 2006 [10a], providing efficient access to a wide array of heterocycles and

Asymmetric Organo-Metal Catalysis: Concepts, Principles, and Applications, First Edition. Liu-Zhu Gong.
© 2022 WILEY-VCH GmbH. Published 2022 by WILEY-VCH GmbH.

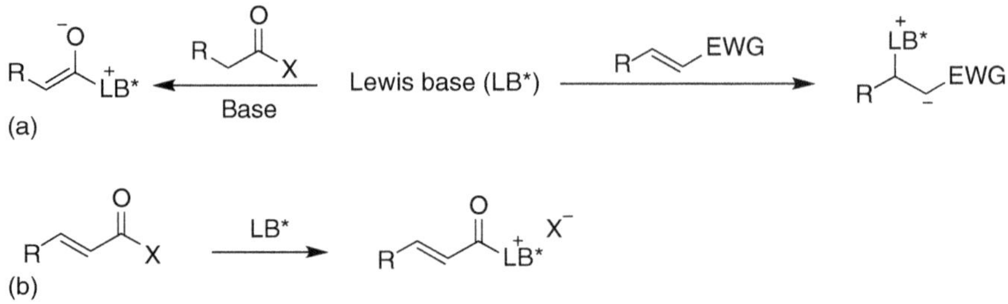

Scheme 7.1 Representative examples of substrates activation via n–π* interaction. (a) Activation of nucleophiles by n→π* interaction. (b) activation of electrophiles by n→π* interaction. EWG, electron withdrawing group. Sources: Based on Denmark and Beutner [2a]; List [2b]; Vedejs and Denmark [2c]; Candish et al. [3].

molecules [10]. Several types of active intermediates are currently described in the chiral tertiary amine nucleophilic catalysis [11], including acylammonium species, ammonium enolates, and α,β-unsaturated acylammonium intermediates, that are capable of undergoing a wide range of transformations owing to their ready accessibility and divergent reactivity (Figure 7.1b). In the field of NHC-catalysis, the pioneering work was introduced by Ukai [12a] and Breslow [12b] on the benzoin reaction. Since these seminal findings, NHCs have been gradually recognized as one of the most powerful and effective types of Lewis base catalysts. The principally diverse activation intermediates [13], such as Breslow intermediates, homoenolate intermediates, azolium enolates, azolium dienolates, acyl azoliums, and α,β-unsaturated acylazoliums, catalytically generated from n–π* interaction of NHCs with carbonyl functionalities, open a large space to create new reactions (Figure 7.1b).

Notwithstanding the great success being achieved in the field of asymmetric Lewis base catalysis, in some cases, the use of Lewis base catalyst alone is insufficient to drive a reaction or to control the selectivity. Thus, the introduction of an additional activation mode seems necessary to address these unsolved issues in the Lewis base catalysis. Lewis acids (LAs), which act as electron pair-acceptors, are capable of enhancing the electrophilicity of substrates, stabilizing the nucleophilic intermediates, and providing coordinative organization of the transition state (Scheme 7.2a) [14]. Over the past decades, synergistic activation of electrophiles (E) and nucleophiles (Nu) via the combined use of Lewis base (LB) and Lewis acid (LA) has become a versatile strategy in the area of asymmetric catalysis (Scheme 7.2b) [15]. In this cooperative catalysis, Lewis acid catalysts are able to activate the electrophiles by lowering LUMO (**I**, **V**, and **VI** Scheme 7.2c), stabilize the nucleophilic intermediates formed in the catalytic cycles (**II**, Scheme 7.2c), or participate in the simultaneous operation of both alternatives and organize the transition state (**III** and **VI**, Scheme 7.2c). The Lewis base–Lewis acid cooperative catalysis has gradually become a robust strategy to enhance the reactivity, to improve the stereoselectivity, and even to change the reaction pathway. The synergistic activation of nucleophiles and electrophiles allows for a more diverse range of bond-breaking and forming

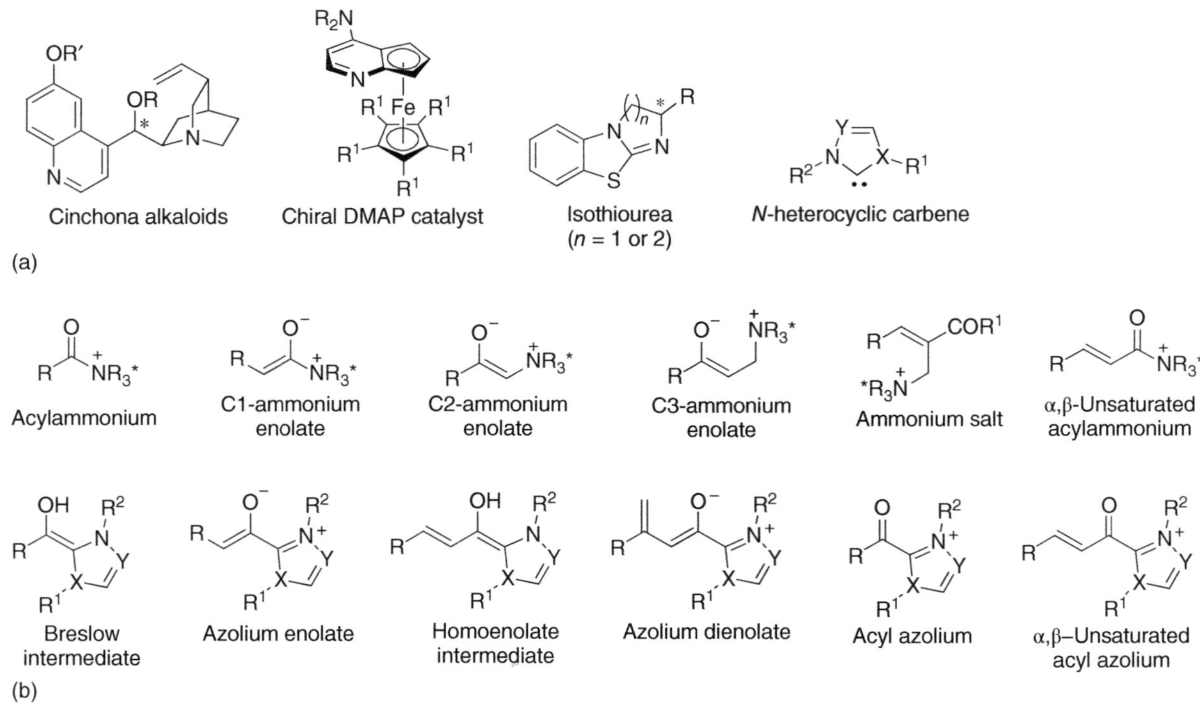

Figure 7.1 Representative chiral Lewis bases and activation intermediates. (a) Representative chiral Lewis base catalysts. (b) Representative activation intermediates.

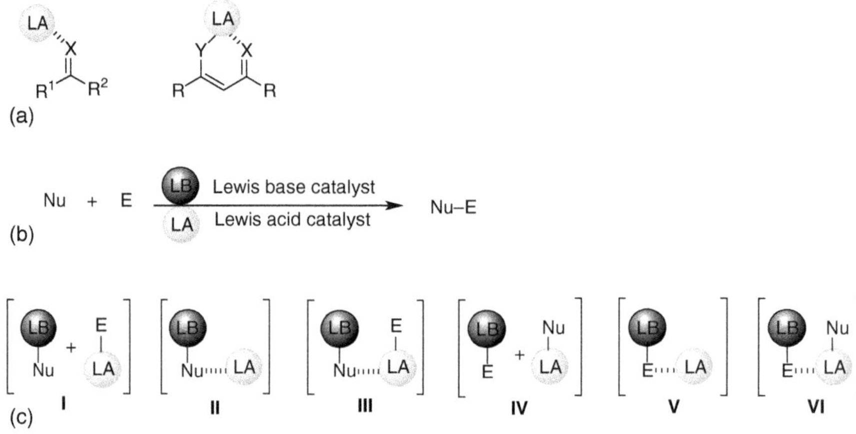

Scheme 7.2 Lewis base–Lewis acid cooperative catalysis. (a) Lewis acid activation modes. Source: Based on Yamamoto [14]. (b) Lewis base–Lewis acid cooperative catalysis. (c) Representative activation modes. Sources: Based on Paull et al. [15a]; Cohen and Scheidt [15b]; Stegbauer et al. [15c]; Wang and Scheidt [15d]; Jia et al. [15e]; Nagao and Ohmiya [15f].

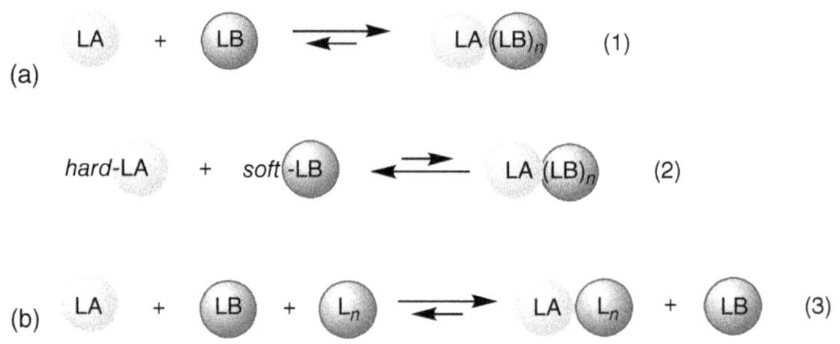

Scheme 7.3 Catalyst "self-quenching" interactions and solutions. (a) Catalyst "self-quenching". Source: Based on Paull et al. [15a]. (b) solutions.

events and thereby leads to larger space for the creation of a broader scope of enantioselective reactions.

Theoretically, due to the nature of the Lewis base and Lewis acid catalysts, the catalyst self-quenching is the most concerned and inevitable problem in the Lewis base–Lewis acid cooperative catalytic systems [15a]. The Lewis base–Lewis acid interaction usually attenuates the nucleophilicity of the Lewis base and the electrophilicity of the Lewis acid, or even leads to the formation of a Lewis base–Lewis acid complex ($LA(LB)_n$) (Eq. (1), Scheme 7.3a), accompanying deceleration or even inhabitation of reactions. Over the past decades, continuous efforts have been directed toward the creation of alternative solutions via fine-tuning of reaction conditions and evaluating compatible Lewis acid and base catalyst pairs based on fundamental theory [16], such as the use of a hard metal ion with a suitable soft base (Eq. (2), Scheme 7.3b), or the addition of ligands (L_n) to refrain the catalysts from or weaken their "self-quenching" (Eq. (3), Scheme 7.3b).

7.1.1 Early Examples in Lewis Base–Lewis Acid Cooperative Catalysis

An early example in the Lewis base–Lewis acid cooperative catalysis was introduced by Aggarwal et al. in 1996. Aggarwal devoted his effort to improve the efficiency of 1,4-diaza[2.2.2]bicyclooctane (DABCO)-catalyzed Baylis–Hillman reaction by using Lewis acid as co-catalyst (Scheme 7.4) [17]. The "hard Lewis acids," lanthanides, and group III metal triflates, were all proven to be eligible co-catalysts combined with DABCO for the reaction acceleration (Eq. (1), Scheme 7.4). The Lewis acid catalyst is generally believed to increase the reaction rate by either activating electronically deficient reaction partners to facilitate the nucleophilic addition of tertiary amine catalysts or stabilizing the generated enolate intermediates, or both (see transition state [**TS-1**] in Scheme 7.4) [17b]. No reaction occurs in the presence of 5 mol% $La(OTf)_3$ until 10 mol% of DABCO is added, possibly because all the DABCO associates with the Lewis acid to give metal complexes (Eq. (2), Scheme 7.4). Further reaction rate acceleration can be obtained upon the addition of diol ligands (such as binol) to displace the DABCO ligand (Eq. (3), Scheme 7.4).

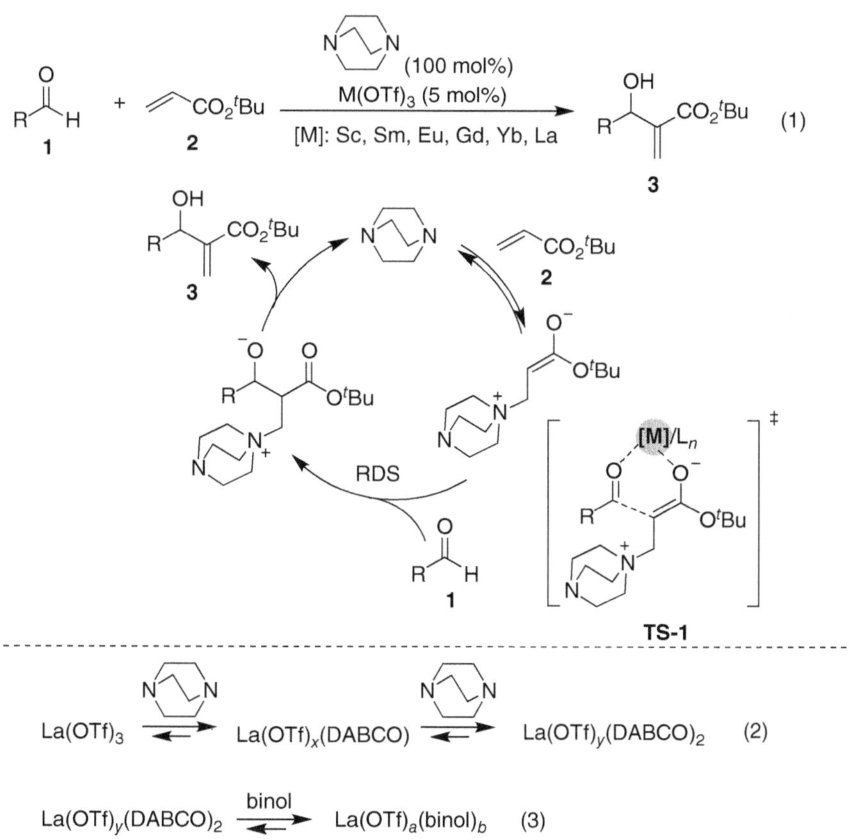

Scheme 7.4 Lewis acid-accelerated catalysis of the Baylis–Hillman reaction.

Barrett et al. found that the addition of a stoichiometric amount of a Lewis acid co-catalyst was able to improve the stereoselectivity of the chiral pyrrolizidine base-catalyzed asymmetric Baylis–Hillman reactions [18]. Lectka and coworkers reported the first cinchona alkaloid/Lewis acid cooperatively catalyzed enantioselective Staudinger-type [2+2] cycloaddition to access β-lactams. The presence of 10 mol% indium(III) triflate significantly enhances the chemical yields [19]. Shi and coworker also indicated that the addition of catalytic amounts of lithium salt was also able to speed up the cinchona alkaloid derivative-catalyzed Baylis–Hillman reaction [20]. The success of these early studies provides the possibility and feasibility to integrate Lewis bases and Lewis acids in cooperative catalysis.

7.2 Asymmetric Reactions Driven by Tertiary Amine-Mediated Ammonium Enolates

C1- and C3-ammonium enolates (**Int-1** and **Int-2**), catalytically generated from reactions with electrophiles such as ketenes or activated carboxylates (Eq. (1), Scheme 7.5a) and α,β-unsaturated carbonyl compounds, respectively (Eq. (2), Scheme 7.5a), have found widespread applications in asymmetric reactions via the HOMO-raising activation [11a, c, e]. In the field of Lewis base–Lewis acid cooperative catalysis involving C1 and C3 ammonium enolates, the Lewis acid catalyst can enhance the electrophilicity of electrophiles (**I**), or stabilize ammonium enolates (**II**), or both through a metal-organized transition-state (**III**) (Scheme 7.5b). It is worth mentioning that the integration of hard metal ions with tertiary amine catalysts (Eq. (2), Scheme 7.3b), or addition of competitive ligands (Eq. (3), Scheme 7.3b) is principally able to address the catalyst self-quenching issue (Eq. (1), Scheme 7.3a) [15a]. The combination of tertiary amine Lewis basic catalysts (including cinchona alkaloids, pyridine-derived nucleophiles, and ITUs) with different Lewis acid catalysts, including lithium, copper, indium, and scandium salts, will be described in this section.

7.2.1 Asymmetric Baylis–Hillman Reactions

Barrett et al. developed a chiral pyrrolizidine-base catalyzed asymmetric Baylis–Hillman reaction of ethyl vinyl ketone and electron-deficient aromatic aldehydes (Scheme 7.6) [18]. The addition of stoichiometric amounts of alkali metal salts, in particular, sodium tetrafluoroborate, remarkably improves the enantioselectivity. The Baylis–Hillman products **6** were obtained in moderate to good yield (up to 93%) and with moderate enantioselectivity (up to 72% ee). The transition state (**TS-2**) proposed by Barrett has documented the key role that the sodium ion plays in the substrate activation process by coordination with hydroxy of the organocatalyst **5** and the aldehyde **1**.

Shi and coworker described a lithium-assisted asymmetric Baylis–Hillman reaction catalyzed by quinidine-derived chiral amines (Scheme 7.7) [20]. The addition of a lithium salt to some degree exerts an impact on the chiral induction rather than

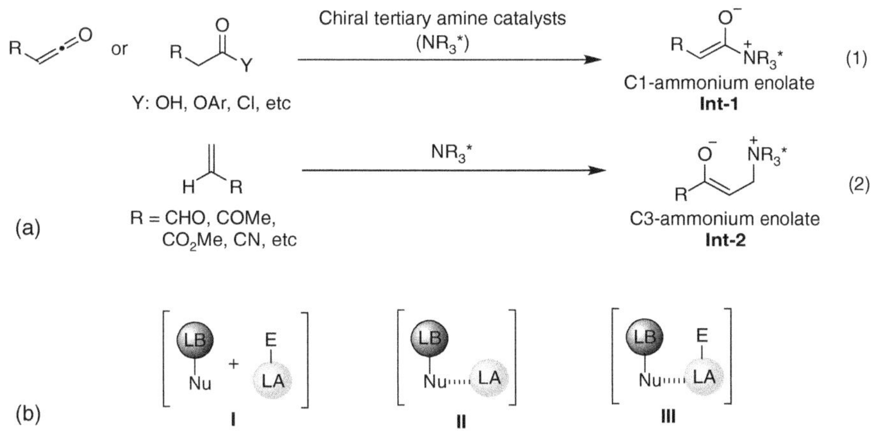

Scheme 7.5 Lewis base–Lewis acid cooperative catalysis involving ammonium enolates. (a) Generation of C1- and C3-ammonium enolates. (b) Representative activation modes. Source: Based on Aggarwal et al. [17b].

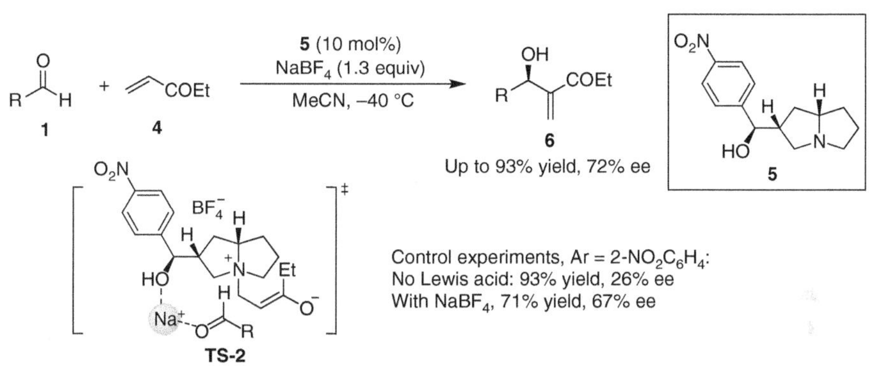

Scheme 7.6 Sodium-assisted asymmetric Baylis–Hillman reactions. Source: Based on Barrett et al. [18].

the reaction rate. In the reaction with *p*-bromobenzaldehyde **1a**, the enantioselectivity of **9** can be enhanced to 49% ee by the addition of 20 mol% lithium perchlorate. The Li^+ acts as a Lewis acid to bridge the carbonyl group of the aldehyde and the phenolic hydroxy group of the catalyst as shown in the **TS-3**, or stabilizes the enolate intermediate formed in the final Baylis–Hillman reaction step as shown in the **TS-3'**. It is believed that the lithium cation renders a more rigid transition state for the chiral induction and thus enhances the enantioselectivity.

In 2009, Connell and coworker reported a MgI_2 and DMAP cooperatively catalyzed Baylis–Hillman reaction [21a]. One year later, the same group utilized Fu's "planar-chiral" DMAP catalyst **11** in concert with MgI_2 to catalyze an enantioselective Baylis–Hillman reaction of cyclopentenone **10** with aromatic and aliphatic aldehydes **1** [21b] (Scheme 7.8). The Morita–Baylis–Hillman (MBH) product **12** was

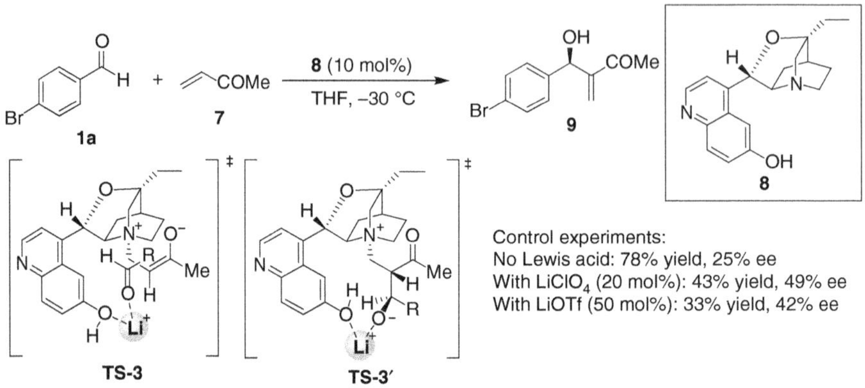

Scheme 7.7 Lithium-assisted asymmetric Baylis–Hillman reactions. Source: Based on Shi and Jiang [20].

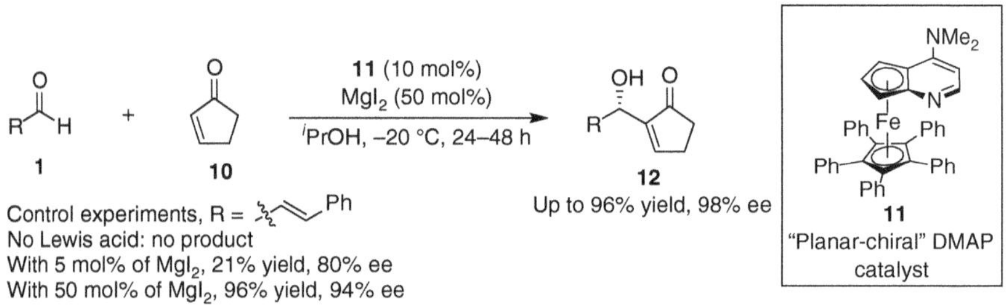

Scheme 7.8 Magnesium-assisted asymmetric Baylis–Hillman reactions.

obtained in good to excellent yields (up to 96%) and with moderate to excellent enantioslectivities (up to 98% ee). The use of MgI_2 as the Lewis acid co-catalyst leads to a significant increase in the reaction rate. Under the standard reaction conditions, no reaction was observed between *trans*-cinnamaldehyde and cyclopentenone in the absence of MgI_2. Variation of the amounts of MgI_2 from 5 to 50 mol% dramatically elevates both the yield (from 21% to 96%) and the enantioselectivity (from 80 to 94% ee).

7.2.2 Asymmetric [2+2] Reactions

In 2002, Lectka and coworkers investigated a cinchona alkaloid/$In(OTf)_3$ collaboratively catalyzed enantioselective [2+2] cycloaddition between substituted acetyl chlorides **13** and imino esters **14** for the synthesis of β-lactams (Scheme 7.9) [19]. The evaluation of the co-catalysts reveals that other metals, including $Rh(PPh_3)_3OTf$ and $Cu(PPh_3)_2ClO_4$, are able to vary the reaction performance, mainly due to the coordination between the metal center and the bridgehead nitrogen atom of quinuclidine core in benzoylquinine (BQ, **15**) (proven by ultraviolet–visible [UV–vis] spectroscopy) [22], and this "self-quenching" interaction deactivates the catalytic activity

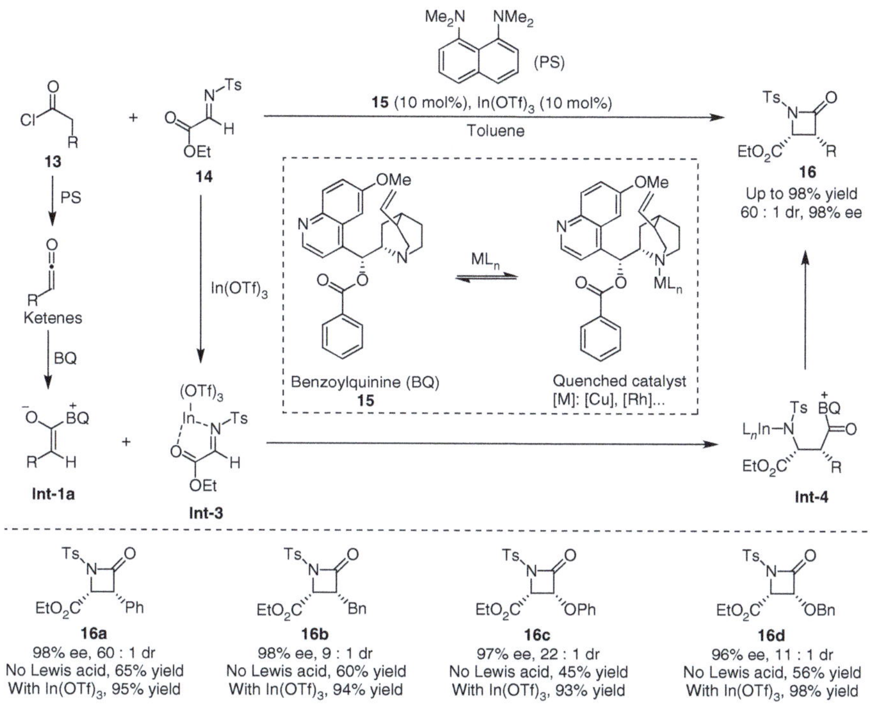

Scheme 7.9 Cinchona alkaloid/In(OTf)$_3$ catalyzed asymmetric [2+2] cycloaddition. Source: France et al. [19].

of the nucleophilic organocatalyst. In sharp contrast, the use of the harder, less azaphilic Lewis acid (In(OTf)$_3$) can minimize the catalyst quenching. The yield of the annulation product **16** was greatly improved by a maximum 98% by using In(OTf)$_3$as the Lewis acid co-catalyst. Mechanistic studies suggest that the metal chelation to the imino ester **14** is the most probable scenario for the enhancement of reactivity. The nonmetal-coordinated zwitterionic enolate **Int-1a** generated from the acetyl chloride **13** with BQ and base undergoes Mannich-type reaction with the In(III) bound imino ester **Int-3** to yield an intermediate **Int-4**. Finally, an intramolecular amide bond formation then delivers β-lactams **16** and regenerates the catalyst.

The cinchona alkaloid-catalyzed [2+2] cycloaddition of aldehydes with ketenes, first reported by Wynberg, represents one of the most efficient methods to access β-lactones [23a]. However, only highly electron-deficient aldehydes (chloral) are amenable. In 2004, Nelson and coworkers found that in the presence of a lithium salt, the Lewis base catalyst *O*-trimethylsilylquinine (TMSQ, **17**) allows the enantioselective [2+2] cycloaddition reaction to tolerate an even broader scope of substrates, leading to a wide range of *cis*-β-lactones **18** in good yields and with high levels of stereoselectivity (Scheme 7.10) [23b]. However, excess amounts of $LiClO_4$ (up to 300 mol%) are required for enabling a smooth reaction involving less reactive and bulky aldehydes. In terms of mechanism, the lithium-stabilized ammonium enolate **Int-5** is the key intermediate and undergoes the asymmetric aldol reaction via a metal-constrained Zimmerman–Traxler transition state (**TS-4**), wherein the aldehyde is activated by the coordination with the lithium cation, providing both enthalpic and entropic activation to the ensuring enolate-aldehyde addition event.

Lanthanide Lewis acid/quinidine cooperatively catalyzed asymmetric [2+2] cycloaddition reactions were subsequently reported (Scheme 7.11) [24]. The Lewis acid shows an obvious synergistic effect on the catalytic process with the chiral Lewis base. The presence of the Lewis acid is able to significantly change the reaction performance. The *O*-trimethylsilylquinidine (TMSQD, **20**) itself affords a sluggish [2+2] cycloaddion of α-phenoxyacetyl chloride with highly reactive 4-nitrobenzaldehyde with poor diastereoselectivity (51 : 49 dr), whereas the addition of $Er(OTf)_3$ dramatically elevates the conversion to preferentially give *cis*-β-lactone in improved diastereoselectivity (92 : 8 dr) (Scheme 7.11a) [24a]. Significantly, a broader scope of aromatic aldehydes **1** can participate in the reaction to result in excellent enantioselectivities of up to >99% ee. In this case, two molecules of $Er(OTf)_3$ are presumably bonded to the aldehyde and the ammonium enolate, respectively, and the aldol-type reaction occurs via a gauche and open transition state **TS-5**, leading to *cis*-β-lactone **21**. Aliphatic acid chlorides are even less reactive and are unable to undergo the cycloaddition catalyzed by the Lewis base **20**, alone. In contrast, in collaboration with 15 mol% of $Sc(OTf)_3$, the asymmetric cycloaddition proceeds cleanly to favorably give *trans*-β-lectones **22** in high yields and excellent stereoselectivity (Scheme 7.11b) [24a]. The trans-selectivity probably arises from the aldol reaction occurring via an open transition-state **TS-6**, in which two molecules of $Sc(OTf)_3$ also bond to the aldehyde and the ammonium enolate. Under the combined catalysis of a silyl cinchona alkaloid **20** and $Yb(N(TMS)_2)_3$, the asymmetric Staudinger-type [2+2] cycloaddition of phenoxyacetyl chloride **19a** and aryl imines **23** undergoes smoothly to afford α-phenoxy-β-aryl-β-lactams **24** in excellent stereoselectivity (Scheme 7.11c) [24b]. Different from the dual activation proposed in the precedents [24a], in this case, the Lewis acid catalyst does not coordinate to the imine functionality but works as a counter cation to stabilize the ammonium enolate (**TS-7**).

In the presence of a base, the sulfonyl chloride **25** undergoes an elimination to generate a sulfene, which can be captured by a chiral Lewis base (NR_3*) to give a zwitterionic intermediate **Int-6** (Scheme 7.12a). Principally, this intermediate **Int-6** has similar reactivity to the ketene and probably undergoes the [2+2] cycloaddition with either aldehydes or imines. The first attempt by Borrmann and Wegler to establish the tertiary amine catalyzed [2+2] cycloaddition of the sulfonyl chloride with highly electron-deficient aldehydes (chloral) indicated that the reaction is indeed workable catalyzed by a Lewis base [26]. In 2007, Peters and coworker described the first application of sulfene intermediates in asymmetric catalysis (Scheme 7.12b) [27]. However, the use of a sterically hindered cinchona alkaloid as the catalyst leads to much diminished conversion. Notably, the addition of a proper Lewis acid as a co-catalyst significantly speeds up the reaction and is beneficial to the stereochemical control. Based on the cooperative catalysis of a chiral nucleophilic tertiary amine $(DHQ)_2PYR$ (diydroquinine 2,5-diphenyl-4,6-pyrimidinediyl diether, **26**) and $In(OTf)_3$ or $Bi(OTf)_3$, Peters and coworkers developed a highly

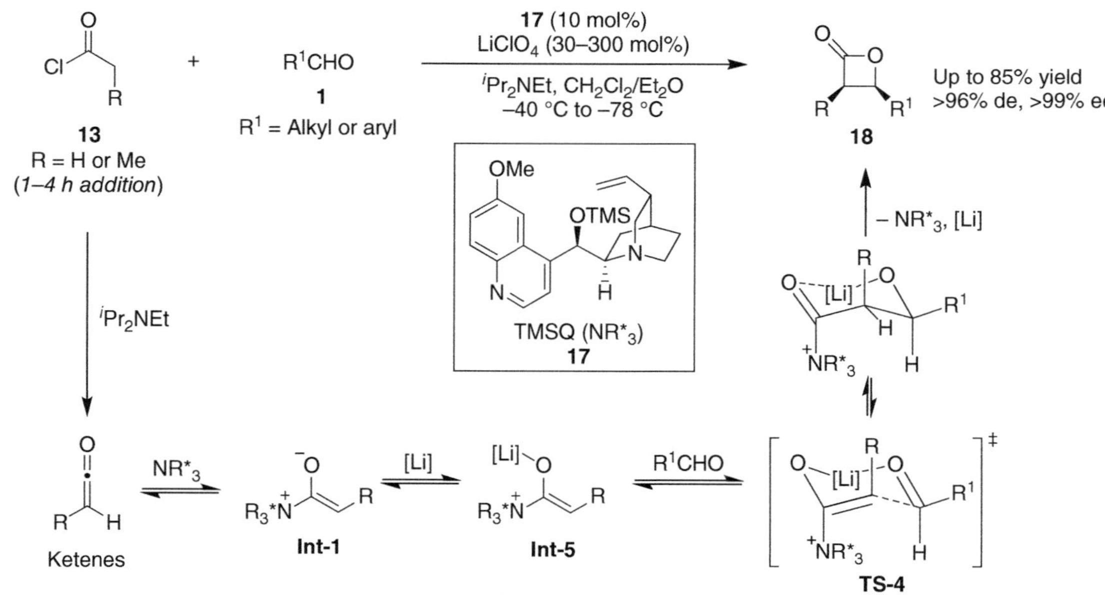

Scheme 7.10 Cinchona alkaloid/lithium-catalyzed asymmetric [2+2] cycloaddition. Source: Zhu et al. [23b].

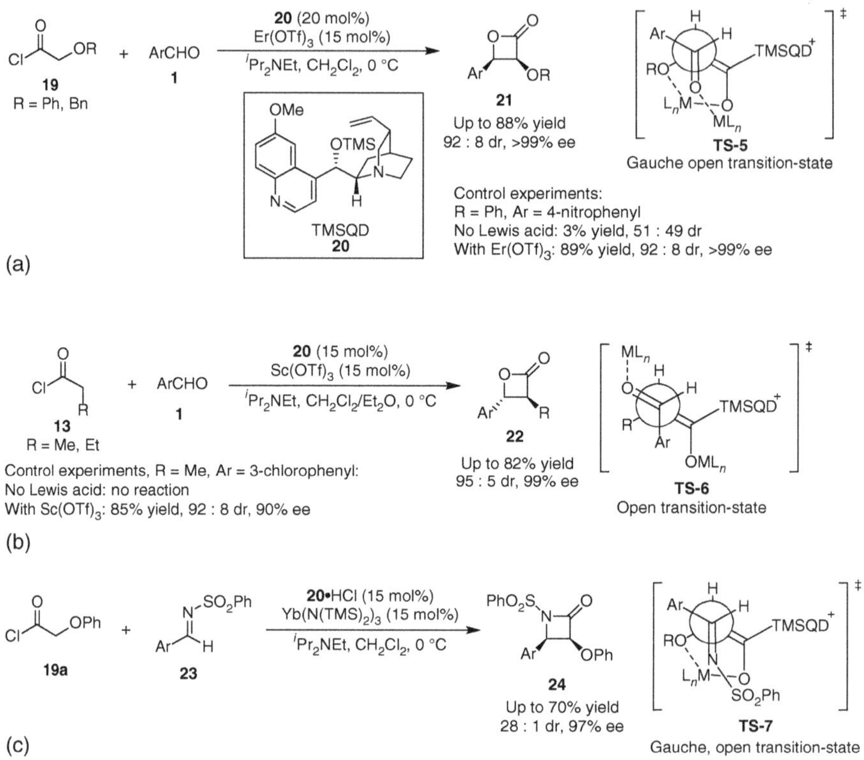

Scheme 7.11 Asymmetric [2+2] cycloaddition reactions cooperatively catalyzed by cinchona alkaloid and lanthanide Lewis acid. (a) $Er(OTf)_3$/TMSQD cooperative catalysis. (b) $Sc(OTf)_3$/TMSQD cooperative catalysis. Source: Based on Calter et al. [24a]. (c) $Yb(N(TMS)_2)_3$ cooperative catalysis. Source: Huang and Calter [24b].

enantioselective [2+2] cycloaddition of alkyl sulfonyl chlorides and aldehydes, providing 3,4-disubstituted β-sultones **27** in high yields (up to 87%) and with excellent stereoselectivities (up to >100 : 1 dr and >99% ee). Deuteration experiments suggest that the Lewis acid co-catalyst does not influence the generation of sulfene intermediates. Presumably, the activation of the aldehyde by coordination of the metal triflate co-catalyst with the carbonyl group and/or stabilization of the sensitive sulfene–amine adduct leads to a well-defined transition state **TS-8**, thereby providing the enhanced enantioselectivity.

The asymmetric [2+2] cycloaddition of imines and alkyl sulfonyl chlorides for the synthesis of enantioenriched β-sultams smoothly occurs under the catalysis of the quinine derivatives, but the substrate scope is limited to the highly electrophilic imines, such as chloral derived *N*-tosyl imine and *N*-tosyl imino ester [28]. The Lewis acid and chiral Lewis base cooperative catalysis enable the asymmetric [2+2] cycloaddition of 2-pyridylsulfonyl imines **28** with sulfonyl chlorides **25** to accommodate a broader scope of substrates (Scheme 7.13). Principally, the 2-pyridylsulfonyl group offers the additional activation by bidentate coordination

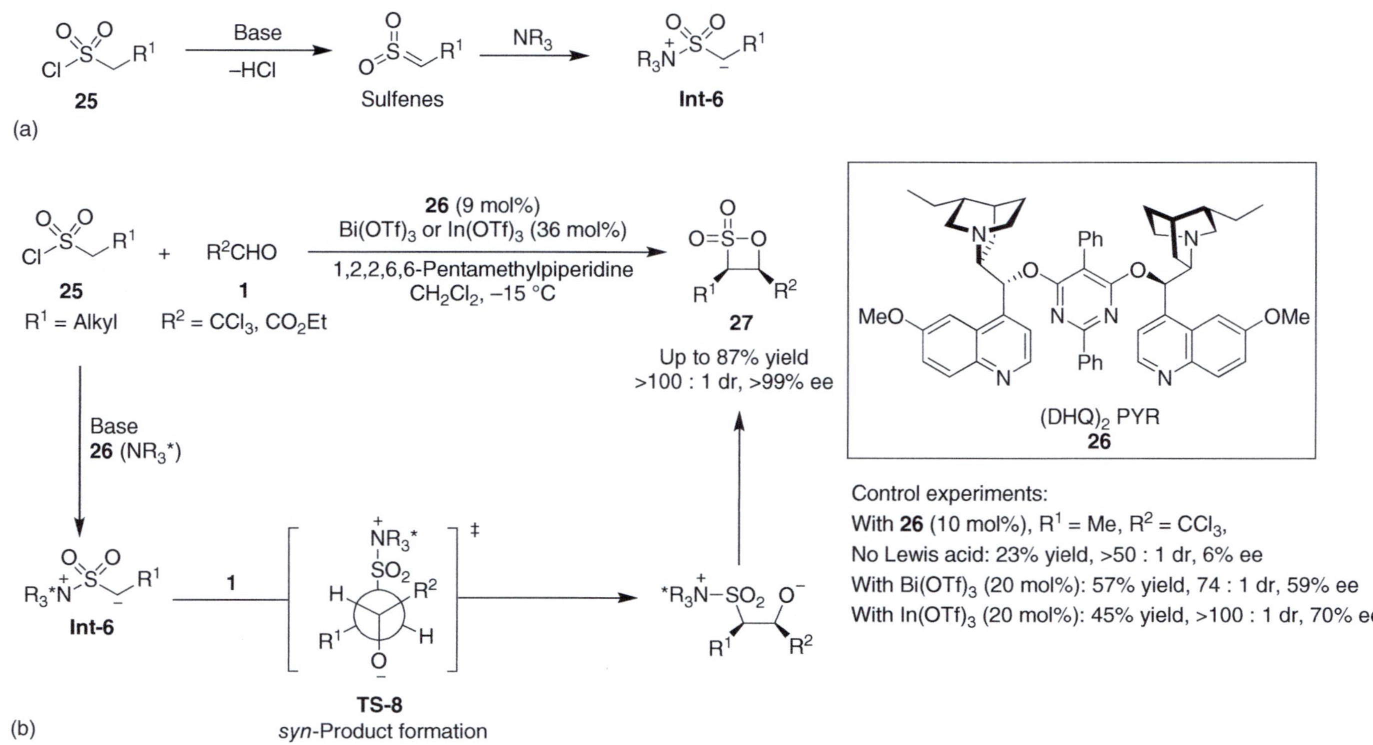

Scheme 7.12 Asymmetric [2+2] cycloaddition of alkyl sulfonyl chlorides and aldehydes. (a) Generation of a nucleophilic zwitterion **Int-6**. (b) Stereoselective formation of β-sultones. Source: Based on Murauski et al. [25].

to a Lewis acid co-catalyst. Thus, the presence of $Yb(OTf)_3$ allows the cinchona alkaloid-catalyzed [2+2] cycloaddition to tolerate a broad scope of 2-pyridylsulfonyl arylaldimines **28**. Deuteration experiments indicate that a zwitterionic aminal **Int-7** turns out to be a key intermediate, and no sulfene formation is involved at the reaction temperature (−80 °C). In the proposed transition state **TS-9**, the Lewis acid co-catalyst activates the zwitterionic aminal **Int-7** through a five-membered chelate ring. The coordination of the sulfonyl moiety in **25** with the ytterbium(III) is likely and would facilitate the N-sulfonylation step.

The chiral ITU and its structural analogues, have recently been recognized as versatile Lewis base organocatalysts that hold a unique capacity for the development of asymmetric transformations [10, 11]. The cooperative catalysis of chiral ITU and Lewis acid was first introduced by Romo and coworkers in 2010 [29]. The chiral ITU and LiCl cooperatively catalyzed enantioselective intramolecular aldol lactonization of keto acids **31** provides bi- and tricyclic β-lactones **33** in good to high yields (up to 93%) and with high enantioselectivities (up to 98% ee, single diastereomer) (Scheme 7.14a). The addition of stoichiomeric amounts of LiCl enables the desired products **33** to be furnished in higher yields than those without LiCl, while in some cases with slightly dropped enantiopurity. The control reaction in the absence of ITU catalyst confirms that the LiCl promotes a non-stereoselective pathway to yield a racemic product. Moreover, as shown in the proposed transition state **TS-11**, the addition of LiCl inhibits the $n_O \rightarrow \sigma^*_{C\text{–}S}$ non-bonding interaction [31] between the enolate and the ITU sulfur atom (**TS-10**) and thus is able to affect the stereocontrol. The expansion of the ITU and LiCl cooperative catalysis to an intramolecular [2+2] cycloaddition of **34** successfully gives rise to a β-lactone **35** in a much higher yield and enantioselectivity than the protocol without LiCl under otherwise identical conditions (Scheme 7.14b) [30].

7.2.3 Asymmetric [4+2] Reactions

The cinchona alkaloid/Lewis acid catalysis has been intensively applied to drive asymmetric hetero-Diels–Alder (HDA) reactions [32]. The asymmetric [4+2] cycloaddition of *o*-benzoquinone diimides **36** and acid chlorides **13** proceeds cleanly under the cooperative catalysis of $Zn(OTf)_2$ and benzoylquinidine, to generate quinoxalinones **38** in almost perfect enantioselectivity (Scheme 7.15a) [32a]. The Lewis acid $Zn(OTf)_2$ coordinates to the quinone diimide **36** to form a more electrophilic complex **Int-8**, importantly, does not show detriment to the activity of the nucleophilic catalyst or the nucleophilic enolate **Int-1** that normally happens with self-quenching, thereby speeds up the reaction and enhances the chemical yield. A similar catalytic concept successfully enables a [4+2] cycloaddition of *o*-benzoquinone imides **40** and acid chlorides **13** to produce optically active 1,4-benzoxazinones **41** (Scheme 7.15b) [32b]. The addition of $Sc(OTf)_3$ obviously improves the catalytic efficiency by activating the quinone imide **40** and thus elevates the chemical yields while preserving the remarkable enantioselectivity.

trans-$(Ph_3P)_2PdCl_2$ was revealed to improve the performance of *O*-benzoylquinidine-catalyzed hetero Diels–Alder reaction of *o*-chloranil **43** and

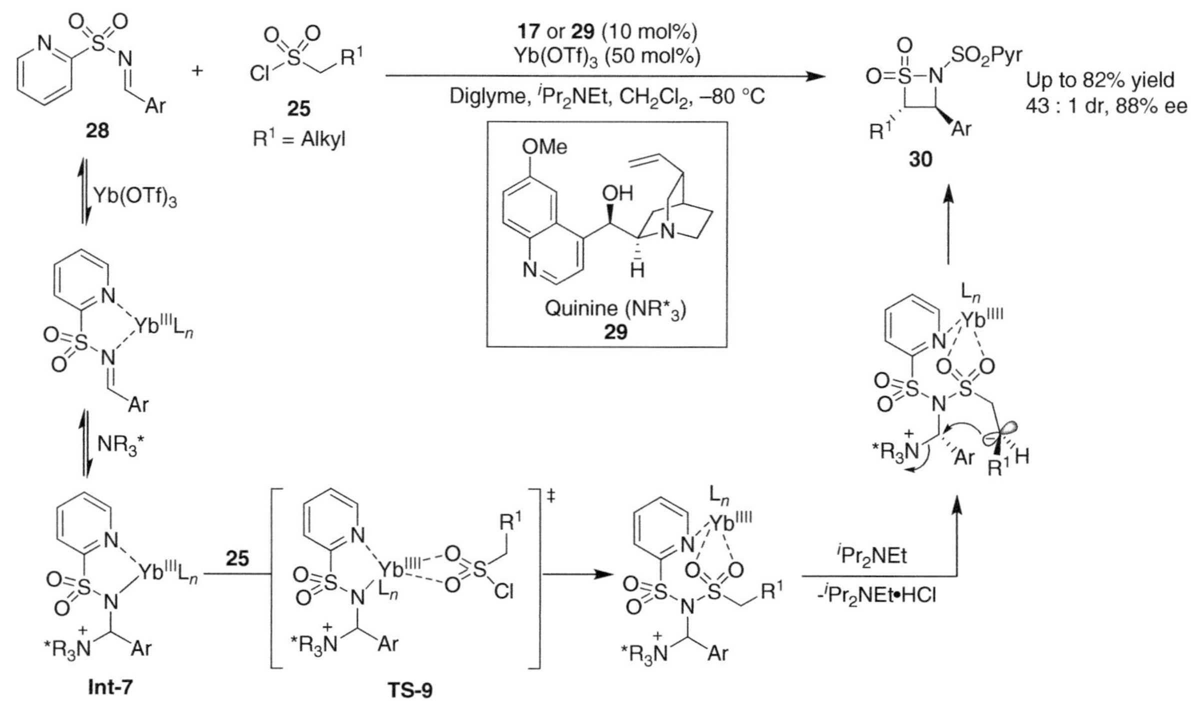

Scheme 7.13 Asymmetric [2+2] cycloaddition reactions of alkyl sulfonyl chlorides and imines.

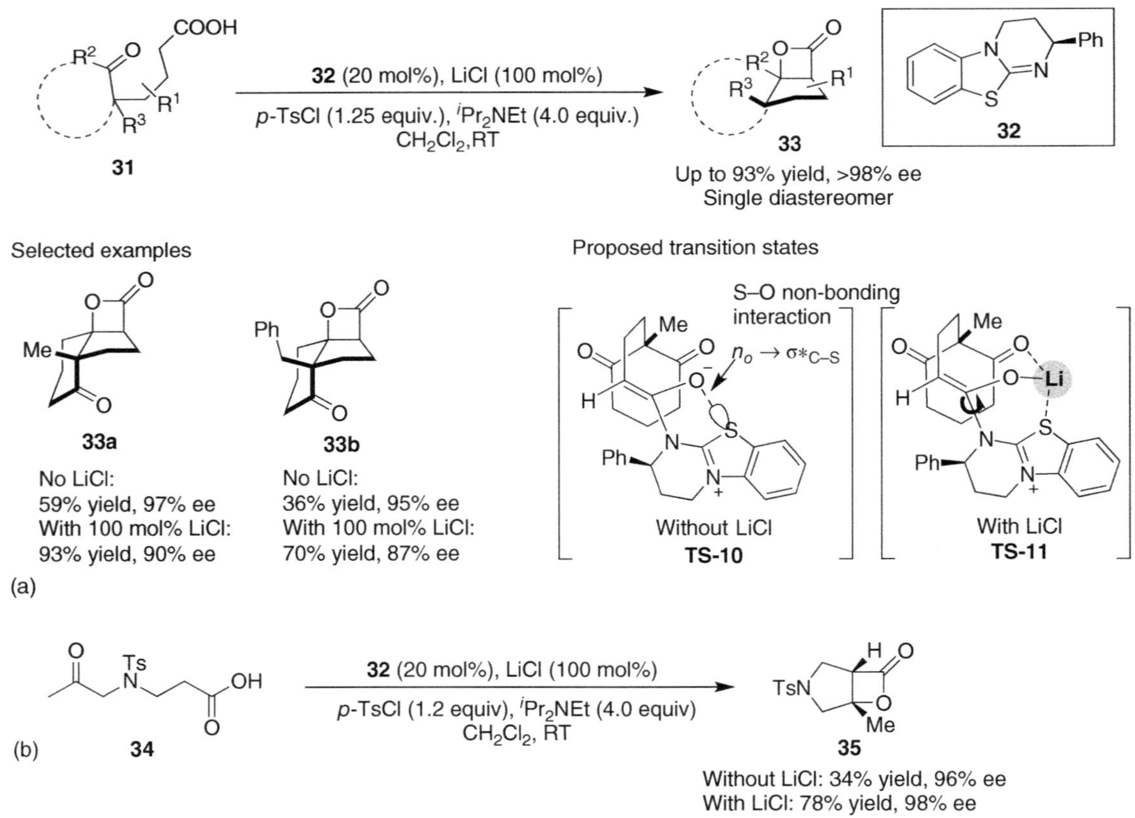

Scheme 7.14 Isothiourea/LiCl catalyzed intramolecular aldol lactonizations. Source: Kong and Romo [30].

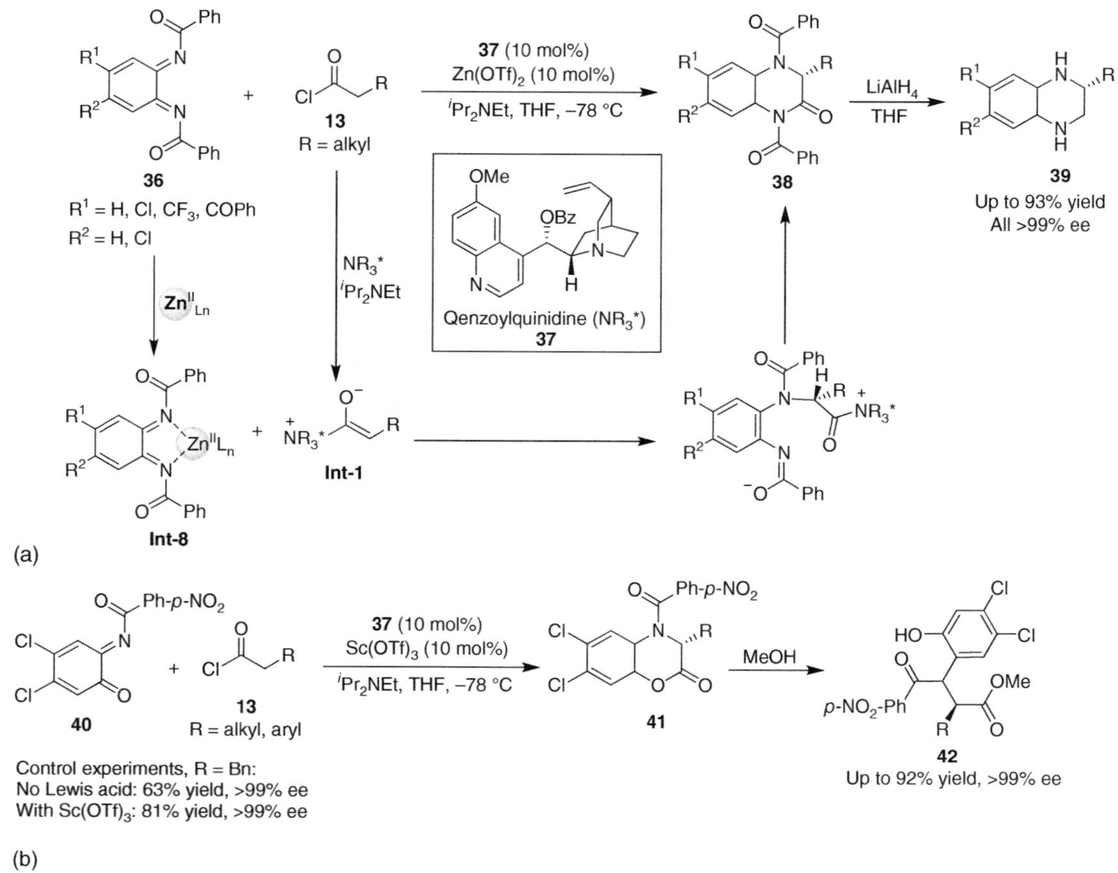

Scheme 7.15 Cinchona alkaloid/Lewis acid catalyzed asymmetric hetero-Diels–Alder reactions. Source: Abraham et al. [32a]; Paull et al. [32b].

acyl chlorides **13**, leading to α-oxygenated products in excellent yields and enantioselectivity (Scheme 7.16) [33]. Both computational and experimental studies indicate that the palladium(II) complex leverages the reaction performance by forming an ammonium palladium enolate as shown in **Int-9** to increase its relative concentration and provides enhanced chemoselectivity toward the quinone **43** (Scheme 7.16), rather than by the classical Lewis acid-activation of the quinone as observed in precedent HDA reactions of *o*-quinone derivatives (as shown in **Int-8**, Scheme 7.15a) [32].

Nelson and coworker created an asymmetric [4+2] cycloaddition of acyl chlorides and *N*-thioacyl imines by using lithium(I) and alkaloid combined catalysis (Scheme 7.17) [34]. The exposure of α-amido sulfone **46** and acid chlorides **13** to diisopropylethylamine simultaneously generates *N*-thioacyl imine and ketenes, which then undergo the enantioselective [4+2] cycloaddition catalyzed by *O*-TMSQ (**17**) in concert with $LiClO_4$ to deliver cis-disubstituted thiazinones **47** with almost perfect enantioselectivity. The high diastereoselectivity is attributed to the enolate-imine addition proceeding through either an open transition state **TS-12** or lithium coordinated and confined transition states **TS-13**.

A [4+2] cycloaddition reaction between trichloromethylacetaldehyde **1b** and vinylketene, *in situ* generated from α,β-unsaturated acid chloride **48** proceeds under the catalysis of cinchona alkaloid, but generates δ-lactones **49** in a low yield (Scheme 7.18a) [35a, c]. The addition of a Lewis acid, such as $Er(OTf)_3$, $Nd(OTf)_3$, and $Sn(OTf)_2$, is able to dramatically increase the yield, whereas in all cases, no change happens in the enantioselectivity ($R^1 = {}^iPr$). In the catalytic cycle shown in Scheme 7.18a, the Lewis acid catalyst is not involved in the cycloaddition step but exerts an impact on the reaction by simply facilitating the dehydrochlorination of **48**. The vinylketene is trapped by the chiral tertiary amine catalyst **20** to form a zwitterionic dienolate **Int-10**. The *cis*-**Int-10** is supposed to undergo [4+2] cycloadditions with aldehydes by either a stepwise or a concerted mechanism. However, the combination of TMSQD **20** with the Lewis acid is still limited to the electron-poor aldehyde (chloral, **1b**). However, the scope of aldehydes becomes even more general by using norephedrine/$Er(OTf)_3$combined catalyst system (Scheme 7.18b) [35b, c]. A wide range of aromatic and heteroaromatic aldehydes **1** are accommodated, leading to the δ-lactones **49** with excellent enantioselectivities (up to 98% ee). Presumably, the ligand and substrates all bind to the same metal center as shown in a proposed transition state **TS-14**, as implied by the absence of a nonlinear effect [35c].

7.2.4 Asymmetric α-Functionalization of Carbonyl Compounds

Lectka and coworkers found that the ammonium enolate generated from *O*-benzoyl quinidine (BQD, **37**) and acetyl chloride can undergo asymmetric fluorination with *N*-fluorodibenzenesulfonimide (NFSI) (Scheme 7.19) [36]. Notably, the presence of either *trans*-$(Ph_3P)_2PdCl_2$ or (dppp)$NiCl_2$ renders a much cleaner reaction of aryl acetyl chlorides **13** and NFSI **51** to deliver α-fluorinated carboxylic acid derivatives in high yields (up to 91%) and excellent enantioselectivities (up to 99% ee) (Scheme 7.19a) [36a]. The metal-bound zwitterionic ketene enolate **Int-11**

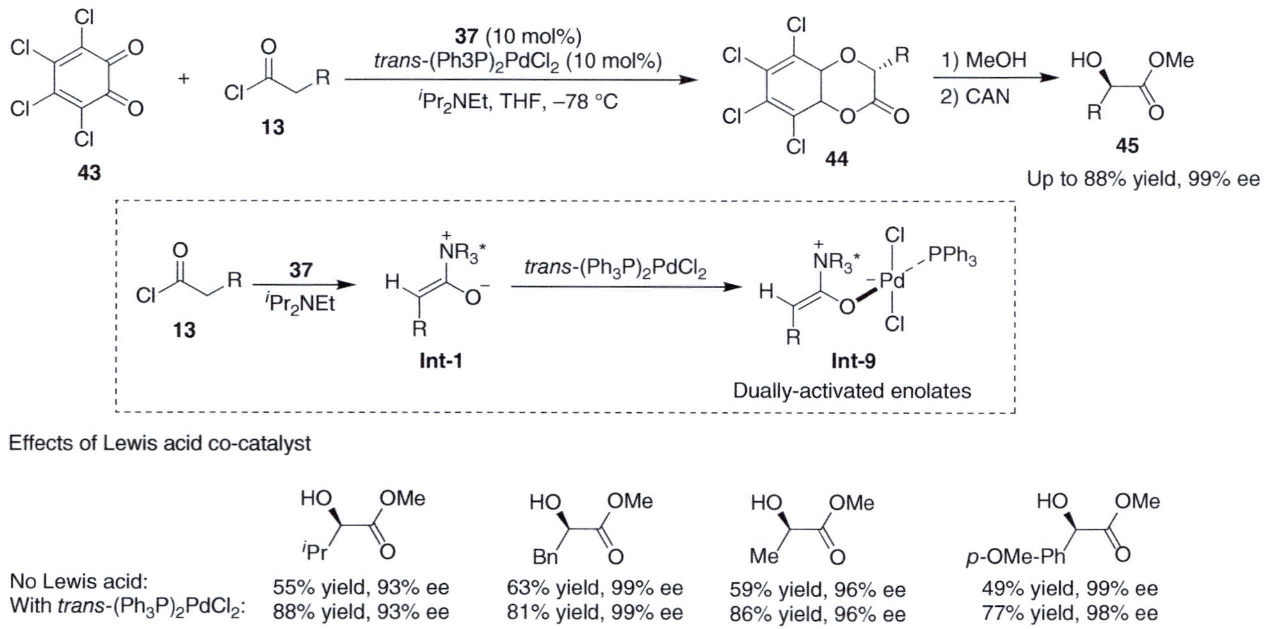

Scheme 7.16 Cinchona alkaloid/palladium catalyzed HDA reaction. Source: Abraham et al. [33].

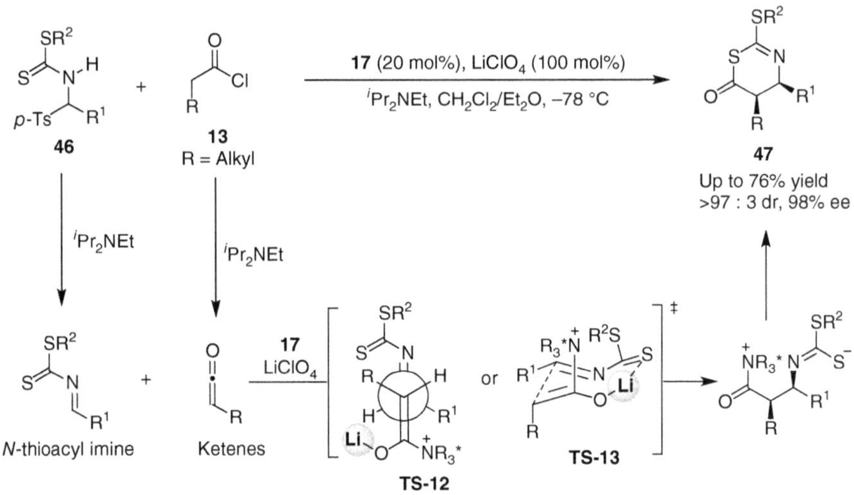

Scheme 7.17 Cinchona alkaloid/$LiClO_4$ catalyzed asymmetric [4+2] cycloadditions. Source: Xu et al. [34].

has been identified as the key intermediate that shows high reactivity toward the fluorinating reagent NFSI to give a chiral α-fluoro amide **52**, which then couples with nucleophiles to afford a variety of carboxylic acid derivatives **53**, such as α-fluorocarboxylic acids, amides, esters, and even peptides. However, aliphatic acid chlorides are seemingly less reactive; thus, one additional Lewis acid, $LiClO_4$, is required to activate both electro- and nucleophiles with chiral Lewis base **37** and the *trans*-$(Ph_3P)_2PdCl_2$ (Scheme 7.19b) [36b]. Mechanistic studies by kinetics, isotopic labeling, spectroscopic measurements, and theoretical calculations suggest that the lithium salt specifically coordinates with the NFSI as shown in **TS-15** and thereby enhances its electrophilicity toward the aliphatic ammonium enolates to make the enantioselective α-fluorination faster and cleaner. Each individual catalyst activates the substrates independently, but all catalysts propel the reaction in an interactive manner.

7.3 Asymmetric Reactions Driven by NHC-Mediated Homoenolates

The homoenolate intermediates (**Int-12**), catalytically generated from the reaction of NHCs with α,β-unsaturated aldehydes, have found widespread applications in asymmetric synthesis via HOMO elevation (Scheme 7.20a) [13a, d]. In the field of Lewis base/Lewis acid cooperative catalysis [15b, d–f] involving NHC-mediated generation of homoenolate intermediates, the Lewis acid catalyst is principally able to enhance the electrophilicity of electrophile (**I**), or leverages both the nucleophile and electrophile through the metal-organized transition-state (**III**) (Scheme 7.20b). However, NHCs routinely act as strongly σ-donating ligands of late transition metals and have been prevalently applied to homogeneous catalysis [37]. The strong

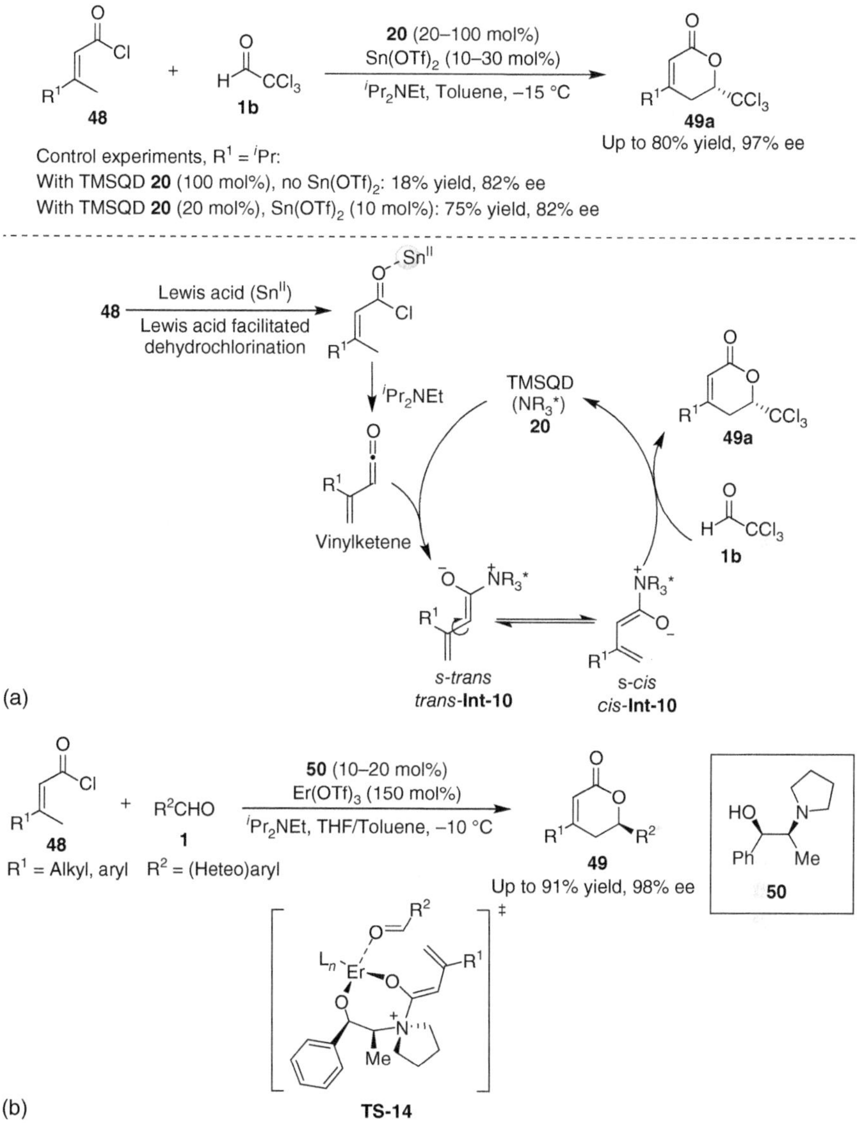

Scheme 7.18 Chiral tertiary amine/Lewis acid catalyzed asymmetric [4+2] cycloadditions of vinylketenes. Sources: Based on Tiseni and Peters [35a]; Tiseni and Peters [35c].

coordination of NHCs to transition metals actually poses constraints on the development of NHC and metal cooperative catalysis due to unavoidable self-quenching events (Scheme 7.20c, right-hand side). The NHCs tend to be soft Lewis bases and thus are compatible with hard Lewis acids. The weak and reversible coordination between the NHC and hard Lewis acid essentially minimizes the self-quenching to principally make their cooperative catalysis workable for the creation of asymmetric transformations (Scheme 7.20c, left-hand side).

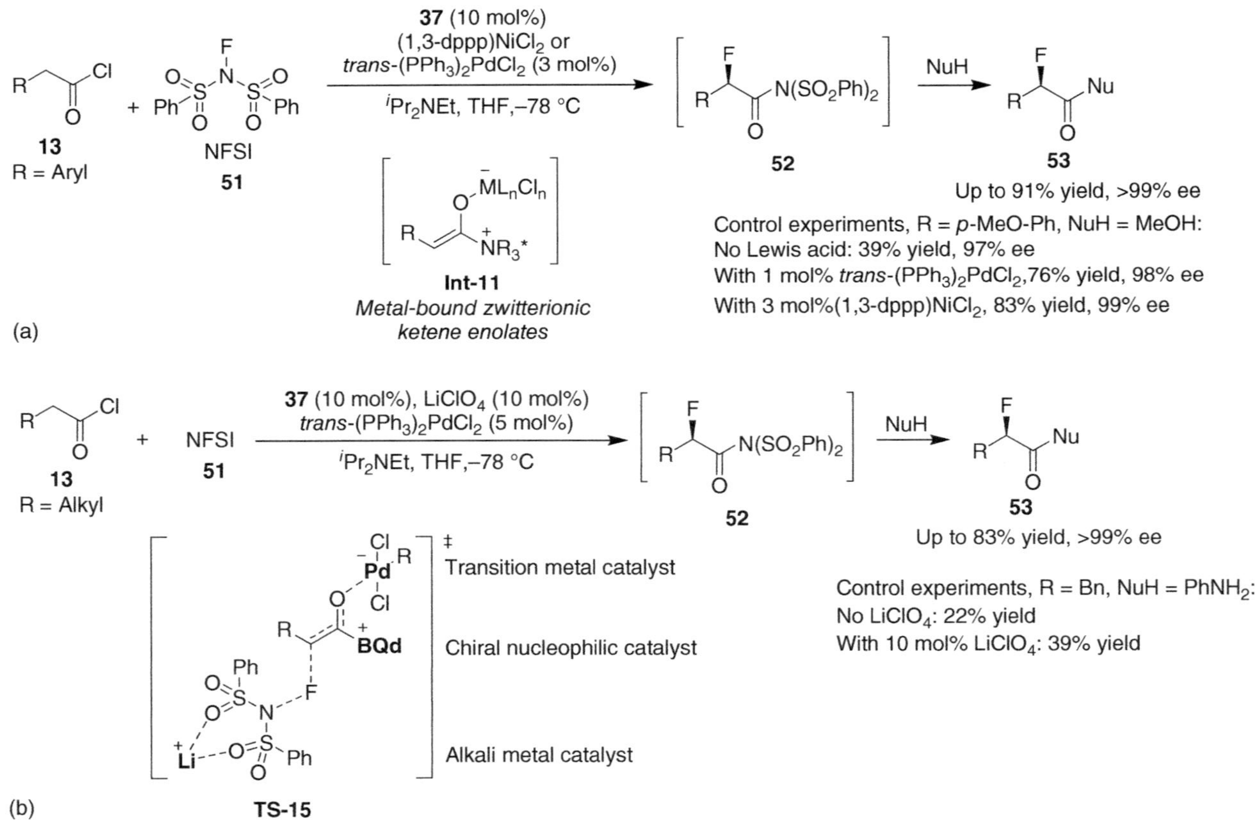

Scheme 7.19 Cinchona alkaloid/Lewis acid catalyzed enantioselective α-fluorination reactions. Sources: Based on Paull et al. [36a]; Erb et al. [36b].

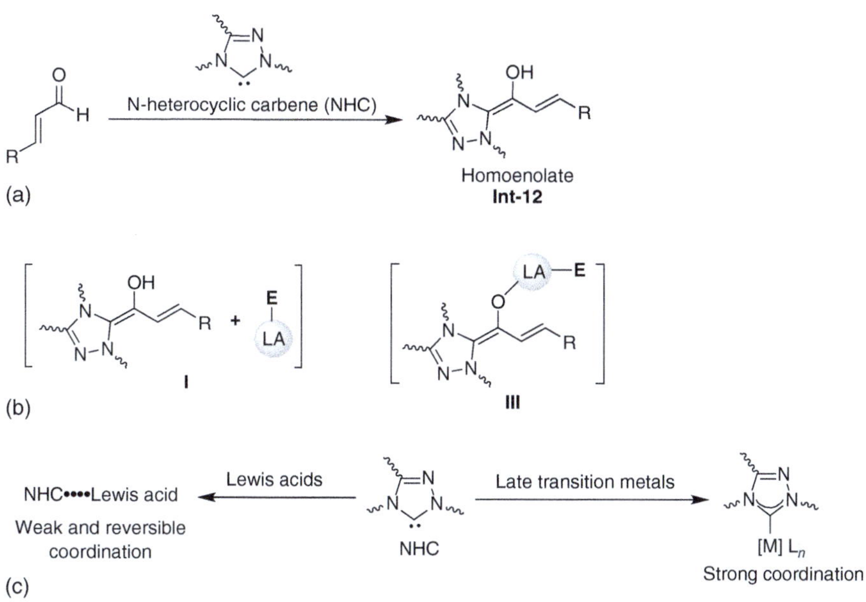

Scheme 7.20 NHC-Lewis acid cooperative catalysis. (a) NHC-mediated generation of homoenolate intermediates. (b) Activation modes in NHC–Lewis acid cooperative catalysis. (c) Interactions between NHCs and metals. Sources: Based on Nair et al. [13a]; Menon et al. [13d].

7.3.1 Asymmetric Annulation Reactions

The proof of concept in the NHC–Lewis acid cooperative catalysis was described by Scheidt and coworkers in 2010 [38]. They elaborated an asymmetric annulation of enals by NHC/Lewis acid cooperative catalysis (Scheme 7.21). In this case, the hard Lewis acids appear to be the choice of co-catalysts for NHCs. Other Lewis acids surveyed, including lanthanide, zinc, copper, and nickel salts, inhibit the reaction, possibly due to the strong and irreversible coordination of NHCs to these metals. The use of Lewis acids with more donating ligands (e.g. metal alkoxides) might attenuate this potentially strong binding interaction of the NHC to the metal center, and thus, the titanium(IV) tetraisopropoxide ($Ti(O^iPr)_4$) was proven to be the optimal co-catalyst. Under the promotion of the combined catalyst of $Ti(O^iPr)_4$ and NHC precursor **56**, the annulation occurs very smoothly to deliver substituted *cis*-cyclopentenes **57** with excellent stereoselectivity. More importantly, the use of $Ti(O^i\text{-}Pr)_4$ results in a reversal of diastereoselectivity, in comparison with the results obtained in the absence of a Lewis acid co-catalyst (see control experiments). From the mechanistic viewpoint, the titanium(IV) Lewis acid activates both the α,β-unsaturated aldehyde **54** and the enone **55**, and holds the homoenolate in close proximity to the β-carbon of the enone **55** as shown in **TS-16**, aligning it for cis-stereoselective C—C bond formation. Computational studies by Domingo et al. support the hypothesis that the Ti(IV) catalyst gets involved in the key bond-forming event [39].

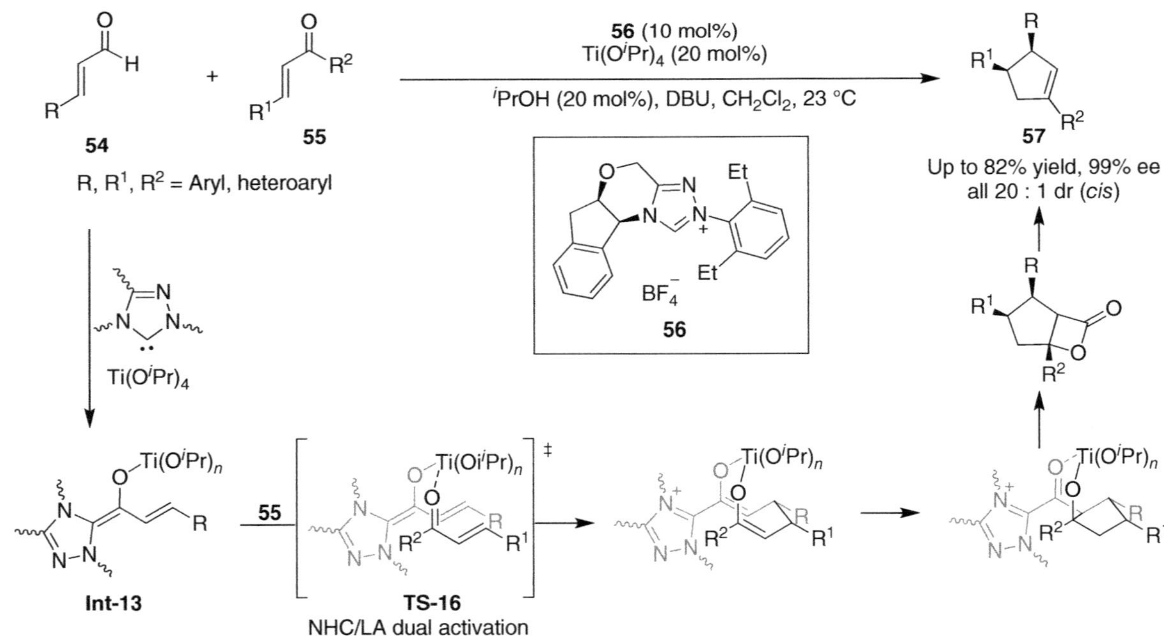

Control experiments, R = R^1 = R^2 = Ph:
Without Lewis acid: 84% yield, *cis* : *trans* = 1 : 6, 99% ee
With Ti(O^{i}Pr)$_4$ (20 mol%): reversal of diastereoselectivity
74% yield, *cis* : *trans* = 20 : 1, 99% ee

Scheme 7.21 NHC/Ti-catalyzed annulation reactions.

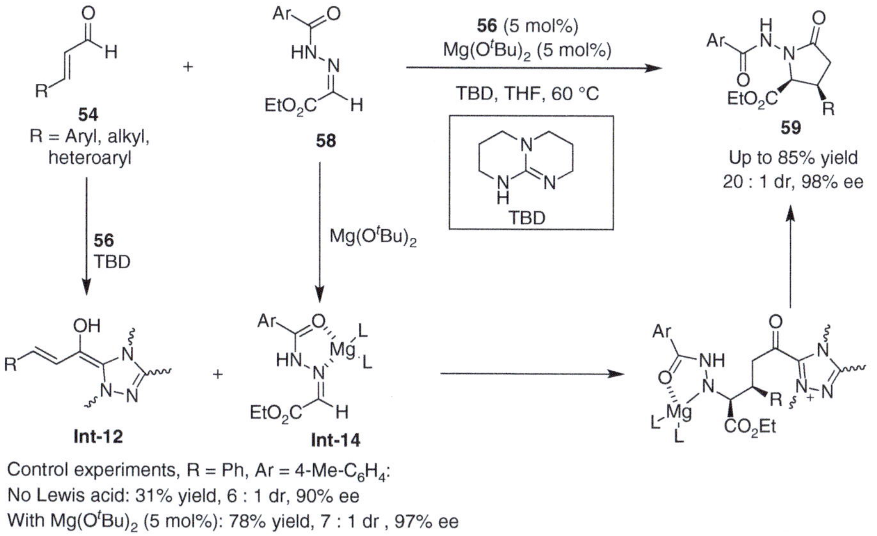

Scheme 7.22 NHC/Mg-catalyzed annulation reactions. Source: Based on Raup et al. [40].

Almost at the same time, the same group established an asymmetric annulation reaction of α,β-unsaturated aldehydes **54** with hydrazones **58** by chiral NHC and Lewis acid cooperative catalysis (Scheme 7.22) [40]. The hard Lewis acid Mg(O^{t}Bu)$_2$ was identified as a proper co-catalyst that nicely collaborated with chiral NHC **56** to enable the annulation delivering highly substituted γ-lactams **59** in moderate to good yields (up to 85%) and with high levels of diastereo- and enantioselectivity (up to 20 : 1 dr, 98% ee). The α,β-unsaturated aldehydes **54** condense with NHCs to generate transient NHC-bound homoenolates **Int-12**, raising the HOMO. Simultaneously, the hydrazone substrate **58** chelates to the magnesium ion with the carbonyl oxygen and imine nitrogen to form **Int-14**, lowering the LUMO. Such dual activation greatly enhances the reactivity of both the nucleophile and electrophile and facilitates the whole process.

The NHC-catalyzed asymmetric dimerization of enals via a 1,2-addition of homoenolates was reported by Bode and coworkers in 2004 [41]. Scheidt and coworkers shifted the reactivity from the 1,2- to a 1,4-addition by exploiting NHC/Lewis acid cooperative catalysis (Scheme 7.23) [42]. The combined use of a chiral azolium salt **60** and metal alkoxides, such as Mg(O^{t}Bu)$_2$, Ba(O^{i}Pr)$_2$, and Sr(O^{i}Pr)$_2$, leads to the 1,2-addition products **62** or to the decomposition of the starting material. In contrast, the oxophilic titanium(IV) isopropoxide is able to greatly enhance the electrophilicity of C—C double bond of the enal **54** by coordination and makes the titanium-bonded homoenolate intermediate, generated from the condensation of NHC and another molecule of enal, preferentially undergo 1,4-addition via **TS-17**. In addition to switching the reaction mode, the presence of Ti(O^{i}Pr)$_4$ is necessary for the reactivity because no desired product **61** is detected without this co-catalyst.

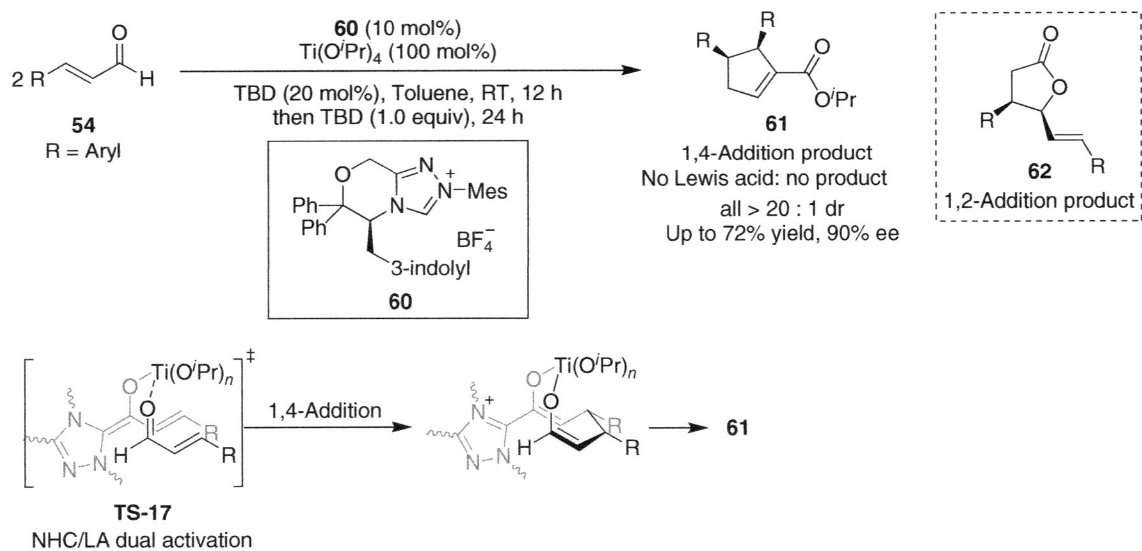

Scheme 7.23 NHC/Ti-catalyzed dimerization of enals. Source: Cohen et al. [14].

The NHC/Ti(O^{i}Pr)$_4$ cooperative catalysis is workable for the 1,4-addition of homoenolate equivalents to α,β-unsaturated α-ketoesters **63** (Scheme 7.24) [43]. The Lewis acid-activated α-ketoesters **63** prefer to undergo the 1,4-addition with the NHC-bonded homoenolates via the transition state **TS-18**, delivering the cyclopentanol products **64** in up to 85% yield with up to 20 : 1 dr and 99% ee. The reaction does not work in the absence of the Lewis acid. Although 20 mol% of NHC organocatalyst is sufficient, large excess amounts of Ti(O^{i}Pr)$_4$ are required to ensure a clean reaction, presumably owing to the strong coordination affinity with the keto ester substrate to inhibit the Lewis acid turnover.

In 2007, Scheidt and coworkers reported an NHC-catalyzed intramolecular Michael addition and lactonization cascade reaction of 2-aroylvinylcinnamaldehydes **65** (Scheme 7.25a) [44a]. Interestingly, Cheng and coworkers found that the addition of Ti(O^{i}Pr)$_4$ as a Lewis acid co-catalyst switched the reaction pathway from the intramolecular cyclization to an intermolecular dimerization, preferentially giving rise to indeno[1,2-c]furan-1-one derivatives **69** in good yields (up to 88%) and with excellent diastereo- and enantioselectivities (up to 22 : 1 dr, >99% ee) (Scheme 7.25b) [44b]. The titanium(IV) catalyst is believed to activate the formyl functionality of **65** and to make the NHC-bound homoenolate preferentially undergo the intermolecular aldol reaction with another molecule of the substrate (via **TS-19**), rather than the intramolecular Michael addition.

Catalyst-controlled divergent annulation of enals **54** with heterocyclic enones **70** was developed by Zhao and coworkers (Scheme 7.26) [45]. The backbone of azolium catalysts has a dramatic effect on the chemo-selectivity of the annulation. In this case, heterocyclic enones **70** prefer to undergo a [4+3] cyclization reaction with enals **54** to generate ε-lactones **71** catalyzed by the NHC **66**. The presence of 10 mol% of Ti(O^{i}Pr)$_4$ greatly enhances the reaction efficiency, but with sacrificed chemo-selectivity (see control experiments). A completely different reaction pathway was observed under the catalysis of NHC **73**, which yielded spirocycle **72** as the major product instead of **71**.

The impact that alkali metal salt exerts on carbene-catalyzed reactions has been observed by Scheidt and coworkers [46a]. In 2012, the same group explored the potential of lithium salts in the context of NHC/Lewis acid asymmetric catalysis. The asymmetric annulation of NHC-bonded homoenolate with isatins proceeds readily under the promotion of excess amounts of lithium chloride (Scheme 7.27) [46b]. Similarly, the addition of LiCl greatly improves the reaction performance. Control experiments reveal that the NHC **56** shows high catalytic activity for the asymmetric annulation of the cinnamaldehyde **54a** and *N*-methyl isatin **74a**, but with poor diastereoselectivity and modest enantioselectivity (1.1 : 1 dr, 34% ee). The addition of two equivalents of LiCl leads to a significant improvement in stereochemical outcome (2.5 : 1 dr, 90% ee). The presence of a large excess of [12] crown-4 completely inhibits the coordination of LiCl with the substrates, and thus the lithium cation is sequestrated from the reaction mixture to result in similar results to those observed in the reaction without LiCl. As shown in **TS-20**, the lithium chloride activates both substrates through the coordinating interactions

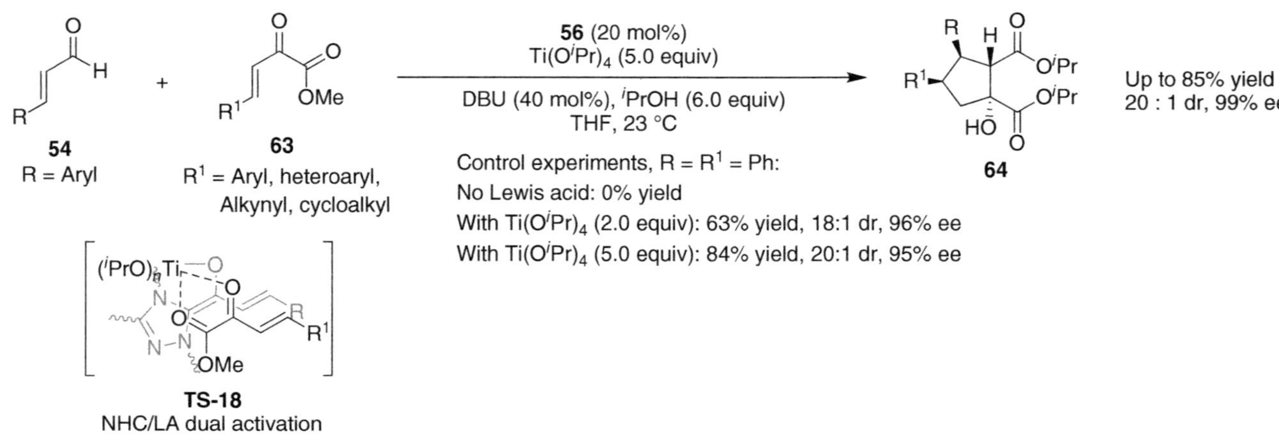

Scheme 7.24 NHC/Ti-catalyzed asymmetric annulation of enals and unsaturated ketoesters. Source: Cohen et al. [43].

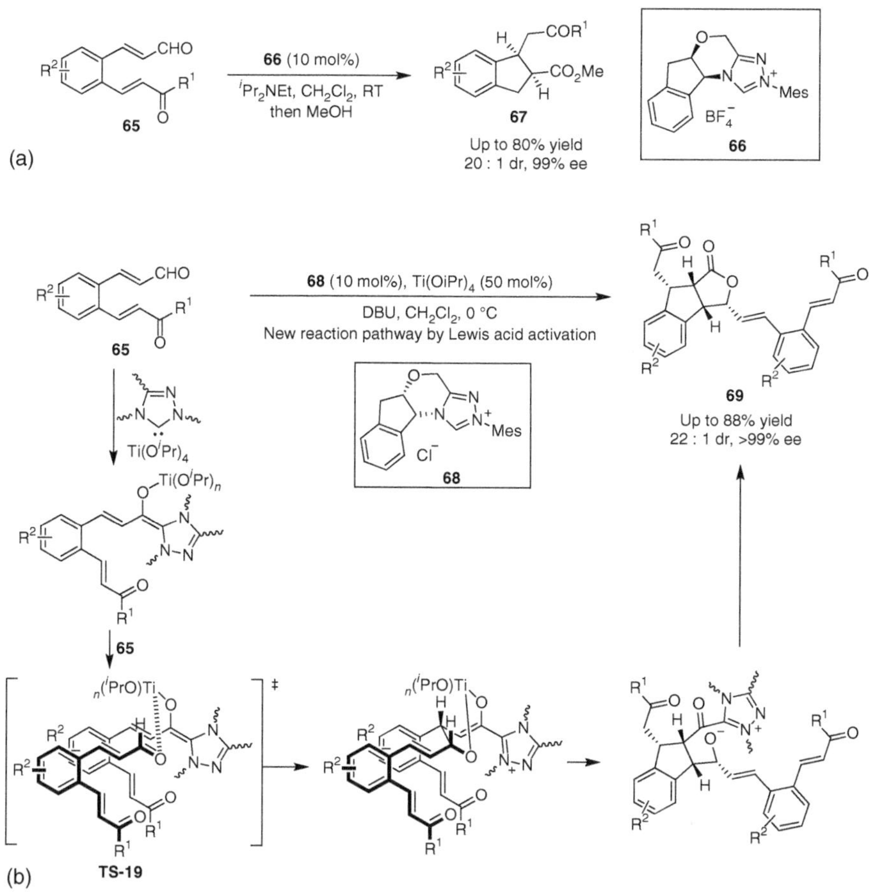

Scheme 7.25 Intramolecular Michael addition and lactonization cascade reactions. (a) Scheidt's work: NHC catalysis. Source: Based on Phillips et al. [44a]. (b) Cheng's work: NHC–Lewis acid cooperative catalysis. Source: Wang et al. [44b].

with the enol oxygen of the NHC-bonded homoenolate and the 1,2-dicarbonyl of the isatin **74**, respectively, to synergistically drive the reaction.

Scheidt and coworkers introduced a ternary catalyst system combining chiral NHCs with achiral Lewis acid and hydrogen-bond donor catalyst to the enantioselective annulation reaction of unsaturated aldehydes **54** with α-ketoesters **76** (Scheme 7.28) [25]. Only if the three individual catalysts are all presented, the reaction proceeds cleanly to give the desired product **79** in high yields and enantioselectivity. As shown in the transition state **TS-21**, density functional theory (DFT) computations and nuclear magnetic resonance (NMR) studies suggest that the Ca^{2+} ion chelates to the carbonyl groups of the α-ketoester electrophile **76** and the oxygen atom of the anionic homoenolate nucleophile. The methoxide counterion, which forms H-bonding interaction with the thiourea N–H groups to stabilize its negative charge, binds with Ca^{2+} to form a distorted tetrahedral metal center. In addition, the thiourea phenyl group engages in a C–H–π interaction. Such multiple non-covalent interactions allow the thiourea to act as a relay auxiliary,

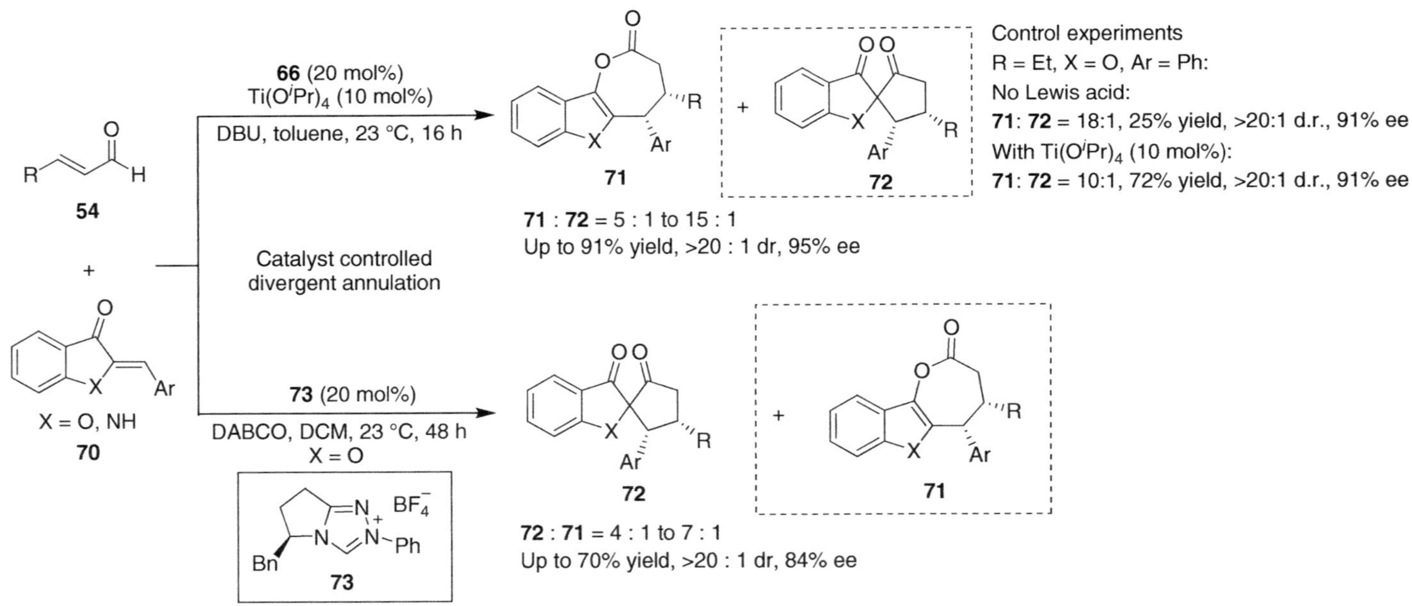

Scheme 7.26 NHC/Ti(IV) cooperatively catalyzed [3+4] annulation reaction. Source: Wang et al. [45].

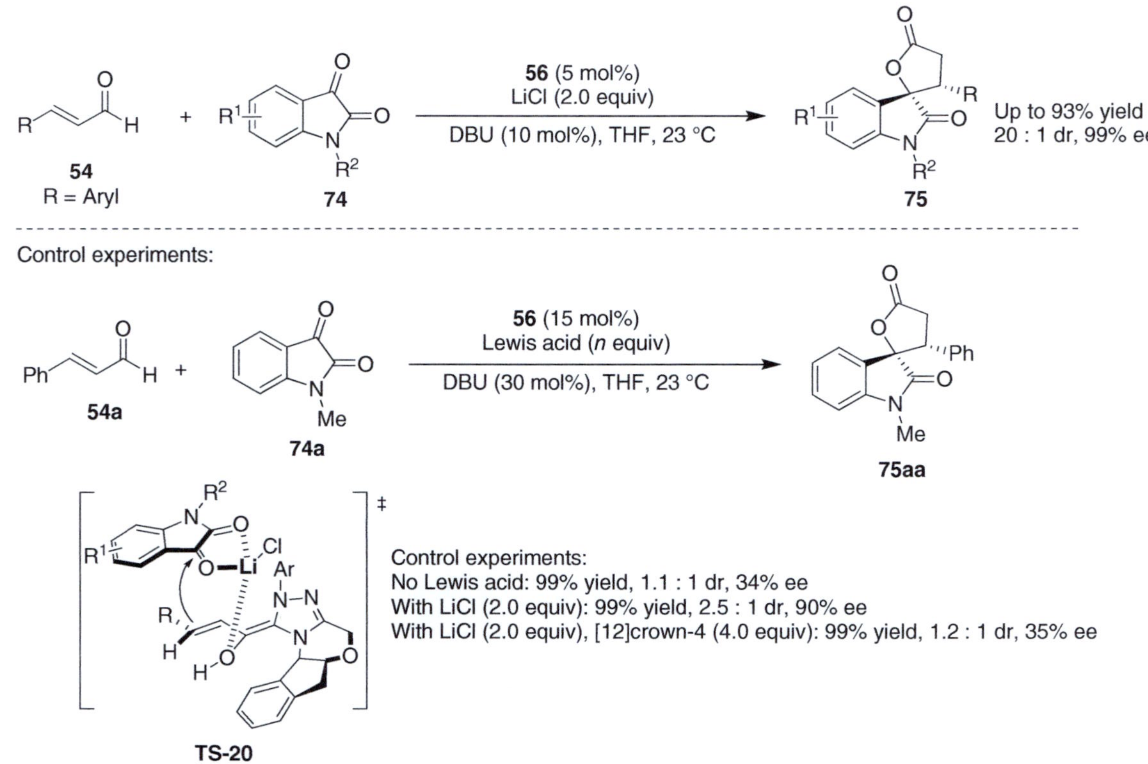

Scheme 7.27 NHC/Li dually catalyzed [3+2] annulation reactions. Source: Dugal-Tessier et al. [46b].

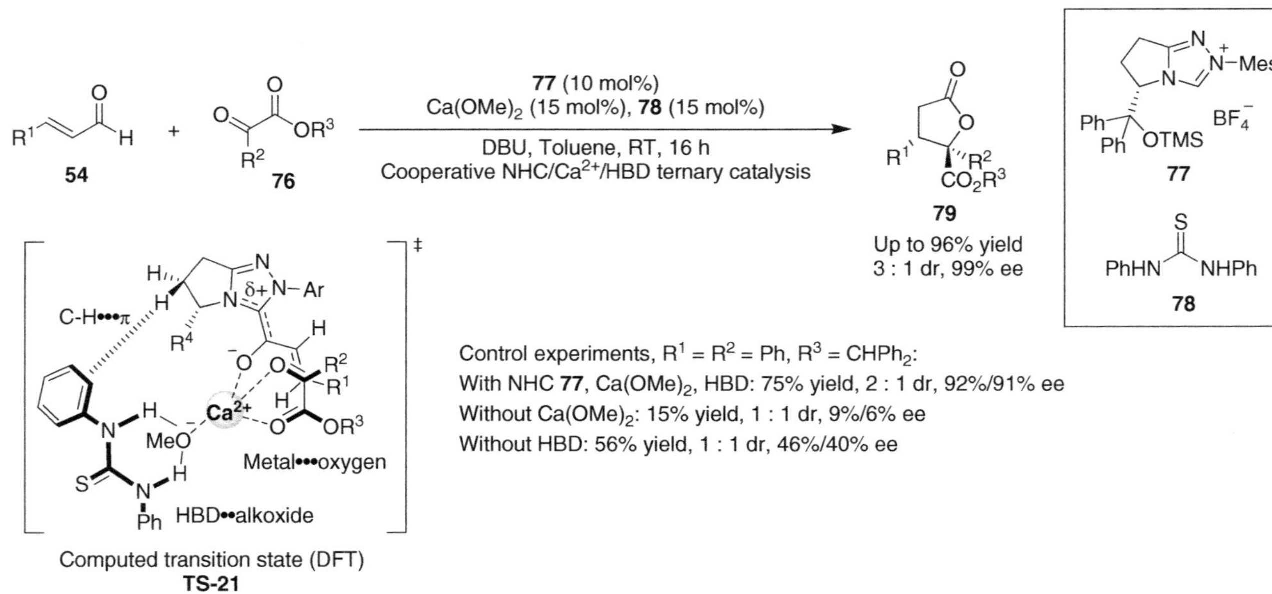

Scheme 7.28 Cooperative NHC/Ca^{2+}/thiourea ternary catalysis. Source: Based on Murauski et al. [25].

with the phenyl groups transferring the chiral information of the catalyst to the distal reactive center.

Under the promotion of LiCl, NHC undergoes a nucleophilic addition to the potassium 2-oxo-3-enoate **80** and followed by a decarboxylation to generate a transient NHC-bonded homoenolate **Int-12** (Scheme 7.29a) [47]. Based on this transformation, Fu, Huang, and coworkers developed an NHC/LiCl cooperatively catalyzed asymmetric [3+2] annulation reaction between potassium 2-oxo-3-enoates **80** and enones **55** or isatins **74**, affording cyclopentenes **57** or spirooxindole lactones **75** in high yields and with excellent levels of stereoselectivity (Scheme 7.29b) [47].

She and coworkers established a [3+2] annulation reaction between alkynals **82** and unsaturated keto esters **63′** on the basis of the NHC/LiCl cooperative catalysis to efficiently assemble butenolide **84** (Scheme 7.30) [48]. The NHC-bounded "allenolate equivalent" **Int-15** undergoes 1,2-addition rather than 1,4-addition with the Lewis acid (LiCl) coordinated intermediate **Int-16** (via **TS-22**). The addition of LiCl would dramatically enhance the reaction rate, and the annulation process would complete in a few minutes, whereas in the absence of the Lewis acid, a trace amount of the target product was obtained. An asymmetric version by using chiral NHC catalyst **85** provides the corresponding butenolide **84aa** in 36% yield and 53% ee.

As reported by Scheidt and coworker, an even more highly enantioselective [3+2] annulation reaction of alkynals and α-ketoesters can be accessed via cooperative catalysis of chiral NHC and chiral lithium phosphate (Scheme 7.31) [49]. No reaction was observed in the absence of chiral phosphoric acid. The ^{31}P NMR study suggests that the chiral phosphoric acid **87** is initially deprotonated under basic conditions, and the phosphate acts as a counterion of the lithium salt. The lithium phosphate turns out to be the Lewis acid catalyst to activate the keto ester substrate **76′**. The chiral NHC, generated from the deprotonation of azolium salt **86** with lithium *tert*-butoxide, mediates the formation of enolate **Int-17**. The enantioselective 1,2-addition between the two intermediates via transition state **TS-23** generates butenolides **88**. However, the configuration of phosphoric acid **87** seems to have little impact on the enantioselectivity, as indicated by the control experiments with racemate and its enantiomer.

7.3.2 Asymmetric β-Protonation Reactions

The enantioselective β-protonation of enals is highly challenging because the chiral auxiliary locates far from the reaction site, and the protons are small in size. Huang and coworkers proposed to use a Lewis acid to bridge the chirality transfer from an NHC to the distal β-carbon (Scheme 7.32) [50], wherein the enol functionality serves as a directing group by coordinating to the Lewis acid (**TS-24**, Scheme 7.32a), which is beneficial to the stereochemical control of the protonation event. The $Cu(OTf)_2$ turns out to be the optimal Lewis acid catalyst, which in collaboration with chiral NHC *ent*-**66** enables the β-protonation of β-aryl and vinyl enals **89** with thiol **90** to deliver chiral thioesters **91** in up to 99% yield and 99% ee (Scheme 7.32b) [50a]. On the contrary, the use of NHC only provides even much diminished yield and enantioselectivity. Alternatively, the copper would be possible to facilitate the

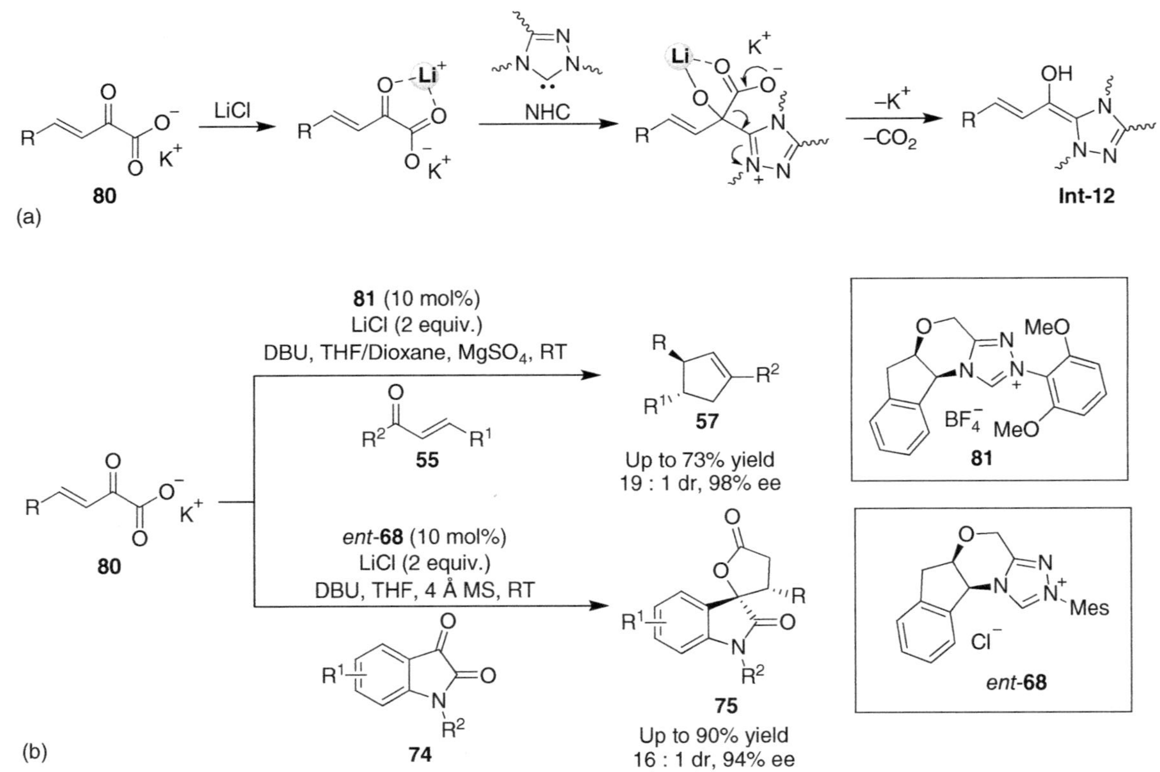

Scheme 7.29 NHC/LiCl-mediated annulation reactions of potassium 2-oxo-3-enoates. Source: Gao et al. [47].

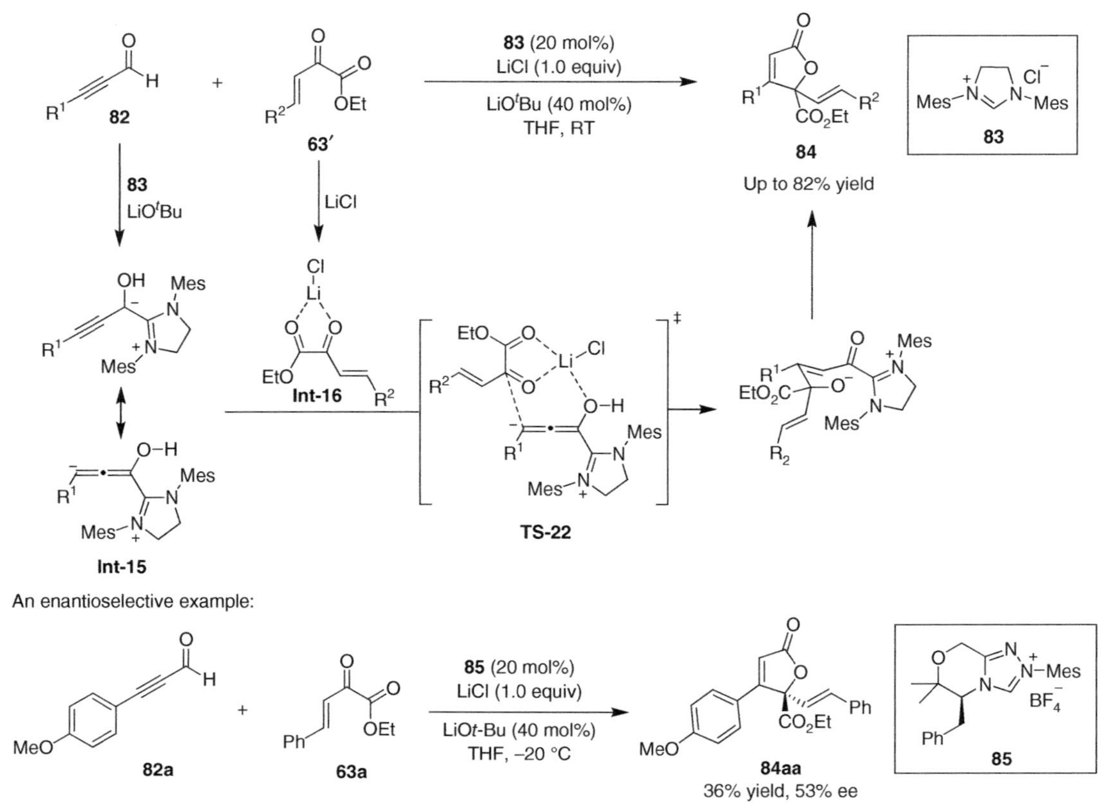

Scheme 7.30 NHC/Li-catalyzed [3+2] annulation reactions of alkynals. Source: Qi et al. [48].

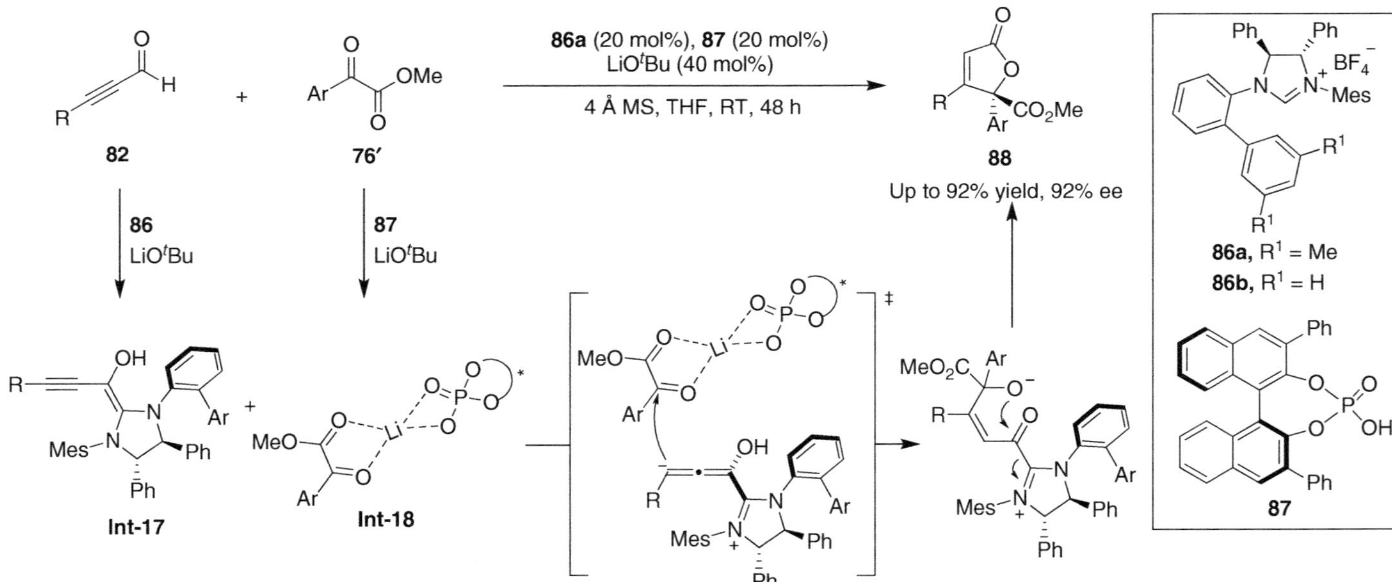

Control experiments, R = Ar = Ph:
With **86b** (20 mol%), no phosphoric acid = no reaction
With **86a** (20 mol%), **87** (20 mol%): 85% yield, 86% ee
ent-**87** (20 mol%): 84% yield, 86% ee
rac-**87** (20 mol%): 74% yield, 82% ee

Scheme 7.31 NHC/lithium phosphate cooperative catalysis. Source: Lee and Scheidt [49].

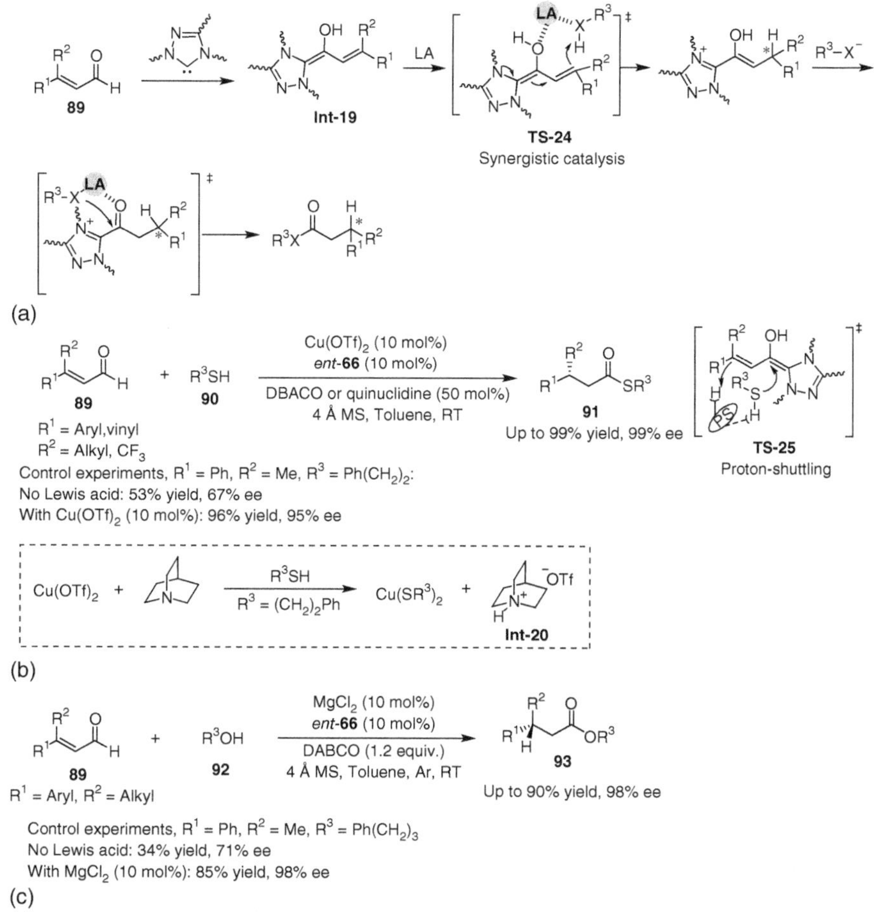

Scheme 7.32 Asymmetric β-protonation of homoenolate intermediates. Source: Chen et al. [50a]; Wang et al. [50b].

generation of a quinuclidinium ion **Int-20**, a more effective and selective proton source, from quinuclidine and thiol, which participates in the highly stereoselective protonation of homoenolates via an organocatalytic proton-shuttling strategy (**TS-25**, Scheme 7.32b). The extension of the Lewis acid assisted concept to create an enantioselective hydroesterification of β-alkyl cinnamaldehydes **89** with alcohols **92** is also successful (Scheme 7.32c) [50b]. In this case, the addition of $MgCl_2$ as Lewis acid co-catalyst significantly elevates both yield and enantioselectivity of the NHC-catalyzed process.

7.3.3 Asymmetric Kinetic Resolutions

Gong, Song, and coworkers established an NHC/copper cooperatively catalyzed highly efficient kinetic resolution (KR) and dynamic kinetic asymmetric transformation (DyKAT) of racemic *N*-tosylaziridines **95** by [3+3] annulation with isatin-derived enals **94** [51] (Scheme 7.33). A large library of enantioenriched *N*-tosylaziridine derivatives (*R*)-**95** (up to >99% ee) and spirooxindole derivatives **96** were obtained with high structural diversity and stereoselectivity (up to >95 : 5 dr, >99% ee). The chemoselection between KR and DyKAT that probably occur can be switched by tuning the dose of the chiral NHC **98** (KR, Cu^{I}: **97**:**98** = 1 : 2 : 2, Scheme 7.33a; DyKAT, Cu^{I}: **97**:**98** = 1 : 2 : 0.5, Scheme 7.33b), therefore regulating the catalytic activity of the copper complex, and the absence or presence of the Cu complex of diphosphine [Cu^{I}(**97**)] governs the reaction pathway. Based on mechanistic studies, a plausible catalytic cycle was proposed for the KR of aziridines (Scheme 7.33a). The copper catalyst [Cu^{I}(**97**)(**98**)] is formed from the coordination reaction of $Cu(CH_3CN)_4PF_6$, diphosphine **97**, and NHC precatalyst **98** under basic conditions. The addition of NHC **98** to isatin-based enal **94a** generates Breslow intermediate **Int-21**, which represents azolium homoenolate species **Int-22** in mesomeric form, and transiently releases some catalytically active [Cu^{I}(**97**)] species via the sequestration of NHC **98** from the [Cu^{I}(**97**)(**98**)] complex. The aziridine **95** coordinates with the metal center of the active [Cu^{I}(**97**)] species to deliver electrophilic **Int-23** and **Int-23′**. Thereafter, **Int-23** undergoes a much faster S_N2 type substitution with NHC-bound nucleophile **Int-22** than its diastereomeric complex **Int-23′**, and further delivers final product **96** with the recovered aziridine (*R*)-**95**. In the DyKAT of electron-rich aryl-substituted aziridines, the use of excess amount of cooperative catalyst to guarantee the existence of the active [Cu^{I}(**97**)] species, which is responsible for the rapid racemization of aziridines, is the key to success (Scheme 7.33b). On the other hand, in the KR of electron-rich aryl-substituted aziridines, mechanistic studies reveal that the transient [Cu^{I}(**97**)] species cannot be detected in the reaction system, thereby enabling KR (Scheme 7.33a).

7.4 Asymmetric Reactions Driven by NHC-Mediated Azolium Enolates

Azolium enolates **Int-24** and dienolates **Int-25**, commonly generated through the reaction of NHCs with carboxylic acid (or aldehyde) derivatives or with α,β-unsaturated carboxylic acid (or aldehyde) derivatives, respectively, have been versatile intermediates generally applicable to the creation of asymmetric transformation to allow efficient α- or γ-functionalization of carbonyl molecules (Scheme 7.34) [13b, c, f].

The Lewis acid/NHC cooperatively catalyzed asymmetric reaction driven by NHC-mediated azolium enolates was initially reported by Chi and coworkers who attempted to develop an oxidative γ-functionalization of enals **99** by trapping the azolium dienolates with aryltrifluoroacetones **100** (Scheme 7.35) [52]. However, an unsatisfactory reaction that generates an unsaturated δ-lactone **102a** ($R^1 = R^2$ = Ph) in a moderate yield (67%) and low enantioselectivity (16% ee) was observed with chiral NHC **66**. Remarkable improvement in both reaction conversion and stereochemical control was accessed by introducing either $Mg(OTf)_2$ or $Sc(OTf)_3$ as a co-catalyst. In particular, the hybrid heterometallic combination of $Mg(OTf)_2$

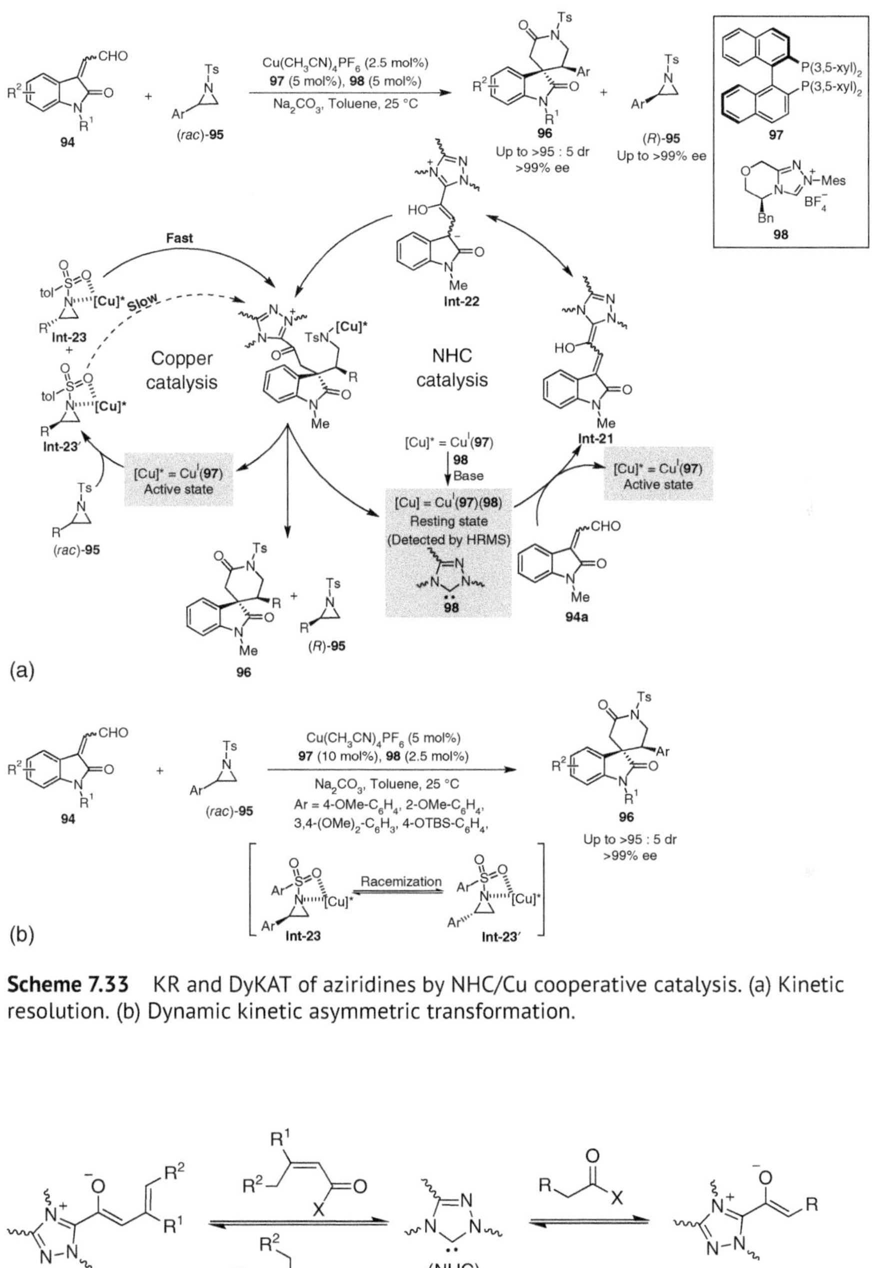

Scheme 7.33 KR and DyKAT of aziridines by NHC/Cu cooperative catalysis. (a) Kinetic resolution. (b) Dynamic kinetic asymmetric transformation.

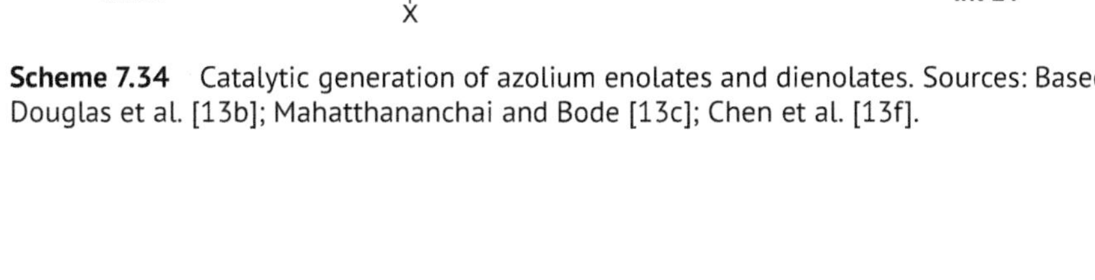

Scheme 7.34 Catalytic generation of azolium enolates and dienolates. Sources: Based on Douglas et al. [13b]; Mahatthananchai and Bode [13c]; Chen et al. [13f].

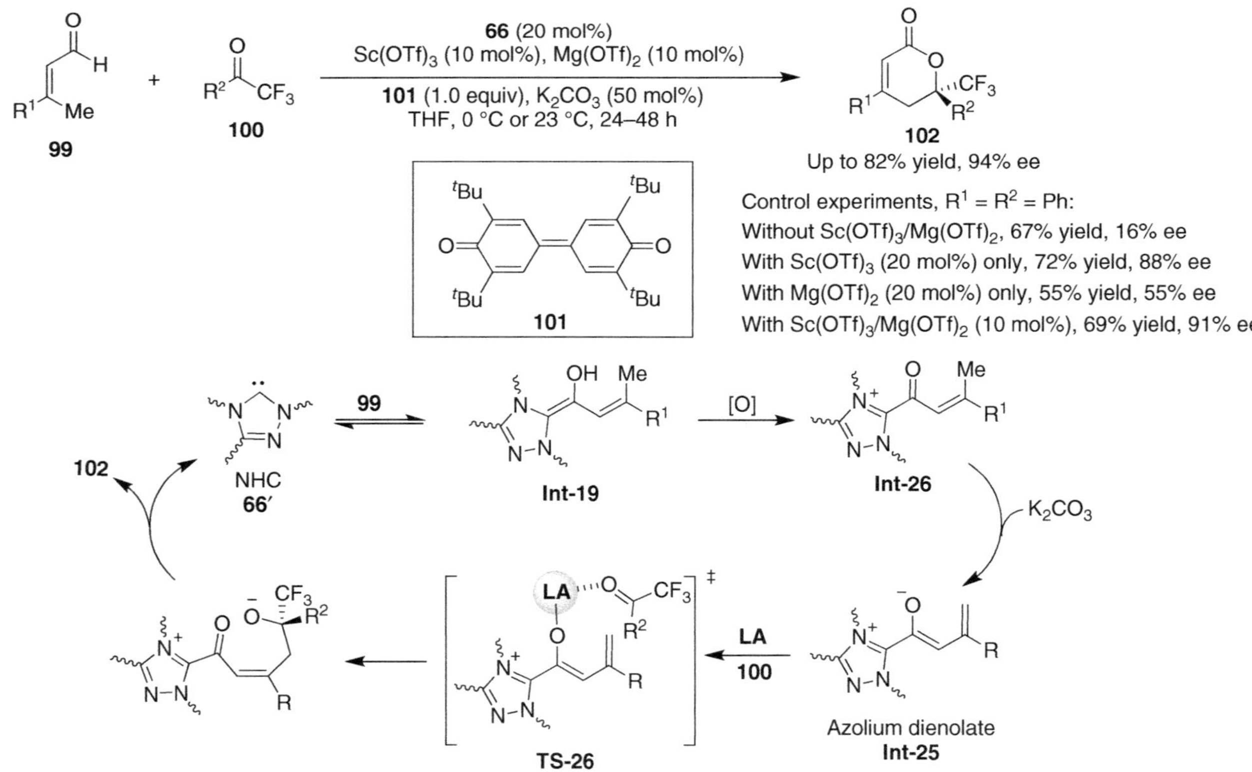

Scheme 7.35 Oxidative γ-addition of enals to trifluoromethyl ketones. Source: Mo et al. [52].

and $Sc(OTf)_3$ with NHC **66** turns out to be the best for the oxidative γ-addition of enals **99** to trifluoromethyl ketones **100**, giving rise to the unsaturated δ-lactones **102** in high yields (up to 82%) and excellent enantioselectivities (up to 94% ee). In the stereogenic step, the NHC-bound homoenlate intermediate **Int-19**, generated from the nucleophilic addition of NHC catalyst to enal **99**, undergoes the oxidation process to deliver unsaturated acyl azolium **Int-26**, which later transforms into a vinyl enolate intermediate **Int-25** through γ-deprotonation. The Lewis acid likely gets involved in multisite coordination to make the NHC-bound acyl azolium dienolate **Int-25** close enough to undergo nucleophilic addition to the ketones **100** (via **TS-26**), and followed by an intramolecular lactonization to afford unsaturated δ-lactones **102**.

A similar Lewis acid/NHC cooperative catalysis was adopted to create the γ-addition of α,β-unsaturated aldehydes to isatins for the synthesis of spirooxindole-dihydropyranone scaffolds [53]. Yao and coworkers established an asymmetric oxidative annulation of 3-methylbut-2-enal **99a** and 1-methylindoline-2,3-dione **74a** under the NHC/$Yb(OTf)_3$ cooperative catalysis to provide a chiral spirooxindole–dihydropyranone **104aa** in 85% yield, but with moderate enantioselectivity (33% ee) (Eq. (1), Scheme 7.36) [53a]. Much more stereoselective synthesis of spirooxindole–dihydropyranone derivatives has been accessed by the asymmetric annulation of 2-bromoenals **105** and isatins **74** by cooperative activation of NHC **106** and $La(OTf)_3$ (Eq. (2), Scheme 7.36) [53b]. $La(OTf)_3$ is proposed to act as a Lewis acid and may have a good affinity for carbonyl oxygen and carboxylates in multisite coordination to bring the electrophile into close proximity with the acyl azolium dienolate, as shown in **TS-27**. An asymmetric [4+2] annulation of 4-((methoxycarbonyl)oxy) butenal **107** with isatins **74** proceeds cleanly under the cooperative catalysis of NHC **66** and $Mg(O^tBu)_2$ to generate spirocyclic oxindolodihydropyranones **104′** in up to 76% yield and 78% ee (Eq. (3), Scheme 7.36) [53c]. In this case, the azolium dienolate **Int-25**, generated from the elimination of vinyl Breslow intermediate **Int-27**, undergoes an asymmetric γ-addition to the isatin activated by $Mg(O^tBu)_2$, presumably via a transition state similar to those in precedents; otherwise, the Lewis acid is different. α,β-Unsaturated carboxylic acid **108** activated by forming an anhydride **111** with pivaloyl chloride couples with the NHC, *in situ* generated from **103**, to give a vinyl Breslow intermediate Int-**25** that participates in $La(OTf)_3$ co-catalyzed [4+2] annulation reaction with isatin derived imines **109** (Eq. (4), Scheme 7.36) [53d].

A dynamic kinetic resolution of α-ketoesters **112** by trapping the azolium dienolates catalytically generated from enals **99** with NHC *ent*-**56** in the presence of the oxidant **101** delivers enantioenriched δ-lactones **113** in good to high yields (up to 88%) and with excellent enantio- and diastereocontrol (up to 96% ee, >20 : 1 dr) (Scheme 7.37) [54]. $Sc(OTf)_3$ turns out to be an optimal co-catalyst and can considerably enhance the enantioselectivity (see control experiments, from 79 to 94% ee). The Lewis acid $Sc(OTf)_3$ is proposed to get involved in multisite coordination to bring the ketone electrophile **112** into close proximity with the azolium dienolate intermediate to enable the highly enantioselective nucleophilic addition via **TS-28**.

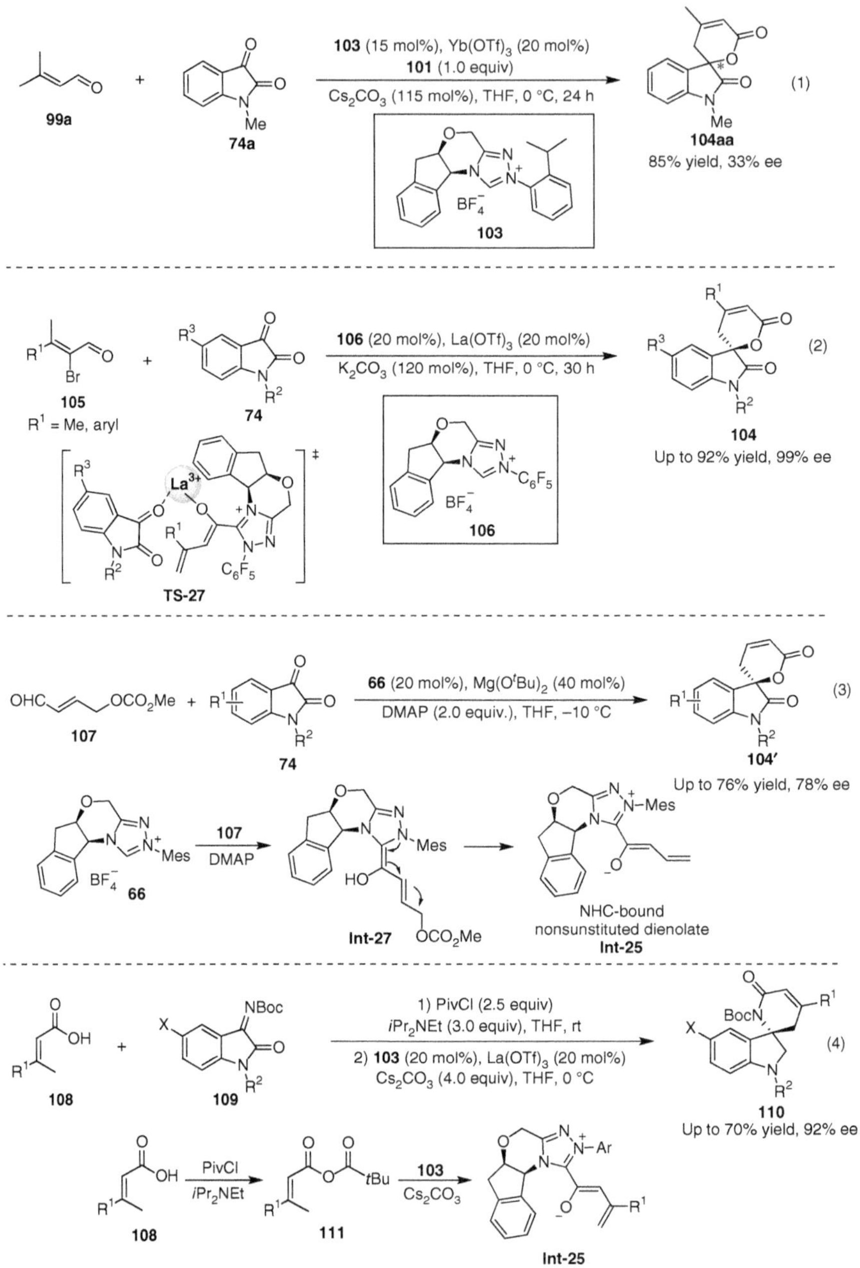

Scheme 7.36 NHC/Lewis acid-catalyzed γ-addition of azolium dienolates to isatins. Sources: Based on Liu et al. [53a]; Xiao et al. [53b]; Cheng et al. [53c]; Jia et al. [53d].

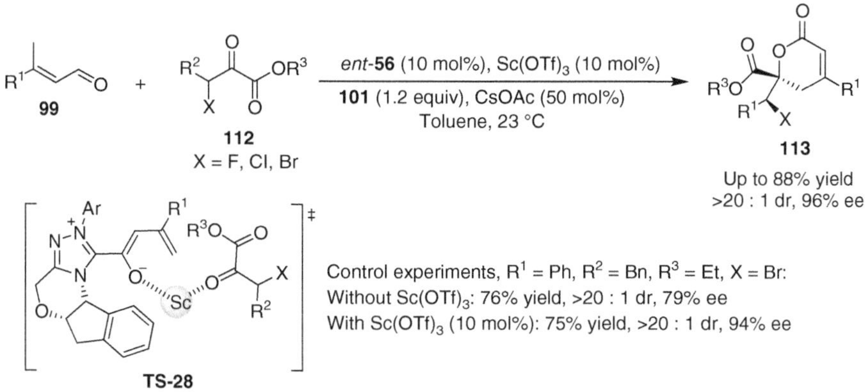

Scheme 7.37 NHC/$Sc(OTf)_3$-catalyzed oxidative dynamic kinetic resolution reaction. Source: Wu et al. [54].

The NHC/LiCl cooperatively catalyzed enantioselective aldol lactonization of acyclic keto acids **114** via a dynamic kinetic resolution furnishes cyclopentane-fused β-lactones **115** with three contiguous stereogenic centers in good yields and high stereocontrol (Scheme 7.38) [55]. The more reactive intermediate **Int-28** *in situ* formed from the keto acids **114** and hexafluorophosphate azabenzotriazole tetramethyl uronium (HATU) **116** reacts with a chiral NHC generated from **68** to furnish two diastereomers **Int-29a** and **Int-29b** that quickly interconvert, leading to the subsequent enantioselective aldol reaction to give a favored diastereomeric intermediate **Int-30**, which undergoes the lactonization to give final product **115**.

7.5 Asymmetric Reactions Driven by Ammonium Salts

Ammonium salts **Int-31**, generated under the catalysis of tertiary amines with MBH carbonates, bear an internal alkene activated by an electron withdrawing group (COR^1) as well as a formally positively charged, allylic leaving group, and represent the most important electrophilic intermediates in chiral tertiary amine catalysis (Scheme 7.39a) [56]. α,β-Unsaturated acylammonium salts **Int-32**, generated from α,β-unsaturated acyl chlorides, carboxylic acids, anhydrides, and esters with chiral tertiary amines, basically lower the energetic potential of the lowest unoccupied molecular orbital (LUMO) of the electrophiles [11d]. In the realm of Lewis base–Lewis acid cooperative catalysis involving ammonium salts, the Lewis acid catalyst can chelate with the nucleophiles to favor a closed conformation (**IV**) or orchestrate both substrates in the transition state through coordinating with both the nucleophile and acylammonium carbonyl group (**V**), and thereby enhances the electrophilicity of α,β-unsaturated acylammoniums (Scheme 7.39b).

A highly enantioselective allylic monofluoromethylation of Morita–Baylis–Hillman carbonates with fluorobis(phenylsulfonyl)methane (FBSM, **118**) proceeds readily under the combined catalysis of bis(cinchona alkaloid) ((DHQD)$_2$AQN, **120**)

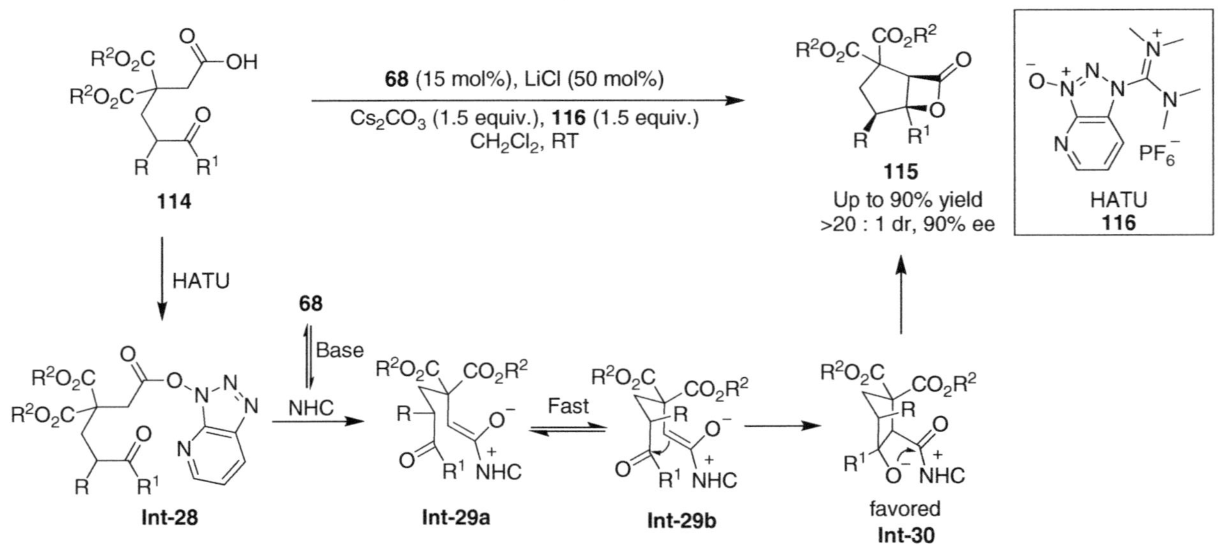

Scheme 7.38 NHC/Li-catalyzed aldol-lactonization. Source: Mondal et al. [55].

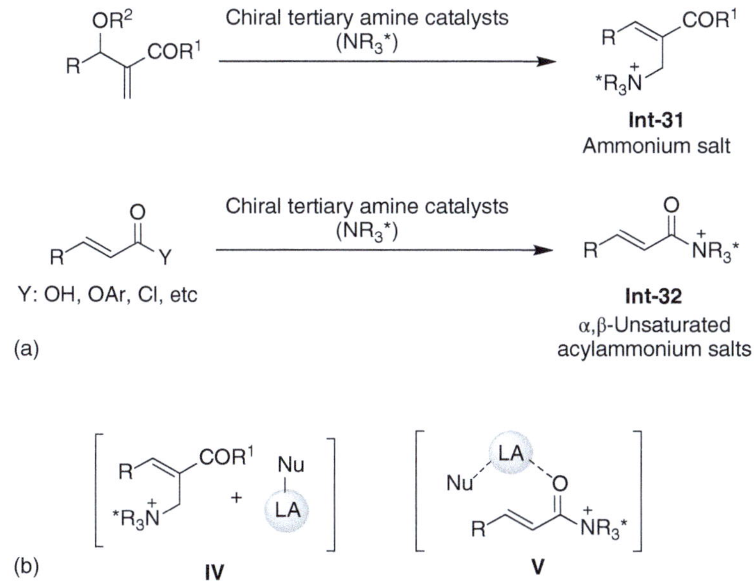

Scheme 7.39 Lewis base–Lewis acid cooperative catalysis involving ammonium salts. (a) Generation of ammonium salts. Source: Based on Liu et al. [56]. (b) Activation modes in Lewis base–Lewis acid cooperative catalysis.

and a Lewis acid (Scheme 7.40) [57]. The evaluation of different Lewis acids, including $FeCl_3$, $FeBr_2$, $Ti(O^iPr)_4$, $AlCl_3$, and $Y(OTf)_3$, reveal that both $FeCl_2$ and $Ti(O^iPr)_4$ are efficient co-catalysts in the cooperative catalysis with $(DHQD)_2AQN$ for this transformation. The additional use of Lewis acid catalyst would improve the yield and the enantioselectivity. The allylic monofluoromethylation product **121** was obtained in good to excellent yields (up to 99%) and with excellent enantioslectivities (up to 97% ee). The asymmetric allylic methylation reaction was also achieved using bis(phenylsulfonyl)methane (BSM, **119**) instead of FBSM **118** under otherwise identical conditions. The ee value for product **122a** was improved from 92 to 96% ee by using $FeCl_2$ as the Lewis acid co-catalyst. It is believed that the bidentate chelation of FBSM with the Lewis acid catalyst locks the FBSM conformation and leads to a more rigid transition state (**TS-29**) for the chiral induction.

In 2013, Romo and coworkers reported an asymmetric [3+3] annulation of an α,β-unsaturated acyl chloride **123a** and enamine **124** enabled by cooperative catalysis of *O*-TMSQD (**20**) and LiCl, leading to dihydropyridinones **125** in 78% yield and with 92% ee (Scheme 7.41) [58].

Later, the same group presented a Michael addition/proton transfer/lactamization cascade process proceeding through the intermediacy of the chiral α,β-unsaturated acylammonium salt, providing feasible access to medium-sized lactams **127** in high yields and with excellent levels of enantioselectivity (Scheme 7.42) [59]. The addition of stoichiometric amounts of LiCl dramatically improves yield and enantioselectivity. The lithium cation assists the chiral Lewis base (TMSQ, **17**) to accelerate

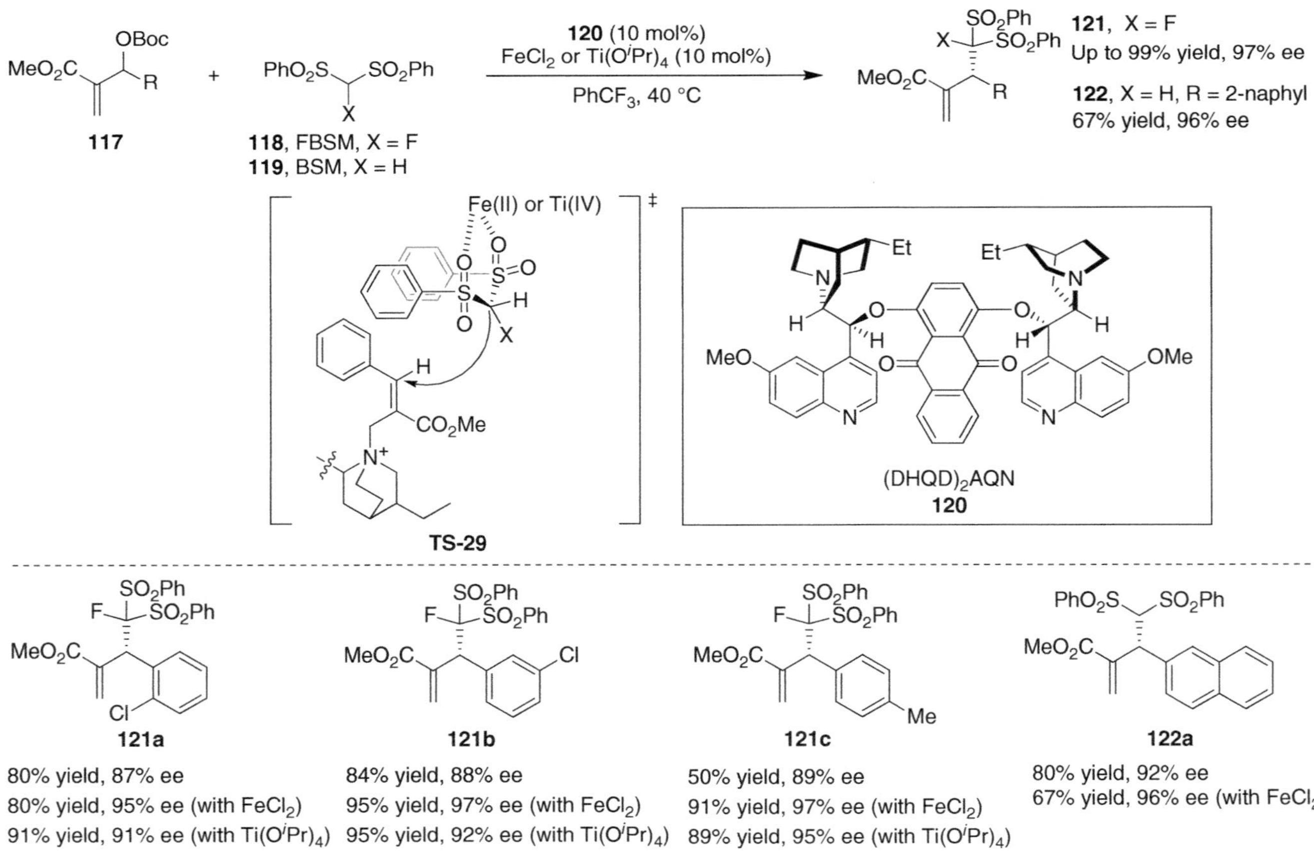

Scheme 7.40 Cinchona alkaloid/Lewis acid co-catalyzed enantioselective methylation of Morita–Baylis–Hillman carbonates. Source: Furukawa et al. [57].

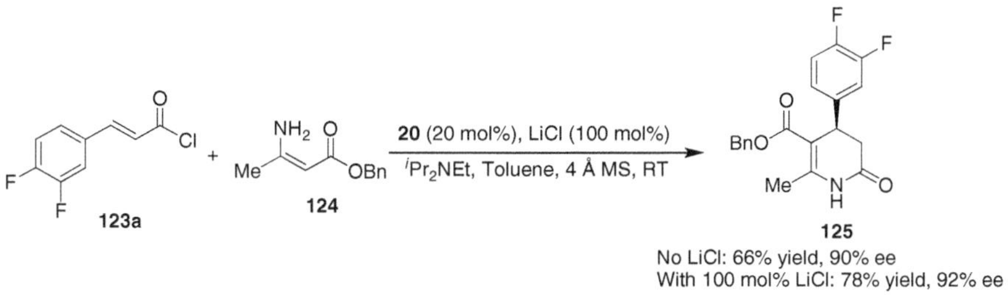

Scheme 7.41 TMSQD/LiCl catalyzed cascade reactions via chiral α,β-unsaturated acylammonium salts. Source: Vellalath et al. [58].

the reaction and control the stereoselectivity by multiple coordinating interactions with the enolate and acylammonium carbonyl group to bring both substrates close enough and thereby to provide a more ordered transition state **TS-30**.

7.6 Asymmetric Reactions Driven by NHC-Mediated α,β-Unsaturated Acyl Azoliums

α,β-Unsaturated acyl azoliums **Int-33**, catalytically generated from α,β-unsaturated acyl chlorides, carboxylic acids, anhydrides, and esters with NHCs, have extensively been employed in asymmetric catalysis via LUMO-lowering activation strategy (Scheme 7.43a) [13e, g]. In Lewis acid/NHC cooperative catalysis involving α,β-unsaturated acyl azoliums, the Lewis acid can activate the Michael acceptor **Int-30** (**V**), or bring both substrates to form a more confined transition state, as shown in **VI** (Scheme 7.43b).

7.6.1 Asymmetric [3+3] Reactions

You and coworkers reported an NHC-catalyzed enantioselective oxidative Michael addition reaction of dicarbonyl compounds to α,β-unsaturated aldehydes (Scheme 7.44a) [60a]. The homoenolate **Int-12** can be readily oxidized by an external organic oxidant **101** to α,β-unsaturated acyl azolium **Int-33**, which can undergo Michael addition with dicarbonyl compounds **128** to deliver substituted dihydropyranones **130** in up to 95% yield and 96% ee. Control experiments, conducted between cinnamaldehyde **54a** (R^1 = Ph) and 1,3-diphenylpropane-1,3-dione **128a** (R^2 = R^3 = Ph), revealed that the addition of catalytic amounts of $NaBF_4$ was essential to achieving enhanced yield and stereocontrol. In contrast, the use of other metal salts, such as $LiBF_4$ and KBF_4, led to diminished results. In 2013, Chi and coworkers established a similar asymmetric [3+3] annulation reaction of saturated aldehydes **131** with dicarbonyl compounds **128** using excess amounts of an oxidant via oxidative NHC catalysis (Scheme 7.44b) [60b]. In this case, LiCl turns out to be the optimal co-catalyst and allows the enantioselectivity to be considerably improved (from 70% to 90% ee). Zheng and coworkers reported an asymmetric synthesis of 4-aryl-dihydropyran-4-carbonitrile

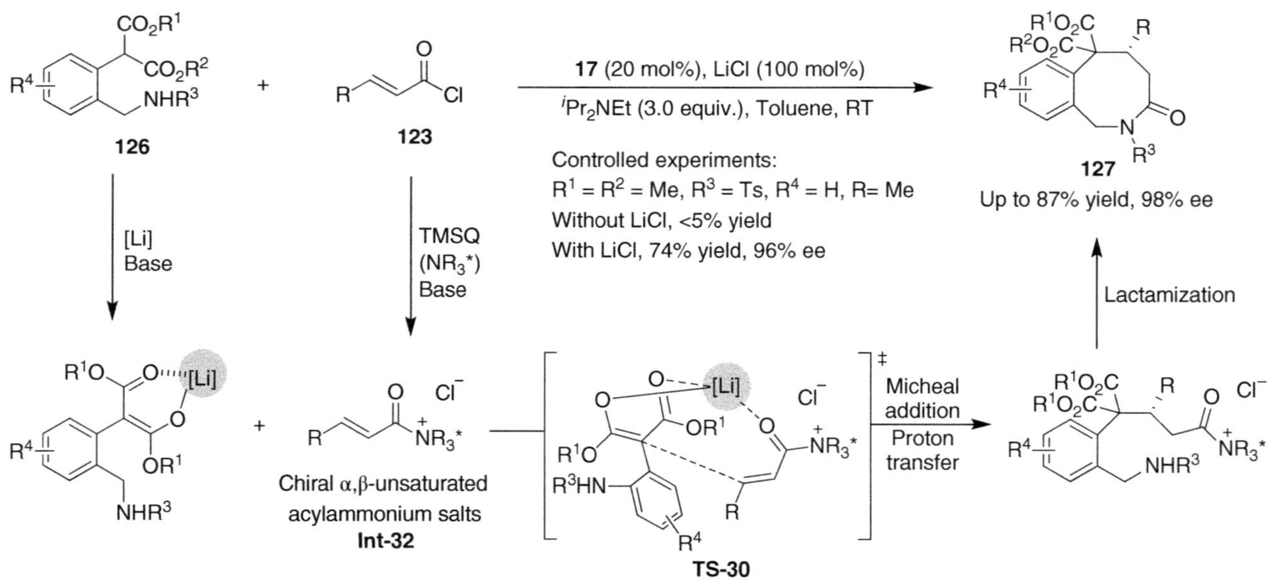

Scheme 7.42 TMSQ/LiCl catalyzed enantioselective synthesis of medium-sized lactams. Source: Kang et al. [59].

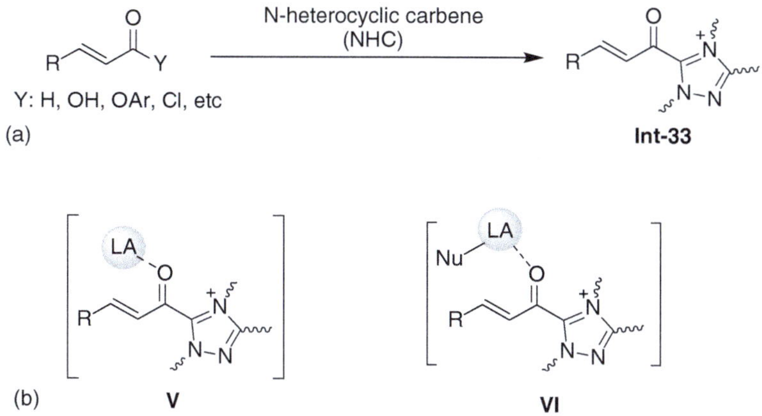

Scheme 7.43 Lewis base–Lewis acid cooperative catalysis involving α,β-unsaturated acyl azolium. (a) Generation of α,β-unsaturated acyl azoliums. Sources: Based on Nair et al. [13a]; Douglas et al. [13b]; Mahatthananchai and Bode [13c]; Menon et al. [13d]; Zhang et al. [13e]; Chen et al. [13f]; Mondal et al. [13g]. (b) Activation modes in Lewis base–Lewis acid cooperative catalysis.

134 through the NHC-catalyzed oxidative cyclization reaction of β-keto esters **133** to β-cyano-substituted α,β-unsaturated aldehydes **132**. Again, the LiCl co-catalyst significantly improves both the reaction efficiency and stereoselectivity (Scheme 7.44c) [60c]. Huang and coworkers developed a similar NHC/LiCl catalyzed [3+3] annulation reaction using potassium 2-oxo-3-enoates **80** as the precursor of α,β-unsaturated acyl azoliums (Scheme 7.44d) [47].

NHC/Li catalyzed [3+3] annulation reaction of 1,3-carbonyl compounds **128** with α,β-unsaturated acyl azoliums generated from aromatic 3-bromoenals **136** and α,β-unsaturated carboxylic acids **137** was investigated by Ma and Yao groups, respectively (Scheme 7.45a,b) [61]. In both cases, the addition of LiCl is able to improve the yield and stereochemical control. A highly chemoselective and enantioselective synthesis of 3,4-2*H*-pyrindin-2-ones **140** was achieved by an NHC-Catalyzed [3+3] cyclization (Scheme 7.45c) [61c]. The NHC catalyst undergoes a nucleophilic addition to the γ-chloroenals **138** followed by C—Cl bond cleavage to generate a transient NHC-bonded dienolate **Int-25**, which participates in a [4+2] cycloaddition reaction with an imine to afford a 5,6-2*H*-pyrindin-2-one derivative **141** as a minor product. The dienolate **Int-25′** can rearrange via a 1,5-H shift to generate α,β-unsaturated acyl azolium **Int-26**, which subsequently undergoes a [3+3] cycloaddition reaction with the imine **139** to afford a 5,6-2*H*-pyrindin-2-one derivative **140** as the major product. The LiCl Lewis acid promotes the formation of the enamine **139′** and accelerates the [3+3] cyclization.

Wang and coworkers reported an enantioselective atroposelective [3+3] annulation of ynals **142** with cyclic 1,3-diones **143** cooperatively catalyzed by NHC and $Mg(OTf)_2$ (Scheme 7.46) [62]. The axially chiral α-pyrone–aryls **145** were obtained in moderate to good yields (up to 76%) and with high enantioselectivities (up to 94% ee). The addition of $Mg(OTf)_2$ makes the reaction cleaner by inhibiting the side reaction

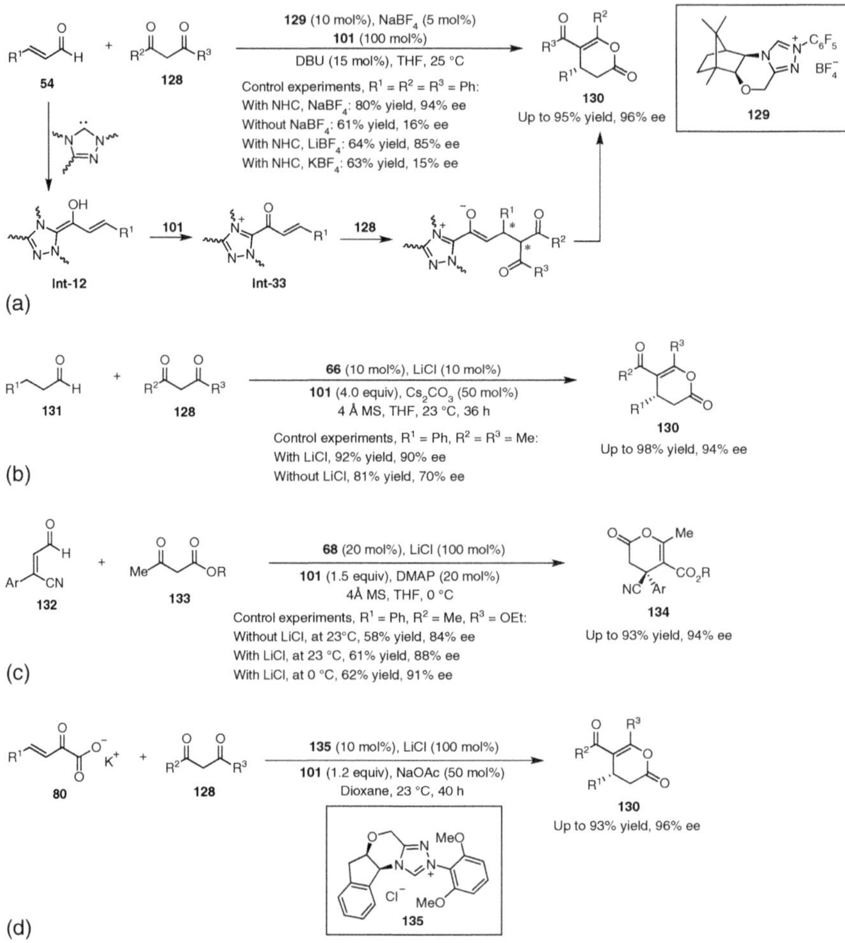

Scheme 7.44 NHC/Lewis acid catalyzed oxidative [3+3] annulation reactions. (a) You's work. Source: Rong et al. [60a]. (b) Chi's work. Source: Mo et al. [60b]. (c) Zheng's work. Source: Wu et al. [60c]. (d) Huang's work. Source: Gao et al. [47].

to generate byproduct **146**. In the cooperative catalytic cycle, the NHC-bonded Breslow intermediate **Int-34**, generated through the nucleophilic addition of the NHC to ynal **142**, is oxidized to deliver alkynyl acyl azolium **Int-35**, which then undergoes the Michael addition with cyclic 1,3-diones **143** via transition state **TS-31** to form the allenolate intermediate **Int-36**. The Lewis acid $Mg(OTf)_2$ actually inhibits "O" attack of the ketoenolate via the transition state **TS-31** but instead promotes the "C" attack via transition state **TS-32**. The subsequent proton transfer and intramolecular lactonization deliver the axially chiral product **145** and regenerate the NHC catalyst.

Chi and coworkers reported an NHC-catalyzed [3+3] annulation reaction between α-bromoenals **147** and thioamides **148** assisted by $Cu(OTf)_2$ (Scheme 7.47) [63]. Both the yield and ee value dropped without the addition of $Cu(OTf)_2$ under otherwise identical reaction conditions. The formation of Cu–NHC complex, as indicated by

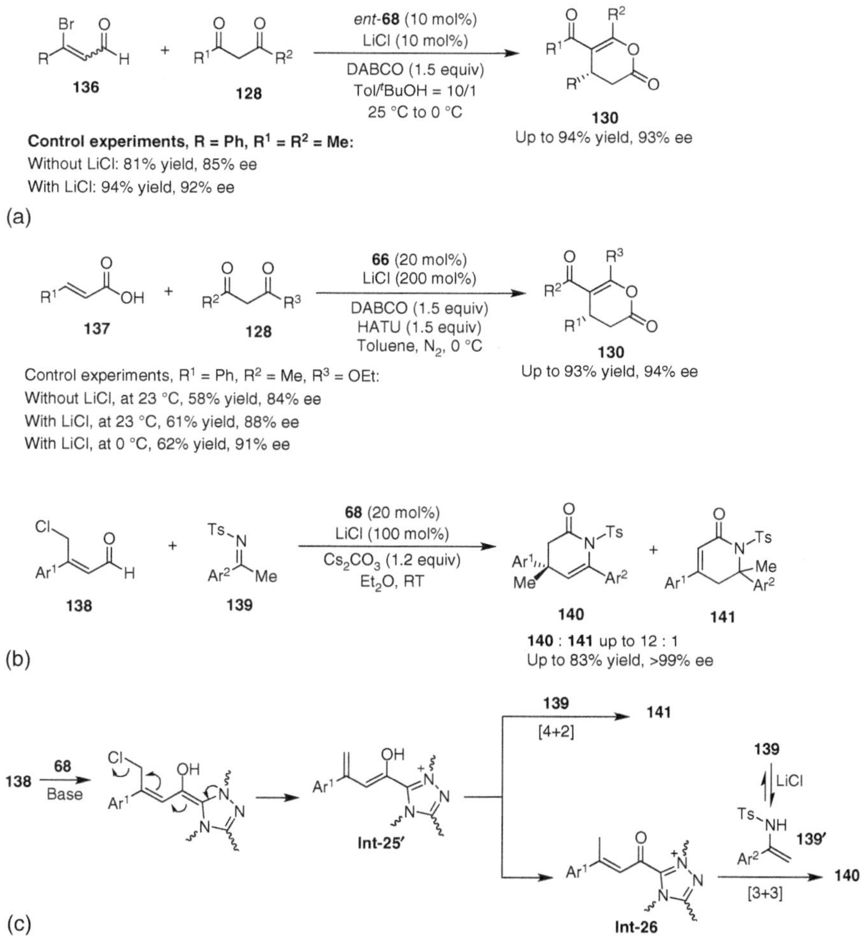

Scheme 7.45 NHC/Li-catalyzed [3+3] annulation reactions. (a) Ma's work. Source: Wang et al. [61a]. (b) Yao's work. Source: Que et al. [61b]. (c) Zhong and Zeng's work. Source: Yan et al. [61c].

high-resolution mass spectrometry analysis of the reaction mixture, consumes part of the active NHC catalyst, and thus the presence of more than 5 mol% of $Cu(OTf)_2$ slows down the reaction.

7.6.2 Asymmetric Cascade Reactions

Studer and coworkers established an oxidative asymmetric Michael addition/[2+2] cyclization reaction of β-keto malonates (**151**) with enals cooperatively catalyzed by NHC and LiCl, allowing for the enantioselective synthesis of highly substituted β-lactones **152** in up to 97% yield and with 99% ee (Scheme 7.48) [64]. The coordination of Li with the O-atom of the α,β-unsaturated acyl azolium, generated from the oxidation of vinyl-Breslow intermediate **Int-12**, makes this Michael acceptor

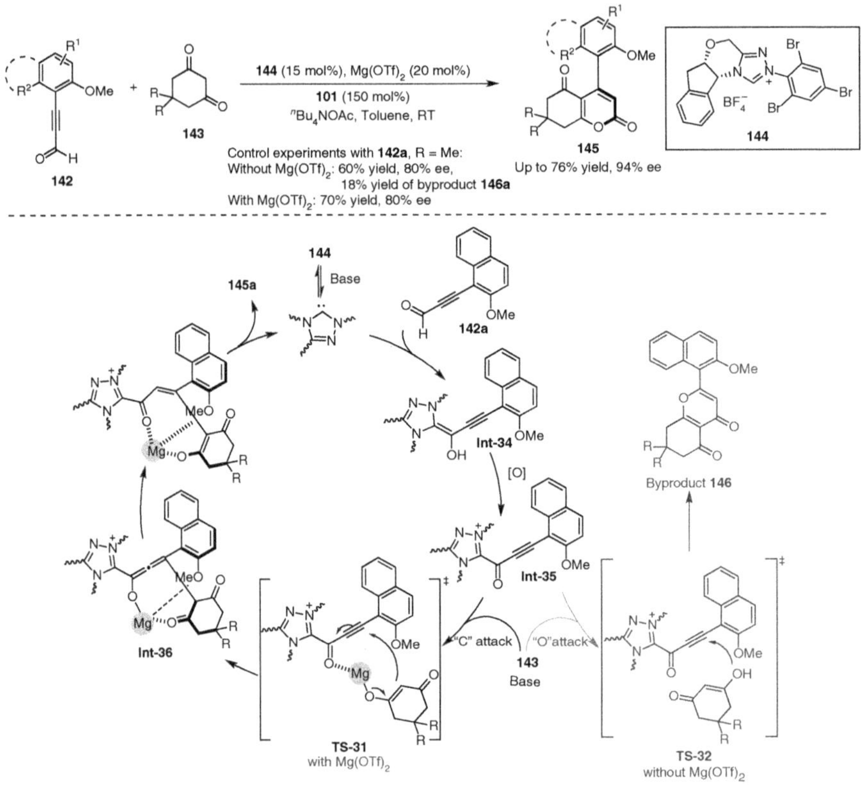

Scheme 7.46 NHC/Mg-catalyzed enantioselective [3+3] atroposelective annulation reaction. Source: Zhao et al. [62].

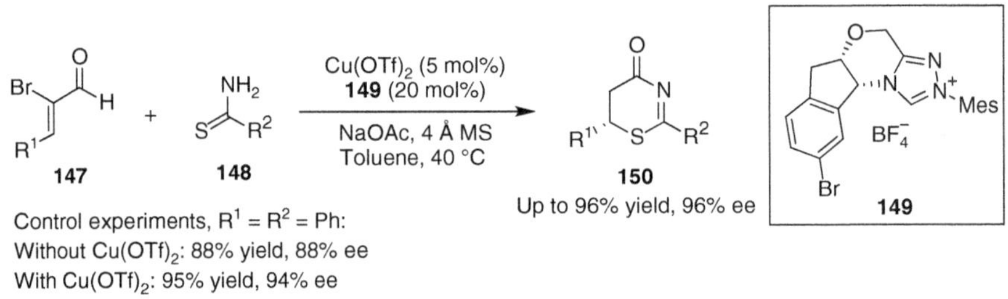

Scheme 7.47 NHC/Cu-mediated [3+3] annulation reaction. Source: Liu et al. [63].

even more reactive toward the nucleophile, leading to an intermediate Int-**37**. The intramolecular formal [2+2] aldol lactonization of **Int-37** affords β-lactones **152**.

The similar catalyst system was extended to an asymmetric Michael addition/[4+2] cyclization reaction of enals **54** and ε-oxo-γ,δ-malonates **153**, providing an efficient entry to cylcopentane- or cyclohexane-fused δ-lactones **154** (Scheme 7.49) [65]. Three contiguous stereogenic centers in **154** are simultaneously built up in great yields and with excellent levels of stereoselectivity. As introduced by Studer and

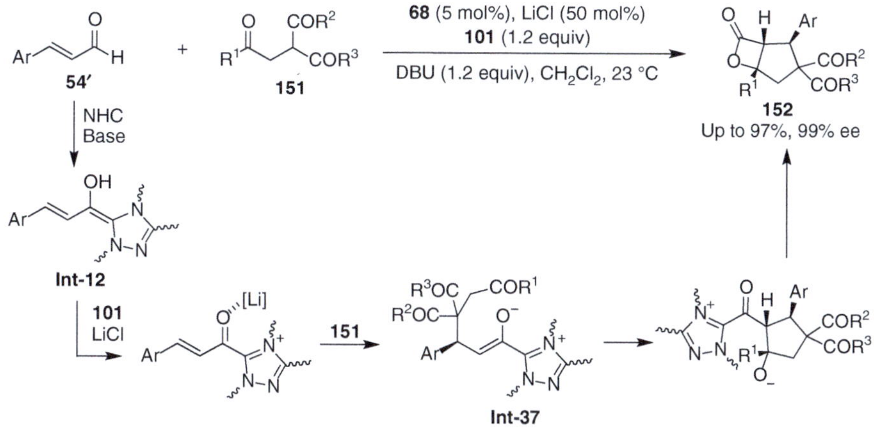

Scheme 7.48 NHC/Li-catalyzed Michael addition/[2+2] annulation reaction. Source: Bera et al. [64].

coworkers, the presence of LiCl Lewis acid greatly enhances the reaction efficiency as only a trace amount of desired product was obtained with NHC catalysis alone (Scheme 7.49a) [65a]. In addition, dropping the loading of LiCl results in much diminished yield and stereoselectivity. The multiple coordination interaction of LiCl with both the α,β-unsaturated acyl azolium intermediate and malonate as shown in **TS-33** accounts for the highly enantioselective Michael addition event to give an intermediate **Int-38**. The sequential cascade diastereoselective Michael addition and lactonization gives the bicyclic products **154** (shown in Scheme 7.49a). Ye and coworkers reported a similar work and found the Lewis acid co-catalyst actually assists both the reaction yield and stereocontrol (Scheme 7.49b) [65b].

A domino Michael/aldol/lactonization/dehydration reaction of enals **54** with four folds of malonates **155** occurs smoothly under the cooperative catalysis of NHC and LiCl to furnish functionalized tricyclic coumarin derivatives **157** in excellent yields and enantioselectivity (Scheme 7.50) [66]. The enolate **Int-39**, generated from the malonate **155** in the presence of 1,8-diazabicyclo[5.4.0]undec-7-ene (DBU) and LiCl, undergoes asymmetric Michael addition to α,β-unsaturated acyl azolium intermediate **Int-33** to give the azolium enolate **Int-40**. The subsequent cascade intramolecular aldol reaction and proton shift processes give rise to **Int-41**. The following δ-lactonization and dehydration form the final product **157**.

7.6.3 Asymmetric Kinetic Resolutions

A kinetic resolution of 3-hydroxy-3-substituted oxindoles **158** with cinnamaldehyde **54a** was successfully accessed by NHC/Lewis acid cooperative catalysis (Scheme 7.51) [67]. The general concept for the kinetic resolution basically relies on the enantioselective acylation of the alcohol **158** with α,β-unsaturated acyl azolium intermediate **Int-33a**. The presence of $Mg(OTf)_2$ and $NaBF_4$ is able to tremendously enhance the resolution efficiency. The $Mg(OTf)_2$ is believed to coordinate with carbonyls of the oxindole and acyl azolium intermediate **Int-33a**

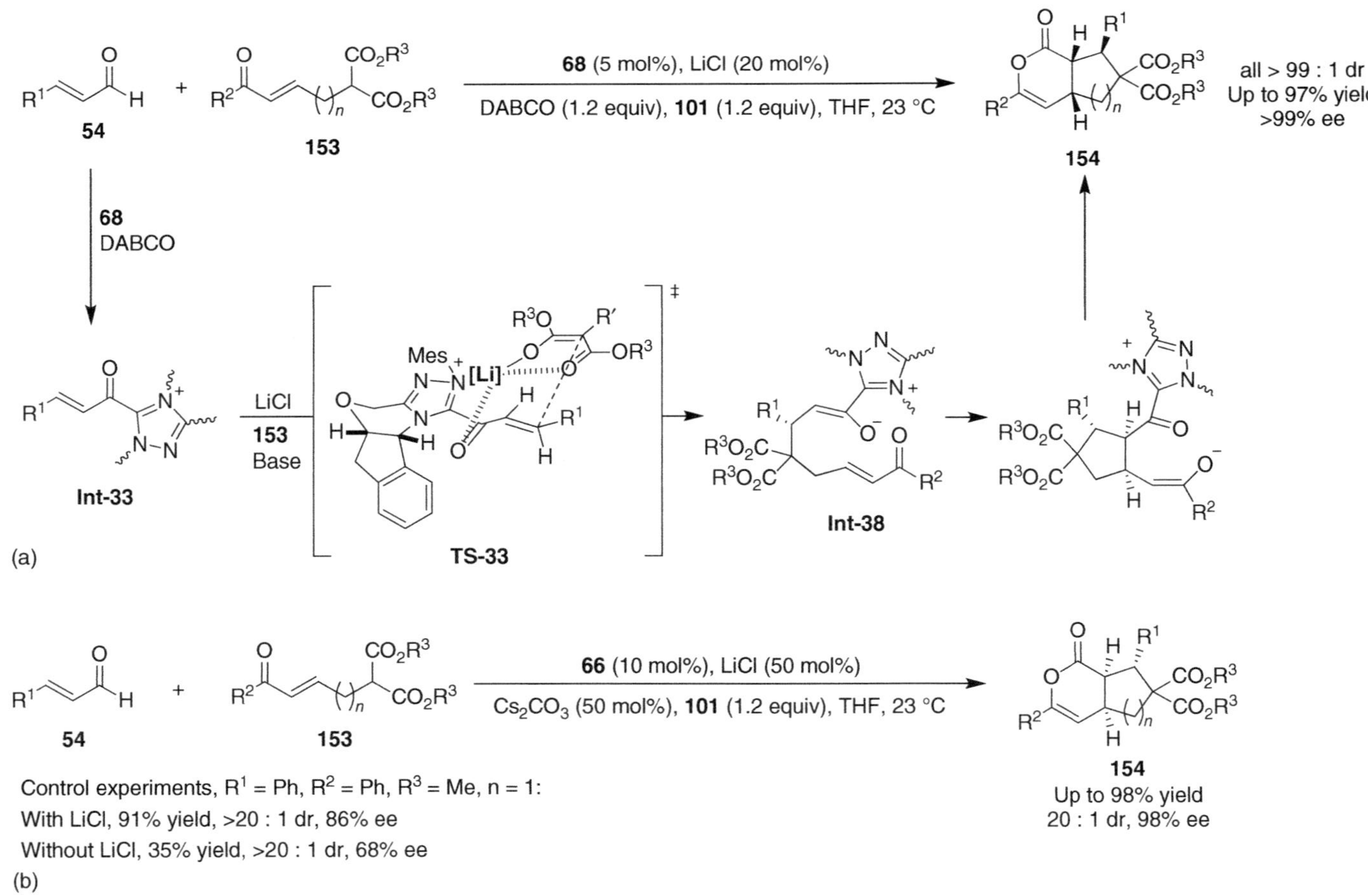

Scheme 7.49 NHC/Li-catalyzed cascade reactions. (a) Studer's work, Source: Bera et al. [65a]. (b) Ye's work. Source: Liang et al. [65b].

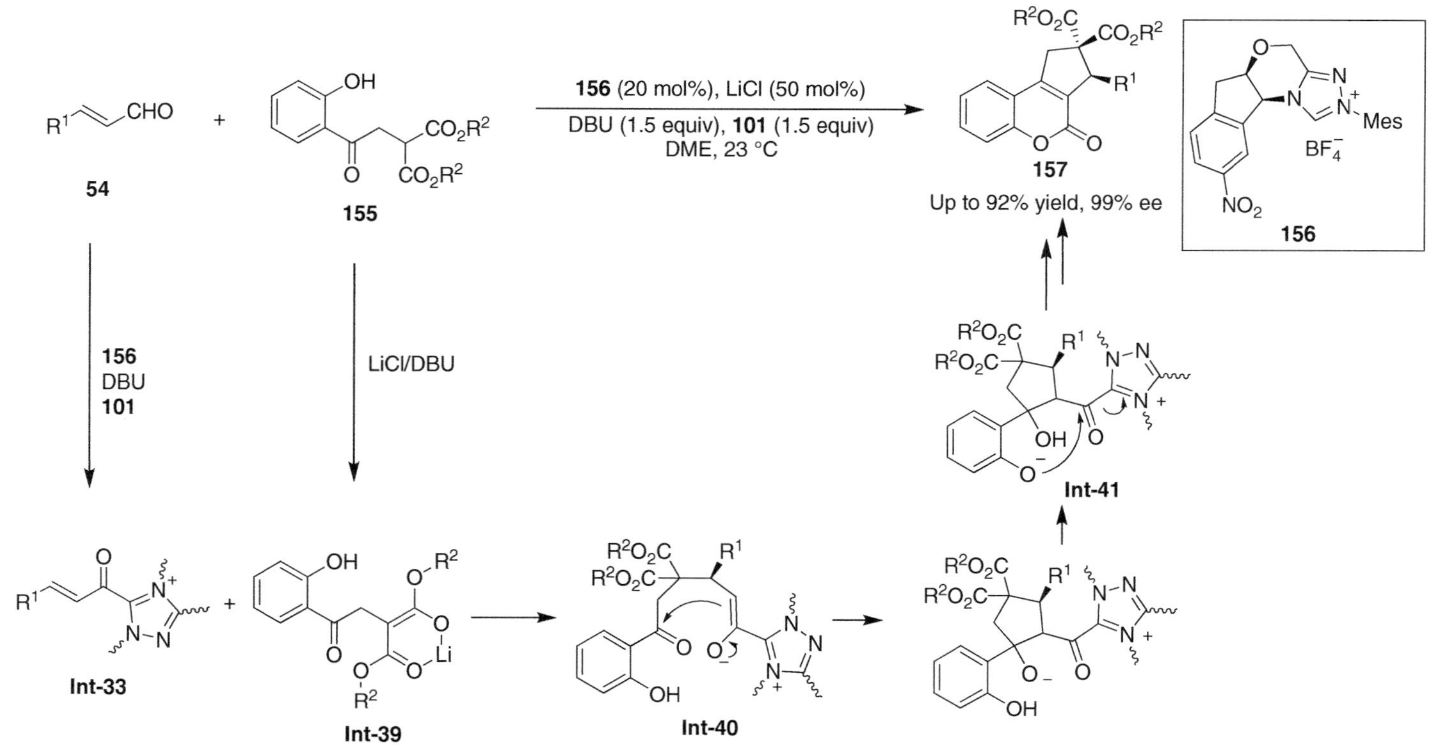

Scheme 7.50 NHC/LiCl-catalyzed cascade reaction. Source: Liu et al. [66].

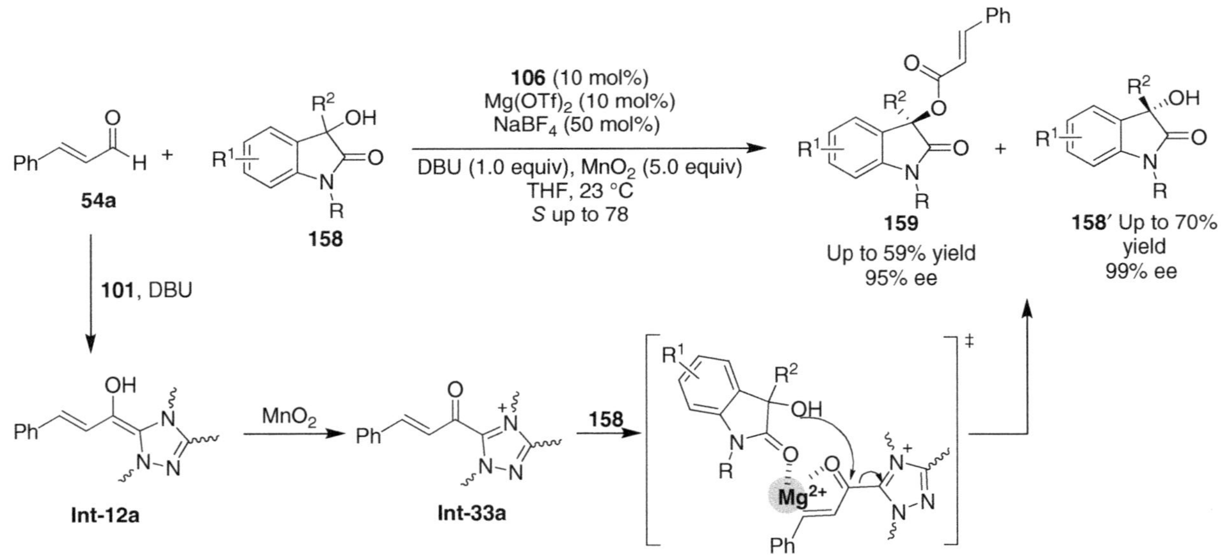

TS-34

Control experiments, R^1 = H, R^1 = Allyl, R = Bn:

With $Mg(OTf)_2$, $NaBF_4$, Conv. = 53%, S = 70

Without $Mg(OTf)_2$, without $NaBF_4$, Conv. = 47%, S = 36

Without $NaBF_4$, Conv. = 49%, S = 48

Scheme 7.51 Kinetic resolution of tertiary alcohols. Sources: Based on Lu et al. [67a]; Lu et al. [67b].

(via **TS-34**) to facilitate the acylation; however, the role of $NaBF_4$ is yet to be understood.

7.7 Conclusion

Lewis base/Lewis acid cooperative catalysis has emerged as a general and productive concept for the development of a wide scope of asymmetric transformations, including Morita–Baylis–Hillman reactions, annulation reactions, cascade reactions, kinetic resolutions, and others. Among the cooperative catalysis of this type, the Lewis acids basically enhance the reactivity of either of reaction components or both by coordinating interactions, and in most cases, improve the stereochemical control. Cinchona alkaloid-based tertiary amines turn out to be compatible with hard Lewis acids, for example, alkali and alkaline earth metal salts. In contrast, NHCs are soft Lewis bases and can survive well with a broad scope of Lewis acids with minimization of "self-quenching," and therefore, the combination of Lewis acids with NHC provides more activation mode for the establishment of new asymmetric transformations. However, the potential of this concept has far less been demonstrated by the known achievements described in this chapter. The future development of enantioselective reactions by Lewis acid/Lewis base cooperative catalysis is still more than anticipated.

References

1 (a) Jensen, W.B. (1980). *The Lewis Acid-Base Concepts*. New York: Wiley-Interscience. (b) Jensen, W.B. (1978). *Chem. Rev.* 78: 1–22.
2 (a) Denmark, S.E. and Beutner, G.L. (2008). *Angew. Chem.* 120: 1584–1663; *Angew. Chem. Int. Ed.* 2008, 47, 1560–1638. (b) List, B. (2012). *Asymmetric Organocatalysis 1 Lewis Base and Acid Catalysts*. New York: Georg Thieme Verlag KG Stuttgart. (c) Vedejs, E. and Denmark, S.E. (2016). *Lewis Base Catalysis in Organic Synthesis*. Wiley-VCH.
3 Candish, L., Nakano, Y., and Lupton, D.W. (2014). *Synthesis* 46: 1823–1835.
4 (a) Erdmann, H. and Huth, P. (1897). *J. Prakt. Chem.* 56: 6–27. (b) Einhorn, A. and Hollandt, F. (1898). *Justus Liebigs Ann. Chem.* 301: 95–115. (c) Verley, A. and Bölsing, F. (1901). *Ber. Dtsch. Chem. Ges.* 34: 3354–3358.
5 Wegler, R. (1932). *Liebigs Ann. Chem.* 498: 62–76.
6 France, S., Guerin, D.J., Miller, S.J., and Lectka, T. (2003). *Chem. Rev.* 103: 2985–3012.
7 For selected reviews, see:(a)Enders, D., Niemeier, O., and Henseler, A. (2007). *Chem. Rev.* 107: 5606–5655. (b) Marion, N., Díez-González, S., and Nolan, S.P. (2007). *Angew. Chem. Int. Ed.* 46: 2988–3000. (c) Grossmann, A. and Enders, D. (2012). *Angew. Chem. Int. Ed.* 51: 314–325. (d) Hopkinson, M.N., Richter, C., Schedler, M., and Glorius, F. (2014). *Nature* 510: 485–496. (e) Flanigan, D.M.,

Romanov-Michailidis, F., White, N.A., and Rovis, T. (2015). *Chem. Rev.* 115: 9307–9387.

8 (a) Pracejus, H. and Mätje, H. (1964). *J. Prakt. Chem. 4. Reihe* 24: 195. (b) Samtleben, R. and Pracejus, H.J. (1972). *Prakt. Chem.* 314: 157–169. (c) Tian, S.-K., Chen, Y., Hang, J. et al. (2004). *Acc. Chem. Res.* 37: 621–631. (d) Marcelli, T. and Hiemstra, H. (2010). *Synthesis* 2010: 1229–1279.

9 (a) Ruble, J.C. and Fu, G.C. (1996). *J. Org. Chem.* 61: 7230–7231. (b) Fu, G.C. (2004). *Acc. Chem. Res.* 37: 542–547. (c) Wurz, R.P. (2007). *Chem. Rev.* 107: 5570–5595. (d) Sun, L., Zhang, Z.-F., Xie, F., and Zhang, W.-B. (2008). *Chin. J. Org. Chem.* 28: 574–587.

10 (a) Birman, V.B. and Li, X. (2006). *Org. Lett.* 8: 1351–1354. (b) Birman, V.B., Jiang, H., Li, X. et al. (2006). *J. Am. Chem. Soc.* 128: 6536–6537. (c) Taylor, J.E., Bull, S.D., and Williams, J.M.J. (2012). *Chem. Soc. Rev.* 41: 2109–2121. (d) Merad, J., Pons, J.-M., Chuzel, O., and Bressy, C. (2016). *Eur. J. Org. Chem.* 2016: 5589–5610. (e) Birman, V.B. (2016). *Aldrichim. Acta* 49: 23–33.

11 (a) Gaunt, M.J. and Johansson, C.C.C. (2007). *Chem. Rev.* 107: 5596–5605. (b) Müller, C.E. and Schreiner, P.R. (2011). *Angew. Chem. Int. Ed.* 50: 6012–6042. (c) Morrill, L.C. and Smith, A.D. (2014). *Chem. Soc. Rev.* 43: 6214–6226. (d) Vellalath, S. and Romo, D. (2016). *Angew. Chem. Int. Ed.* 55: 13934–13943. (e) McLaughlin, C. and Smith, A.D. (2021). *Chem. Eur. J.* 27: 1533–1555.

12 (a) Ukai, T., Tanaka, R., and Dokawa, T. (1943). *J. Pharm. Soc. Jpn.* 63: 296–300. (b) Breslow, R. (1958). *J. Am. Chem. Soc.* 80: 3719–3726.

13 (a) Nair, V., Menon, R.S., Biju, A.T. et al. (2011). *Chem. Soc. Rev.* 40: 5336–5346. (b) Douglas, J., Churchill, G., and Smith, A.D. (2012). *Synthesis* 44: 2295–2309. (c) Mahatthananchai, J. and Bode, J.W. (2014). *Acc. Chem. Res.* 47: 696–707. (d) Menon, R.S., Biju, A.T., and Nair, V. (2015). *Chem. Soc. Rev.* 44: 5040–5052. (e) Zhang, C., Hooper, J.F., and Lupton, D.W. (2017). *ACS Catal.* 7: 2583–2596. (f) Chen, X.-Y., Liu, Q., Chauhan, P., and Enders, D. (2018). *Angew. Chem. Int. Ed.* 57: 3862–3873. (g) Mondal, S., Yetra, S.R., Mukherjee, S., and Biju, A.T. (2019). *Acc. Chem. Res.* 52: 425–436.

14 Yamamoto, H. (ed.) (2000). *Lewis Acids in Organic Synthesis*. Weinheim: Wiley VCH.

15 (a) Paull, D.H., Abraham, C.J., Scerba, M.T. et al. (2008). *Acc. Chem. Res.* 41: 655–663. (b) Cohen, D.T. and Scheidt, K.A. (2012). *Chem. Sci.* 3: 53–57. (c) Stegbauer, L., Sladojevich, F., and Dixon, D.J. (2012). *Chem. Sci.* 3: 942–958. (d) Wang, M.H. and Scheidt, K.A. (2016). *Angew. Chem. Int. Ed.* 55: 14912–14922. (e) Jia, Q., Li, Y., Lin, Y., and Ren, Q. (2019). *Catalysts* 9: 863–885. (f) Nagao, K. and Ohmiya, H. (2019). *Top. Curr. Chem.* 377: 35.

16 Pearson, R.G. (1963). *J. Am. Chem. Soc.* 85: 3533–3539.

17 (a) Aggarwal, V.K., Tarver, G.J., and McCague, R. (1996). *J. Chem. Soc., Chem. Commun.*: 2713–2714. (b) Aggarwal, V.K., Mereu, A., Tarver, G.J., and McCague, R. (1998). *J. Org. Chem.* 63: 7183–7189. (c) Aggarwal, V.K., Dean, D.K., Mereu, A., and Williams, R. (2002). *J. Org. Chem.* 67: 510–514.

18 Barrett, A.G.M., Cook, A.S., and Kamimura, A. (1998). *Chem. Commun.*: 2533–2534.

19 France, S., Wack, H., Hafez, A.M. et al. (2002). *Org. Lett.* 4: 1603–1605.
20 Shi, M. and Jiang, J.-K. (2002). *Tetrahedron: Asymmetry* 13: 1941–1947.
21 (a) Bugarin, A. and Connell, B.T. (2009). *J. Org. Chem.* 74: 4638–4641. (b) Bugarin, A. and Connell, B.T. (2010). *Chem. Commun.* 46: 2644–2646.
22 France, S., Shah, M.H., Weatherwax, A. et al. (2005). *J. Am. Chem. Soc.* 127: 1206–1215.
23 (a) Wynberg, H. and Staring, E.G.J. (1982). *J. Am. Chem. Soc.* 104: 166–168. (b) Zhu, C., Shen, X., and Nelson, S.G. (2004). *J. Am. Chem. Soc.* 126: 5352–5353.
24 (a) Calter, M.A., Tretyak, O.A., and Flaschenriem, C. (2005). *Org. Lett.* 7: 1809–1812. (b) Huang, Y. and Calter, M.A. (2007). *Tetrahedron Lett.* 48: 1657–1659.
25 Murauski, K.J.R., Walden, D.M., Cheong, P.H.-Y., and Scheidt, K.A. (2017). *Adv. Synth. Catal.* 359: 3713–3719.
26 Borrmann, D. and Wegler, R. (1966). *Chem. Ber.* 99: 1245–1251.
27 (a) Koch, F.M. and Peters, R. (2007). *Angew. Chem. Int. Ed.* 46: 2685–2689. (b) Koch, F.M. and Peters, R. (2011). *Chem. Eur. J.* 17: 3679–3692.
28 Zajac, M. and Peters, R. (2009). *Chem. Eur. J.* 15: 8204–8222.
29 Leverett, C.A., Purohit, V.C., and Romo, D. (2010). *Angew. Chem. Int. Ed.* 49: 9479–9483.
30 Kong, W. and Romo, D. (2017). *J. Org. Chem.* 82: 13161–13170.
31 (a) Nagao, Y., Hirai, T., Goto, S. et al. (1998). *J. Am. Chem. Soc.* 120: 3104–3110. (b) Minkin, V.I. and Minyaev, R.M. (2001). *Chem. Rev.* 101: 1247–1266. (c) Brameld, K.A., Kuhn, B., Reuter, D.C., and Stahl, M. (2008). *J. Chem. Inf. Model.* 48: 1–24. (d) Liu, P., Yang, X., Birman, V.B., and Houk, K.N. (2012). *Org. Lett.* 14: 3288–3291.
32 (a) Abraham, C.J., Paull, D.H., Scerba, M.T. et al. (2006). *J. Am. Chem. Soc.* 128: 13370–13371. (b) Paull, D.H., Alden-Danforth, E., Wolfer, J. et al. (2007). *J. Org. Chem.* 72: 5380–5382. (c) Paull, D.H., Wolfer, J., Grebinski, J.W. et al. (2007). *Chimia* 61: 240–246.
33 Abraham, C.J., Paull, D.H., Bekele, T. et al. (2008). *J. Am. Chem. Soc.* 130: 17085–17094.
34 Xu, X., Wang, K., and Nelson, S.G. (2007). *J. Am. Chem. Soc.* 129: 11690–11691.
35 (a) Tiseni, P.S. and Peters, R. (2007). *Angew. Chem. Int. Ed.* 46: 5325–5328. (b) Tiseni, P.S. and Peters, R. (2008). *Org. Lett.* 10: 2019–2022. (c) Tiseni, P.S. and Peters, R. (2010). *Chem. Eur. J.* 16: 2503–2517.
36 (a) Paull, D.H., Scerba, M.T., Alden-Danforth, E. et al. (2008). *J. Am. Chem. Soc.* 130: 17260–17261. (b) Erb, J., Paull, D.H., Dudding, T. et al. (2011). *J. Am. Chem. Soc.* 133: 7536–7546.
37 (a) Nolan, S.P. (2006). *N-Heterocyclic Carbenes in Synthesis*. Weinheim, Germany: Wiley-VCH. (b) Díez-González, S., Marion, N., and Nolan, S.P. (2009). *Chem. Rev.* 109: 3612–3676. (c) Janssen-Müller, D., Schlepphorst, C., and Glorius, F. (2017). *Chem. Soc. Rev.* 46: 4845–4854.
38 Cardinal-David, B., Raup, D.E.A., and Scheidt, K.A. (2010). *J. Am. Chem. Soc.* 132: 5345–5347.

39 (a) Domingo, L.R., Zaragozá, R.J., and Arnó, M. (2011). *Org. Biomol. Chem.* 9: 6616–6622. (b) Domingo, L.R., Sáez, J.A., and Arnó, M. (2014). *Org. Biomol. Chem.* 12: 895–904.
40 Raup, D.E.A., Cardinal-David, B., Holte, D., and Scheidt, K.A. (2010). *Nat. Chem.* 2: 766–771.
41 Sohn, S.S., Rosen, E.L., and Bode, J.W. (2004). *J. Am. Chem. Soc.* 126: 14370–14371.
42 Cohen, D.T., Cardinal-David, B., Roberts, J.M. et al. (2011). *Org. Lett.* 13: 1068–1071.
43 Cohen, D.T., Cardinal-David, B., and Scheidt, K.A. (2011). *Angew. Chem. Int. Ed.* 50: 1678–1682.
44 (a) Phillips, E.M., Wadamoto, M., Chan, A., and Scheidt, K.A. (2007). *Angew. Chem. Int. Ed.* 46: 3107–3110. (b) Wang, Z.-Y., Ding, Y.-L., Wang, G., and Cheng, Y. (2016). *Chem. Commun.* 52: 788–791.
45 Wang, M., Rong, Z.-Q., and Zhao, Y. (2014). *Chem. Commun.* 50: 15309–15312.
46 (a) Phillips, E.M., Riedrich, M., and Scheidt, K.A. (2010). *J. Am. Chem. Soc.* 132: 13179–13181. (b) Dugal-Tessier, J., O'Bryan, E.A., Schroeder, T.B.H. et al. (2012). *Angew. Chem. Int. Ed.* 51: 4963–4967.
47 Gao, Y., Ma, Y., Xu, C. et al. (2018). *Adv. Synth. Catal.* 360: 479–484.
48 Qi, J., Xie, X., Han, R. et al. (2013). *Chem. Eur. J.* 19: 4146–4150.
49 Lee, A. and Scheidt, K.A. (2014). *Angew. Chem. Int. Ed.* 53: 7594–7598.
50 (a) Chen, J., Yuan, P., Wang, L., and Huang, Y. (2017). *J. Am. Chem. Soc.* 139: 7045–7051. (b) Wang, L., Wang, Q., Chen, J., and Huang, Y. (2018). *Acta Chim. Sinica* 76: 850–856.
51 Zhang, Z.-J., Wen, Y.-H., Song, J., and Gong, L.-Z. (2021). *Angew. Chem. Int. Ed.* 60: 3268–3276.
52 Mo, J., Chen, X., and Chi, Y.R. (2012). *J. Am. Chem. Soc.* 134: 8810–8813.
53 (a) Liu, R., Yu, C., Xiao, Z. et al. (2014). *Org. Biomol. Chem.* 12: 1885–1891. (b) Xiao, Z., Yu, C., Li, T. et al. (2014). *Org. Lett.* 16: 3632–3635. (c) Cheng, J.-T., Chen, X.-Y., Gao, Z.-H., and Ye, S. (2015). *Eur. J. Org. Chem.* 2015: 1047–1053. (d) Jia, W.-Q., Zhang, H.-M., Zhang, C.-L. et al. (2016). *Org. Chem. Front.* 3: 77–81.
54 Wu, Z., Li, F., and Wang, J. (2015). *Angew. Chem. Int. Ed.* 54: 1629–1633.
55 Mondal, S., Mukherjee, S., Das, T.K. et al. (2017). *ACS Catal.* 7: 3995–3999.
56 Liu, T.-Y., Xie, M., and Chen, Y.-C. (2012). *Chem. Soc. Rev.* 41: 4101–4112.
57 Furukawa, T., Kawazoe, J., Zhang, W. et al. (2011). *Angew. Chem. Int. Ed.* 50: 9684–9688.
58 Vellalath, S., Van, K.N., and Romo, D. (2013). *Angew. Chem. Int. Ed.* 52: 13688–13693.
59 Kang, G., Yamagami, M., Vellalath, S., and Romo, D. (2018). *Angew. Chem. Int. Ed.* 57: 6527–6531.
60 (a) Rong, Z.-Q., Jia, M.-Q., and You, S.-L. (2011). *Org. Lett.* 13: 4080–4083. (b) Mo, J., Shen, L., and Chi, Y.R. (2013). *Angew. Chem. Int. Ed.* 52: 8588–8591. (c) Wu, Q., Li, C., Wang, W. et al. (2017). *Org. Chem. Front.* 4: 2323–2326.

61 (a) Wang, G., Chen, X., Miao, G. et al. (2013). *J. Org. Chem.* 78: 6223–6232. (b) Que, Y., Lu, Y., Wang, W. et al. (2016). *Chem. Asian J.* 11: 678–681. (c) Yan, J., Song, Z., Zhao, C. et al. (2020). *Org. Lett.* 22: 3329–3334.

62 Zhao, C., Guo, D., Munkerup, K. et al. (2018). *Nat. Commun.* 9: 611–620.

63 Liu, C., Wu, S., Xu, J. et al. (2019). *Org. Lett.* 21: 9493–9496.

64 Bera, S., Samanta, R.C., Daniliuc, C.G., and Studer, A. (2014). *Angew. Chem. Int. Ed.* 53: 9622–9626.

65 (a) Bera, S., Daniliuc, C.G., and Studer, A. (2015). *Org. Lett.* 17: 4940–4943. (b) Liang, Z.-Q., Wang, D.-L., Zhang, H.-M., and Ye, S. (2015). *Org. Lett.* 17: 5140–5143.

66 Liu, Q., Chen, X.-Y., Puttreddy, R. et al. (2018). *Angew. Chem. Int. Ed.* 57: 17100–17103.

67 (a) Lu, S., Poh, S.B., Siau, W.-Y., and Zhao, Y. (2013). *Angew. Chem. Int. Ed.* 52: 1731–1734. (b) Lu, S., Poh, S.B., Siau, W.-Y., and Zhao, Y. (2013). *Synlett* 24: 1165–1169.

8

Lewis Base-Transition Metal Cooperative Catalysis

8.1 Introduction

As summarized in Chapter 7, ternary amines, phosphines and related trivalent phosphorus, and N-heterocyclic carbenes (NHCs) are among the most important Lewis base catalysts [1–4]. Both phosphines and NHCs are typical and are among the most popular dative ligands in organotransition metal chemistry and homogeneous catalysis [5, 6]. The coordination of these Lewis bases and transition metals principally attenuates the nucleophilicity, or at least the amounts of the catalytically active Lewis base. The unavoidable and disliked "self-quenching" event also happens widely in the Lewis base–Lewis acid combined catalysis (see Chapter 7). However, the presence of either phosphines or NHCs in some cases improves the catalytic activity of transition metal complexes. In this context, by elaborately tuning the ratio of the Lewis bases and transition metals the optimal combined catalyst systems in most cases can be achieved [7]. Because the C—N bond is shorter than the C—P bond and the C—N—C bond angle is larger than the C—P—C bond, transition metal complexes of tertiary amines are more sterically congested than those of tertiary phosphines. The structural difference between amines and phosphines, together with the hard–soft mismatch with most of the transition metals, makes ternary amines coordinate with transition metals more weakly than phosphines and NHC. In this regard, the ternary amines are more compatible with transition metals to offer much easier access to the Lewis base-transition metal combined catalysis [8]. Similar to the principle demonstrated in the proof of concept, the Lewis base basically activates nucleophiles by covalent interactions while the transition metal activates the other reaction partner by undergoing elementary organometallic reactions, respectively.

In 2003, Krische and coworkers first demonstrated the robustness of cooperative catalysis of transition metal complexes and Lewis base catalysts in the creation of unusual reactions showcased by an enone cycloallylation of mono-enone mono-allylic acetate **1** (Scheme 8.1) [9]. The enolate generated from the nucleophilic addition of tributyl phosphine to the enone moiety of **1** undergoes Tsuji–Trost reaction with concomitantly formed electrophilic π-allyl palladium species and followed by the elimination of tributyl phosphine to give cyclopentenyl methanone derivatives **2**. The binary catalyst system nicely integrates the nucleophilicity of the

Asymmetric Organo-Metal Catalysis: Concepts, Principles, and Applications, First Edition. Liu-Zhu Gong.
© 2022 WILEY-VCH GmbH. Published 2022 by WILEY-VCH GmbH.

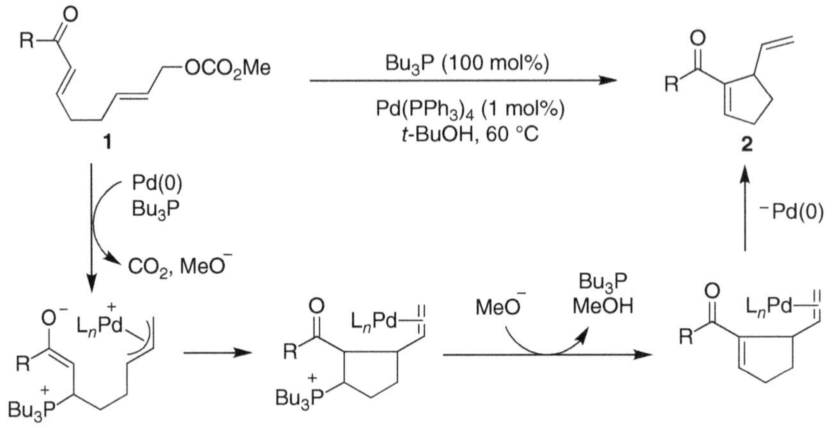

Scheme 8.1 The first combination of Lewis base catalysis and transition metal catalysis. Source: Modified from Jellerichs et al. [9].

Morita–Baylis–Hillman reaction and the electrophilicity of the Trost–Tsuji reaction to allow a unique transformation that is otherwise unable to be accessed. This work initiates the field of combining transition metal and Lewis base catalysis and puts forward the concept of Lewis base and transition metal cooperative catalysis for the first time.

Since this seminal event, the phosphines, tertiary amines, and NHCs, in concert with transition metal complexes, provide a diverse range of catalytic activity and have been frequently applied for the development of asymmetric transformations [7, 8]. The coupling of Lewis base-bonded nucleophiles, including enolate zwitterion [10], and NHC-mediated nucleophilic intermediates [11] (Scheme 8.2a), with the transition metal-mediated electrophilic intermediates, including π-(allyl)palladium, π-(allyl)iridium, alkynyl-copper, alkynyl-gold, copper-allenylidene, copper-amide, and acyl-palladium as well (Scheme 8.2b), turns out to be the key event in the Lewis

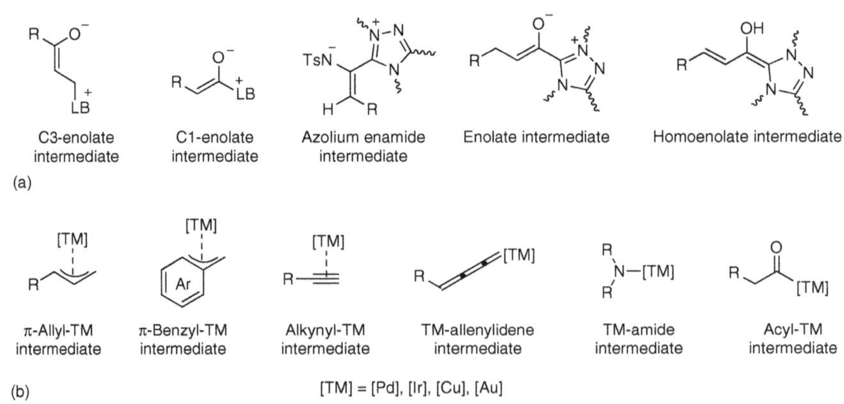

Scheme 8.2 Lewis base-transition metal cooperative catalysis. (a) Lewis base (LB)-bonded nucleophilic intermediates. (b) Transition metal activated electrophilic intermediates.

base and transition metal cooperatively catalyzed reactions. This chapter focuses on summarizing the asymmetric transition metal/Lewis base combined catalysis that integrates the reactivity of these intermediates.

8.2 Phosphine and Transition Metal Cooperative Catalysis

Nucleophilicity-featured chiral phosphines [3] have been receiving increasing attention in asymmetric organocatalysis arising from pioneering contributions by Rauhut, Currier, Lu, and others [12]. In phosphine and transition metal combined catalyst systems recently developed, the phosphine catalyst not only acts as a Lewis base organocatalyst which is compatible with the metal catalyst, on some occasions but also acts as a ligand toward the metal center.

Gong and coworkers reported a highly efficient sequential C–H functionalization/asymmetric allylation or Michael addition reaction of indoles **3** and 3-diazooxindoles **4** with an allenoate **6** or vinyl ketones **9** to afford 3,3-indolyloxindole derivatives **7** or **10** enabled by Rh(II)/chiral phosphine combined catalysis (Scheme 8.3) [13]. The one-pot reaction principally encompasses two catalytic cycles. Mechanistically, the 3-diazooxindole **4a** reacts with the rhodium catalyst to afford rhodium carbene species, and the following C–H insertion reaction of indole **3** generates a nucleophilic intermediate **Int-1**. In the sequential C–H functionalization/allylation reaction, lithium-stabilized phosphonium enolate **Int-2** is the key intermediate and acts as a stronger base to facilitate the enolization of **Int-1**. The ternary system synergistically integrates the Lewis acid, hydrogen bonding, and Lewis base activations, as shown in the transition state **TS-I**. The oxindole lithium enolate interacts with a chiral thiourea–chloride complex through Li–Cl coordination to deliver relatively higher levels of stereochemical control than in the absence of LiCl.

A cascade annulation of Morita–Baylis–Hillman (MBH) carbonates prepared from isatins **11** and allylic carbonates **12** enabled by phosphine/palladium cooperative catalysis proceeds efficiently to assemble spirooxindoles **13** incorporating a 4-methylene-2-cyclopentene motif (Scheme 8.4) [14]. The enantioselective version of this reaction by using chiral phosphine catalyst **14** provides the corresponding spirooxindole **13aa** in 86% yield and with 65% ee (Eq. (2), Scheme 8.4). In the stereogenic step, the E-configured allylic [P*]-ylide species **Int-3**, generated from nucleophilic addition of chiral phosphine catalyst **14** to **11a**, undergoes the allylic alkylation to deliver **Int-5**. The subsequent oxidative addition of olefin-coordinated Pd(0) to carbon–phosphonium bond in **Int-5** (Pd-decomplexation cation was detected by high-resolution mass spectrometry [HRMS] analysis), turns out to be the key step to deliver **Int-6**. Finally, the automatic tandem intramolecular Heck coupling of **Int-6** occurs to afford the spirooxindole **13aa**.

An asymmetric [5+4] annulation of ortho-quinone methides **15** and vinylethylene carbonates (VECs) **16** cooperatively catalyzed by 2,2′-bis(diphenylphosphino)-1,1′-binaphthyl (BINAP) and palladium complex yields optically active nine-membered benzo-heterocycles **18** (Scheme 8.5) [15]. BINAP **17** undergoes a nucleophilic attack

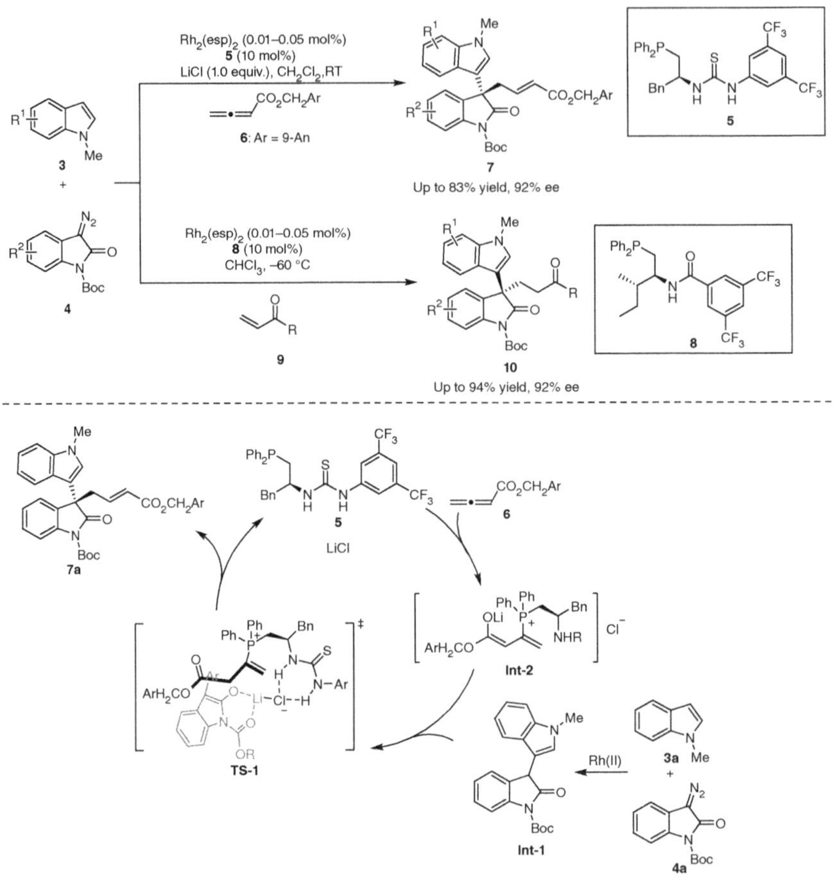

Scheme 8.3 Rh(II)/chiral phosphine combined catalysis. Source: Modified from Chen et al. [13].

to ortho-quinone methides **15** to form a zwitterionic phosphonium **Int-8**, which may exist in equilibrium with a five-membered ring phosphorane **Int-8′**. Simultaneously, Pd(0) reacts with **16** to form a π-allyl palladium complex **Int-9** through decarboxylation. The following *O*-allylation and intramolecular annulation deliver the nine-membered ring product **18** and regenerate the chiral phosphine catalyst **17**. BINAP acts as an organocatalyst in this reaction, as indicated by the fact that the BINAP-palladium complex Pd(**17**)$(PPh_3)_2$ prepared *in situ* could not control the stereoselectivity.

8.3 N-Heterocyclic Carbene and Transition Metal Cooperative Catalysis

NHCs are very strong ligands of transition metals, in particular, late transition metals, owing to their electron-rich σ-donor nature [4, 6, 16]. The coordination

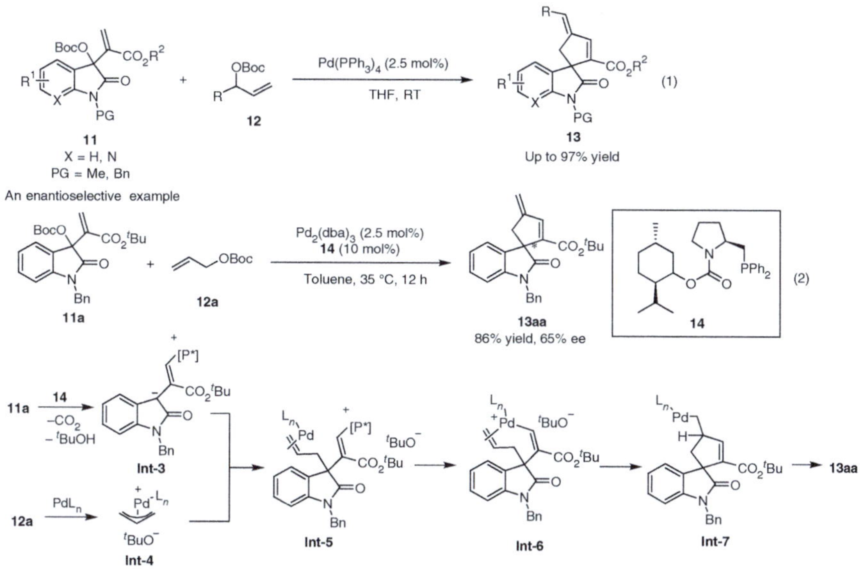

Scheme 8.4 Asymmetric annulation of MBH carbonates and allylic carbonates enabled by Pd(0)/chiral phosphine catalysis. Source: Modified from Chen et al. [14].

interaction between transition metals and NHCs varies and commonly enhances the catalytic activity of transition metals. Therefore, the transition metal complexes of chiral NHCs have been frequently applied to asymmetric catalysis [6]. However, in the combined catalysis of transition metal and NHC [7], such a coordination event unavoidably consumes some amounts of NHC and thereby to some degree quenches the nucleophilicity of NHC to slow down the nucleophilic Lewis base catalysis. Accordingly, the solutions to this issue can be deliberating appropriate catalyst combinations based on fundamental coordination chemistry, such as introducing extra ligands and by means of dynamic and tunable dissociation events. Either of these ways would be able to provide enough active metal complexes and NHC organic catalysts to enable them work together synergistically on driving a reaction (Scheme 8.6).

8.3.1 π-Allyl Metal Mediated Transformations

π-Allyl metal intermediates, generated from the oxidative addition of transition metal complexes to allylic substrates equipped with a leaving group such as allylic halides, esters, and carbonates, are commonly electrophiles capable of undergoing substitution reaction with a broad scope of nucleophiles to form chemical bonds [17] (Scheme 8.7). Palladium, iridium, and many others have been widely investigated for catalyzing regio-, stereo-, and enantioselective allylic substitution reactions.

The combination of NHC and transition metal catalysis was initially attempted to enable a cascade processes that proceed via π-allyl metal intermediates [18]. Harada first developed a cascade allylic substitution and Stetter reaction sequentially catalyzed by a palladium complex and an NHC *in situ* generated from thiazolium salt

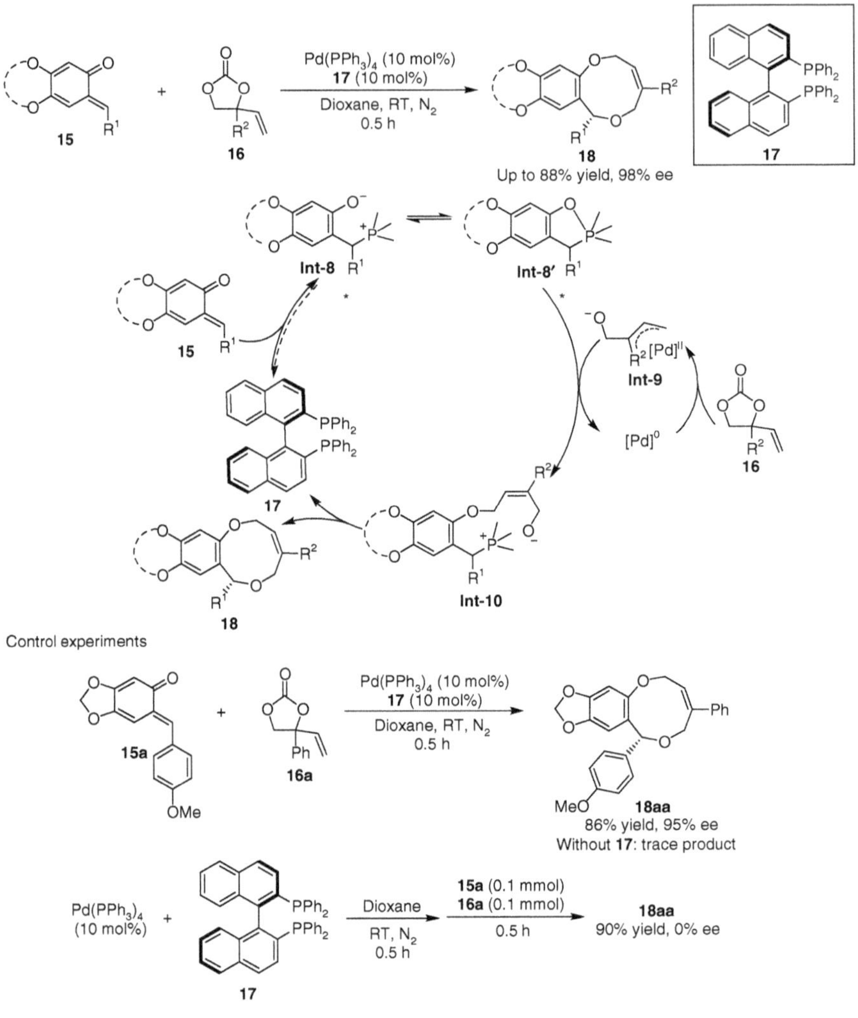

Scheme 8.5 Pd(0)/BINAP cooperatively catalyzed [5+4] annulation. Source: Modified from Xia et al. [15].

[18a]. Glorius [18b] and She [18c] successively reported a cascade benzoin reaction and allylic substitution enabled by sequential catalysis of NHC and palladium complex. In these cases, the NHC catalyzes a transformation to generate a nucleophilic product, rather than an NHC-bonded nucleophile to couple with π-allyl palladium species.

The first example describing the coupling of the NHC-mediated nucleophile and π-allyl-palladium intermediate was reported by Liu and coworkers who found an unusual coupling reaction of picolinaldehyde **19** with allyl acetate **20** occurring under the NHC and palladium cooperative catalysis (Scheme 8.8) [19]. The allylic substitution between Breslow intermediate **Int-11** formed from the picolinaldehyde **19** with NHC **21′** and π-allyl-palladium intermediate **Int-12** proceeds via

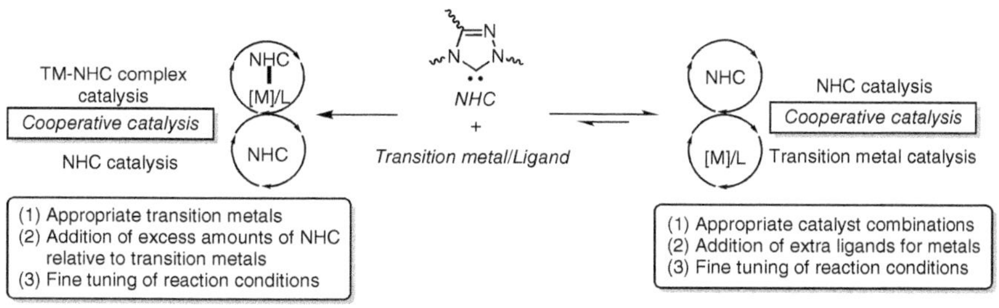

Scheme 8.6 General concept in the NHC and transition metal cooperative catalysis.

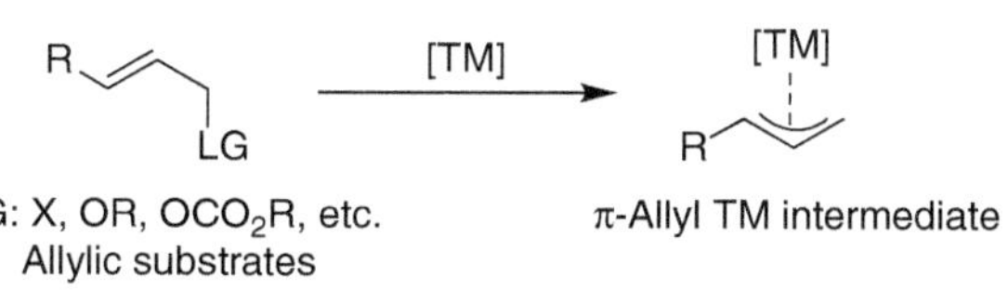

Scheme 8.7 The generation of π-allyl metal intermediates.

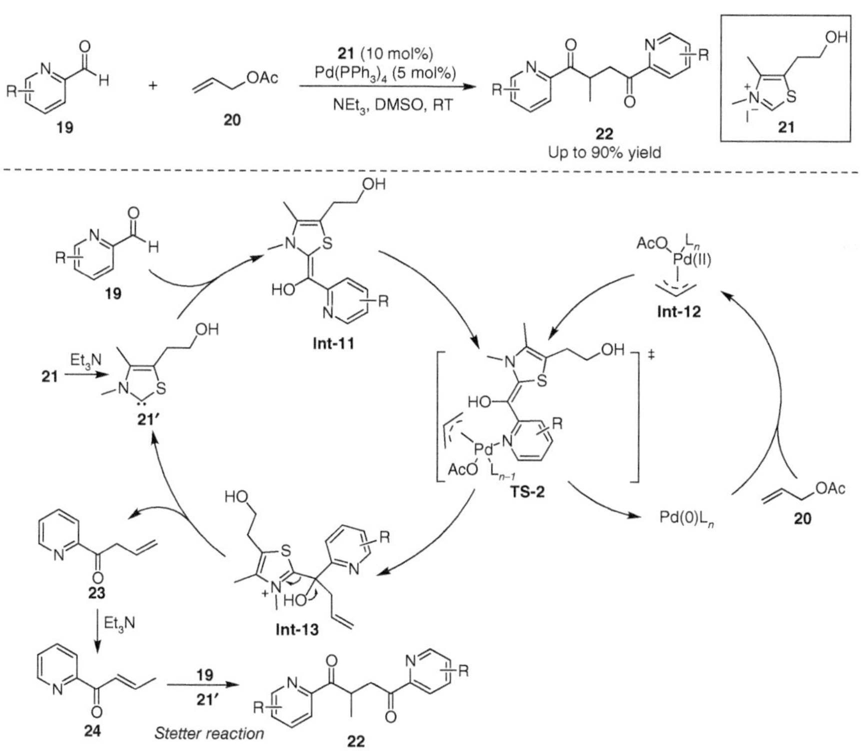

Scheme 8.8 Pd(0)/NHC relay catalytic benzoin and allylic substitution cascade. Source: Modified from Bai et al. [19].

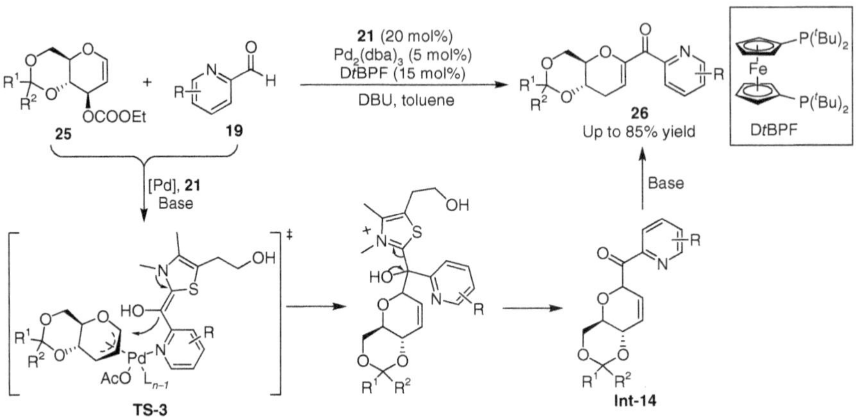

Scheme 8.9 Pd(0)/NHC cooperatively catalyzed C-glycosylation. Source: Modified from Bai et al. [20].

transition state **TS-2** to generate an intermediate **Int-13**, wherein the coordination of palladium(II) with the pyridinyl nitrogen holds the two intermediates and makes the reaction approximate enough to undergo the substitution. This hypothesis is proven by the fact that other aldehydes only gave the typical NHC-catalyzed benzoin products. Elimination event happens in the intermediate **Int-13** regenerates NHC **21′** and gives β,γ-unsaturated ketones **23**, which then undergoes a proton transfer under the basic reaction conditions to generate α,β-unsaturated ketone **24**. A classical NHC-catalyzed Stetter reaction between **24** and the picolinaldehydes **19** gives final products **22**.

Afterward, they established a coupling reaction of glycals **25** with picolinaldehyde **19** by using a similar combined catalyst system, leading to C-glycosylation products **26** (Scheme 8.9) [20]. D*t*BPF turns out to be the optimal ligand and allows the palladium to form a highly reactive π-allyl-palladium intermediate, which preferentially undergoes the allylic substitution at the O-bonded terminus (via **TS-3**). In the presence of a base, double bond migration of **Int-14** occurs to give final product **26**.

Scheidt and coworkers reported a palladium-catalyzed allylation of NHC-activated 2-*O*-alloc cinnamaldehydes (Scheme 8.10) [21]. The oxidative addition of palladium to the 2-*O*-alloc cinnamaldehyde **27**, followed by subsequent decarboxylation, generates an aldehyde **Int-15** or an intermediate **Int-16**, either of which then undergoes the addition reaction with NHC **28′** to give an extended Breslow intermediate **Int-17** or **Int-18**. Subsequently, the NHC-bound homoenolate **Int-17** undergoes a β-protonation to generate catalytic enol intermediate **Int-19**, which can alternatively be generated from the concurrent events of β-protonation and palladium insertion of **Int-18**. The allylation between the NHC-enol and cationic π-allyl palladium species generates acyl azolium **Int-20** and releases the palladium catalyst. Finally, a lactonization of the acyl azolium **Int-20** proceeds to furnish the desired product **30** and regenerates the NHC catalyst **28′**. The bidentate phosphine ligands are very important for the success of this cooperative catalysis because they can strongly coordinate to the palladium and minimize the undesired Pd–NHC ligation.

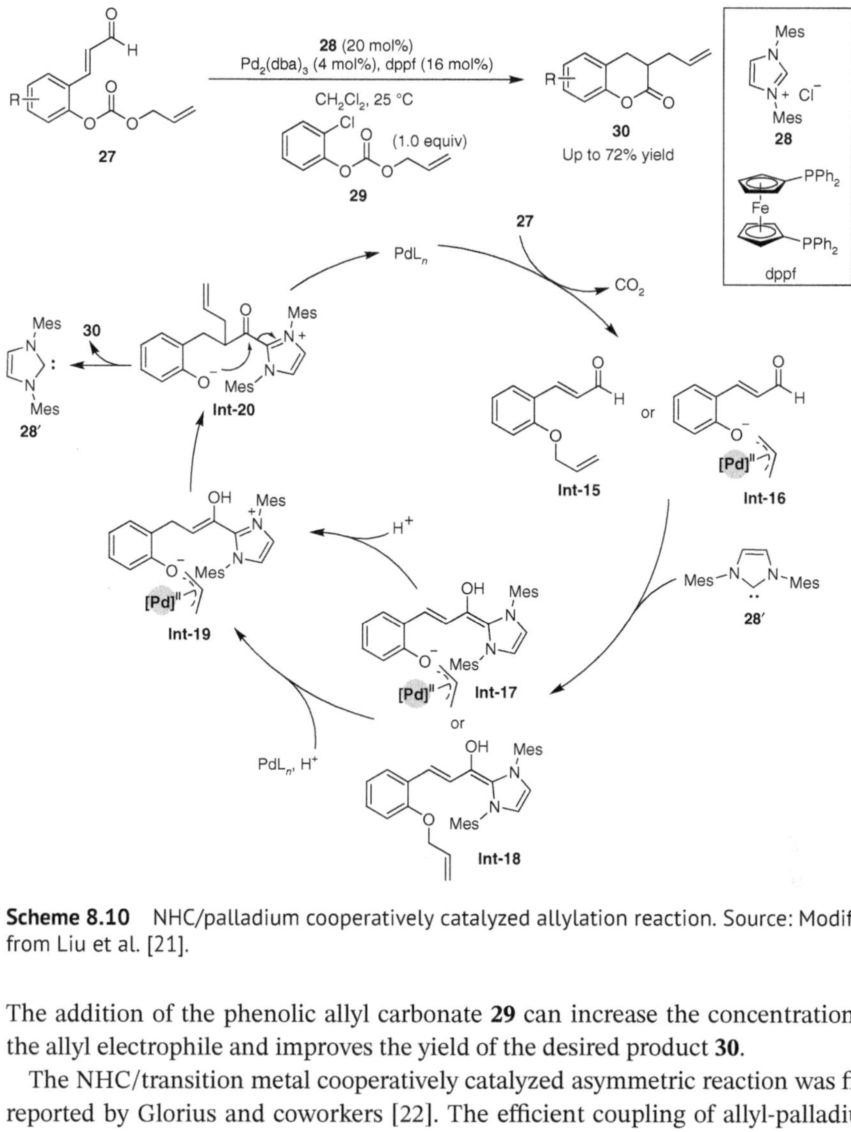

Scheme 8.10 NHC/palladium cooperatively catalyzed allylation reaction. Source: Modified from Liu et al. [21].

The addition of the phenolic allyl carbonate **29** can increase the concentration of the allyl electrophile and improves the yield of the desired product **30**.

The NHC/transition metal cooperatively catalyzed asymmetric reaction was first reported by Glorius and coworkers [22]. The efficient coupling of allyl-palladium complex and NHC-bound homoenolate allows an enantioselective umpolung annulation of vinyl benzoxazinanones **31** and enals **32** to deliver a diverse set of benzazepine derivatives **34** in good yields (up to 98%) and with excellent stereoselectivities (up to 20 : 1 dr, 99% ee) (Scheme 8.11) [22a]. Mechanistic studies reveal that the NHC **33′** not only acts as a Lewis base catalyst but also as a chiral ligand to coordinate with palladium [22b]. Initially, the hybrid palladium complex bound with NHC and PPh_3 sequentially undergoes oxidative addition and decarboxylation with vinyl benzoxazinanones **31a** to generate a hybrid π-allyl-Pd/NHC/phosphine intermediate **Int-21**. Simultaneously, the condensation of the enal **32a** and the NHC **33′** gives a chiral NHC-bound homoenolate **Int-22**. Afterward, the nucleophilic substitution of the allylpalladium species **Int-21** with the NHC-bound homoenolate

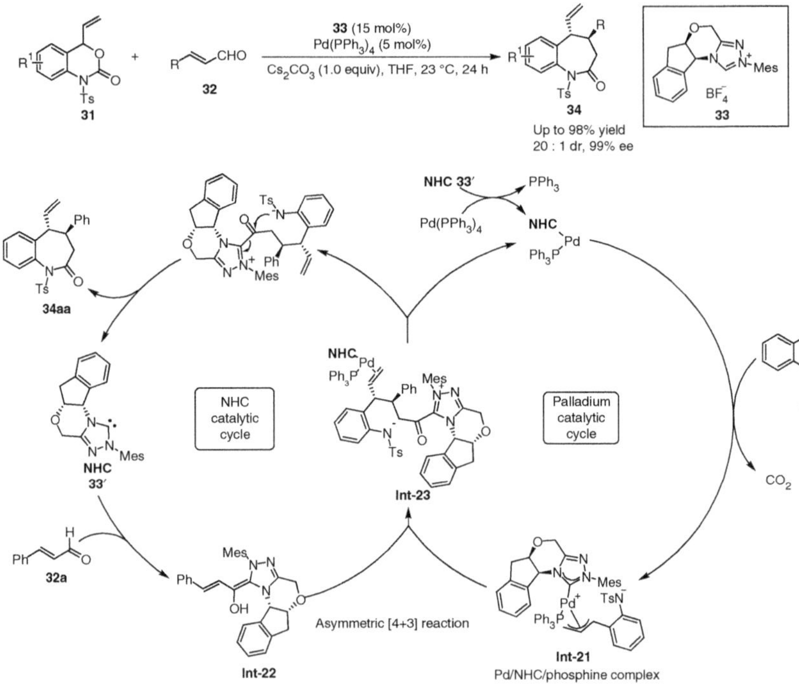

Scheme 8.11 NHC/palladium cooperatively catalyzed asymmetric [4+3] annulation. Source: Modified from Guo et al. [22a].

Int-22 occurs to form an intermediate **Int-23**, which then undergoes a lactamization to furnish enantioenriched benzazepines **34aa**. Although NHC plays a key role in the stereochemical control, the presence of the achiral triphenylphosphine can tune the catalytic efficiency of the palladium complex and thus dramatically enhances the performance of the process.

Shortly after this event, the same group extended this concept to develop an enantioselective [5+2] annulation of VECs **16** with enals **32** (Scheme 8.12) [23]. High levels of stereochemical control are enabled by the synergistic action of chiral ligand and NHC. As indicated by control experiments, the (*R*)-tol-BINAP **35** matches the chirality of the NHC **36** and their combination allows the reaction to give the highest enantioselectivity. Interestingly, the combination of NHC **36** and racemic BINAP also gives excellent enantioselectivity, but with a diminished yield. The (*R*)-tol-BINAP **35** in concert with an achiral NHC **38** also offers an excellent enantiomeric excess, but with a low yield, indicating that the bidentate phosphine ligand, (*R*)-tol-BINAP, strongly bonds to the palladium and can control the stereoselectivity efficiently. Unlike the case of $Pd(PPh_3)_4$ and NHC combined catalysis wherein a hybrid palladium complex of NHC and PPh_3 turns out to be the true catalyst [22], in this case, the bidentate phosphine strongly coordinates with the palladium to form a highly catalytically active metal complex whereas the NHC does not coordinate to the palladium and completely acts as Lewis base catalyst. Mechanistically, a chiral π-allyl-palladium intermediate **Int-9**, generated from the

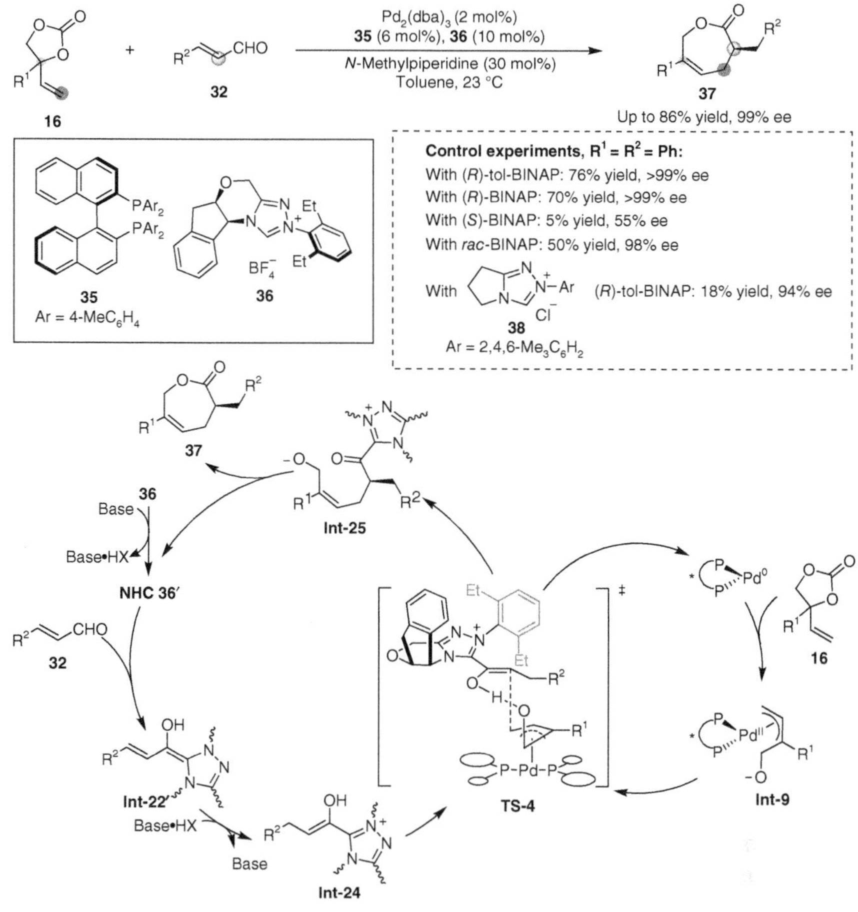

Scheme 8.12 NHC/palladium cooperatively catalyzed asymmetric [5+2] annulations. Source: Modified from Singha et al. [23].

decarboxylation of a VEC **16** with (*R*)-tol-BINAP-Pd complex, couples with the NHC-bound enolate **Int-24**, concurrently formed from the facile β-protonation of **Int-22′**, and further leads to an intermediate **Int-25** via a transition state **TS-4**. Finally, the intramolecular lactonization gives the seven-membered lactone **37** and regenerates the free NHC and Pd catalyst. As can be seen in **TS-4**, both the chiral ligand and NHC exert an impact on the chiral induction of allylic alkylation event.

An enantioselective coupling of a π-allyl-iridium intermediate with NHC-homoenolate was reported by Glorius and coworkers in 2020 (Scheme 8.13) [24]. The iridium complex of chiral phosphoramidite **40** works compatibly with chiral NHC to enable a stereoselective [3+2] annulation reaction of VEC **16a** with enals **32** or α-chloro aldehyde **39** (Scheme 8.9a). Again, a hybrid π-allyliridium complex Ir(**40**)(**42′**)Cl bound with NHC **42′** and phosphoramidite **40** appears to be an active catalyst (detected by HRMS). The Ir–π-allyl intermediate **Int-26** generated from the Ir-mediated decarboxylation of VEC **16a** was also identified by the electrospray

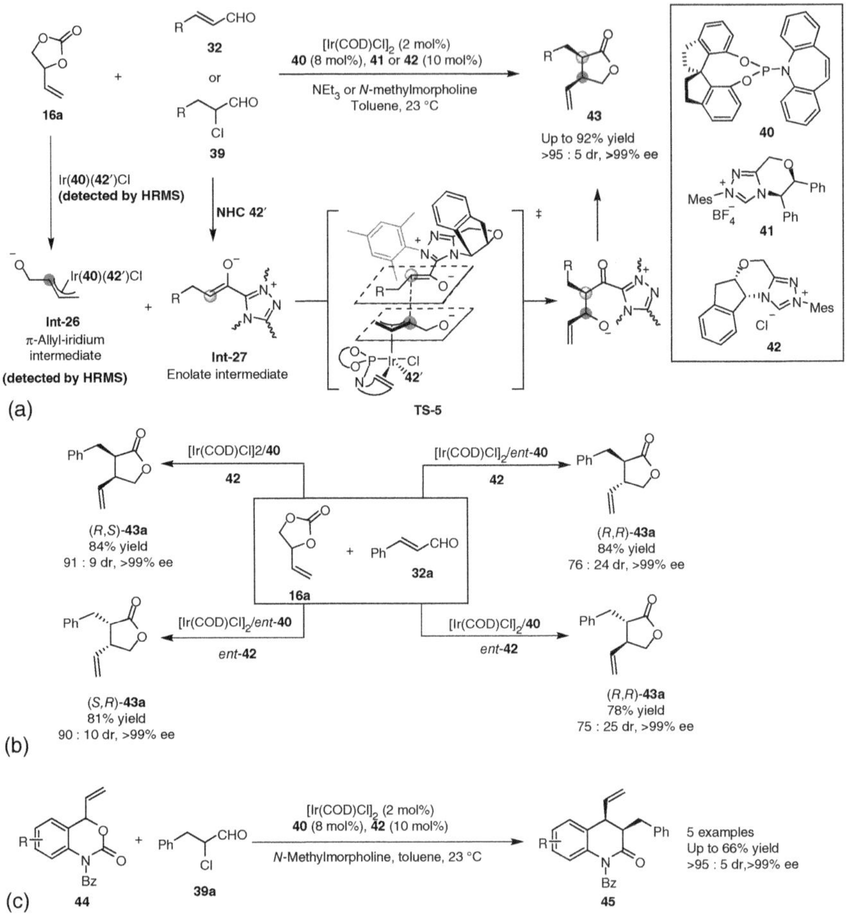

Scheme 8.13 NHC/iridium cooperatively catalytic system. Source: Modified from Singha et al. [24].

ionization and mass spectrometry analysis of the crude reaction mixture. The NHC-enolate **Int-27**, generated from umpolung-type reaction of NHC **42′** with an α,β-unsaturated aldehyde **32** or α-chloro aldehyde **39**, attacks the Ir–π-allyl intermediate **Int-26** via an open transition state **TS-5**, preferring branched to the linear substitution [25], and subsequently undergoes cyclization to produce the five-membered lactone **43**. The successful coupling between the chiral enolate and π-allyl-iridium intermediate allows stereodivergent dual catalysis, wherein two distinct and highly face-selective catalytic cycles are merged (Scheme 8.13b) [26]. As such, the full array of stereoisomers of α,β-disubstituted γ-butyrolactones **43a** are obtained in good yields (78%–84%) and superb enantioselectivity (all >99% ee) and high diastereoselectivity (76 : 24 to 90 : 10 dr) from the same set of starting materials **16a** and **32a** by an orthogonal combination of different enantiomers of phosphoramidite ligand and NHC-precatalyst. Such a dual NHC/Ir catalyst system is also amenable to the [4+2] annulation of *N*-benzoyl-vinyl benzoxazinanones **44** with α-chloro aldehyde **39a**, leading to *cis*-δ-lactams **45** in good yields and with high levels of stereoselectivity (Scheme 8.13c).

8.3.2 Alkynyl-metal Mediated Transformations

Recently, transition metal-catalyzed electrophilic activation of alkynes through π-alkyne complex formation has been studied extensively [27], and has proven compatible with the Lewis base mediated nucleophilic addition process. In 2014, Chi and coworkers disclosed a copper/NHC relay catalytic asymmetric three-component annulation reaction of alkynes, TsN_3, and imines (Scheme 8.14) [28]. As shown in Scheme 8.14a, the ketenimine **Int-28**, generated from the

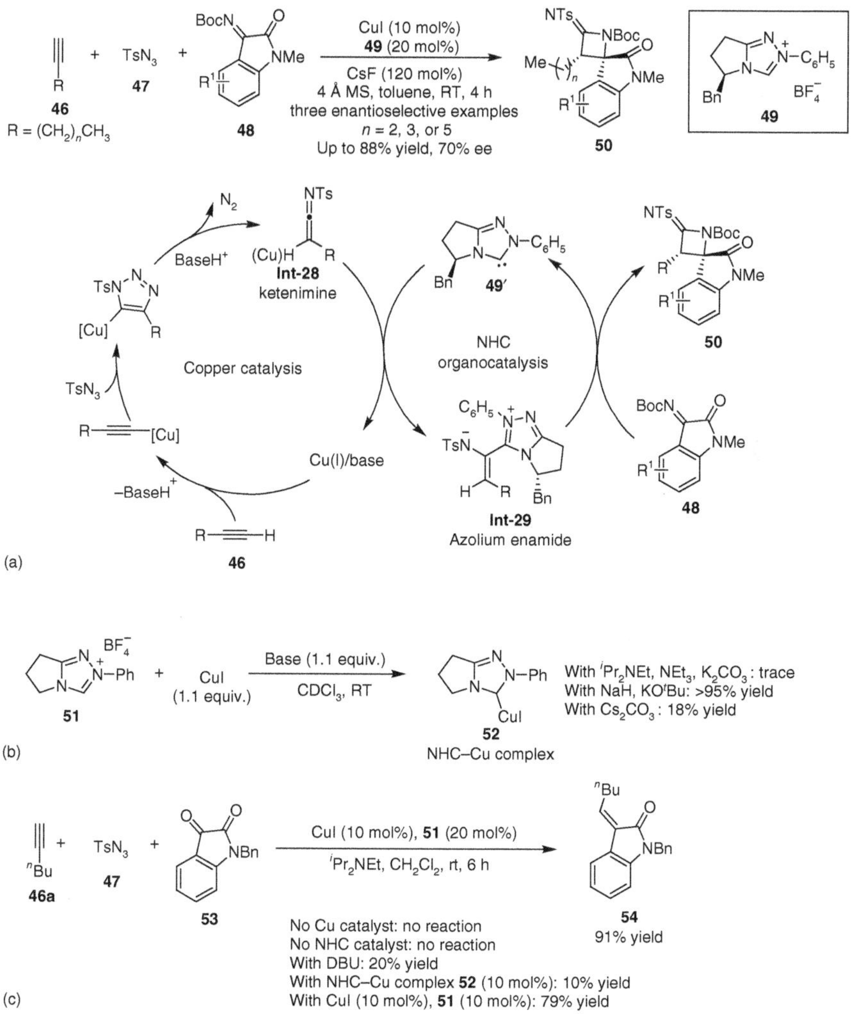

Scheme 8.14 Copper and NHC relay catalysis. (b) Coordination effect of NHC/Cu. (c) Control experiments.

copper-catalyzed 1,3-dipolar cycloaddition of an alkyne **46** with TsN_3, undergoes the subsequent denitrogenation with NHC **49′** to form a highly nucleophilic azolium enamide **Int-29**, which then undergoes an [2+2] annulation reaction with isatin–imines **48** to afford the desired products **50**. Three examples of enantioenriched spirocyclic lactam products were obtained in good yields and moderate ee (up to 88% yield, 70% ee). Careful investigation of the compatibility between CuI and NHC catalyst **51** reveals that strong bases, such as NaH, KO^tBu, and 1,8-diazabicyclo[5.4.0]undec-7-ene (DBU), allow the formation of NHC–Cu complex **52**, while the use of weak bases (such as $^iPr_2NEt$, NEt_3, and K_2CO_3) is essential to modulate the formation of free carbene catalyst and NHC–Cu complex **52** with a meaningful level of concentration (Scheme 8.14b). In addition, using the preformed NHC–Cu complex **52** alone in the three-component reaction diminishes the yield significantly (Scheme 8.14c). Instead, the combination of 20 mol% of NHC **51** and 10 mol% of CuI allows the reaction to proceed completely under otherwise identical conditions. These facts suggest that the Cu(I) catalyst and triazolium NHC organic catalyst work independently to accomplish their mission without quenching each other, in particular, under conditions with weak bases.

Based on gold and NHC relay catalysis, Chi and coworkers developed a cascade cycloisomerization/asymmetric cycloaddition of ynamides and enals (Scheme 8.15) [29]. The gold-catalyzed intramolecular annulation of ynamide **55** generates an **Int-30** and then isomerizes into an azadiene intermediate **Int-31**. Concurrently, the azolium homoenolate **Int-32** generated from the coupling reaction of the enal **56** and NHC **57′** undergoes a proton transfer to give an enolate **Int-33**. The asymmetric [4+2] annulation of ketenimine **Int-31** and azolium enolate **Int-33** occurs to afford a bicyclic lactam product **58**. NHC is principally a dative ligand capable of strongly coordinating with gold(I), and hence its catalytic activity would be weakened by the coordination interaction with the metal. Control experiments reveal that the combination of 10 mol% gold(I) and 20 mol% NHC **57** is highly efficient to drive the desired relay catalytic cascade reaction. The presence of NHC/Au catalyst in a ratio of 1/1 (10 mol% each) enables the conversion of ynamide to the α,β-unsaturated imine **59** via the protodemetallation of **Int-31**, but the organocatalytic reaction pathway is not involved. As a consequence, the variation of the molar ratio of NHC/gold(I) from 1/1 to 2/1 or the addition of a competing ligand, such PPh_3, renders the carbene organic catalyst **57′** to coexist at a reasonable concentration enough to activate the enal **56**, and hence allows the reaction to efficiently generate the product **59**. In addition, the combination of a preformed chiral gold-catalyst [Au(**57′**)Cl] **60** and an achiral NHC precatalyst **61** still provides moderate stereocontrol (**58a**, 43% ee), indicating that the chiral gold complex participates in catalyzing the [4+2] cycloaddition event. In terms of the facts, the process is a combination of gold/carbene relay and cooperative catalysis.

8.3.3 Metal-allenylidene Mediated Transformations

Over the past decades, great advances have been made on the transition metal-catalyzed propargylic substitution of propargylic alcohol derivatives **62**

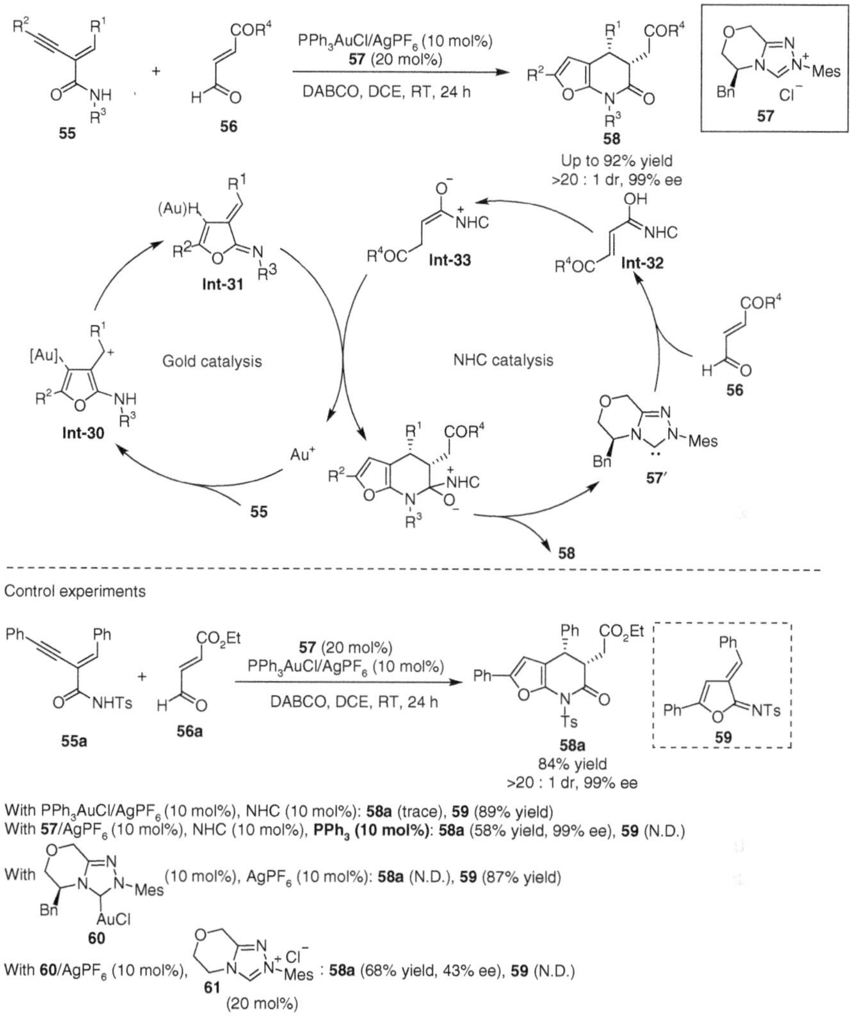

Scheme 8.15 Gold and NHC cooperatively catalyzed [4+2]-annulation. Source: Modified from Zhou et al. [29].

with various nucleophiles, wherein electrophilic metal-allenylidene complexes **Int-34** served as key intermediates (Scheme 8.16a) [30]. The enantioselective catalytic processes have been achieved by using chiral metal complexes, for example, chiral thiolate bridged diruthenium complexes and chiral Cu–pybox complexes. On the other hand, Lewis base-bound enolates are typical nucleophiles and are principally able to participate in a nucleophilic addition to metal-allenylidene complexes **Int-34** (Scheme 8.16b). As such, the combination of transition metal catalysis and chiral Lewis base catalysis would be a complementary concept for the development of asymmetric propargylation reactions.

Ethynylethylene carbonates (EECs) **63**, first introduced by Zhang and coworkers [31a], undergoes decarboxylation to deliver nucleophile/copper-allenylidene

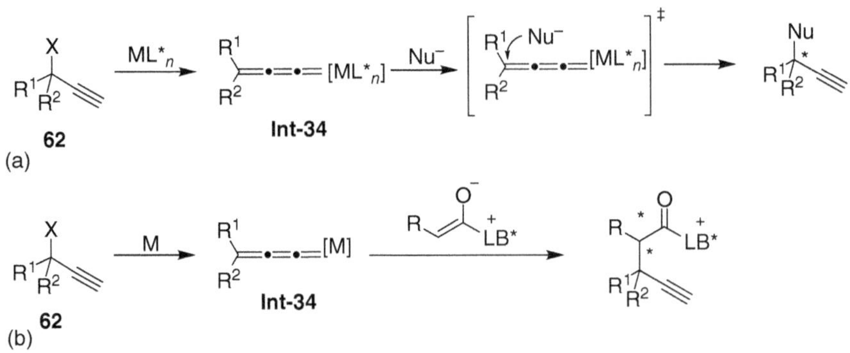

Scheme 8.16 General concept in transition metal-catalyzed propargylic substitution. (a) Chiral transition metal catalysis. Sources: Miyake et al. [30a]; Ljungdahl and Kann [30b]; Detz et al. [30c]; Ding and Hou [30d]; Nishibayashi [30e]; Zhang and Hu [30f]; Hu et al. [30g]. (b) Transition metal/chiral Lewis base cooperative catalysis.

bifunctional intermediates under cooper catalysis, which can undergo asymmetric reactions to produce functionalized propargylic products [31]. A [3+3] annulation reaction between EECs **63** and isatin-derived enals **64** was developed by cooperative catalysis of copper complex and chiral NHC **65** to generate spirooxindole δ-lactones **66** in up to 77% yield, >95 : 5 dr, and 99% ee (Scheme 8.17). As reported by Chi and coworkers [28], the NHC acts as a strong σ-donor ligand and can form a stable complex with copper(I). In this case, the nonlinear effects and (electrospray ionization) ESI-MS studies suggest that the chiral copper complex of **NHC 65′** forms and turns out to be an active species to undergo decarboxylation with **63** to afford the key dipolar intermediates **Int-35** or **Int-36** (Scheme 8.17, left cycle). Chiral homoenolates **Int-37**, concurrently generated from the addition of **NHC 65′** to the enals **64** (Scheme 8.17, right cycle), then undergo nucleophilic addition to the transient copper-allenylidene intermediates (**Int-35** or **Int-36**) and followed by lactonization to access enantioenriched spirooxindoles **66** [32].

Ethynyl benzoxazinanones **67**, first introduced by Lu and Xiao [33a], undergo decarboxylation upon exposure to copper(I) complex to generate 1,4-dipoles containing an electrophilic Cu-allenylidene and a nucleophilic sulfonamide. Such intermediates are also highly reactive toward a broad scope of nucleophiles, culminating in a number of transition-metal-catalyzed cyclization reactions [33]. The extension of the NHC and copper cooperative catalysis to the asymmetric [3+4] annulation of isatin-derived enals **64** with ethynyl benzoxazinanones **67** is also successful (Scheme 8.18) [32]. In this case, the copper allenylidene intermediates **Int-38** formed from the decarboxylation of ethynyl benzoxazinanones **67** with copper(I) complex couple with the chiral homoenolates **Int-37**, and followed by the lactamization to yield the final products. The combination of NHC **68** and $Cu(OTf)_2$ enables the [3+4] annulation reaction to deliver chiral spirobenzazepinones **69** in up to 99% yield and with >94 : 6 dr, and 97% ee.

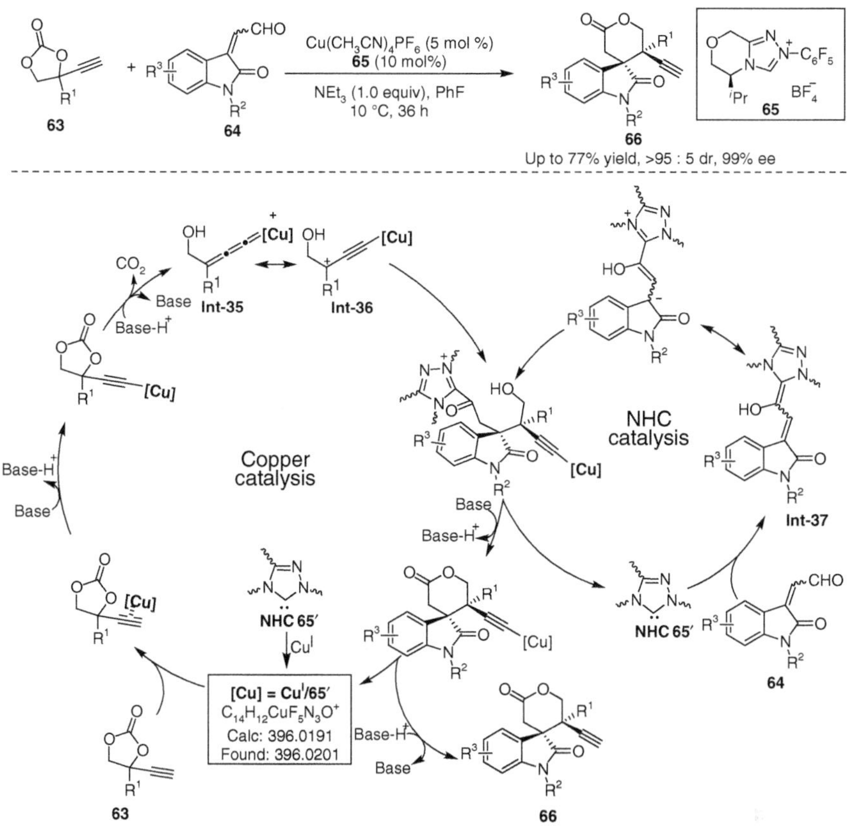

Scheme 8.17 [3+3] Annulation of ethynylethylene carbonates with enals.

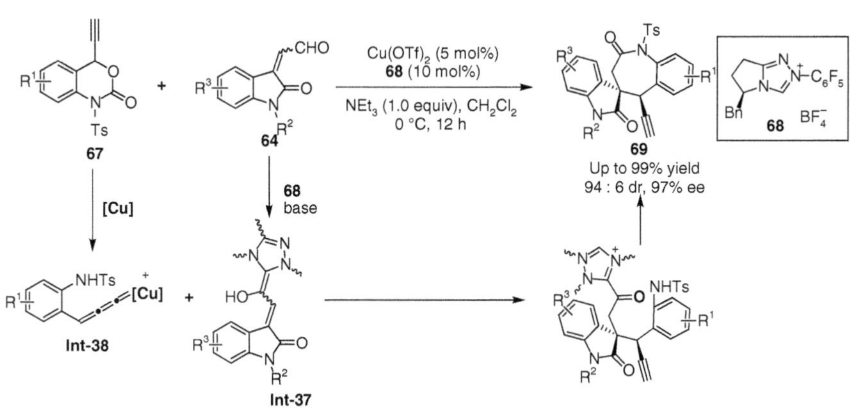

Scheme 8.18 [4+3] Annulation of ethynyl benzoxazinanones with enals. Source: Modified from Zhang et al. [32].

8.4 Tertiary Amine and Transition Metal Cooperative Catalysis

8.4.1 π-Allyl Metal Mediated Transformations

C1-ammonium enolates can be accessed via the nucleophilic attack of a tertiary amine catalyst on ketenes either preformed or *in situ* generated from acid derivatives [10] and have emerged as versatile nucleophiles capable of participating in a broad scope of asymmetric reaction by capturing different electrophiles. The possibility of integrating the reactivity of the π-allyl-metal electrophiles with chiral C1-ammonium enolates in a synergistic fashion was first introduced by Snaddon and coworkers in 2016 (Scheme 8.19) [34]. An asymmetric α-allylation of aryl acetates **70** with either allyl methylsulfonates or cinnamyl *tert*-butylcarbonates **71** proceeds cleanly under the chiral isothiourea/palladium cooperative catalysis to produce a variety of highly enantioenriched α-branched esters **73** in excellent yields and enantioselectivity. In this reaction, the C1-ammonium enolate **Int-39** is formed from a sequential *N*-acylation and deprotonation process of the perfluorophenyl 2-arylacetate **70** with the chiral benzotetramisole catalyst (BTM, **72**) and adopts an extended s-*cis* conformation with a stabilized *syn*-coplanar electrostatic 1,5-S···O interaction (n_0 to σ^*_{C-S}) between the acylammonium oxygen and the sulfur atom of **72** [35]. The enantioselective allylic allylation of the C1-ammonium enolate **Int-39** with a π-allyl-palladium complex **Int-40 (40′)** formed from allylic ester **71** occurs, wherein the *Si*-face of the enolate double bond is blocked by the pendant phenyl group of the nucleophilic catalyst and thus the π-allyl-palladium electrophile is pushed to the *Re*-face distal to the phenyl, furnishing the (*R*)-configured acyl

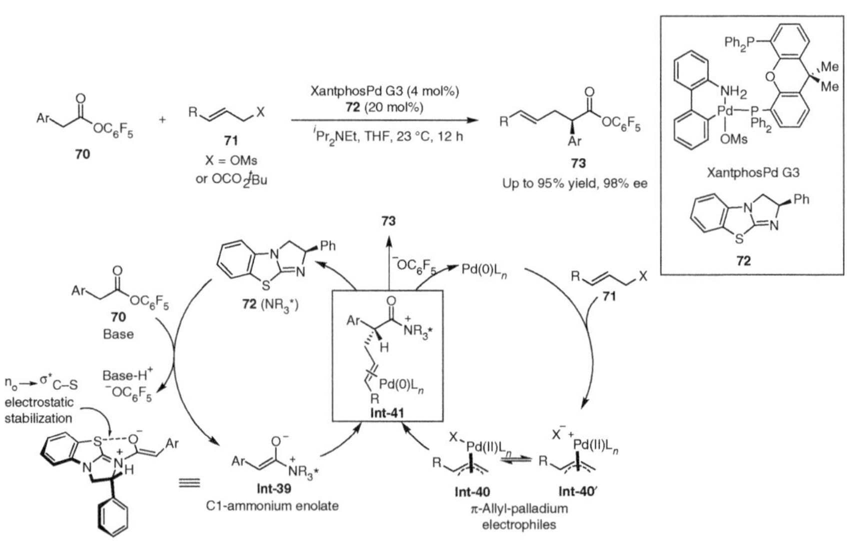

Scheme 8.19 Chiral ITU/palladium cooperatively catalyzed asymmetric allylic alkylation. Source: Modified from Schwarz et al. [34].

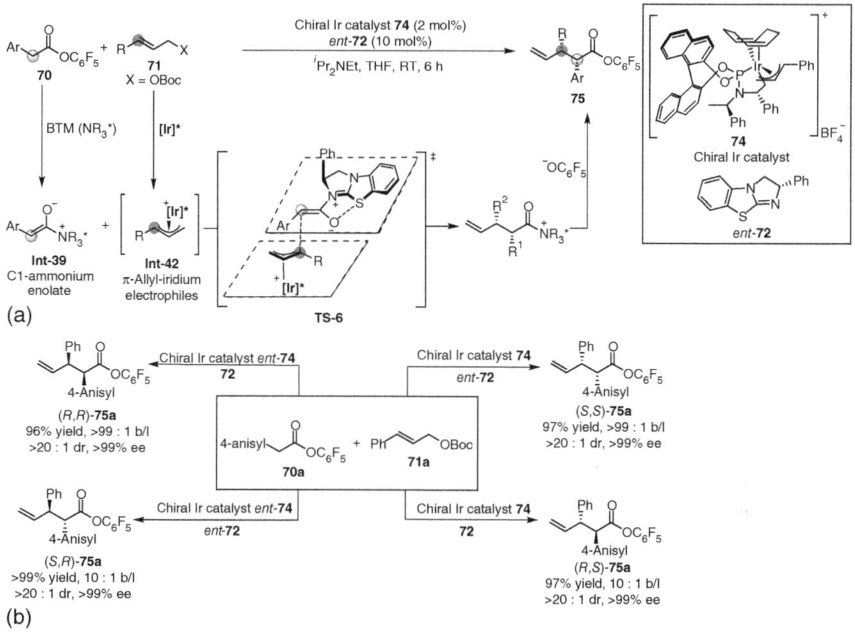

Scheme 8.20 Stereodivergent allylic alkylation enabled by chiral ITU/iridium cooperative catalysis. Source: Modified from Jiang et al. [36].

ammonium ions **Int-41**. Finally, the decomplexation and phenolate ($C_6F_5O^-$) rebound processes simultaneously proceed in the acyl ammonium intermediate **Int-41** to afford the final product **73** and regenerate the BTM **72** and the palladium catalyst.

The branch-selective allylic alkylation of aryl acetates with either allyl or cinnamyl *tert*-butylcarbonates **71** has been accessed by the cooperative catalysis of chiral isothiourea Lewis base and chiral iridium complex, leading to α-allyl pentafluorophenyl ester products **75** in high yields and excellent stereoselectivities (Scheme 8.20) [36]. Both the iridium complex **74** and chiral isothiourea catalyst *ent*-**72** can govern the stereochemistry of two stereogenic centers of the product arising from enantioselective coupling of electrophile **Int-42** and nucleophile **Int-39** happening via **TS-6**; therefore, a stereodivergent reaction can be realized through simple permutations of the enantiomers [26] of the iridium complex and the isothiourea catalyst, allowing all four isomers of **75a** to be obtained from the same set of starting materials **70a** and **71a** in excellent yields and with excellent levels of diastereo- and enantioselectivities (Scheme 8.20b).

Smith and coworkers found that the allylic ammonium salt *in situ* formed from the Pd-catalyzed allylic amination between allylic phosphate **71** and *N*,*N*-dialkylglycine esters **76** smoothly undergoes an enantioselective [2,3]-rearrangement reaction catalyzed by the chiral isothiourea **72** (Scheme 8.21) [38]. In this case, a bench-stable succinimide-based palladium complex **77** stands out among the palladium precatalysts tested and enables the allylic phosphate **71** to undergo allylic amination

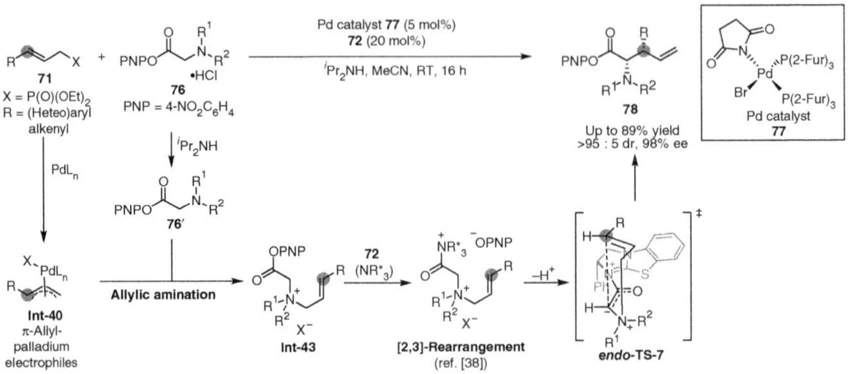

Scheme 8.21 Chiral ITU/palladium relay catalytic system. Source: Modified from Spoehrle et al. [38].

with **76′** to give an ammonium salt **Int-43**. The subsequent chiral isothiourea (ITU)-catalyzed stereoselective [2,3]-sigmatropic rearrangement of **Int-43**, a similar process reported by the same group [37], affords *syn*-α-amino acid derivatives **78** with high diastereo- and enantioselectivities (up to >95 : 5 dr, 98% ee). A variety of substituted allylic ethyl phosphates **71** and *N*,*N*-dialkylglycine esters **76** are all tolerant in this reaction, leading to a diverse range of highly enantioenriched β,β-disubstituted amino esters **78**.

Snaddon and coworkers developed an enantioselective α-allylation of aryl acetates **70** with 2-substituted allyl methylsulfonates **79** based on the chiral isothiourea/palladium cooperative catalysis (Scheme 8.22) [39]. The electronic and steric features of phosphine ligands have a significant impact on the reaction. The combination of the palladium-Xantphos and chiral isothiourea, which has successively

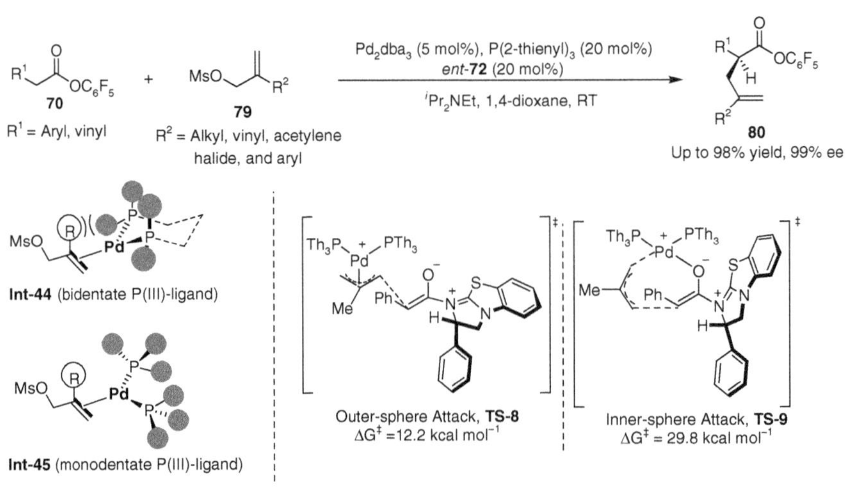

Scheme 8.22 Chiral isothiourea/palladium-catalyzed allylic alkylation of 2-substituted allyl electrophiles. Source: Modified from Schwarz et al. [39].

enabled asymmetric allylic alkylation of allyl methylsulfonates by the same group [34]; however, fails to drive this reaction, presumably due to the adverse nonbonding interactions between the 2-substituent of **79** and a pseudo-axially-positioned benzene ring of the Xantphos ligand (**Int-44**). None of the other bidentate phosphines tested are able to enable the palladium-catalyzed reaction. In partnership with chiral isothiourea catalyst *ent*-**72**, the palladium adorned with 2-thienylphosphine ligands turns out to be a particularly effective catalyst for this transformation, which would relieve the adverse nonbonding interactions and reengage the cooperative process, probably arising from the low steric demand and significant π-accepting character of these ligands (**Int-45**). A wide range of 2-substituted allyl electrophiles **79**, including alkyl, vinyl, acetylene, halide, and aryl moieties, can be tolerated. Variation of the substituent and substitution pattern of the benzene ring of **70** is also allowed, illustrating the insensitivity of the nucleophile. Significantly, α-vinyl acetic acid esters are also competent nucleophiles. The density functional theory (DFT) calculations reveal that the π-(allyl)Pd complex attacks the less hindered face of the C1-ammonium enolate. The outer-sphere pathway (**TS-8**) features a relatively lower barrier in comparison with the inner-sphere pathway (**TS-9**), which is free of steric repulsion between the C1-ammonium enolate and the palladium ligands.

The chiral isothiourea/palladium cooperative catalysis is generally amenable to the asymmetric allylic alkylation with different kinds of functionalized allylic substrates (Scheme 8.23). For instance, an asymmetric α-allylation of pentafluorophenyl acetic esters **70** with Si-substituted allyl methylsulfonates **81** catalyzed by the palladium complex of tri(2-furanyl)phosphine and BTM furnishes the corresponding allylation products **82** in up to 95% yield and 88% ee (Scheme 8.23a) [40]. The presence of silyl group provides additional space for molecular modulation through simple chemical transformations. The expansion of a similar catalyst system to an allylic alkylation between electron-deficient electrophiles and pentafluorophenyl acetic esters **70** works well, but aldehyde, ketone, ester, and amide substituents of **83** greatly affects the reaction efficiency and the *E*/*Z* selectivity, especially in the case of amides (Scheme 8.23b) [41a]. The *E*/*Z* selectivity in palladium-catalyzed allylic alkylation originates from the relative stability and reactivity of (*syn or anti*) π-allylPd(II) intermediates, which can be directed by electronic and steric effects dispatched by both the substrate and the supporting ligand [41b]. Tris[tri(2-thienyl)phosphine]Pd(0) complex is a generally effective catalyst for all substrates, and the alkylation products **84** are obtained exclusively as the linear regiomer, with high levels of *E*-selectivity (or exclusively *E*) and enantiocontrol. An enantioselective α-allylation of the esters **70** with B(pin)-substituted allyl methylsulfonates **85** also undergoes under the promotion of palladium complex and isothiourea Lewis base (Scheme 8.23c) [42]. The chiral BTM *ent*-**72** dominantly controls the enantioselectivity, whereas the phosphine ligand of Pd catalyst exerts considerable impact on the alkene geometry of the allylation products and BINAP offers the highest E/Z selectivity. The use of (*S*)-BINAP to replace racemic BINAP, in concert with (*S*)-BTM (*ent*-**72**), slightly elevates the enantioselectivity from 94% ee to 96% ee. In contrast, the employment of (*R*)-BINAP results in diminished enantioselectivity under otherwise identical conditions, indicating the existence

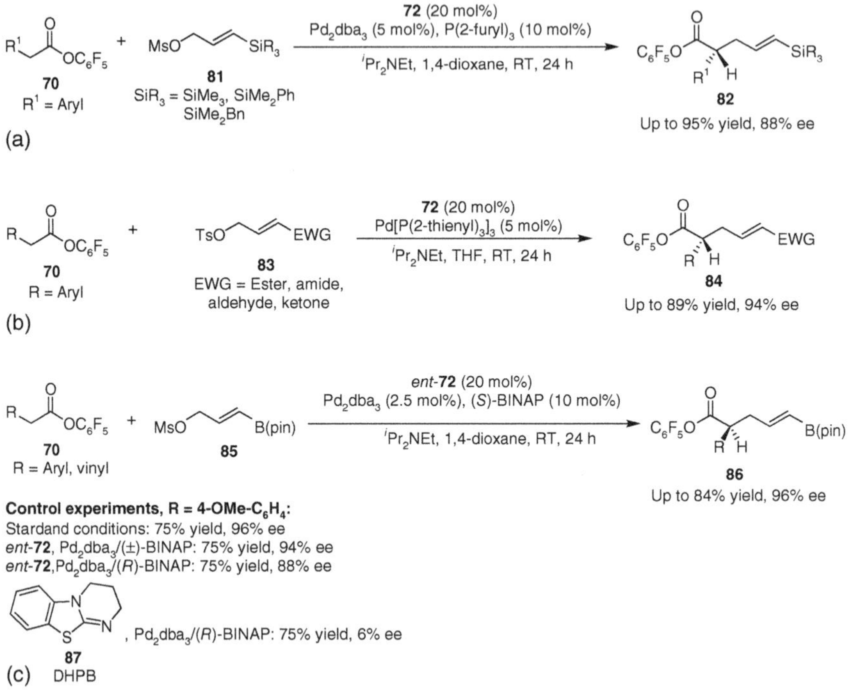

Scheme 8.23 Chiral isothiourea/palladium-catalyzed allylic alkylation reactions. (a) Enantioselective α-allylation with silicon-substituted electrophiles. (b) Enantioselective α-allylation with EWG-substituted electrophiles. (c) Enantioselective α-allylation with B(pin)-substituted electrophiles.

of the stereochemically matched effect between these two chiral catalysts. The combination of the achiral Lewis base DHPB **87** and (*R*)-BINAP gave the desired product in only 6% ee. These facts demonstrate that the BTM catalyst controls the absolute stereochemistry of the bond-forming event, while the phosphine ligand exerts little impact on the stereochemical control.

A chiral isothiourea/Pd cooperatively catalyzed enantioselective α-allylation of 2-pyrrolylacetates **88** provides highly enantioenriched products **89** with excellent generality for both coupling partners (Scheme 8.24) [43]. The α-alkylated pyrroles **89** have been readily elaborated in a manner, which points towards potential applications in target-directed synthesis.

Snaddon and coworkers introduced a sequential one-pot chiral isothiourea/transition-metal catalyzed allylic alkylation/Hofmann rearrangement reaction (Scheme 8.25) [44]. The treatment of chiral homoallylic esters generated from the asymmetric allylic alkylation with ammonia yields chiral amides **Int-46**, which then undergo a stereospecific Hofmann rearrangement reaction by exposing to an oxidant PIFA ([bis(trifluoroacetoxy)iodo]benzene). The final alcoholysis reaction of **Int-47** gives a range of linear homoallylic amines **90** in almost perfect optical purity.

Similarly, branched homoallylic amines **91** can be synthesized from the sequential branch-selective allylic alkylation enabled by chiral isothiourea/iridium cooperative

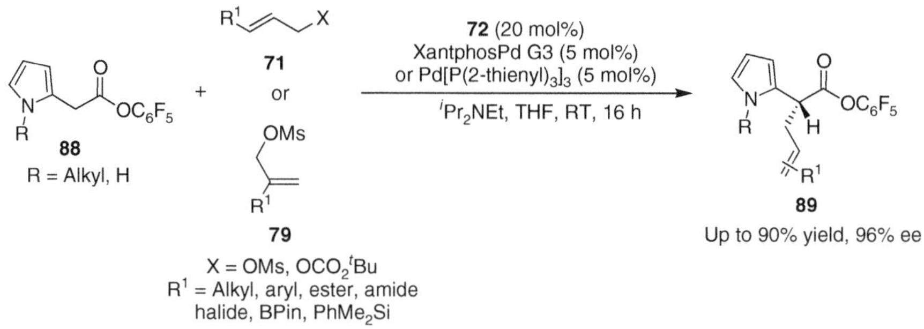

Scheme 8.24 Chiral isothiourea/palladium-catalyzed allylic alkylation of 2-pyrrolylacetates. Source: Modified from Scaggs et al. [43].

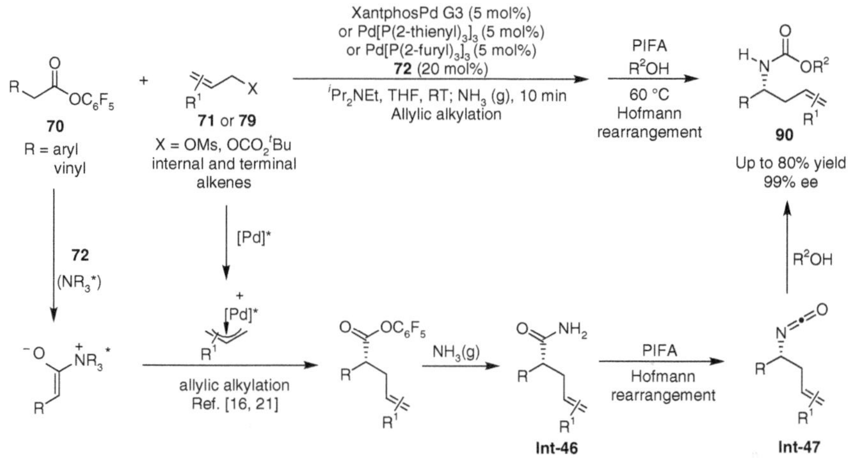

Scheme 8.25 Linear-selective allylic alkylation/Hofmann rearrangement process. Source: Modified from Pearson et al. [44].

catalysis [36] and Hofmann rearrangement process (Scheme 8.26) [44]. In addition, all four stereomers of branched homoallylic amine **91** have been accessed by a similar procedure based on the stereodivergent asymmetric allylic allylation of **70** and **71** enabled by the cooperative catalysis of isothiourea with iridium catalysts and followed by Hofmann rearrangement.

8.4.2 π-Benzyl-metal Mediated Transformations

Since the first report of palladium complex-catalyzed benzylic substitution reaction by Fiaud and Legros in 1992 [45], the π-benzyl palladium species (**Int-48**) has been used as a versatile intermediate in catalytic benzylation reactions for the assembly of carbon–carbon and carbon–heteroatom bond (Scheme 8.27) [46]. In comparison with the formation of π-allyl-palladium complexes, the generation of π-benzyl-palladium electrophiles requires relatively higher energy for the dearomatization of the aryl system to bring extraordinary challenges. Two possible pathways

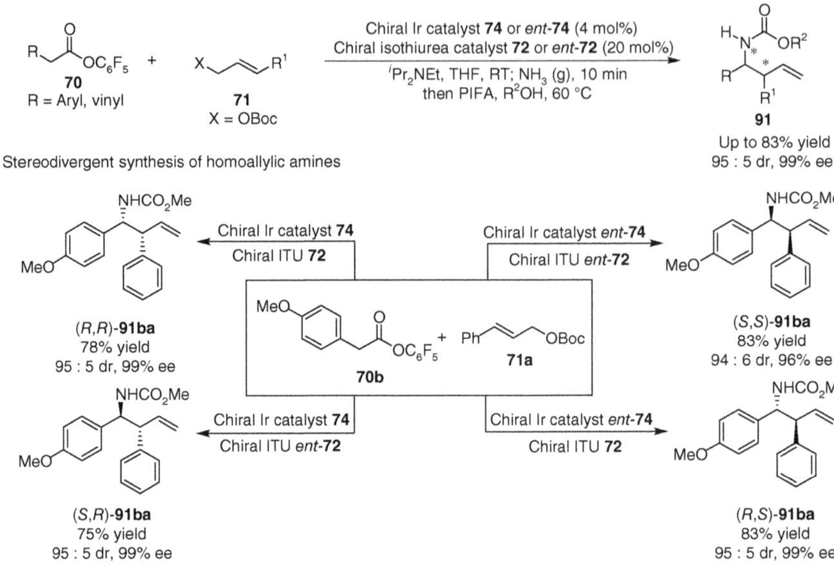

Scheme 8.26 Branch-selective allylic alkylation/Hofmann rearrangement process. Source: Modified from Pearson et al. [44].

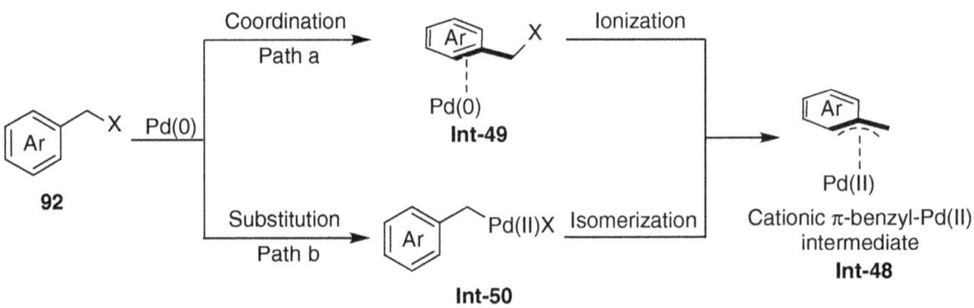

Scheme 8.27 Generation of cationic π-benzyl-palladium intermediates. Sources: Kuwano [46a]; Trost and Czabaniuk [46b]; Le Bras and Muzart [46c]; Trost and Czabaniuk [47].

are outlined in the generation of cationic π-benzyl-palladium (η^3-benzyl-palladium) complexes, including coordination of Pd(0) to the aromatic nucleus of **92** followed by ionization (Scheme 8.27, path a, analogous to that of allylic alkylation), and an S_N2 displacement of the leaving group of **92** by Pd(0) followed by isomerization (Scheme 8.27, path b) [47]. Either of the pathways principally depends on the nature of the leaving and aryl groups.

An asymmetric benzylation reaction of α-aryl acetates enabled by the cooperative catalysis of chiral isothiourea and palladium complex was reported by Snaddon and coworkers (Scheme 8.28) [48]. 2-Naphthalenylmethyl diphenyl phosphate **93**, because the barrier of its dearomatization is lower than that of the monocyclic benzyl counterparts [47], is easier to generate a relatively more stable cationic π-benzyl-Pd(II) **Int-48**, which then reacts with a stereodefined C1-ammonium enolate **Int-39**, leading to the products **94**. A variety of pentafluorophenyl

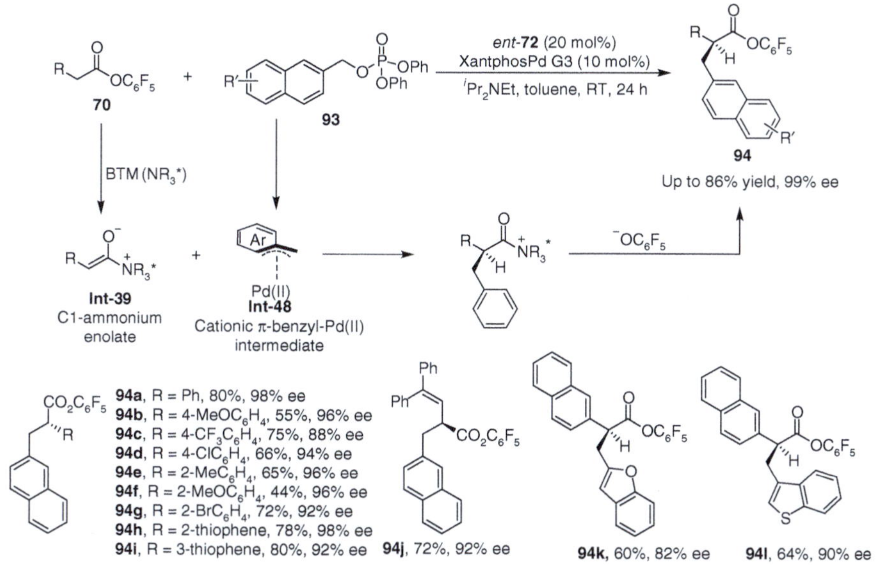

Scheme 8.28 Chiral isothiourea/palladium-catalyzed benzylic alkylation. Source: Modified from Schwarz et al. [48].

2-arylacetates participate in the reaction to provide benzylation products **94a–94i** in excellent enantioselectivity of up to 98% ee. 3,3′-Diaryl butenoic acid pentafluorophenyl esters perform well as nucleophiles, as exemplified by **94j**. In addition to 2-naphthalenylmethyl diphenyl phosphates, 3-benzothiophenylmethyl and 2-benzofuranylmethyl diphenyl phosphates are also excellent electrophiles to undergo the benzylic alkylation, as shown by the cases to give **94k** and **94l**. However, the extension to simple monocyclic benzylic phosphates or esters failed, presumably due to the high energy required for dearomatization to generate reactive π-benzyl-Pd(II) intermediate.

Notably, this method is highly valuable in organic synthesis and has successfully been applied to the concise synthesis of optically pure thrombin inhibitor DX-9065A (Scheme 8.29) [48].

8.4.3 Metal-allenylidene Mediated Transformations

The asymmetric addition of chiral isothiourea-bound enolates to electrophilic Cu-allenylidenes was first described by Gong and coworkers [49]. Based on this elementary event, they established an asymmetric decarboxylative [4+2] annulation of ethynyl benzoxazinanones **67** and carboxylic acids **95** enabled by the cooperative catalysis of the copper complex and chiral isothiourea Lewis base (Scheme 8.30). The copper–allenylidene complex **Int-38′**, generated from the decarboxylation of **67** with the copper catalyst, smoothly couples with the transient chiral C1-ammonium enolate **Int-39** generated from chiral isothiourea Lewis base **72** and carboxylic acids **95** that is activated by the formation of 2-arylacetic tosyl anhydrides, after

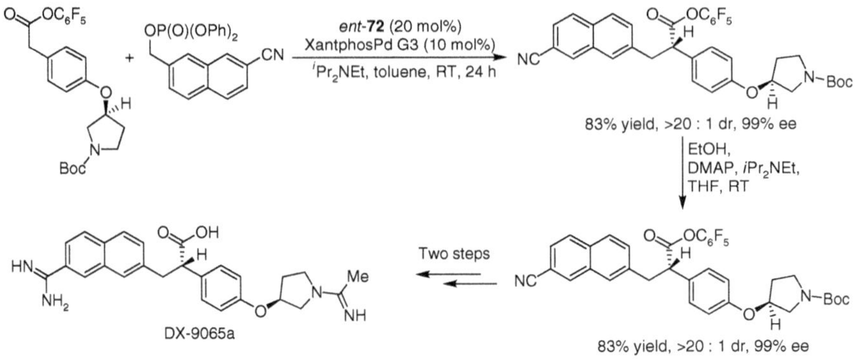

Scheme 8.29 Asymmetric synthesis of DX-9065A. Source: Modified from Schwarz et al. [48].

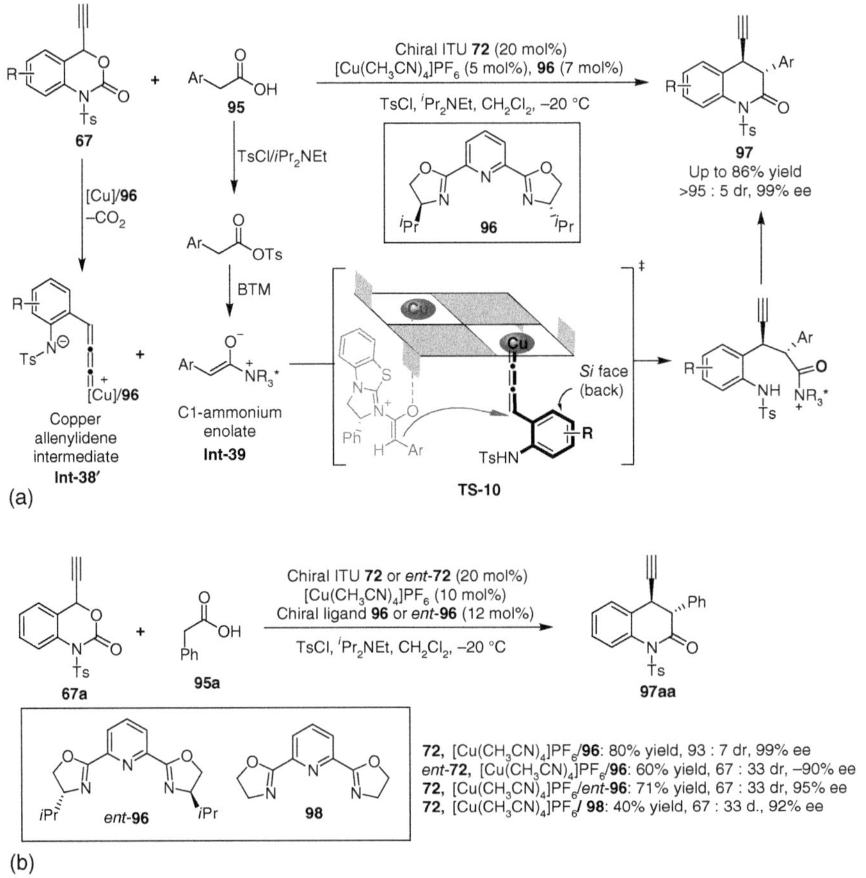

Scheme 8.30 Chiral isothiourea/copper cooperatively catalyzed [4+2] annulation. (a) [4+2] Annulation and the reaction pathway. (b) Identifying the key elements to control stereochemistry.

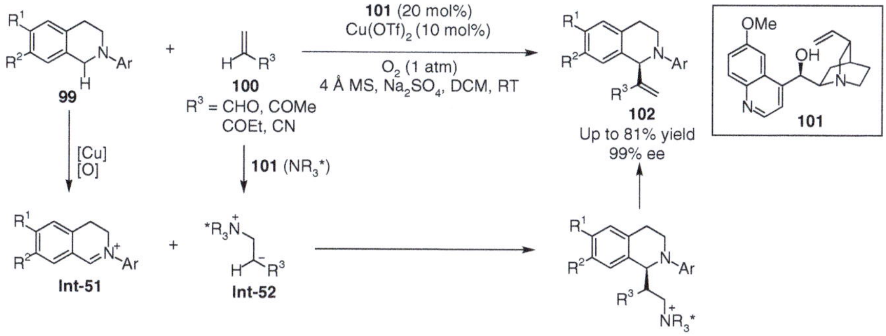

Scheme 8.31 Quinine/copper-catalyzed asymmetric oxidative aza-Morita–Baylis–Hillman reaction. Source: Modified from Zhang et al. [51a].

lactamization, leading to 3,4-dihydroquinolin-2-one derivatives **97** in high yield (up to 86%) and with excellent levels of stereoselectivity (up to 95 : 5 dr, 99% ee). In contrast, the *ent*-**72** enables the reaction to give the enantiomer of **97aa** with considerably lower diastereo- and enantioselectivities (Scheme 8.30b). Only switching the configuration of the chiral ligand does not change the stereochemistry of the product, but slightly varies the diastereoselectivity. Even the achiral ligand **98**, in concert with chiral ITU **72**, still offers excellent enantioselectivity. These results lead to a conclusion that the chiral ITU catalyst controls the absolute stereochemistry of the bond-forming event, and the matched chirality of pybox ligand **96** and isothiourea catalyst **72** is mainly beneficial to controlling diastereoselectivity rather than enantioselectivity. As shown in the proposed transition state **TS-10**, one copper center might serve as an activator of the copper–allenylidene complex **Int-38′**, and the other acts as a counterion to the C1-ammonium (*Z*)-enolate **Int-39**. The propargylation step favors the *Si*-face of the copper–allenylidene complex to give the experimentally observed product as the major isomer. Coincidently, Cao and Wu group independently reported a similar reaction at nearly the same time, wherein the pivaloyl chloride was the activator for carboxylic acids [50].

8.4.4 Other Transition Metal Mediated Transformations

In 2012, Wang and coworkers explored a quinine/copper-catalyzed asymmetric oxidative aza-Morita–Baylis–Hillman reaction of *N*-aryl tetrahydroisoquinolines and electron-deficient olefins (Scheme 8.31) [51a]. The evaluation of metals salts, such as AgOTf, $Pd(OAc)_2$, and $Cu(OTf)_2$ shows that cooper salt is the best co-catalyst with quinine **101**. The copper-catalyzed dehydrogenative process gives an imine-type intermediate (**Int-51**), which reacts with the chiral tertiary amine-activated ammonium intermediates **Int-52** to afford C1-alkene tetrahydroisoquinoline derivatives **102** in satisfactory yields (up to 81%) and excellent enantioselectivities (up to 99% ee).

In 2012, Fu group reported a dynamic kinetic resolution (DKR) of aryl alkyl carbinols **103** via relay catalysis of a planar-chiral 4-dimethylaminopyridine (DMAP) derivative and a ruthenium complex (Scheme 8.32) [52]. $(C_5Ph_5)Ru(CO)_2(O^tBu)$

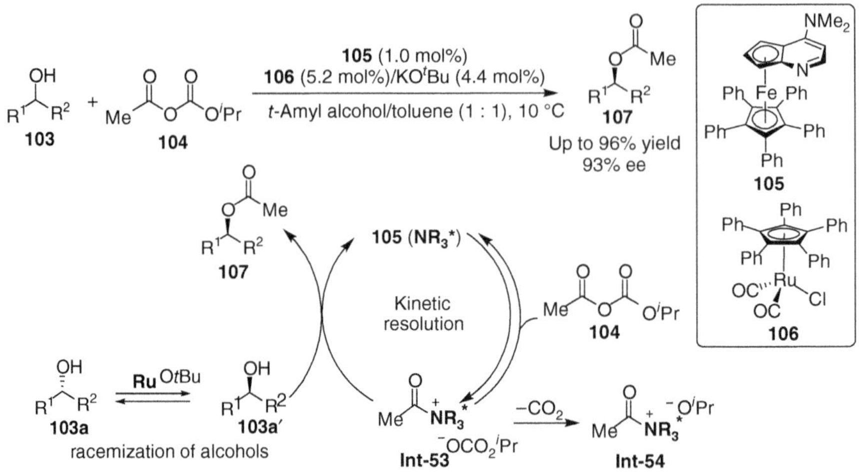

Scheme 8.32 Chiral DMAP/ruthenium-catalyzed DKR of secondary alcohols. Source: Modified from Lee et al. [52].

(**Ru**$^{\mathbf{O}t\mathbf{Bu}}$), generated *in situ* from $(C_5Ph_5)Ru(CO)_2Cl$ (RuCl) **106** and KO^{t}Bu, is a highly efficient catalyst for the racemization of aryl alkyl carbinols **103** at room temperature [53], while the planar-chiral DMAP derivative **105** serves as an effective catalyst for the kinetic resolution (KR) of carbinols. The racemization and KR processes operate independently and compatibly, and thus the DKR tolerates a broad scope of aryl alkyl carbinols to result in high yields and great enantioselectivity (up to 96% yield and 93% ee). Mechanistic studies reveal that the use of less electrophilic acyl carbonates **104** is essential for preventing the deactivation of the metal catalyst, and the acyl transfer process from ion pair **Int-53**, rather than **Int-54**, to the alcohol, is likely the rate-determining step of the DKR.

In 2008, Shi and coworkers established a copper-catalyzed α-amination of esters, a state to art method to access hydantoins that are frequently encountered as core structural elements in natural products and pharmaceuticals [54]. However, the enantioselective version has not been reported until Gong described an enantioselective α-amination of esters enabled by the chiral isothiourea/copper cooperative catalysis (Scheme 8.33) [55]. Although the copper complex of tributylphosphine can catalyze the non-stereoselective α-amination, the use of chiral ITU **109** as the co-catalyst still offers a highly enantioselective version tolerant of a broad scope of the perfluorophenyl acetates **70**. The mechanistic studies by the electron paramagnetic resonance (EPR) spectroscopy reveal that the cleavage of the N—N bond of *N*,*N*-diaziridinone **108** by the Cu complex generates a Cu(II) radical species **Int-55**. Then, the chiral C1-ammonium enolate **Int-39**, concurrently generated from the perfluorophenyl acetates **70** and chiral ITU **109**, couples with the copper-bound nitrogen radical intermediate **Int-55** via transition state **TS-11**. Probably, the isopropyl group of the chiral ITU catalyst effectively shields the *Si* face of the C1-ammonium (*Z*)-enolate **Int-39** to permit the *Re* face open for the nitrogen radical intermediate **Int-55** to undergo the enantioselective radical addition, and

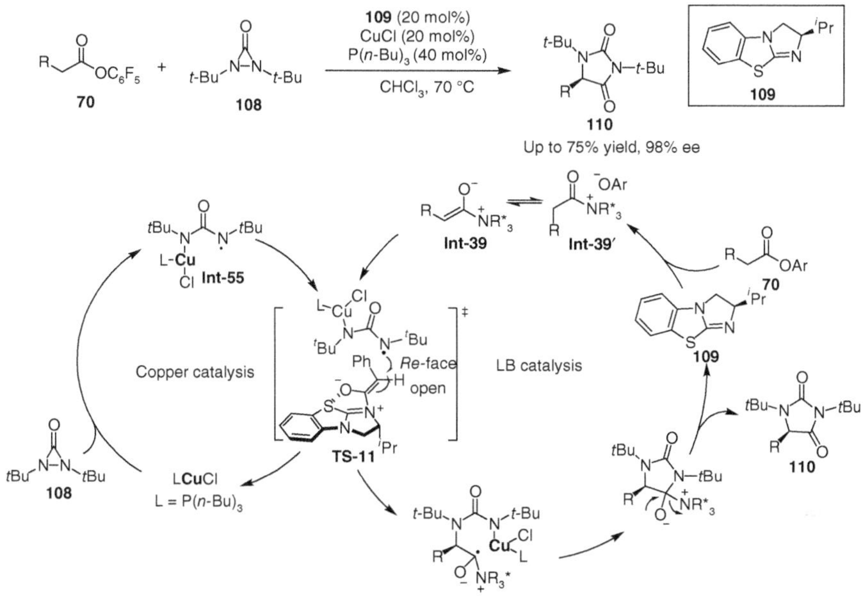

Scheme 8.33 Chiral BTM/copper cooperatively catalyzed α-amination of esters. Source: Modified from Song et al. [55].

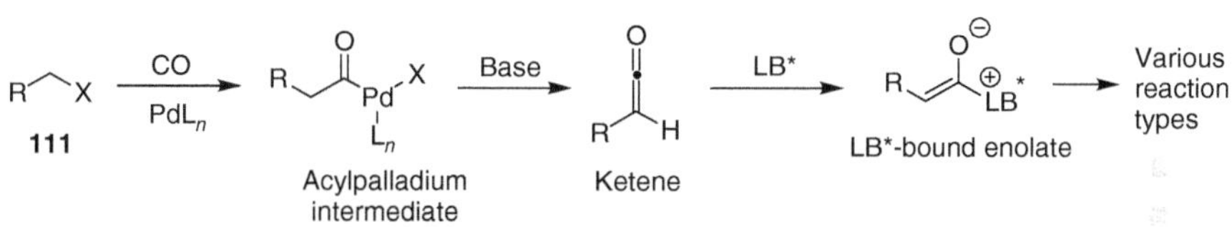

Scheme 8.34 Catalytic generation of Lewis base-bound enolate from Pd-catalyzed carboxylation

then followed by the subsequent lactamization to give experimentally observed product **110** (up to 75% yield, 98% ee).

The palladium-catalyzed carboxylation of an organohalide **111** with carbon monoxide (CO) generates an acylpalladium intermediate bearing an α-proton that can be transformed to a ketene. The addition of a Lewis base to the ketene furnishes a Lewis base-bound enolate, which is a versatile intermediate capable of participating in various types of reactions (Scheme 8.34).

In 2019, Gong, Han, and coworkers reported asymmetric formal [1+1+4] and [1+1+2] annulation reactions of benzyl bromides **111**, CO, and *N*-tosylimines (**112** or **116**) to respectively give chiral dihydropyridones **115** and β-lactams **117** in high yields and with excellent enantio- and diastereoselectivities (Scheme 8.35) [56]. Mechanistically, the oxidative addition of the palladium complex to benzyl bromide **111a** generates an intermediate **Int-56**, which can undergo an insertion reaction with CO to afford acyl-palladium **Int-57**. β-Hydride elimination from the

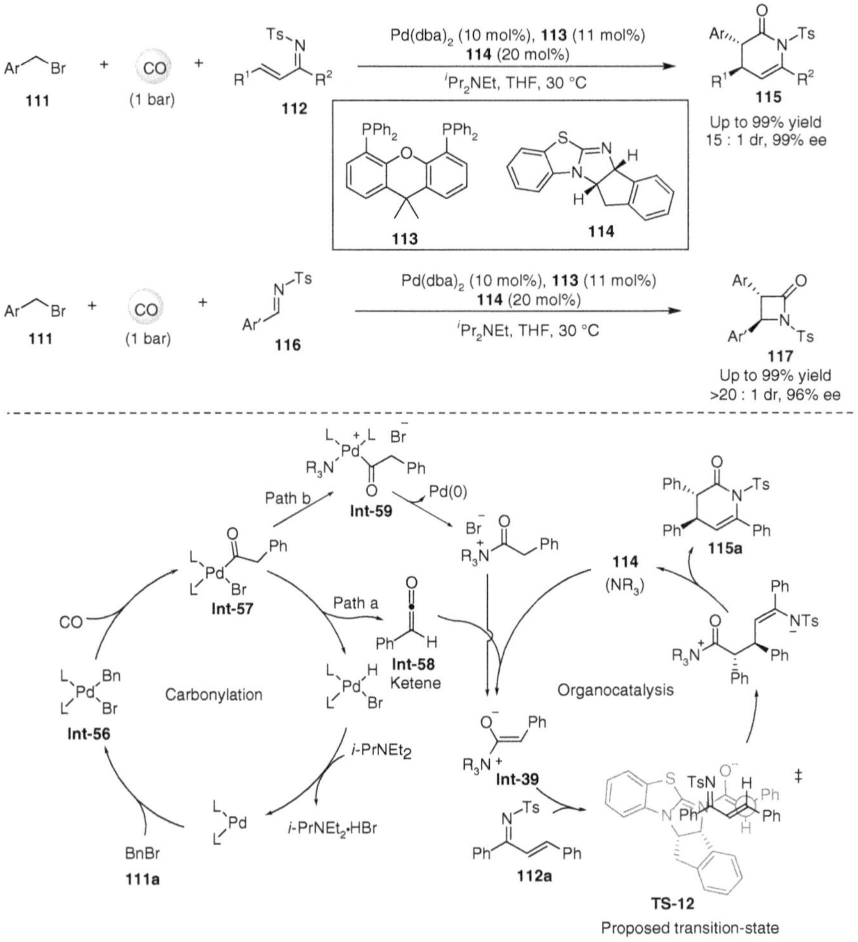

Scheme 8.35 Palladium and chiral ITU relay catalytic system. Source: Modified from Li et al. [56].

acyl-palladium intermediate **Int-57** affords ketene **Int-58** (path a), which then reacts with the chiral Lewis base **114** to form the key C1-ammonium enolate intermediate **Int-39a**. A series of kinetic studies reveal that the C–H cleavage is likely involved in the rate-limiting step. Alternatively, the reaction pathway b, leading to the generation of **Int-59** from acylpalladium **Int-57**, is also possible [57]. Due to the slow generation and *in situ* capture of the unstable ketene intermediate, side and racemic background reactions are essentially evaded. Once formed, the C1-ammonium enolate intermediate **Int-39a** undergoes a stereoselective Michael addition with ketimine **112** via a proposed transition state **TS-12** [58], followed by the intramolecular cyclization, to generate the corresponding dihydropyridine **115a**. For the formal [1+1+2] reaction, the intermediate **Int-39a** and imine **116** undergo a similar cascade Mannich-type reaction/intramolecular cyclization process (Staudinger [2+2] reaction) to give β-lactam **117.**

8.5 Conclusions

Although the unavoidable coordination interaction commonly happened between the Lewis base and transition metal causes the "self-quenching" to weaken or even kill the catalytic activity of either or both of them, the combined catalysis of these principally distinct catalysts has gradually become more and more visible in the asymmetric synthetic chemistry since the proof concept and recent breakthroughs appeared. The Lewis base/transition metal cooperative catalysis has shown remarkable compatibility, outstanding catalytic activity, and excellent stereocontrol ability, leading to different types of enantioselective catalytic transformations, including C–C and C–heteroatom bond-forming reactions. Despite the "self-quenching" occasionally encountered in some transition metal and Lewis base combined catalysis, this issue has been partially figured out by strategies based on fundamental coordination chemistry, such as the addition of external competing ligands, the presence of more amounts of Lewis base relative to the transition metal to allow the organocatalyst coexisting at a catalytically active concentration. However, some limitations remain in the cooperative catalysis of this type, including: (i) the Lewis base catalysts are restricted to NHCs and chiral isothioureas, and the activation modes are also limited to the Lewis base-mediated nucleophilic species. In sharp contrast, Lewis base-mediated electrophilic intermediates are yet underdeveloped; (ii) the transition metal catalysts are restricted to palladium, iridium, copper, and gold. Many other transition-metal catalysts widely used in the creation of new reactions, such as rhodium, platinum, ruthenium, etc., and related activation modes are yet to be combined with chiral Lewis bases for the cooperative catalysis; (iii) the readily available protocols are mainly constituted of allylation, propargylation, and annulation reactions, while a more diverse range of synthetically useful reactions needs to be explored; (iv) the potential of combining catalysis events with subsequent value-added transformations in stereoselective chemical synthesis should be further developed.

References

1 (a) Denmark, S.E. and Beutner, G.L. (2008). *Angew. Chem. Int. Ed.* 47: 1560–1638. (b) List, B. (2012). *Asymmetric Organocatalysis 1 Lewis Base and Acid Catalysts*. New York: Georg Thieme Verlag KG Stuttgart. (c) Vedejs, E. and Denmark, S.E. (2016). *Lewis Base Catalysis in Organic Synthesis*. Wiley-VCH.

2 For reviews on ternary amine Lewis base catalysis, see: France, S., Guerin, D.J., Miller, S.J., and Lectka, T. (2003). *Chem. Rev.* 103: 2985–3012.

3 For reviews on phosphine Lewis base catalysis, see: (a) Lu, X.-Y., Zhang, C.-M., and Xu, Z.-R. (2001). *Acc. Chem. Res.* 34: 535–544. (b) Ye, L.-W., Zhou, J., and Tang, Y. (2008). *Chem. Soc. Rev.* 37: 1140–1152. (c) Wang, S.X., Han, X.Y., Zhong, F.R. et al. (2011). *Synlett* 2011: 2766–2778. (d) Chai, Z. and Zhao, G. (2012). *Catal. Sci. Technol.* 2: 29–41. (e) Wei, Y. and Shi, M. (2014). *Chem. Asian J.* 9: 2720–2734.

4 For reviews on NHC Lewis base catalysis, see: (a) Enders, D., Niemeier, O., and Henseler, A. (2007). *Chem. Rev.* 107: 5606–5655. (b) Hopkinson, M.N., Richter, C., Schedler, M., and Glorius, F. (2014). *Nature* 510: 485–496. (c) Flanigan, D.M., Romanov-Michailidis, F., White, N.A., and Rovis, T. (2015). *Chem. Rev.* 115: 9307–9387. (d) Chen, X., Wang, H., Jin, Z., and Chi, Y.R. (2020). *Chin. J. Chem.* 38: 1167–1202.

5 (a) Beller, M. and Bolm, C. (2004). *Transition Metals for Organic Synthesis: Building Blocks and Fine Chemicals*. 2nd revised and enlarged ed. Weinheim: Wiley-VCH. (b) Hartwig, J.F. (2010). *Organotransition Metal Chemistry: From Bonding to Catalysis*. Sausalito, CA: University Science Books.

6 (a) Glorius, F. (2007). *N-Heterocyclic Carbenes in Transition Metal Catalysis*. Berlin: Springer. (b) Díez-González, S., Marion, N., and Nolan, S.P. (2009). *Chem. Rev.* 109: 3612–3679. (c) Fortman, G.C. and Nolan, S.P. (2011). *Chem. Soc. Rev.* 40: 5151–5169. (d) Janssen-Müller, D., Schlepphorst, C., and Glorius, F. (2017). *Chem. Soc. Rev.* 46: 4845–4854.

7 For reviews on N-heterocyclic carbene (NHC)/metal cooperative catalysis, see: (a) Wang, M.H. and Scheidt, K.A. (2016). *Angew. Chem. Int. Ed.* 55: 14912–14922. (b) Lu, H., Liu, J.-Y., Li, H.-Y., and Xu, P.-F. (2018). *Acta Chim. Sinica* 76: 831–837. (c) Nagao, K. and Ohmiya, H. (2019). *Top. Curr. Chem.* 377: 35.

8 For reviews on nucleophilic ternary amine/metal cooperative catalysis, see: (a) Zhou, Q. and Lu, P. (2018). *Acta Chim. Sinica* 76: 825–830. (b) Knox, G.J., Hutchings-Goetz, L.S., Pearson, C.M., and Snaddon, T.N. (2020). *Top. Curr. Chem.* 378: 16.

9 Jellerichs, B.G., Kong, J.-R., and Krische, M.J. (2003). *J. Am. Chem. Soc.* 125: 7758–7759.

10 (a) Gaunt, M.J. and Johansson, C.C.C. (2007). *Chem. Rev.* 107: 5596–5605. (b) Morrill, L.C. and Smith, A.D. (2014). *Chem. Soc. Rev.* 43: 6214–6226. (c) McLaughlin, C. and Smith, A.D. (2021). *Chem. Eur. J.* 27: 1533–1555.

11 (a) Nair, V., Menon, R.S., Biju, A.T. et al. (2011). *Chem. Soc. Rev.* 40: 5336–5346. (b) Douglas, J., Churchill, G., and Smith, A.D. (2012). *Synthesis* 44: 2295–2309. (c) Mahatthananchai, J. and Bode, J.W. (2014). *Acc. Chem. Res.* 47: 696–707. (d) Menon, R.S., Biju, A.T., and Nair, V. (2015). *Chem. Soc. Rev.* 44: 5040–5052.

12 For the pioneering work, see: (a) M. Rauhut and H. Currier. (American Cyanamid Co.), U.S. Patent 3,074,999, 1963; Chem. Abstr. 1963, 58, 11224a; (b) C.-M. Zhang and X.-Y. Lu. *J. Org. Chem.* 1995, 60, 2906–2908; (c) Zhang, C.-M. and Lu, X.-Y. (1995). *J. Org. Chem.* 60: 2906–2908.

13 Chen, D.F., Zhang, C.-L., Hu, Y. et al. (2015). *Org. Chem. Front.* 2: 956–960.

14 Chen, P., Chen, Z.-C., Li, Y. et al. (2019). *Angew. Chem. Int. Ed.* 58: 4036–4040.

15 Xia, C., Wang, D.-C., Qu, G.-R., and Guo, H.-M. (2020). *Org. Chem. Front.* 7: 1474–1480.

16 (a) Strassner, T. (2004). *Top. Organomet. Chem.* 13: 1–20. (b) Cavallo, L., Correa, A., Costabile, C., and Jacobsen, H.J. (2005). *Organomet. Chem.* 690: 5407–5413. (c) Díez-González, S. and Nolan, S.P. (2007). *Coord. Chem. Rev.* 251: 874–883.

17 (a) Trost, B.M. and Van Vranken, D.L. (1996). *Chem. Rev.* 96: 395–422. (b) Trost, B.M. and Crawley, M.L. (2003). *Chem. Rev.* 103: 2921–2943. (c) Sundararaju, B.,

Achard, M., and Bruneau, C. (2012). *Chem. Soc. Rev.* 41: 4467–4483. (d) Butta, N.A. and Zhang, W. (2015). *Chem. Soc. Rev.* 44: 7929–7967.

18 (a) Nemoto, T., Fukuda, T., and Hamada, Y. (2006). *Tetrahedron Lett.* 47: 4365–4368. (b) Lebeuf, R., Hirano, K., and Glorius, F. (2008). *Org. Lett.* 10: 4243–4246. (c) He, J., Tang, S., Tang, S. et al. (2009). *Tetrahedron Lett.* 50: 430–433.

19 Y. Bai, S. Xiang, M. L. Leow, and X.-W. Liu. *Chem. Commun.* 2014, 50, '–6170.

20 Bai, Y., Leng, W.L., Li, Y., and Liu, X.-W. (2014). *Chem. Commun.* 50: 13391–13393.

21 Liu, K., Hovey, M.T., and Scheidt, K.A. (2014). *Chem. Sci.* 5: 4026–4031.

22 (a) Guo, C., Fleige, M., Janssen-Müller, D. et al. (2016). *J. Am. Chem. Soc.* 138: 7840–7843. (b) Guo, C., Janssen-Müller, D., Fleige, M. et al. (2017). *J. Am. Chem. Soc.* 139: 4443–4451.

23 Singha, S., Patra, T., Daniliuc, C.G., and Glorius, F. (2018). *J. Am. Chem. Soc.* 140: 3551–3554.

24 Singha, S., Serrano, E., Mondal, S. et al. (2020). *Nat. Catal.* 3: 48–54.

25 Cheng, Q., Tu, H.-F., Zheng, C. et al. (2019). *Chem. Rev.* 119: 1855–1969.

26 (a) Krautwald, S., Sarlah, D., Schafroth, M.A., and Carreira, E.M. (2013). *Science* 340: 1065–1068. (b) Krautwald, S. and Carreira, E.M. (2017). *J. Am. Chem. Soc.* 139: 5627–5639.

27 Anaya de Parrodi, C. and Walsh, P.J. (2009). *Angew. Chem. Int. Ed.* 48: 4679–4682.

28 Namitharan, K., Zhu, T., Cheng, J. et al. (2014). *Nat. Commun.* 5: 3982–3988.

29 Zhou, L., Wu, X., Yang, X. et al. (2020). *Angew. Chem. Int. Ed.* 59: 1557–1561.

30 (a) Miyake, Y., Uemura, S., and Nishibayashi, Y. (2009). *ChemCatChem* 1: 342–356. (b) Ljungdahl, N. and Kann, N. (2009). *Angew. Chem. Int. Ed.* 48: 642–644. (c) Detz, R.J., Hiemstra, H., and van Maarseveen, J.H. (2009). *Eur. J. Org. Chem.* 2009: 6263–6276. (d) Ding, C.-H. and Hou, X.-L. (2011). *Chem. Rev.* 111: 1914–1937. (e) Nishibayashi, Y. (2012). *Synthesis* 44: 489–503. (f) Zhang, D.-Y. and Hu, X.-P. (2015). *Tetrahedron Lett.* 56: 283–295. (g) Hu, X.-H., Liu, Z.-T., Shao, L., and Hu, X.-P. (2015). *Synthesis* 47: 913–923.

31 (a) Tian, L., Gong, L., and Zhang, X. (2018). *Adv. Synth. Catal.* 360: 2055–2059. (b) Zhang, Y.-C., Zhang, B.-W., Geng, R.-L., and Song, J. (2018). *Org. Lett.* 20: 7907–7911. (c) Gómez, J.E., Cristòfol, À., and Kleij, A.W. (2019). *Angew. Chem. Int. Ed.* 58: 3903–3907.

32 Zhang, Z.-J., Zhang, L., Song, J. et al. (2019). *Angew. Chem. Int. Ed.* 58: 12190–12194.

33 (a) Wang, Q., Li, T.-R., Lu, L.Q. et al. (2016). *J. Am. Chem. Soc.* 138: 8360–8363. (b) Li, T.-R., Wang, Y.-N., Xiao, W.-J., and Lu, L.-Q. (2018). *Tetrahedron Lett.* 59: 1521–1530.

34 Schwarz, K.J., Amos, J.L., Klein, J.C. et al. (2016). *J. Am. Chem. Soc.* 138: 5214–5217.

35 (a) Liu, P., Yang, X., Birman, V.B., and Houk, K.N. (2012). *Org. Lett.* 14: 3288–3291. (b) Greenhalgh, M.D., Smith, S.M., Walden, D.M. et al. (2018). *Angew. Chem. Int. Ed.* 57: 3200–3206. (c) Antúnez, D.-J.B., Greenhalgh, M.D.,

Brueckner, A.C. et al. (2019). *Chem. Sci.* 10: 6162–6173. (d) Young, C.M., Elmi, A., Pascoe, D.J. et al. (2020). *Angew. Chem. Int. Ed.* 59: 3705–3710.
36 Jiang, X., Beiger, J.J., and Hartwig, J.F. (2017). *J. Am. Chem. Soc.* 139: 87–90.
37 (a) West, T.H., Daniels, D.S.B., Slawin, A.M.Z., and Smith, A.D. (2014). *J. Am. Chem. Soc.* 136: 4476–4479. (b) West, T.H., Spoehrle, S.S.M., and Smith, A.D. (2017). *Tetrahedron* 73: 4138–4149. (c) West, T.H., Walden, D.M., Taylor, J.E. et al. (2017). *J. Am. Chem. Soc.* 139: 4366–4375.
38 Spoehrle, S.S.M., West, T.H., Taylor, J.E. et al. (2017). *J. Am. Chem. Soc.* 139: 11895–11902.
39 Schwarz, K.J., Pearson, C.M., Cintron-Rosado, G.A. et al. (2018). *Angew. Chem. Int. Ed.* 57: 7800–7803.
40 Fyfe, J.W.B., Kabia, O.M., Pearson, C.M., and Snaddon, T.N. (2018). *Tetrahedron* 74: 5383–5391.
41 (a) Hutchings-Goetz, L., Yang, C., and Snaddon, T.N. (2018). *ACS Catal.* 8: 10537–10544. (b) Trost, B.M. (1996). *Chem. Rev.* 96: 395–422.
42 Scaggs, W.R. and Snaddon, T.N. (2018). *Chem. Eur. J.* 24: 14378–14381.
43 Scaggs, W.R., Scaggs, T.D., and Snaddon, T.N. (2019). *Org. Biomol. Chem.* 17: 1787–1790.
44 Pearson, C., Fyfe, J., and Snaddon, T.N. (2019). *Angew. Chem. Int. Ed.* 58: 10521–10527.
45 Legros, J.-Y. and Fiaud, J.-C. (1992). *Tetrahedron Lett.* 33: 2509–2510.
46 (a) Kuwano, R. (2009). *Synthesis*: 1049–1061. (b) Trost, B.M. and Czabaniuk, L.C. (2014). *Angew. Chem. Int. Ed.* 53: 2826–2851. (c) Le Bras, J. and Muzart, J. (2016). *Eur. J. Org. Chem.*: 2565–2593.
47 Trost, B.M. and Czabaniuk, L.C. (2012). *J. Am. Chem. Soc.* 134: 5778–5781.
48 Schwarz, K.J., Yang, C., Fyfe, J.W.B., and Snaddon, T.N. (2018). *Angew. Chem. Int. Ed.* 57: 12102–12105.
49 Song, J., Zhang, Z.-J., and Gong, L.-Z. (2017). *Angew. Chem. Int. Ed.* 56: 5212–5216.
50 Lu, X., Ge, L., Cheng, C. et al. (2017). *Chem. Eur. J.* 23: 7689–7693.
51 (a) Zhang, G., Ma, Y., Wang, S. et al. (2012). *J. Am. Chem. Soc.* 134: 12334–12337. (b) Boess, E., Schmitz, C., and Klussmann, M. (2012). *J. Am. Chem. Soc.* 134: 5317–5325.
52 Lee, S.Y., Murphy, J.M., Ukai, A., and Fu, G.C. (2012). *J. Am. Chem. Soc.* 134: 15149–15153.
53 Martín-Matute, B., Edin, M., Bogár, K. et al. (2005). *J. Am. Chem. Soc.* 127: 8817–8825.
54 Zhao, B., Du, H., and Shi, Y. (2008). *J. Am. Chem. Soc.* 130: 7220–7221.
55 Song, J., Zhang, Z.-J., Chen, S.-S. et al. (2018). *J. Am. Chem. Soc.* 140: 3177–3180.
56 Li, L.-L., Ding, D., Song, J. et al. (2019). *Angew. Chem. Int. Ed.* 58: 7647–7651.

57 Lagueux-Tremblay, P.L., Fabrikant, A., and Arndtsen, B.A. (2018). *ACS Catal.* 8: 5350–5354.

58 (a) Simal, C., Lebl, T., Slawin, A.M.Z., and Smith, A.D. (2012). *Angew. Chem. Int. Ed.* 51: 3653–3657. (b) Shi, Q.Q., Zhang, W., Wang, Y. et al. (2018). *Org. Biomol. Chem.* 16: 2301–2311.

9

Chiral Organocatalyst Combined with Transition Metal Based Photoredox Catalyst

9.1 Introduction

For billions of years, nature has been directed by the sunlight through converting solar energy into chemical energy, namely photosynthesis [1], which inspires the flourishment of light-driven synthetic organic chemistry [2]. In the past two decades, photocatalysis utilizing a visible-light absorbing molecule, i.e. photocatalyst (PC), particularly transition metal-based ruthenium(II) and iridium(III) polypyridyl complexes (Figure 9.1) [3], has become a unique and programmable strategy to forge remarkable chemical bonds otherwise inaccessible [4]. Although most ground-state PCs are indolent in performing single-electron transfer (SET), photoirradiation of these complexes leads to the excitation of the highest occupied molecular orbital (HOMO) electron, which can rapidly undergo intersystem crossing to give long-lived excited states. These exciplexes often result in donating one electron to a substrate or taking another electron from somewhere, namely oxidative quenching or reductive quenching. In general, the SET induced by an excited-state PC affords active radical intermediates for further transformations [5]. Alternatively, the excited-state PC may undergo energy transfer (EnT) with an appropriate substrate to afford the excited-state substrate possessing partial di-radical character, which is highly reactive toward radical functionalization [6]. Although a variety of impressive advances in synthetic transformations through visible-light-driven SET or EnT mechanisms have been achieved [7], the enantioselective variants have still met with limited success presumably because the high reactivity of radicals generally brings about uncontrollable background reactions [8].

In the past several decades, the use of an organic small molecule as the chiral catalyst, namely organocatalysis, has been extensively developed and kept abreast of traditional metal catalysis and enzymatic catalysis toward offering powerful solutions for asymmetric catalysis [9]. Chiral organocatalysts, which are usually low-cost, readily accessible, as well as moisture- and oxygen-tolerated, can perform either covalent or noncovalent activation on the substrates (Figure 9.2). In terms of covalent bonding activation modes [10], organocatalysts selectively and reversibly react with specific functionalities of the substrate to generate reactive intermediates, enabling the transformation by lowering the energy barrier. The premier examples of covalent activation include: (i) chiral amine catalysts activate aldehydes and

Asymmetric Organo-Metal Catalysis: Concepts, Principles, and Applications, First Edition. Liu-Zhu Gong.
© 2022 WILEY-VCH GmbH. Published 2022 by WILEY-VCH GmbH.

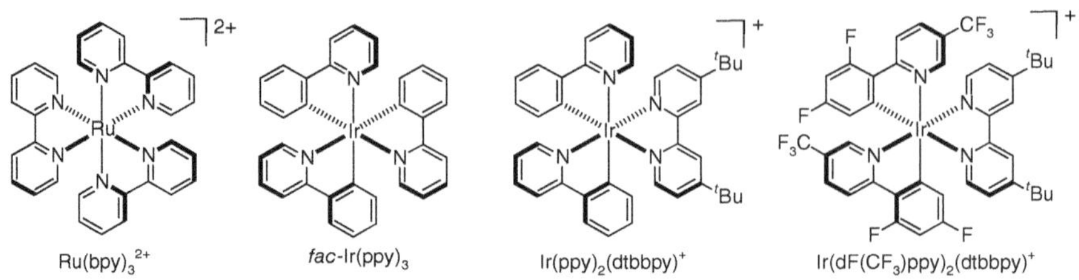

Figure 9.1 Representative transition metal-based photocatalysts.

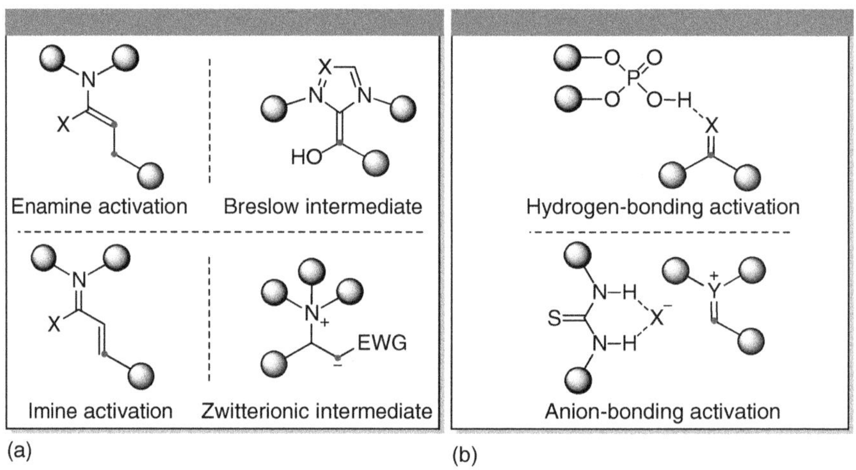

Figure 9.2 Representative activation modes of chiral organocatalysis. (a) Covalent-based activation modes. (b) Non-covalent-based activation modes.

ketones via the formation of chiral enamine and iminium intermediates [11]; (ii) nitrogen-heterocyclic carbene (NHC) catalysts activate aldehydes through the formation of Breslow intermediate [12]; (iii) Lewis base catalysts undergo 1,4-addition to electron-deficient olefin substrates, affording zwitterionic intermediates [13] (Figure 9.2a). On the other hand, the noncovalent bonding activation modes, such as hydrogen-bonding activation and anion-binding activation (Figure 9.2b) [14], rely on the weaker and less directional interactions between the catalyst and specific functionalities of the substrates. By virtue of the unique ability in stereocontrol, chiral organocatalysis, combined with transition metal-based photoredox catalysis, has achieved impressive advances in enantioselective radical transformations, illustrating the great potential of the integration of both types of catalysis to harvest challenging photochemical processes [8c]. In this chapter, we will outline the initiation and development of the combined catalysis of transition metal-based photoredox catalysts and chiral organocatalysts. All of these combined catalyzes will be discussed in activation modes of organocatalysts. In particular, the landmark discoveries in this rapidly growing field will be highlighted to inspire further novel transformations.

9.2 Covalent-Based Organocatalytic Activation in Combination with Transition Metal-Based Photoredox Catalyst

9.2.1 Chiral Amine/Photoredox Combined Catalysis

Chiral amine catalysts have been widely applied to the enantioselective α-functionalization of carbonyl compounds through enamine activation, as exemplified by organocatalytic asymmetric Aldol, Mannich, and Michael reactions [15]. However, the direct α-alkylation of carbonyl compounds with alkyl halides via enamine catalysis has proven to be challenging [16]. The cooperation of chiral amine and photoredox catalysts offers a radical approach to this anionic chemistry. In 2008, MacMillan and coworker established an enantioselective α-alkylation of aldehydes with α-bromocarbonyls by merging $Ru(bpy)_3Cl_2$ and chiral amine catalyst **3** [17], in which α-bromocarbonyls were used as electrophilic radical precursors to overcome the intrinsic high energy barrier in traditional S_N2 pathway (Scheme 9.1). The authors proposed that electron-deficient alkyl radical **5** was generated from α-bromocarbonyl **2** via SET with $Ru(bpy)_3^+$, whereas electron-rich

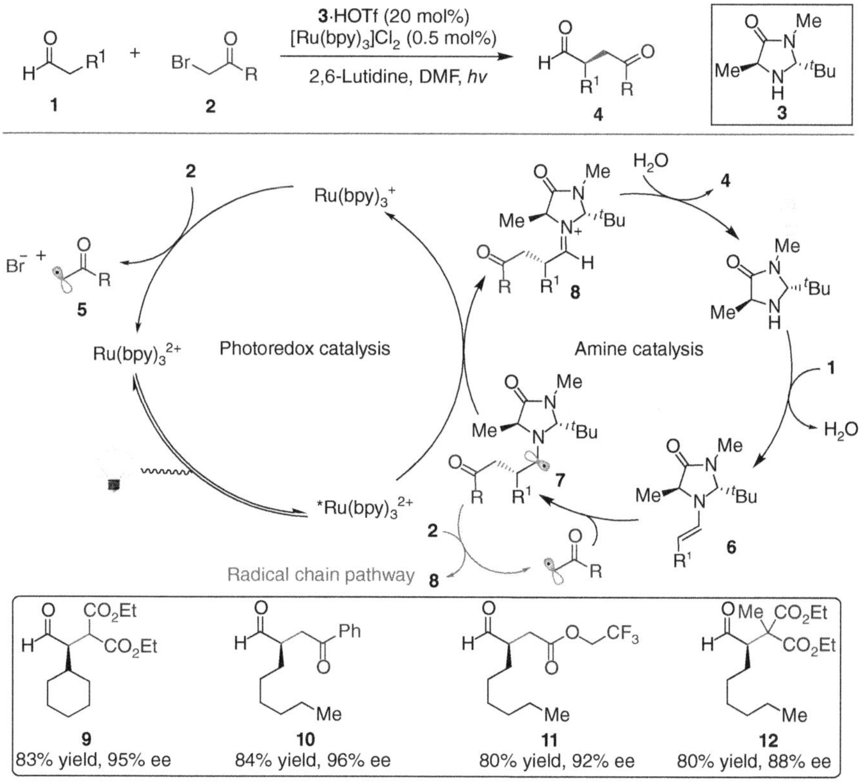

Scheme 9.1 Asymmetric α-alkylation of aldehydes via merging photoredox catalysis with chiral amine catalysis.

chiral enamine **6** was formed through the condensation of aldehyde **1** and chiral amine catalyst **3**. Then electron-deficient radical **5** rapidly added to enamine **6**, affording intermediate **7**, which underwent sequential single-electron oxidation and hydrolysis to release the alkylation product **4** and regenerate chiral amine **3** for further catalytic cycles. A broad range of electron-deficient α-bromo carbonyls, such as diethyl bromomalonate, α-bromoacetophenone, α-bromo ester, and tertiary diethyl methylbromomalonate, are nicely tolerated to afford the corresponding alkylation products (**9–12**) in excellent yields and enantioselectivities. Notably, later mechanistic investigations involving quantum yield and luminescence quenching experiments [18] suggested that a radical chain propagation, which consisted of a SET process of α-bromocarbonyl **2** and α-amino radical **7**, might dominate in the formation of products.

Given a similar mechanism with other alkylating agents, the combination of transition metal-based photoredox catalysts and chiral amine catalysts were also viable to accomplish enantioselective α-alkylation reaction of aldehydes (Scheme 9.2), including trifluoromethylation [19], benzylation [20], and cyanoalkylation [21]. With regard to alkyl halides, such as trifluoromethyl iodide **16**, benzylic bromide **18,** and bromoacetonitrile **20**, electrophilic radicals were smoothly generated via single-electron reduction by either low-oxidation-state or excited PCs, in which spin-center shift (SCS) [20b] may be involved to generate the active electrophilic radical (e.g. the conversion of protonated heterobenzylic alcohol **22** into radical **24**).

Moreover, in terms of photoredox catalysts, the replacement of precious ruthenium(II) and iridium(III) complexes with relatively cheap inorganic semiconductors [22] or iron-based polypyridyl complex [23], is also able to achieve the enantioselective α-alkylation of aldehydes with α-bromocarbonyls (Scheme 9.3), albeit with slightly decreased enantioselectivities when heterogeneous semiconductor photocatalysts are used.

Although chiral amine and photoredox combined catalyst systems have proven to be successful with aldehydes, the similar functionalization of ketone substrates remains challenging. With the use of photoredox/chiral primary amine combined catalysis, in 2014, Luo and coworkers reported an enantioselective α-alkylation of β-ketocarbonyls to generate all-carbon quaternary stereocenters (Scheme 9.4) [24]. Based on the experimental observations, the authors believed that this reaction

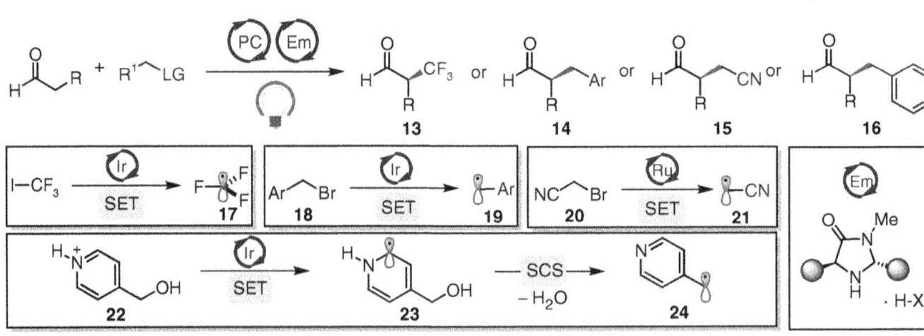

Scheme 9.2 Asymmetric α-functionalization of aldehydes via photoredox/chiral amine combined catalysis.

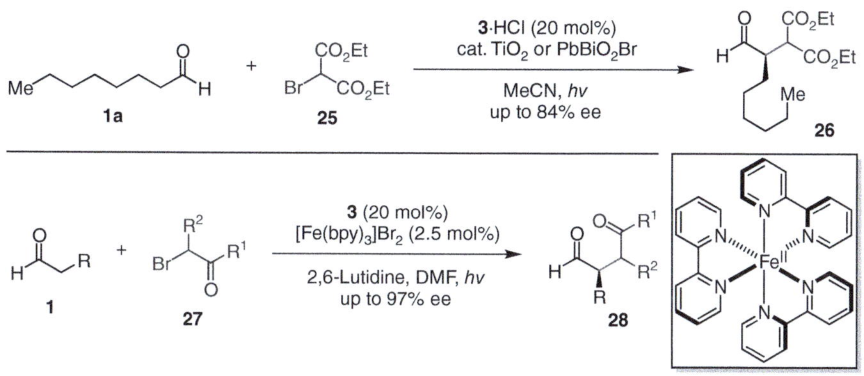

Scheme 9.3 The combination of chiral amine with earth-abundant metal-based photoredox catalysts.

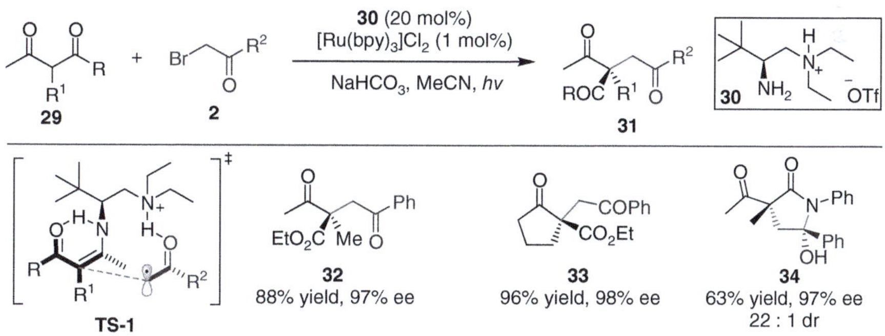

Scheme 9.4 Asymmetric α-alkylation of β-ketocarbonyls. Source: Modified from Zhu et al. [24].

mainly proceeded with a photoredox complex-mediated pathway, and an electron donor-acceptor (EDA) pathway [25] was also possible but should be minor. In the key C–C forming transition state **TS-1**, the addition of phenacyl radical to the chiral enamine was directed by a hydrogen-bonding interaction. Both acyclic and cyclic β-ketoesters can deliver the alkylation products (**32–33**) with high yields and excellent enantioselectivities, while a β-ketoamide substrate with a free N—H bond can afford spiro-γ-lactam **34** through a late-stage intramolecular ketalization.

Chiral amine catalyst can also perform iminium activation toward asymmetric radical conjugate addition to α,β-unsaturated carbonyls. In 2016, Fagnoni and Melchiorre designed a chiral primary amine catalyst **37**, containing a redox-active carbazole moiety, enabling visible-light-promoted enantioselective radical conjugate additions to β,β-disubstituted cyclic enones **35** (Scheme 9.5) [26]. The nucleophilic carbon-centered radicals are generated from bench-stable precursors **42** and **44** via tetrabutylammonium decatungstate (TBADT)-mediated hydrogen atom transfer (HAT) and Ir^{III}-mediated single-electron oxidation, respectively. Radical addition to the cationic iminium ion **39** generates an unstable α-iminyl radical cation **40**, which is subsequently converted to a more stable enamine intermediate **41** through

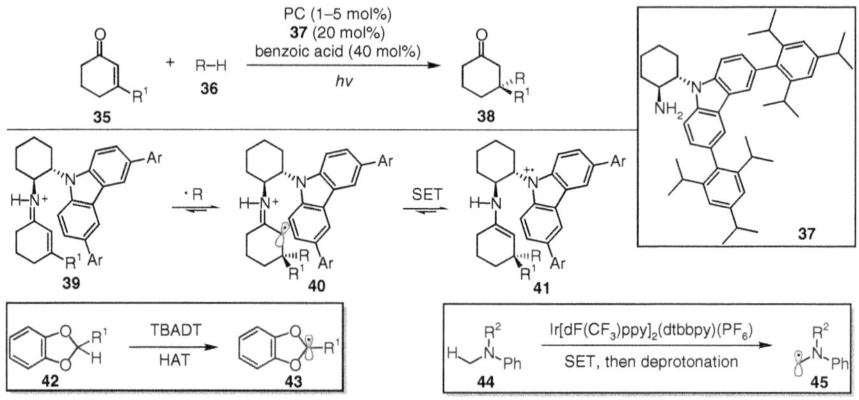

Scheme 9.5 Enantioselective trapping of carbon-centered radicals via the dual photoredox/organo catalysis. Source: Modified from Murphy et al. [25d].

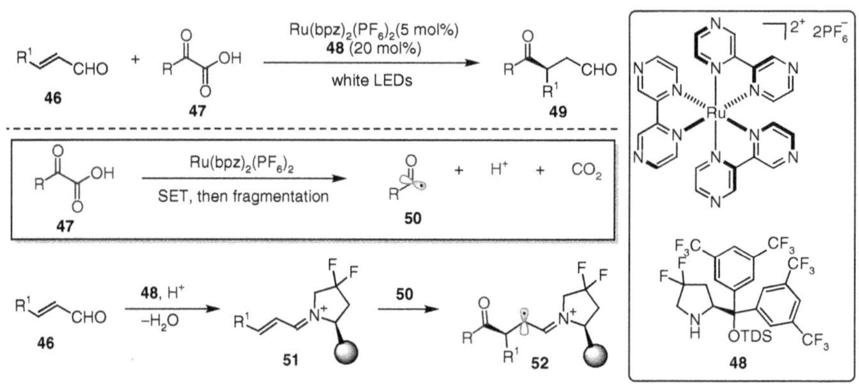

Scheme 9.6 Enantioselective trapping of acyl radicals via the dual photoredox/organocatalysis. Source: Modified from Yang et al. [28].

an electron-relay mechanism. Then **41** undergoes single-electron reduction and hydrolysis to release the amine catalyst **37** and the corresponding product **38**.

In 2019, Yu and coworkers described an enantioselective radical hydroacylation of enals **46** with α-ketoacids **47** in the presence of photoredox/amine dual catalysis [27], providing an efficient way to access chiral 1,4-dicarbonyl compounds (Scheme 9.6). Acyl radical **50** is generated from α-keto acid through single-electron oxidation and subsequent fragmentation. With the assistance of proton, chiral iminium ion **51** is generated from enal **46** and chiral amine catalyst **48**, and then trapped by the acyl radical **50** to deliver the corresponding adduct **52**. **52** is reduced by Ru(I) complex, then undergoes tautomerization and hydrolysis to generate the final product **49**.

In addition, to involve the radical species in the stereoselective bond-forming events, chiral amine/photoredox combined catalysis can also function in a relay or sequential process, taking the advantage of photoredox catalysis in generating ion-type electrophiles from easily accessible raw materials and chiral amine catalysis in stereoselective bond formation. In 2017, Luo and Wu reported an asymmetric

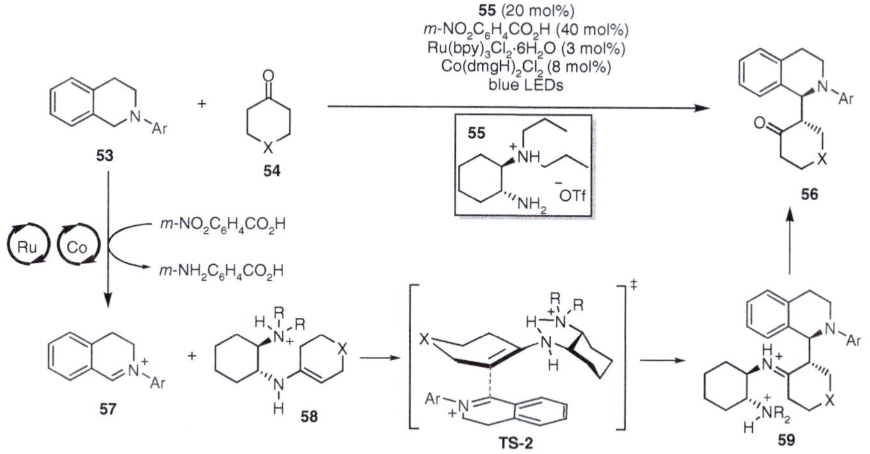

Scheme 9.7 Asymmetric cross-dehydrogenative coupling of tertiary amines with cyclic ketones. Source: Modified from Hou et al. [29].

cross-dehydrogenative coupling (CDC) reaction between tetrahydroisoquinoline **53** and cyclic ketone **54** using a multiple catalyst system, consisting of a chiral primary amine catalyst, $Ru(ppy)_3Cl_2{\cdot}H_2O$ and Co-based hydrogen transfer catalyst (Scheme 9.7) [28]. Upon visible-light irradiation of Ru/Co combined catalysis, tetrahydroisoquinoline **53** is oxidized to an iminium cation intermediate **57** by using m-$NO_2C_6H_4COOH$ as an ultimate hydrogen acceptor. Then the chiral enamine intermediate **58** couples with the iminium cation **57** through transition state **TS-2** to generate the imine intermediate **59**, which undergoes the hydrolysis to provide the desired product **56** and regenerate the chiral amine catalyst **55**.

Later in 2018, Rueping and coworkers described a similar asymmetric α-alkylation of tetrahydroisoquinolines (THIQ) **53** with cyclic ketones **54** using oxygen as an external oxidant (Scheme 9.8) [29]. In comparison with the one-pot reaction, a sequential process for this protocol resulted in better enantioselectivity and diastereoselectivity. Chiral amino acid **60** derived enamine intermediate might also function as the chiral anion of iminium ion intermediate to enhance the enantioinduction of bond-forming process in **TS-3**. Interestingly, in a previous work of the same group, the CDC reaction between **53** and acetone by using L-proline and $Ru(bpy)_3(PF_6)_2$ only provided 8% ee [30].

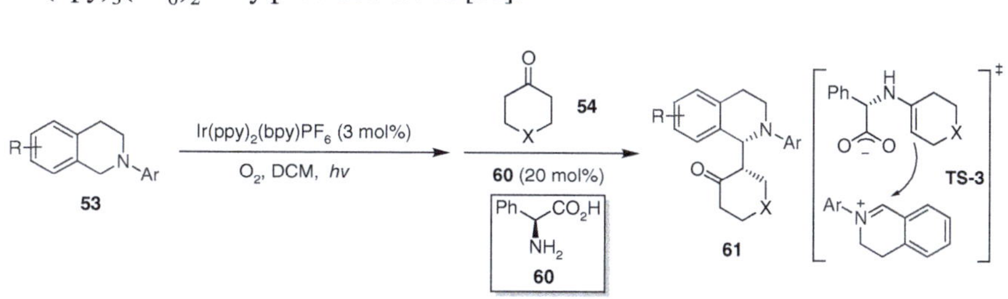

Scheme 9.8 Asymmetric cross-dehydrogenative coupling of tetrahydroisoquinolines and cyclic ketones. Source: Modified from Hou et al. [29].

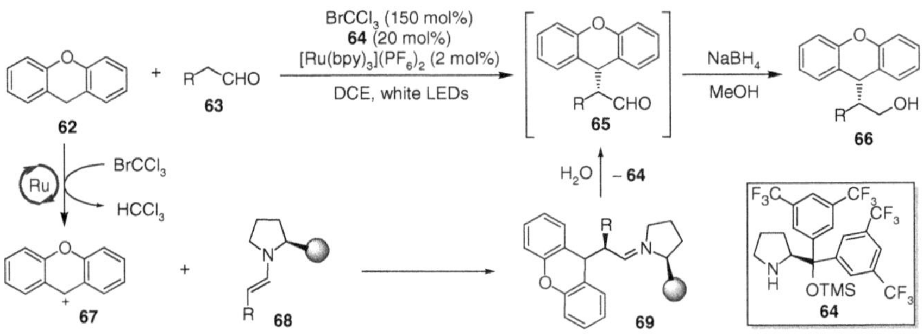

Scheme 9.9 Asymmetric cross-dehydrogenative coupling of aldehydes with xanthenes. Source: Modified from Larionov et al. [31].

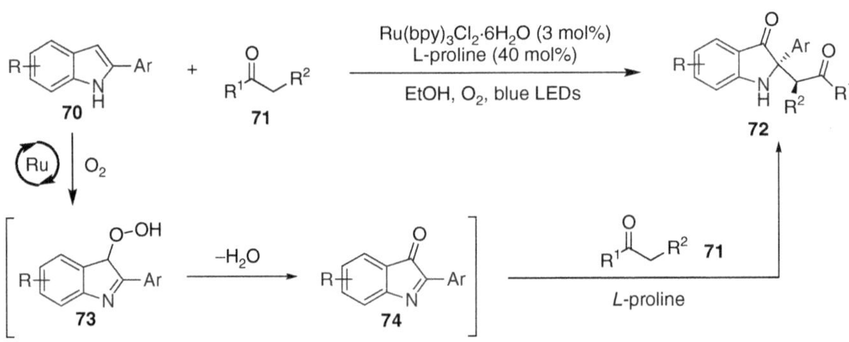

Scheme 9.10 Asymmetric oxidative dearomatization of 2-arylindoles. Source: Modified from Dong et al. [32].

Nonfunctionalized C(sp^3)—H bonds can also participate in asymmetric photoredox CDC reactions. In 2017, Pericàs and coworkers reported an asymmetric coupling between diaryl substituted methylene unit of **62** and aldehydes **63** enabled by the combination of enamine and photoredox catalysis (Scheme 9.9) [31]. Mechanistic studies by experimental and computational methods suggest that this reaction proceeds in an ionic manner. Using $BrCCl_3$ as external oxidant, diarylmethane **62** is photocatalytically oxidized to the corresponding cation, which is then trapped by an enamine intermediate **68** generated *in situ* from aldehyde **63** and chiral amine catalyst **64**.

An alternative type of asymmetric photoredox CDC reaction, direct asymmetric oxidative dearomatization of 2-arylindoles **70** to access C2-quaternary indolin-3-ones **72**, was recently reported by He and Guan (Scheme 9.10) [32]. In this reaction, 2-substituted indoles **70** first undergoes a photocatalyzed aerobic oxidative dearomatization to provide intermediate **74**. Then a proline-catalyzed asymmetric Mannich reaction between **74** and ketones or aldehydes **71** proceeds to afford chiral indolin-3-ones **72**.

9.3 Photoredox-Mediated SOMO Catalysis

Singly occupied molecular orbital (SOMO) organocatalysis [33], in which transient chiral enamine intermediate is converted to a highly reactive radical species via

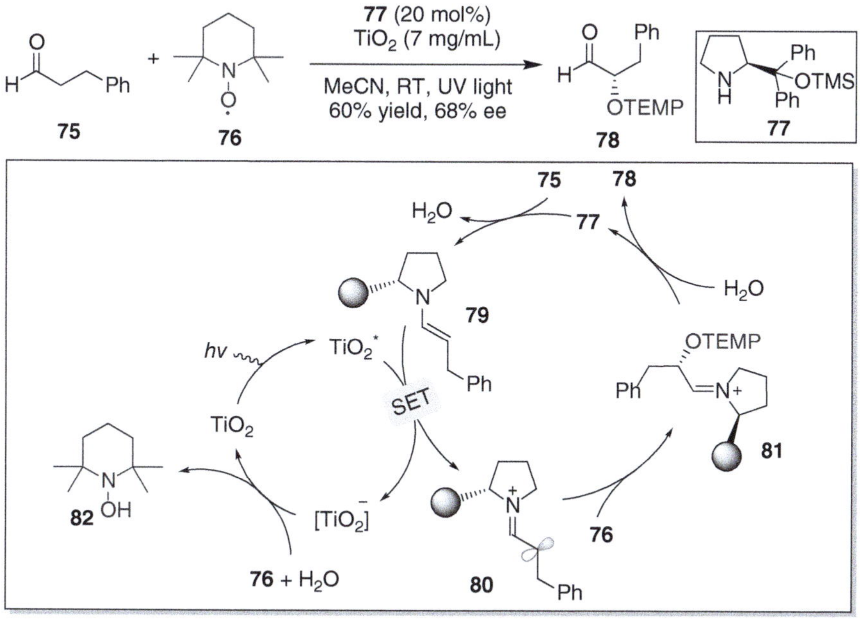

Scheme 9.11 TiO_2-induced enantioselective α-oxyamination of aldehydes with chiral amine. Source: Modified from Ho et al. [37].

chemical single-electron oxidation, has been independently presented by MacMillan [34] and Sibi [35] in 2007. SOMO activation has demonstrated great potential to render a range of asymmetric transformations that are otherwise inaccessible [36]. In 2011, Jang and coworkers established an enantioselective α-oxidation of aldehydes with moderate enantioselectivity by using the combination of heterogeneous TiO_2 photocatalyst and chiral amine catalyst **77** (Scheme 9.11) [37], avoiding the requirement for stoichiometric strong oxidant in previous SOMO catalysis. Based on control experiments, the authors proposed that TiO_2-mediated SET oxidation of enamine intermediate **79** was the major pathway to generate α-iminyl radical cation **80**. Subsequently, radical–radical coupling of **80** with **76**, and followed by hydrolysis, afforded α-oxyaminated aldehyde **78**. Notably, 2,2,6,6-tetramethylpiperidinooxy (TEMPO) **78** not only acted as a coupling partner but also an electron acceptor to regenerate the TiO_2 photocatalyst.

Later in 2012, Jang and coworkers reported an asymmetric difunctionalization of α,β-unsaturated aldehydes **83** under Ru(II)@TiO_2/chiral amine combined catalysis (Scheme 9.12) [38]. This protocol was proposed to proceed through a tandem Michael addition/α-oxyamination of aldehyde process. Iminium intermediate **88**, *in situ* generated from the secondary amine **77** and **83**, undergoes an asymmetric 1,4-addition with malonate **84** to provide the enamine intermediate **89**, which then couples with TEMPO **76** via the Ru(II)-mediated SOMO activation. Notably, TiO_2-supported Ru(II) dye, though in lower catalyst loading, still exhibits much better catalytic activity than that of sole Ru(II) or TiO_2 photoredox catalyst.

MacMillan and coworkers established an enantioselective α-aldehyde alkylation reaction using simple olefins as coupling partners through the combination of chiral enamine catalysis, photoredox-mediated SOMO catalysis, and HAT catalysis

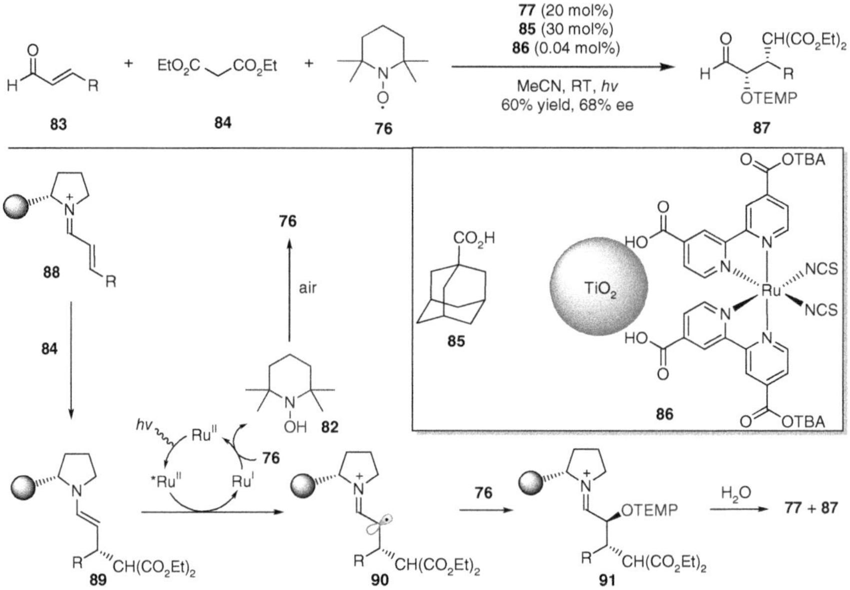

Scheme 9.12 Difunctionalization of α,β-unsaturated aldehyde with Ru(II)@TiO_2/chiral amine combined catalysis. Source: Modified from Yoon et al. [38].

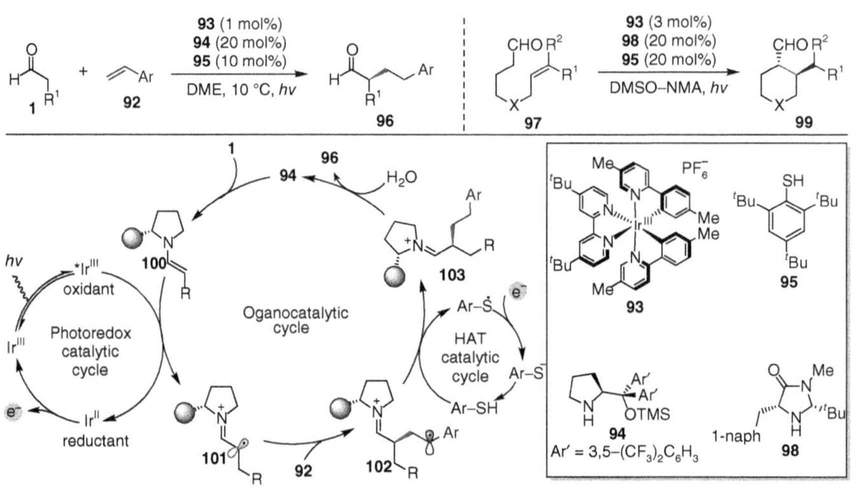

Scheme 9.13 Direct α-alkylation of aldehydes using simple olefins. Source: Modified from Capacci et al. [39].

(Scheme 9.13) [39]. Chiral prolinol **94**, aldehyde **1**, iridium(III) photoredox catalyst **93,** and visible light are merged to deliver an electrophilic α-iminyl radical cation **101**, which rapidly undergoes an enantioselective addition to olefin **92**, generating nucleophilic radical **102**. Then **102** participates in a HAT event with thiophenol

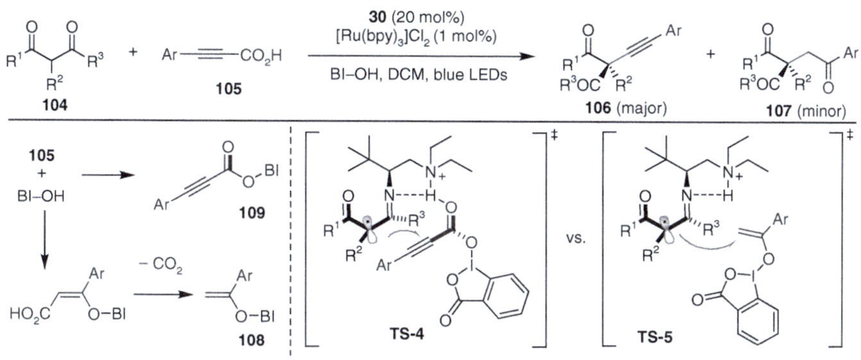

Scheme 9.14 Enantioselective decarboxylative α-alkynylation of β-ketocarbonyls. Source: Modified from Wang et al. [40].

95 to give imine **103**. Finally, the hydrolysis of **103** affords chiral aldehyde derivatives **96** and regenerates **94**. A similar multiple catalytic system also enables an intramolecular α-coupling of aldehydes and olefin to construct cyclic products **99**.

Ketones can also undergo photoredox-mediated SOMO activation in the presence of chiral primary amine catalyst. Luo and coworkers reported an asymmetric α-alkylation and α-alkynylation of β-ketocarbonyls by using combined chiral primary amine catalysis and visible-light photoredox catalysis (Scheme 9.14) [40]. For major α-alkynylation product **106,** the carbon–carbon bond formation proceeds via radical α-addition to **109** through a transition state **TS-4**, in which hydrogen bonding between protonated N–H and propiolate carbonyl accounts for the observed (*R*)-selectivity. The hydroxylation of phenyl propiolic acid **105** with 1-hydroxy-1,2-benziodoxole-3-one (BI–OH) is unavoidable and generates an intermediate **108** after a CO_2 extrusion event. The formation of minor α-alkynylation product **107** is attributed to the radical addition of imino radical to **108** through a transition state **TS-5**.

Another unconventional coupling reaction has also been achieved through the combination of chiral primary amine catalysis and visible-light photoredox catalysis. In 2013, MacMillan and coworkers established a direct β-functionalization of aldehydes and ketones by means of this dual-catalysis. The use of cinchona-derived chiral amine catalyst **111** smoothly affords the β-arylation product **112** with moderate enantioselectivity (Scheme 9.15) [41]. Mechanistic studies suggest that nitrogen-centered radical cation **114** could undergo deprotonation at the allylic position in the presence of an appropriate base, generating a 5π-electron β-enaminyl radical **115**. Then **115** undergoes a radical coupling with radical anion **116**, generated from 1,4-dicyanobenzene **110** through SET reduction mediated with the iridium(III) photoredox catalyst. The resulting cyclohexadienyl anion **117** then undergoes a rapid β,δ-elimination of a cyanide, and hydrolysis to deliver the final product **112**.

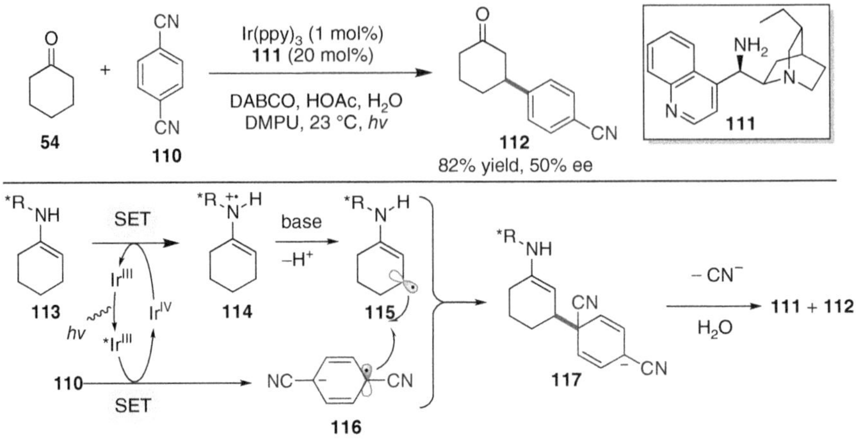

Scheme 9.15 Direct β-arylation of ketone enabled by merging photoredox catalysis with chiral amine catalysis. Source: Modified from Pirnot et al. [41].

9.4 Nucleophilic Organocatalyst in Combination with Photoredox Catalyst

Chiral nucleophilic organocatalysts, such as NHCs and ternary amines, are an important class of organocatalysts to promote versatile umpolung transformations of aldehydes and electron-withdrawing olefins. However, the combination of nucleophilic organocatalysts with photoredox catalysts has rarely been reported. In 2012, Rovis and coworker reported the first examples of chiral NHC/photoredox combined catalysis, enabling an asymmetric α-acylation of tertiary amines **53** with aldehydes **118** (Scheme 9.16) [42]. This approach proceeds through the *in situ* generation of iminium **57** from photoredox-mediated oxidation of tertiary amine **53**. Then the aza-benzoin-type reaction of iminium ion **57** and the chiral Breslow intermediate **121** gives intermediate **122**, which undergoes a 1,2-elimination to afford α-amino ketone **120**.

Asymmetric α-alkylation of THIQ can also be realized by the combination of photoredox catalysts with other nucleophilic organocatalysts. In 2014, Xiao and Lu reported a direct sp^3 C–H acroleination of *N*-aryl-THIQ by the combination of photoredox catalysis and chiral nucleophilic organocatalysis (Scheme 9.17a) [43]. In the catalytic asymmetric variant, THIQ was initially converted to the iminium **57** through photoredox catalysis. Then chiral ternary amine **124**-catalyzed aza-Baylis–Hillman-type reaction of iminium **57** with the zwitterionic intermediate **126** affords the product **125** with moderate enantioselectivity. Similarly, Smith and coworkers found that ammonium enolate derived from an activated aryl ester **128** and chiral isothiourea catalyst **129** was also able to react with the photoredox generated iminium **57** (Scheme 9.17b) [44], leading to the synthesis of substituted chiral tetrahydroisoquinolines in high yield and excellent enantioselectivity.

Recently, Gao and Ye described a direct alkylation reaction of γ-oxidized enals with alkylhalides for the synthesis γ-multisubstituted α,β-unsaturated esters

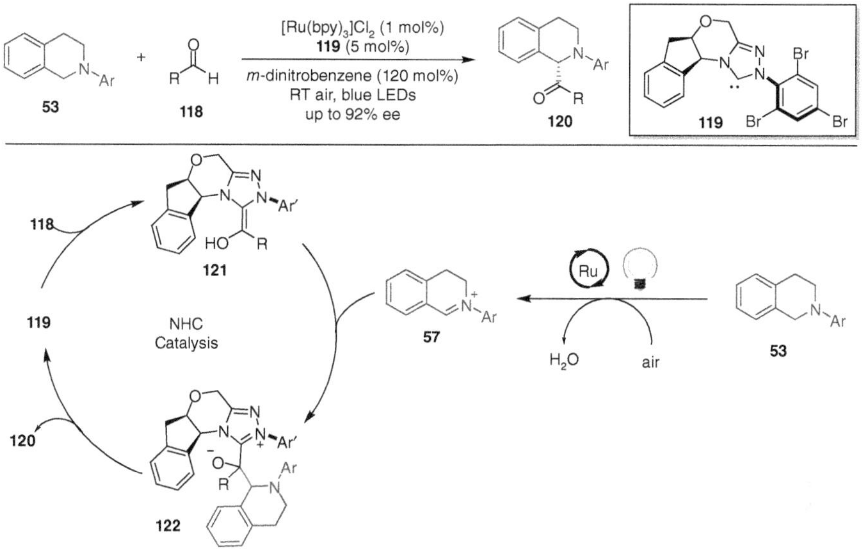

Scheme 9.16 α-Acylation of tertiary amines enabled by N-heterocyclic carbene and photoredox combined catalysis. Source: Modified from DiRocco and Rovis [42].

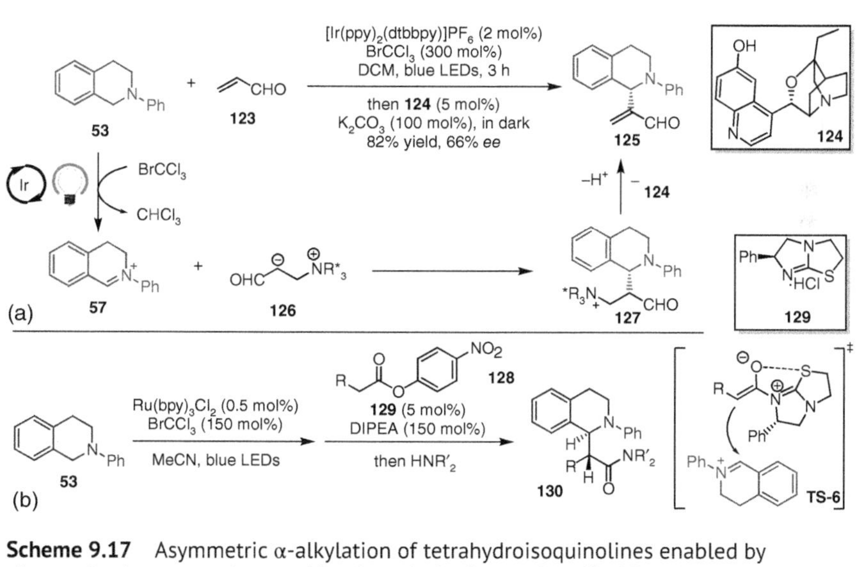

Scheme 9.17 Asymmetric α-alkylation of tetrahydroisoquinolines enabled by photoredox/ternary amine combined catalysis. Source: Modified from Feng et al. [43] and modified from Arokianathar et al. [44].

driven by photoredox/NHC combined catalysis (Scheme 9.18) [45]. The dienolate intermediate **136** generated from the condensation and fragmentation of γ-oxidized enal **131** and chiral NHC catalyst **133** couples with alkyl radical **21** photocatalytically generated from alkyl halide **132** to give a homoenolate radical **137**. Single-electron oxidation of **137** affords acyl azolium intermediate **138**, which is trapped by

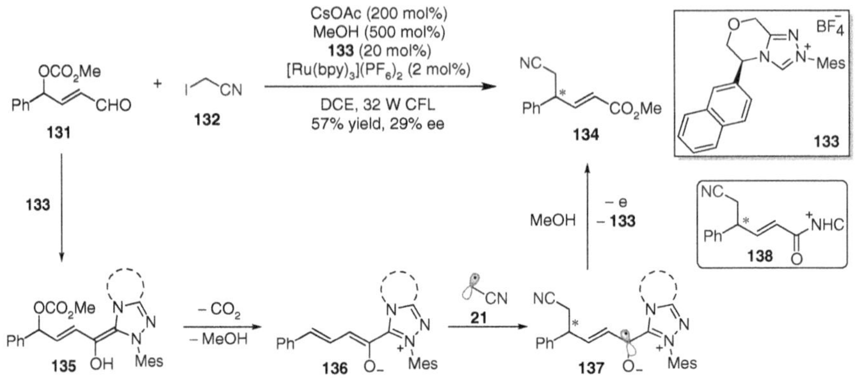

Scheme 9.18 Asymmetric γ-alkylation of enals by merging photoredox catalysis and NHC catalysis. Source: Modified from Dai et al. [45].

methanol to give the final product **134**. Unfortunately, only 29% enantioselectivity is observed upon using chiral NHC precursor **133** as the catalyst.

9.5 Noncovalent-Based Organocatalytic Activation in Combination with Transition Metal-Based Photoredox Catalyst

9.5.1 Chiral Phosphate/Photoredox Combined Catalysis

In the seminal works by Terada [46] and Akiyama [47], chiral phosphoric acids and derivatives have been established as a powerful Brønsted acid tool box in asymmetric catalysis, enabling a wide range of enantioselective reactions [14]. The versatility and good compatibility of chiral phosphoric acid with other catalysts make it suitable for engaging in multi-catalyst systems [14d]. In recent years, the combination of photoredox and chiral phosphoric acids or derivatives has attracted considerable attention to achieve novel asymmetric transformations. In 2013, Knowles and coworkers reported an intramolecular enantioselective reductive coupling of ketones and hydrazones enabled by the combination of a chiral phosphoric acid and an iridium(III) photoredox catalyst (Scheme 9.19) [48]. The reaction proceeds through the formation of the ketyl radical **145** via a concerted proton-coupled electron transfer (PCET) pathway jointly mediated by chiral phosphoric acid **141** and a strongly reducing iridium(II) species. The hydrogen-bonding interaction between the neutral ketyl radical and chiral phosphate anion is critical to the enantioselective carbon–carbon bond-forming process. The resulting hydrazyl intermediate **146** then undergoes HAT from Hantzsch dihydropyridine (hantzsch 1,4-dihydropyridine [HEH], **140**) to furnish the *syn*-1,2-amino alcohol derivatives **142** with high enantioselectivity. Lastly, the iridium(II) species is recycled from the reduction of photoexcited state of the Ir^{III} catalyst with neutral HEH radical **147**, and the resulting acidic pyridinium ion is deprotonated by the phosphate anion to regenerate chiral phosphoric acid **141**.

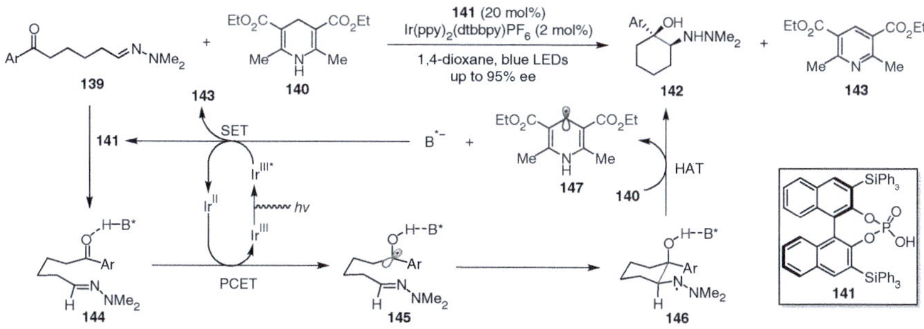

Scheme 9.19 Asymmetric aza-pinacol cyclization by chiral phosphoric acid/photoredox combined catalysis. Source: Modified from Rono et al. [48].

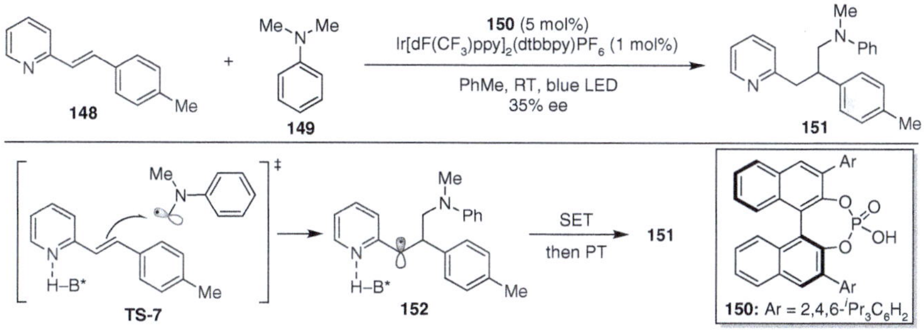

Scheme 9.20 Conjugate addition of photochemically generated α-amino radical to alkenylpyridine. Source: Modified from Hepburn and Melchiorre [50].

This study demonstrates that the great potential of concerted PCET activation to generate radicals can maintain weak interactions inherent to noncovalent organocatalysts, enabling enantioselective radical-based bond-forming transformations [49].

The combination of chiral phosphate and photoredox catalysis can also functionally cooperate in asymmetric radical additions to alkenes and heteroarenes. In 2016, Melchiorre and coworker accomplished a radical conjugate addition to β-substituted alkenylpyridine **148** by merging a chiral phosphoric acid and an iridium-based photoredox catalyst (Scheme 9.20) [50]. The protonated alkenylpyridine is convinced to be a highly reactive and electrophilic pseudo-iminium ion intermediate, which undergoes radical conjugate addition with a nucleophilic amino radical (**TS-7**), *in situ* generated from dimethylaniline **149** via sequential single-electron oxidation and deprotonation. The resulting radical intermediate **152** then undergoes single-electron reduction and protonation to furnish the product **151**. However, the use of chiral phosphoric acid **150** only results in 35% ee. A similar visible-light-induced radical-addition to electron-withdrawing olefins is also feasible in the presence of an equivalent amount of chiral hydrogen-bonding template and catalytic amount of photoredox catalyst [51], albeit with only moderate enantioselectivity.

Phipps and coworkers reported a Minisci-type addition of radical nucleophiles to pyridines and quinolines with excellent control of enantioselectivity (Scheme 9.21a)

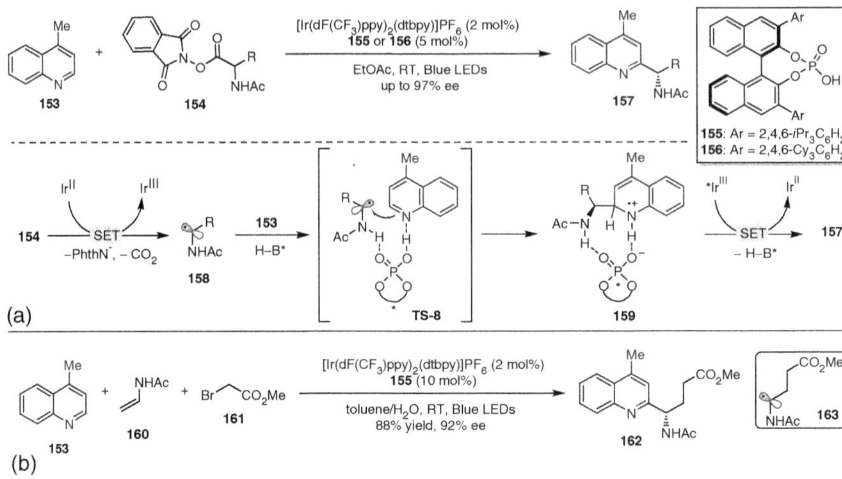

Scheme 9.21 Catalytic enantioselective Minisci-type addition to pyridines and quinolones. Source: Modified from Proctor et al. [52] and modified from Zheng and Studer. [53].

[52]. The prochiral *N*-acyl α-aminoalkyl radical **158** is generated from the fragmentation of amino acid-derived redox-active esters (RAEs, **154**) through a photoredox-mediated single-electron reduction. Chiral phosphate acts as a bifunctional counter anion, engaging in hydrogen bonding interactions with radical nucleophile **158** as well as protonated quinoline. In this manner, stereoselective carbon–carbon bond formation through transition state **TS-8** furnishes the iminium ion intermediate **159**, which then undergoes photoredox-mediated single-electron oxidation and following deprotonation to restore the heteroarene. However, the extension of this protocol to other classes of heteroarenes is still very challenging. Recently, the authors designed a rigorous analysis of the reaction landscape and facilitated reaction data set through multivariate linear regression (MLR) analysis, predicting pyrimidines to be amenable to this enantioselective Minisci reaction [53]. Through *in situ* generating nucleophilic radical **163** from enamide and electrophilic radical, an enantioselective three-component reaction of quinolines or pyridines **153** with enamides **160** and α-bromo carbonyl compounds **161** can also be accomplished to yield **162** by dual photoredox and chiral Brønsted acid catalysis (Scheme 9.21b) [54].

The photoredox/chiral phosphate dual catalysis is also compatible with chiral hydrogen-bonding interaction to facilitate stereoselective radical transformations. In 2018, Knowles and coworkers reported an asymmetric coupling of tryptamine **164** and TEMPO **76** to afford chiral pyrroloindolines **167** (Scheme 9.15) [55]. The reaction is triggered by the single-electron oxidation of hydrogen-bonded adduct **168** with iridium(IV) species to generate the indole radical cation **169**, which then reacts with TEMPO **76** to form an iminium adduct **170**. Intramolecular capture of iminium ion by the nucleophilic carbamate moiety furnishes the desired pyrroloindoline product **167**. The high level of enantioinduction is attributed to the tight hydrogen-bonding interaction of indole radical cation-chiral phosphate ion pair during the carbon–oxygen bond-forming step. TEMPO **76** is not only a reactant

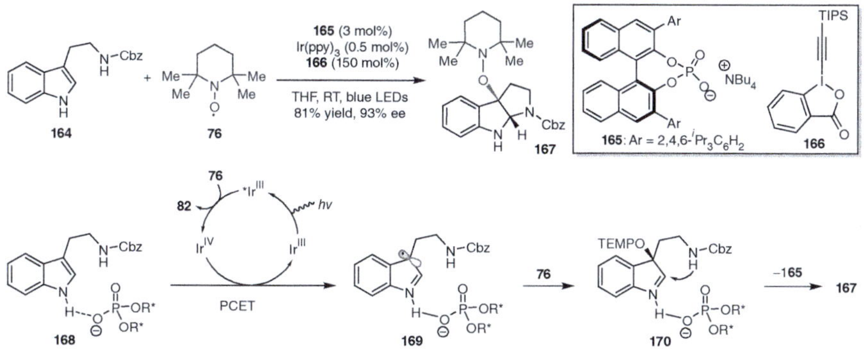

Scheme 9.22 Asymmetric synthesis of substituted pyrroloindolines.

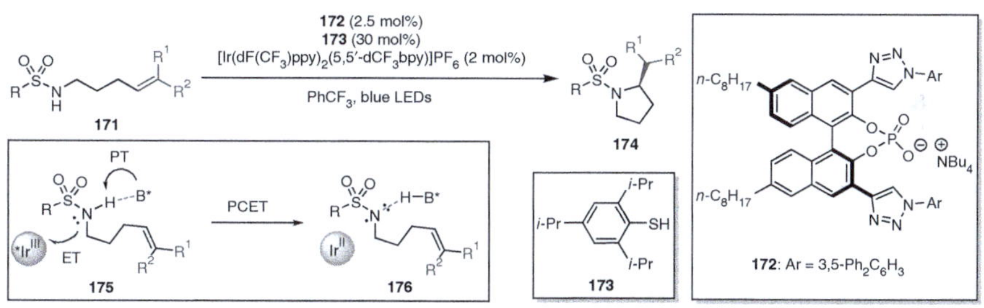

Scheme 9.23 Enantioselective hydroamination of alkenes. Source: Modified from Roos et al. [57].

but also acts as stoichiometric oxidant to regenerate the iridium(IV) species. Interestingly, this reaction also proceeds well with the metal-free visible-light-excited TEMPO-promoted single-electron oxidation [56] (Scheme 9.22).

Recently, Knowles and coworkers found that transient hydrogen-bonding interactions could also enforce stereoselectivity in the intramolecular hydroamination of alkenes with sulfonamides (Scheme 9.23) [57]. This protocol was proposed to proceed via a key sulfonamidyl radical intermediate **176** formed by PCET activation of sulfonamide N—H bonds of **171** with an excited state Ir(III) oxidant and chiral phosphate base **172**. Then the H-bonding interaction between the chiral phosphoric acid and the neutral sulfonamidyl radical, as ahown in **176**, serves as the basis for asymmetric induction in a subsequent C—N bond-forming step. The resulting alkyl radical participates in a HAT event with thiophenol **173** to give **174**.

Moreover, a ternary catalyst system consists of an Ir(III)-based photoredox catalyst, a chiral phosphate base, and a cysteine-containing peptide thiol was recently established by Miller and coworkers to realize the light-driven deracemization of cyclic ureas (Scheme 9.24) [58]. Initiated by an excited-state Ir photocatalyst, the racemic urea substrates **177** is reversibly oxidized to form a mixture of arene radical cations (*R*)-**181** and (*S*)-**181**. Then kinetically controlled deprotonation of **181** with chiral phosphate base **178** resulted in the preferred conversion of fast-reacting (*S*)-**181** to neutral α-amino radicals **182**, and unreacted (*R*)-**181** was able to convert back to the cyclic ureas (*R*)-**177** by charge recombination with the reduced Ir(II).

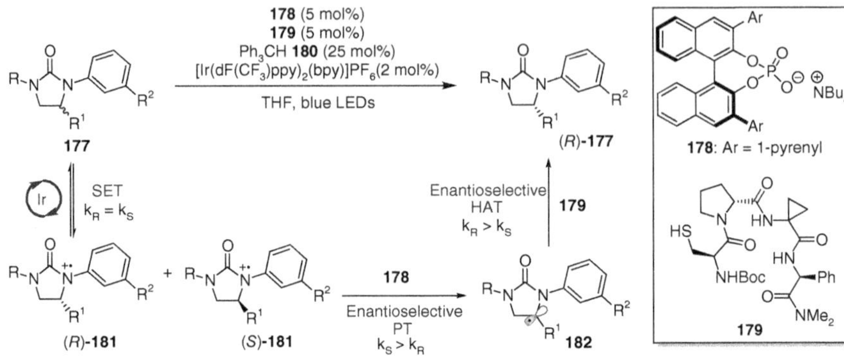

Scheme 9.24 Light-driven deracemization of cyclic ureas. Source: Modified from Shin et al. [58].

Meanwhile, α-amino radicals **182** participates in an enantioselective HAT event with chiral thiol **179** to give (*R*)-**177**. These two independent stereoselective steps allow the optically enriched products (*R*)-**177** to be formed from the racemic starting material **177** in a kinetically controlled manner.

Dual catalysis of photoredox and chiral phosphoric acids can also be applied to creating asymmetric transformations from easily accessible starting materials. In 2014, Lu and Xiao reported a cascade aerobic oxidation and semipinacol rearrangement reaction, allowing efficient access to 2,2-disubstituted indolin-3-ones (Scheme 9.25) [59], the indole substrate **183** is first oxidized to a superoxide intermediate **186** with molecular oxygen in the presence of $Ru(bbp)_3Cl_2{\cdot}6H_2O$ catalyst. Then **186** undergoes an O–O cleavage with another indole **183** to afford a tertiary alcohol **187**, which finally undergoes a chiral phosphoric acid-catalyzed semipinacol rearrangement to furnish indolin-3-one **185** in moderate enantioselectivity.

Wang and coworkers established an asymmetric Friedel–Crafts reaction of indoles with α-amino acid-derived RAEs **154** using visible-light photoredox and chiral phosphate relay catalysis (Scheme 9.26) [60]. The RAEs **154** can serve as *N*-acyl imine equivalents through a light-induced single-electron reduction, fragmentation, and

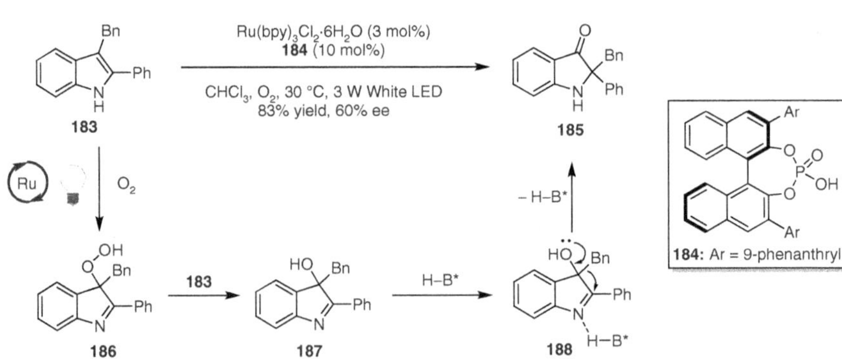

Scheme 9.25 Photocatalytic aerobic oxidation/semipinacol rearrangement cascade reaction. Source: Modified from Ding et al. [59].

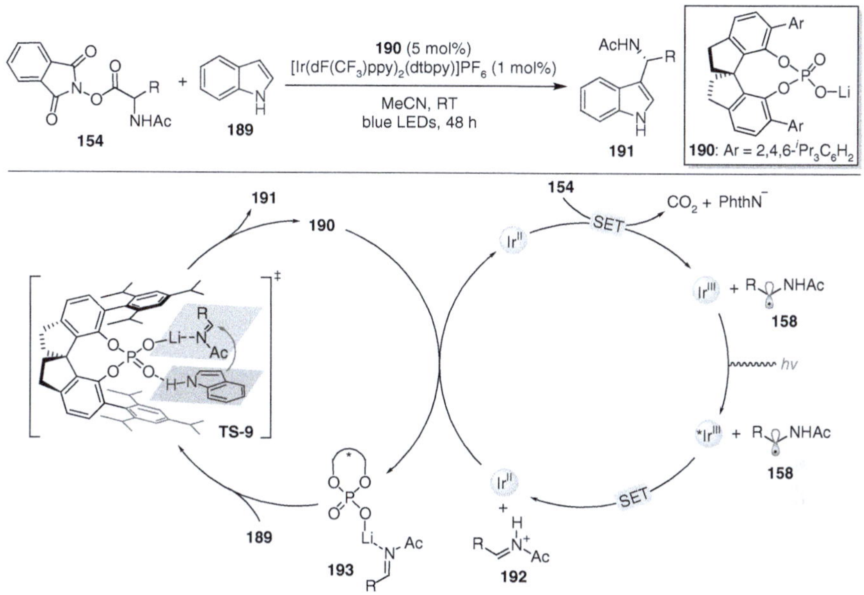

Scheme 9.26 Asymmetric Friedel–Crafts reaction of indoles and α-amino acids derived redox-active esters. Source: Modified from Shen et al. [60].

single-electron oxidation consecutive process. The subsequent trapping of *in situ* generated *N*-acyl imines **192** with indoles through transition state **TS-9** furnishes the corresponding Friedel–Crafts alkylation products **191**. Notably, the use of chiral alkali phosphate **190** significantly improves the enantioselectivity than the corresponding chiral phosphoric acid.

Asymmetric functionalization of inert C—H bonds can also be established with photoredox and chiral phosphoric acid relay catalysis. Recently, Wang and coworkers developed a light-mediated asymmetric C–H functionalization of alkanes with exocyclic enones [61] by using TBADT as HAT catalyst and chiral phosphoric acid **195** as chiral proton-transfer shuttle [62] (Scheme 9.27). This reaction represents a cascade process consisting of a HAT, radical addition, hydrogen abstraction, and enantioselective protonation. A wide range of hydrocarbons, including cycloalkanes, benzylic and allylic hydrocarbons are nicely tolerated to provide chiral ketones **196** with excellent enantioselectivities.

9.6 Asymmetric Ion-Pair/Photoredox Combined Catalysis

In 2014, Jacobsen and coworkers described a sequential amine oxidation and imine addition reaction through the merger of photoredox and asymmetric anion-binding catalysis (Scheme 9.28) [63], providing an efficient approach to

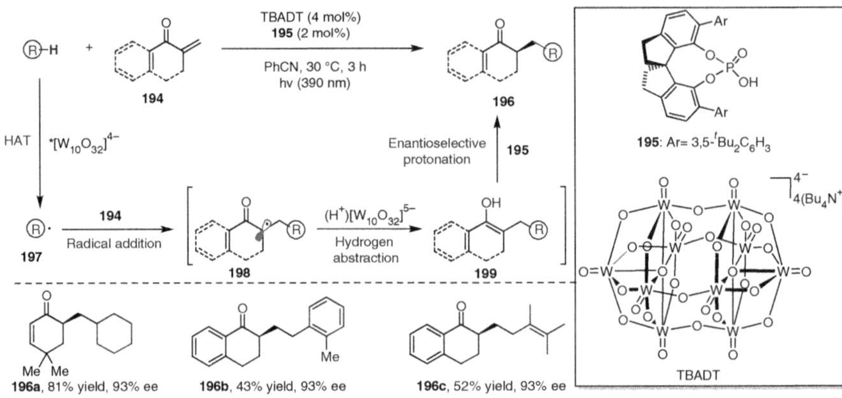

Scheme 9.27 Light-mediated asymmetric C–H functionalization of alkanes with exocyclic enones.

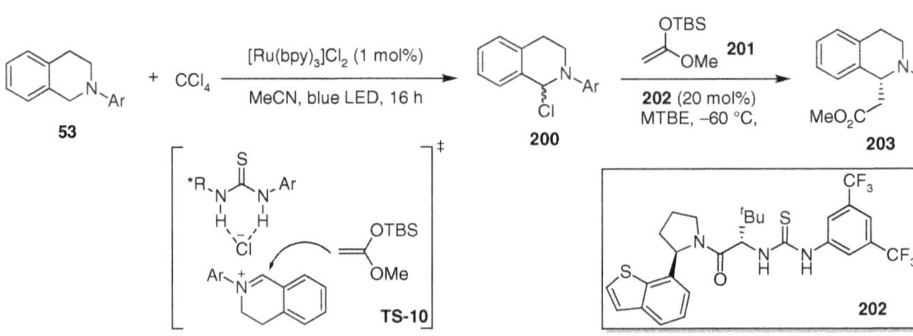

Scheme 9.28 Enantioselective synthesis of β-amino esters. Source: Modified from Bergonzini et al. [63].

access enantiopure β-amino esters. The iminium ion precursor **200** is generated from *N*-aryltetrahydroisoquinoline **53** via photocatalytic oxidation using CCl_4 as the oxidant and chlorinating reagent. The presence of a chiral thiourea catalyst **202** accelerates the extrusion of Cl^- to form a chiral complexed anion-binded iminium, which then engages in stereoselective addition (**TS-10**) with silyl ketene acetal **201** to afford products **203**.

Ooi and coworkers reported a highly enantioselective radical coupling reaction between *N*-sulfonyl aldimines **204** and *N*-arylaminomethanes **205** by using the combination of photoredox and ionic Brønsted acid catalysis (Scheme 9.29) [64]. In this case, the Ir(II)-mediated single-electron reduction of *N*-sulfonyl aldimine **204** generates an anion-radical, which can undergo ion exchange with *P*-spiro aminophosphonium **206** to afford chiral ion pair **208**. Then **208** couples with an aminomethyl radical **209** *in situ* generated from amine **205** via SET oxidation and deprotonation to yield 1,2-diamine derivative **207** with excellent enantioselectivity. A similar process using α-silyl amines **210** as alkyl radical precursors in the presence of $Ir(ppy)_3$ as photoredox catalyst also proceeds, smoothly [65].

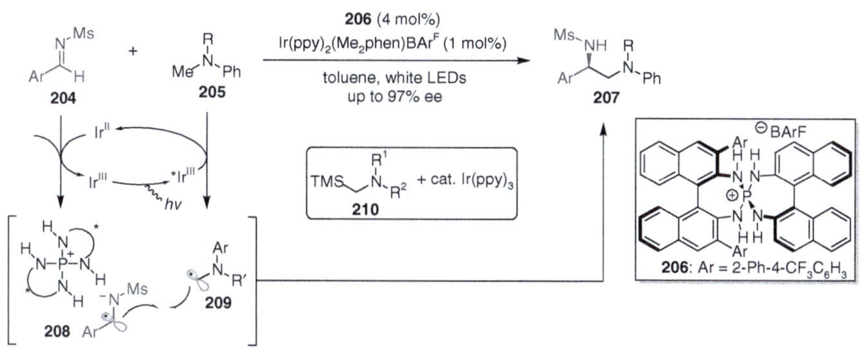

Scheme 9.29 Asymmetric α-coupling of *N*-arylaminomethanes with aldimines. Source: Modified from Uraguchi et al. [64].

9.7 Summary and Outlook

Over the past decade, numerous exciting processes toward the control of stereoselectivity in radical transformations have been achieved by the combination of photoredox catalysis and chiral organocatalysis. By exploiting the intrinsic merits of radical reactivity, these advances offer opportunities for asymmetric transformations that are otherwise challenging to access via conventional two-electron chemistry. Although the concept of this combined catalysis in radical transformations has proven to be successful, this research field is still in its infancy and will continue to grow. Given the great potential of photoredox-catalyzed C–H functionalization, such as PCET and HAT processes, the combined catalysis is expected to glow insights in asymmetric functionalization of inert C—H bonds. Compared to photoredox single-electron transfer process, the energy transfer mechanism is unique to generate highly reactive excited-state molecules, capable of enabling novel stereoselective bond formation in the presence of chiral organocatalysts. Moreover, advances in photochemical radical-mediated cascade reactions are also highly desired to access divergent synthesis of molecular complexity. Lastly, the discovery of earth-abundant metal-based photoredox catalysts and easy-recyclable heterogeneous photocatalysts with applications spacing from laboratory to the industry would be another exciting advancement in this area.

References

1 (a) Ehleringer, J.R., Cerling, T.E., and Helliker, B.R. (1997). *Oecologia* 112: 285–299. (b) Chaves, M.M., Pereira, J.S., Maroco, J. et al. (2002). *Ann. Bot.* 89: 907–916.

2 (a) Meyer, T.J. (1989). *Acc. Chem. Res.* 22: 163–170. (b) Roth, H.D. (1989). *Angew. Chem. Int. Ed. Engl.* 28: 1193–1207. (c) Griesbeck, A.G. and Meierhenrich, U.L. (2002). *Angew. Chem. Int. Ed.* 41: 3147–3154.

3 Prier, C.K., Rankic, D.A., and MacMillan, D.W. (2013). *Chem. Rev.* 113: 5322–5363.

4 (a) Kalyanasundaram, K. (1982). *Coord. Chem. Rev.* 46: 159–244. (b) Juris, A., Balzani, V., Barigelletti, F. et al. (1988). *Coord. Chem. Rev.* 84: 85–277.

5 (a) Wang, C. and Lu, Z. (2014). *Org. Chem. Front.* 2: 179. (b) Buzzetti, L., Crisenza, G.E.M., and Melchiorre, P. (2019). *Angew. Chem. Int. Ed.* 58: 3730–3747.

6 Brimioulle, R., Lenhart, D., Maturi, M.M., and Bach, T. (2015). *Angew. Chem. Int. Ed.* 54: 3872–3890.

7 (a) Lang, X., Zhao, J., and Chen, X. (2016). *Chem. Soc. Rev.* 45: 3026–3038. (b) Shaw, M.H., Twilton, J., and MacMillan, D.W. (2016). *J. Org. Chem.* 81: 6898–6926. (c) Studer, A. and Curran, D.P. (2016). *Angew. Chem. Int. Ed.* 55: 58–102. (d) Liu, Q. and Wu, L.-Z. (2017). *Natl. Sci. Rev.* 4: 359–380. (e) Parasram, M. and Gevorgyan, V. (2017). *Chem. Soc. Rev.* 46: 6227–6240.

8 (a) Wang, D., Zhang, L., and Luo, S. (2017). *Acta Chim. Sin.* 75: 22. (b) Garrido-Castro, A.F., Maestro, M.C., and Alemán, J. (2018). *Tetrahedron Lett.* 59: 1286–1294. (c) Silvi, M. and Melchiorre, P. (2018). *Nature* 554: 41–49.

9 (a) Dalko, P.I. and Moisan, L. (2004). *Angew. Chem. Int. Ed.* 43: 5138–5175. (b) Seayad, J. and List, B. (2005). *Org. Biomol. Chem.* 3: 719–724. (c) Bertelsen, S. and Jorgensen, K.A. (2009). *Chem. Soc. Rev.* 38: 2178–2189. (d) Jacobsen, E.N. and MacMillan, D.W.C. (2010). *Proc. Natl. Acad. Sci. U.S.A.* 107: 20618–20619.

10 Holland, M.C. and Gilmour, R. (2015). *Angew. Chem. Int. Ed.* 54: 3862–3871.

11 (a) List, B. (2004). *Acc. Chem. Res.* 37: 548–557. (b) Mukherjee, S., Yang, J.W., Hoffmann, S., and List, B. (2007). *Chem. Rev.* 107: 5471–5569.

12 (a) Enders, D., Niemeier, O., and Henseler, A. (2007). *Chem. Rev.* 107: 5606–5655. (b) Denmark, S.E. and Beutner, G.L. (2008). *Angew. Chem. Int. Ed.* 47: 1560–1638.

13 (a) Basavaiah, D., Rao, A.J., and Satyanarayana, T. (2003). *Chem. Rev.* 103: 811–892. (b) Basavaiah, D., Reddy, B.S., and Badsara, S.S. (2010). *Chem. Rev.* 110: 5447–5674.

14 (a) Schreiner, P.R. (2003). *Chem. Soc. Rev.* 32: 289–296. (b) Terada, M. (2010). *Synthesis* 12: 1929–1982. (c) Yu, J., Shi, F., and Gong, L.Z. (2011). *Acc. Chem. Res.* 44: 1156–1171. (d) Parmar, D., Sugiono, E., Raja, S., and Rueping, M. (2014). *Chem. Rev.* 114: 9047–9153.

15 (a) Melchiorre, P., Marigo, M., Carlone, A., and Bartoli, G. (2008). *Angew. Chem. Int. Ed.* 47: 6138–6171. (b) Melchiorre, P. (2012). *Angew. Chem. Int. Ed.* 51: 9748–9770.

16 (a) Vignola, N. and List, B. (2004). *J. Am. Chem. Soc.* 126: 450–451. (b) List, B., Coric, I., Grygorenko, O.O. et al. (2014). *Angew. Chem. Int. Ed.* 53: 282–285.

17 Nicewicz, D.A. and MacMillan, D.W. (2008). *Science* 322: 77–80.

18 Cismesia, M.A. and Yoon, T.P. (2015). *Chem. Sci.* 6: 5426–5434.

19 Nagib, D.A., Scott, M.E., and MacMillan, D.W. (2009). *J. Am. Chem. Soc.* 131: 10875–10877.

20 (a) Shih, H.W., Vander Wal, M.N., Grange, R.L., and MacMillan, D.W. (2010). *J. Am. Chem. Soc.* 132: 13600–13603. (b) Nacsa, E.D. and MacMillan, D.W.C. (2018). *J. Am. Chem. Soc.* 140: 3322–3330.

21 Welin, E.R., Warkentin, A.A., Conrad, J.C., and MacMillan, D.W. (2015). *Angew. Chem. Int. Ed.* 54: 9668–9672.

22 Cherevatskaya, M., Neumann, M., Fuldner, S. et al. (2012). *Angew. Chem. Int. Ed.* 51: 4062–4066.

23 Gualandi, A., Marchini, M., Mengozzi, L. et al. (2015). *ACS Catal.* 5: 5927–5931.

24 Zhu, Y., Zhang, L., and Luo, S. (2014). *J. Am. Chem. Soc.* 136: 14642–14645.

25 (a) Arceo, E., Jurberg, I.D., Alvarez-Fernandez, A., and Melchiorre, P. (2013). *Nat. Chem.* 5: 750–756. (b) Arceo, E., Bahamonde, A., Bergonzini, G., and Melchiorre, P. (2014). *Chem. Sci.* 5. (c) Silvi, M., Arceo, E., Jurberg, I.D. et al. (2015). *J. Am. Chem. Soc.* 137: 6120–6123. (d) Murphy, J.J., Bastida, D., Paria, S. et al. (2016). *Nature* 532: 218–222.

26 Murphy, J.J., Bastida, D., Paria, S. et al. (2016). *Nature* 532: 218.

27 Zhao, J.J., Zhang, H.H., Shen, X., and Yu, S. (2019). *Org. Lett.* 21: 913–916.

28 Yang, Q., Zhang, L., Ye, C. et al. (2017). *Angew. Chem. Int. Ed.* 56: 3694–3698.

29 Hou, H., Zhu, S., Atodiresei, I., and Rueping, M. (2018). *Eur. J. Org. Chem.* 2018: 1277–1280.

30 Rueping, M., Vila, C., Koenigs, R.M. et al. (2011). *Chem. Commun.* 47: 2360–2362.

31 Larionov, E., Mastandrea, M.M., and Pericàs, M.A. (2017). *ACS Catal.* 7: 7008–7013.

32 Dong, C.L., Ding, X., Huang, L.Q. et al. (2020). *Org. Lett.* 22: 1076–1080.

33 Meciarova, M., Tisovsky, P., and Sebesta, R. (2016). *New J. Chem.* 40: 4855–4864.

34 Beeson, T.D., Mastracchio, A., Hong, J.-B. et al. (2007). *Science* 316: 582–585.

35 Sibi, M.P. and Hasegawa, M. (2007). *J. Am. Chem. Soc.* 129: 4124–4125.

36 (a) Kim, H. and MacMillan, D.W.C. (2008). *J. Am. Chem. Soc.* 130: 398. (b) Amatore, M., Beeson, T.D., Brown, S.P., and MacMillan, D.W.C. (2009). *Angew. Chem. Int. Ed.* 48: 5121–5124. (c) Conrad, J.C., Kong, J., Laforteza, B.N., and MacMillan, D.W.C. (2009). *J. Am. Chem. Soc.* 131: 11640. (d) Jui, N.T., Lee, E.C.Y., and MacMillan, D.W.C. (2010). *J. Am. Chem. Soc.* 132: 10015–10017. (e) Rendler, S. and MacMillan, D.W.C. (2010). *J. Am. Chem. Soc.* 132: 5027. (f) Um, J.M., Gutierrez, O., Schoenebeck, F. et al. (2010). *J. Am. Chem. Soc.* 132: 6001–6005. (g) Pham, P.V., Ashton, K., and MacMillan, D.W.C. (2011). *Chem. Sci.* 2: 1470–1473. (h) Jui, N.T., Garber, J.A.O., Finelli, F.G., and MacMillan, D.W.C. (2012). *J. Am. Chem. Soc.* 134: 11400–11403.

37 Ho, X.-H., Kang, M.-J., Kim, S.-J. et al. (2011). *Catal. Sci. Technol.* 1: 923.

38 Yoon, H.S., Ho, X.H., Jang, J. et al. (2012). *Org. Lett.* 14: 3272–3275.

39 Capacci, A.G., Malinowski, J.T., McAlpine, N.J. et al. (2017). *Nat. Chem.* 9: 1073–1077.

40 Wang, D., Zhang, L., and Luo, S. (2017). *Org. Lett.* 19: 4924–4927.

41 Pirnot, M.T., Rankic, D.A., Martin, D.B., and MacMillan, D.W. (2013). *Science* 339: 1593–1596.

42 DiRocco, D.A. and Rovis, T. (2012). *J. Am. Chem. Soc.* 134: 8094–8097.

43 Feng, Z.J., Xuan, J., Xia, X.D. et al. (2014). *Org. Biomol. Chem.* 12: 2037–2040.

44 Arokianathar, J.N., Frost, A.B., Slawin, A.M.Z. et al. (2018). *ACS Catal.* 8: 1153–1160.

45 Dai, L., Xia, Z.H., Gao, Y.Y. et al. (2019). *Angew. Chem. Int. Ed.* 58: 18124–18130.

46 Uraguchi, D. and Terada, M. (2004). *J. Am. Chem. Soc.* 126: 5356–5357.
47 Akiyama, T., Itoh, J., Yokota, K., and Fuchibe, K. (2004). *Angew. Chem. Int. Ed.* 43: 1566–1568.
48 Rono, L.J., Yayla, H.G., Wang, D.Y. et al. (2013). *J. Am. Chem. Soc.* 135: 17735–17738.
49 (a) Lin, L., Bai, X., Ye, X. et al. (2017). *Angew. Chem. Int. Ed.* 56: 13842–13846. (b) Li, J., Kong, M., Qiao, B. et al. (2018). *Nat. Commun.* 9: 2445. (c) Liu, Y., Liu, X., Li, J. et al. (2018). *Chem. Sci.* 9: 8094–8098. (d) Cao, K., Tan, S.M., Lee, R. et al. (2019). *J. Am. Chem. Soc.* 141: 5437–5443. (e) Li, F., Tian, D., Fan, Y. et al. (2019). *Nat. Commun.* 10: 1774. (f) Qiao, B., Li, C., Zhao, X. et al. (2019). *Chem. Commun.* 55: 7534–7537.
50 Hepburn, H.B. and Melchiorre, P. (2016). *Chem. Commun.* 52: 3520–3523.
51 Lenhart, D., Bauer, A., Poethig, A., and Bach, T. (2016). *Chem. Eur. J.* 22: 6519–6523.
52 Proctor, R.S.J., Davis, H.J., and Phipps, R.J. (2018). *Science* 360: 419–422.
53 Zheng, D. and Studer, A. (2019). *Angew. Chem. Int. Ed.* 58: 15803–15807.
54 Reid, J.P., Proctor, R.S.J., Sigman, M.S., and Phipps, R.J. (2019). *J. Am. Chem. Soc.* 141: 19178–19185.
55 Gentry, E.C., Rono, L.J., Hale, M.E. et al. (2018). *J. Am. Chem. Soc.* 140: 3394–3402.
56 Liang, K., Tong, X., Li, T. et al. (2018). *J. Org. Chem.* 83: 10948–10958.
57 Roos, C.B., Demaerel, J., Graff, D.E., and Knowles, R.R. (2020). *J. Am. Chem. Soc.* 142: 5974–5979.
58 Shin, N.Y., Ryss, J.M., Zhang, X. et al. (2019). *Science* 366: 364–369.
59 Ding, W., Zhou, Q.Q., Xuan, J. et al. (2014). *Tetrahedron Lett.* 55: 4648–4652.
60 Shen, M.L., Shen, Y., and Wang, P.S. (2019). *Org. Lett.* 21: 2993–2997.
61 Dai, Z.-Y., Nong, Z.-S., and Wang, P.-S. (2020). *ACS Catal.* 10: 4786–4790.
62 Ren, Y.Y., Zhu, S.F., and Zhou, Q.L. (2018). *Org. Biomol. Chem.* 16: 3087–3094.
63 Bergonzini, G., Schindler, C.S., Wallentin, C.J. et al. (2014). *Chem. Sci.* 5.
64 Uraguchi, D., Kinoshita, N., Kizu, T., and Ooi, T. (2015). *J. Am. Chem. Soc.* 137: 13768–13771.
65 Kizu, T., Uraguchi, D., and Ooi, T. (2016). *J. Org. Chem.* 81: 6953–6958.

10

Applications in Organic Synthesis

10.1 Introduction

Due to the increasing demand for natural products or drugs and their related core structures, especially chiral molecules, and when these products or drugs have found a wide scope of applications in pharmaceutical development and production, the efficient synthetic methods steadily gain in more and more concern [1]. The chiral metal-complex catalysis, which can activate a wide range of chemical bonds, has received great attention for half a century because of the prosperous appearance of chiral ligands. Organocatalysis has also been recognized as one of the most powerful strategies to produce chiral molecules with high levels of stereoselectivity during the past decade [2]. Although the synthesis of structurally complex molecules, particularly the total synthesis of natural products, has reached a remarkable level that almost every discovered natural molecule can be synthesized, the rapid construction and the control of stereochemistry of these structurally diverse compounds still need efficient synthetic methods. Recently, the integration of asymmetric organocatalysis and metal catalysis has enabled an amazing creation of unprecedented asymmetric transformations by means of cooperative and sequential activation of chemical bonds, exceeding the performance of either most organocatalysts or some highly active transition-metal catalysts [3]. In some cases, the binary catalyst systems, which can facilitate the step-economical synthesis of key building blocks for building up natural products and drugs with perfect control of stereoselectivity, have shown great potential in organic synthesis.

In this chapter, some representative examples of the applications of asymmetric protocols enabled by organocatalyst and metal combined catalysis to the synthesis of natural products or pharmaceutically significant molecules will be described.

10.2 Applications of Chiral Phosphoric Acid-Metal Cooperative Catalysis

(+)-Cuparene, which possesses adjacent quaternary centers in a cyclopentane skeleton, was first isolated by Enzell and Erdtman [4]. Although some impressive advances on asymmetric synthesis of (+)-Cuparene were achieved [5], List reported

Asymmetric Organo-Metal Catalysis: Concepts, Principles, and Applications, First Edition. Liu-Zhu Gong.
© 2022 WILEY-VCH GmbH. Published 2022 by WILEY-VCH GmbH.

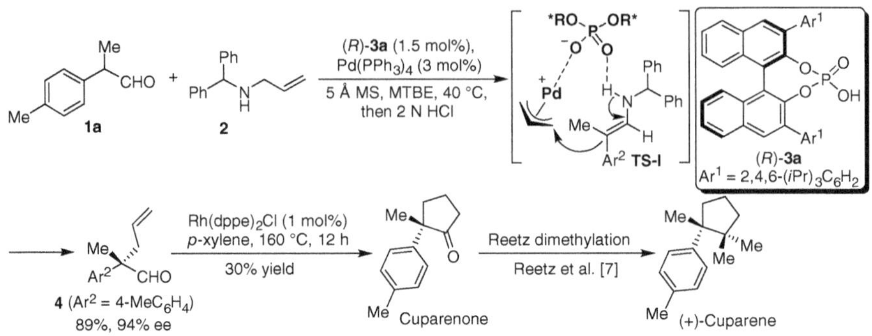

Scheme 10.1 Formal synthesis of (+)-Cuparene. Source: Modified from Mukherjee and List [6].

a more concise approach to this molecule starting with a highly enantioselective α-alkylation of α-branched aldehyde **1a** and allylamine **2a**, synergistically catalyzed by a chiral phosphoric acid [(*R*)-**3a**] and a palladium(0) complex to furnish a chiral aldehyde **14** in 89% yield with 94% ee (Scheme 10.1) [6]. As shown in transition state **TS-I**, phosphoric acid **3a** not only acts as the proton source but functions as a chiral counteranion of the cationic π-allyl-Pd-intermediate to induce the chirality. The formal synthesis of (+)-Cuparene was then accomplished via an Rh-catalyzed hydroacylation of aldehyde **4** in 30% yield, and followed by Reetz dimethylation of Cuparenone [7].

Hu et al. have established a three-component reaction of aryl-/alkyl-diazoacetates, alcohols, and aldimines enabled by cooperative catalysis of a rhodium complex and chiral phosphoric acid **3b**, affording a wide range of chiral β-amino-α-hydroxy ester derivatives with good diastereo- and enantioselectivities [8]. Products **8**, which are generated from asymmetric Mannich-type reaction of diazoacetate **5a**, alcohol **6**, and imine **7**, are important chiral pools in asymmetric synthesis (Scheme 10.2). The taxol side chain and (−)-*epi*-cytoxazone have been prepared in good yields from the *syn*-β-amino-α-hydroxy esters **8a** and **8b**, respectively, through routine manipulations (Scheme 10.2) [9].

Subsequently, Hu and coworkers expanded the concept to a four-component reaction of diazoketone, water, ethyl glyoxylate (**1b**), and anilines in the presence of $Rh_2(OAc)_4$ and chiral phosphoric acid **3c** to produce β-hydroxyl-α-amino ester derivatives in good yields and with high levels of stereoselectivity [10]. The reaction of the linear alkyl vinyl diazoketone **5b** with aniline **9b** can offer the corresponding product **10** in 72% yield (*anti*:*syn* = 70 : 30) with 90% ee, which turns out to be a key intermediate for the synthesis of D-*lyxo*-Phytosphingosine and its analogs (Scheme 10.3).

Zhu and Zhou reported a highly enantioselective N–H insertion reaction of α-diazoketones cooperatively catalyzed by a rhodium complex and chiral spiro-phosphoric acid **3d** [11]. The reaction between α-diazoketone **5c** and $BocNH_2$ gives a ketone **11a** with 79% yield and 94% ee in one minute. The removal of the N-protecting group of **11a** with concentrated HCl affords the key intermediate with maintained enantioselectivity for the synthesis of monoamine oxidase

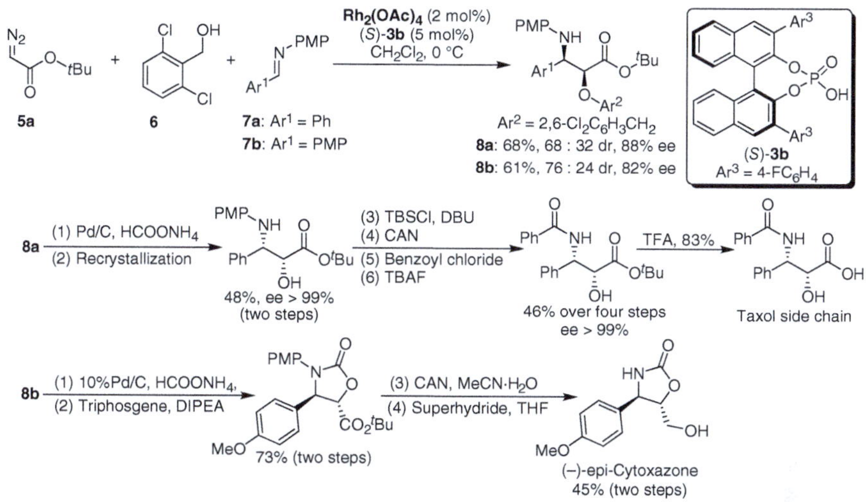

Scheme 10.2 Synthesis of the Taxol side chain and (−)-*epi*-Cytoxazone Source: Modified from Qian et al. [9].

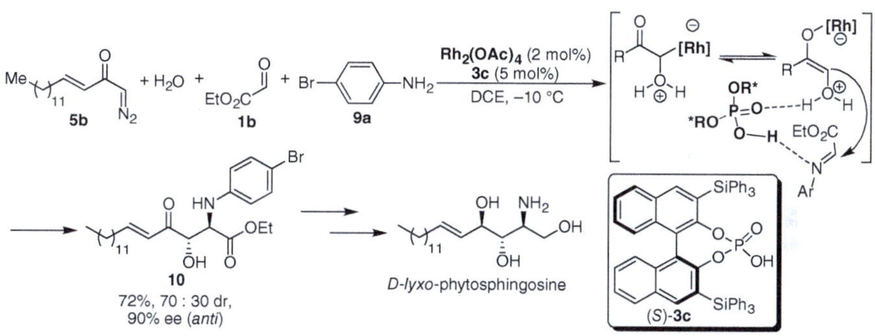

Scheme 10.3 Synthesis of *D*-lyxo-Phytosphingosine.

(MAO) inhibitor (Scheme 10.4) [12]. (L)-(−)-733,061, a selective antagonist for the neurokinin-1 (NK-1) receptor, can be synthesized from chiral α-aminoketone **11b** prepared from the asymmetric N–H insertion reaction of **5d** with $BocNH_2$ by following a known procedure [13].

(*S*)-Eflucimibe (F12511), a new Acyl-CoA cholesterol acyltransferase (ACAT) inhibitor, which is not only able to reduce the plasma cholesterol levels but also has antiatherosclerotic actions in animals' models, has been used as a medicine for the treatment of lipoprotein disorders and atherosclerosis [14]. By means of a similar binary catalyst system [11], Zhou and coworkers developed an asymmetric S—H bond insertion reaction of α-diazo esters. The combination of $Rh_2(TPA)_4$ and chiral spiro-phosphoric acid **3e** allows the reaction of α-phenyl-α-diazoacetate **5e** and *n*-dodecylmercaptan **12** to deliver a chiral α-mercaptoester **11c** in 86% yield and 93% ee [15]. Hydrolysis of the α-mercaptoester **11c** and followed by amidation provides (*S*)-Eflucimibe in 78% yield and with almost maintained optical purity

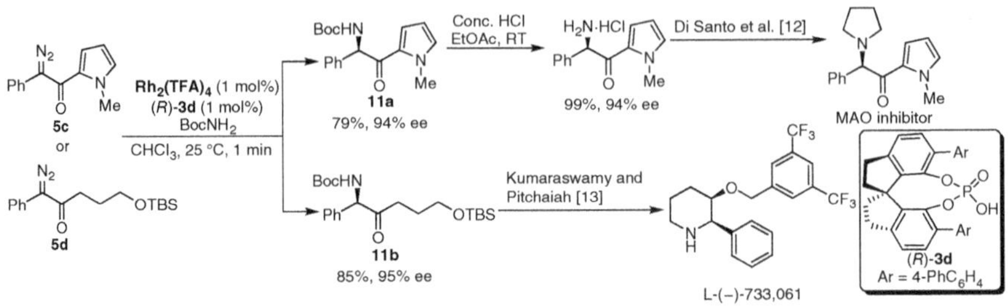

Scheme 10.4 Formal synthesis of MAO inhibitor and (L)-(−)-733,061. Source: Modified from Santo et al. [12].

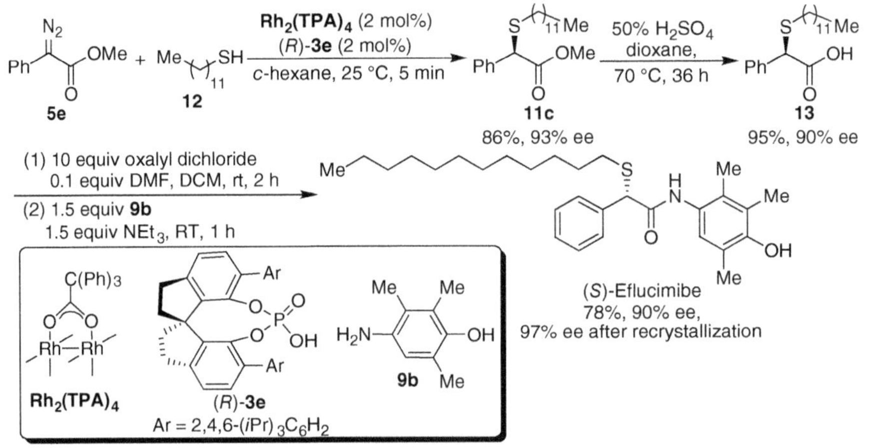

Scheme 10.5 The synthesis of (*S*)-Eflucimibe.

(from 93% to 90% ee). The optical purity of (*S*)-Eflucimibe can be enhanced to 97% ee after a single recrystallization (Scheme 10.5).

(+)-Stephadiamine, which features a norhasubanan skeleton and contains two adjacent α-tertiary amines, was isolated from the snake vine *Stephania japonica* by Taga in 1984 [16]. In 2018, the first total synthesis of racemic Stephadiamine was accomplished by Trauner and coworkers based on the construction of the azapropellane core [17], starting with the key alkylation [18]/decarboxylative Tsuji allylation [19] of tetralone **15** (Scheme 10.6). An asymmetric approach to access chiral tetralone **16** with the benzylic quaternary stereocenter was also investigated. A regio- and enantioselective allylation of unsymmetrical ketone **17** rendered by the cooperative catalysis of palladium complex formed from $Pd_2(dba)_3$ and *t*-BuXPhos and (*R*)-H^8-TRIP (3,3′-bis(2,4,6-triisopropylphenyl)-1,1′-binaphthyl-2,2′-diyl hydrogenphosphate) (**3f**) [20] gives (*S*)-**16** in 63% yield and 93% ee, a key intermediate for the asymmetric synthesis of (+)-Stephadiamine. Moreover, (*R*)-**16** can also be prepared in 81% yield and with 90% ee by switching the configuration of the chiral phosphoric acid. Both the enantiomers of **16** serve as building blocks for the synthesis of Morphine and Hasubanan alkaloids.

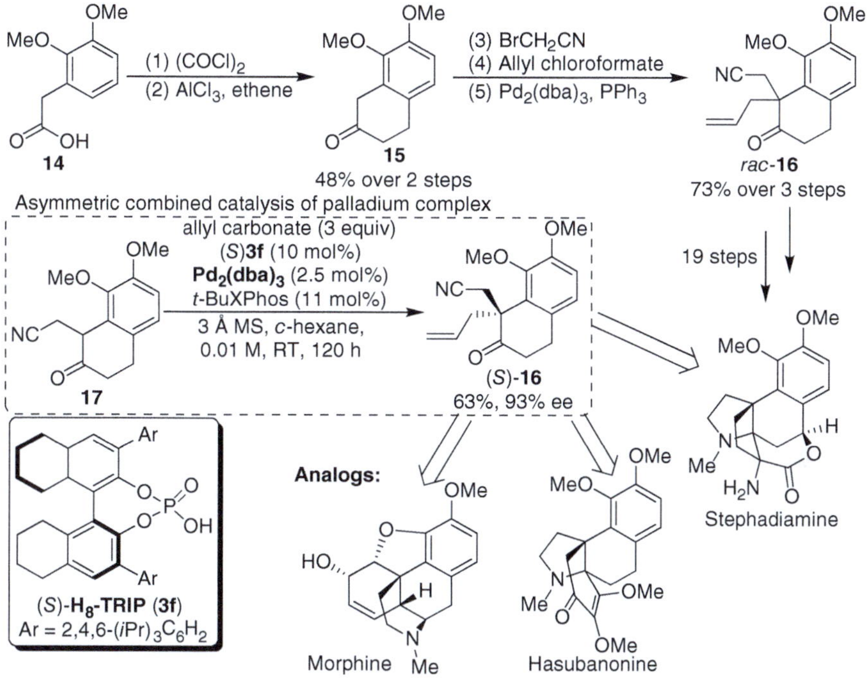

Scheme 10.6 Total synthesis of Stephadiamine and formal asymmetric synthesis of Morphine and Hasubanan alkaloids.

10.3 Application of Transition Metal Catalysis Combined with Secondary Amine Catalysis

In 2009, MacMillan and coworkers established a cascade cross-metathesis/Mukaiyama–Michael addition/intramolecular aldol reaction of crotonaldehyde **1c**, hexene-2-one **18**, and trimethylsilyloxyfuran **20** sequentially catalyzed by Grubbs-II catalyst, chiral secondary amine **19a**, and L-proline **21a**. By the triple-cycle-sequential catalysis, the butenolide **22** was obtained in 64% yield, 5 : 1 dr, and 95% ee through a dicarbonyl intermediate **Int-I** [21]. The hydrogenation catalyzed by palladium/carbon of **22** and followed by reduction with $LiAlH_4$ yields a tetraol **23**. The protection of tetraol **23** with triethylsilyltriflate generates **24**. The Swern oxidation of **24** and followed by Wittig reaction furnishes diene **25**. A reaction sequence of Grubbs metathesis, cyclopropanation with carbene *in situ* generated from $CHBr_3$ and *t*-BuOK, methylation with Me_2CuLi and deprotection leads to (–)-Aromadendranediol (Scheme 10.7).

The bisabolene sesquiterpenes such as (*S*)-(+)-Tumerone, (*S*)-(+)-Curcumene and (*S*)-(+)-Dehydrocurcumene, which exhibit antimicrobial as well as anti-cancer activities and are used as additives in cosmetics, flavors, and perfumes, have been valuable target molecules for the synthetic community [22]. In 2011, a facile strategy for the total synthesis of these sesquiterpenes was developed by Córdova and coworkers, based on the enantioselective conjugate addition enabled

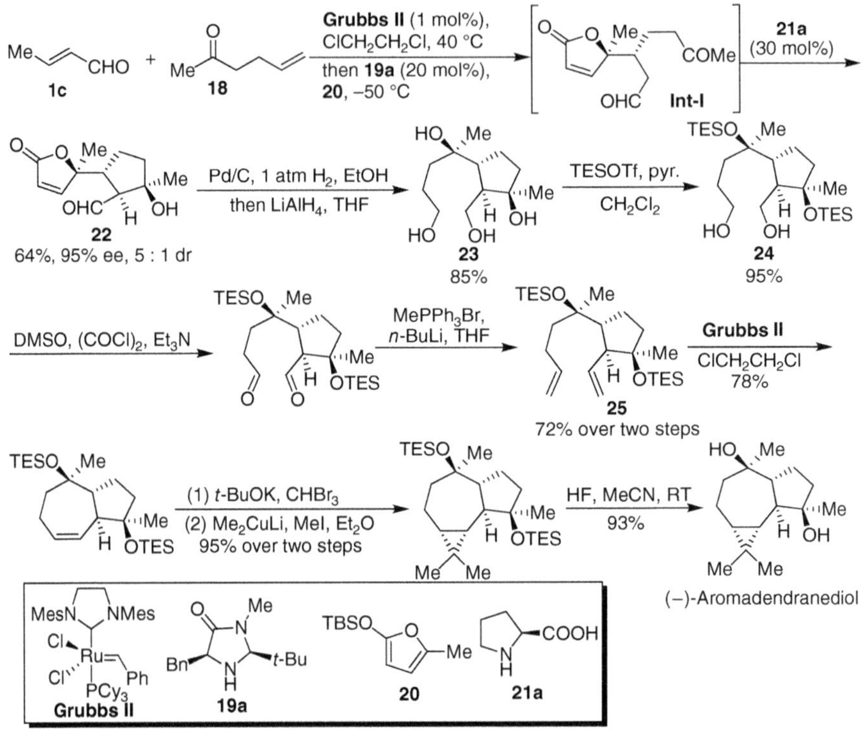

Scheme 10.7 Total synthesis of (−)-Aromadendranediol.

by cooperative catalysis of a chiral amine and copper triflate (Scheme 10.8) [23]. The combination of copper triflate and chiral amine **26a** allows the asymmetric conjugate addition of $ZnMe_2$ to enal **1d** to afford β-methyl aldehyde **27** in 65% yield and with 94% ee. In this case, the activation of the enal by forming transient iminium species with **26a** facilitates the enantioselective addition toward the *in situ* formed L–Cu^{II}–Me complex via transition state **TS-II**. The β-methyl aldehyde **27** is a key chiral building block for the expeditious total synthesis of (*S*)-(+)-Tumerone (51% total yield over two steps), (*S*)-(+)-Curcumene (37% total yield over four steps) and (*S*)-(+)-Dehydrocurcumene (44% yield over two steps).

The biologically active natural (*R*)-(−)-Arundic acid, which has shown efficacy in preventing expansion of cerebral infarction by improving astrocyte function, can be used for the treatment of acute ischemic stroke, as well as clinical development in other neurodegenerative diseases including Alzheimer's disease and Parkinson's disease [24]. Córdova and coworkers reported an intermolecular asymmetric α-allylic alkylation of linear aldehydes, cooperatively catalyzed by an achiral Pd(0) complex and chiral amine **26a** [25]. This method paves a way to the concise and enantioselective total synthesis of Arundic acid (Scheme 10.9). The regiospecific allylic alkylation reaction of allyl acetate **30** and octanal **1e** gives the corresponding alcohol with 93% ee after a one-pot reduction. The total synthesis of (S)-(+)-Arundic acid is accomplished by the sequential hydrogenation over Pd/C and oxidation. Moreover,

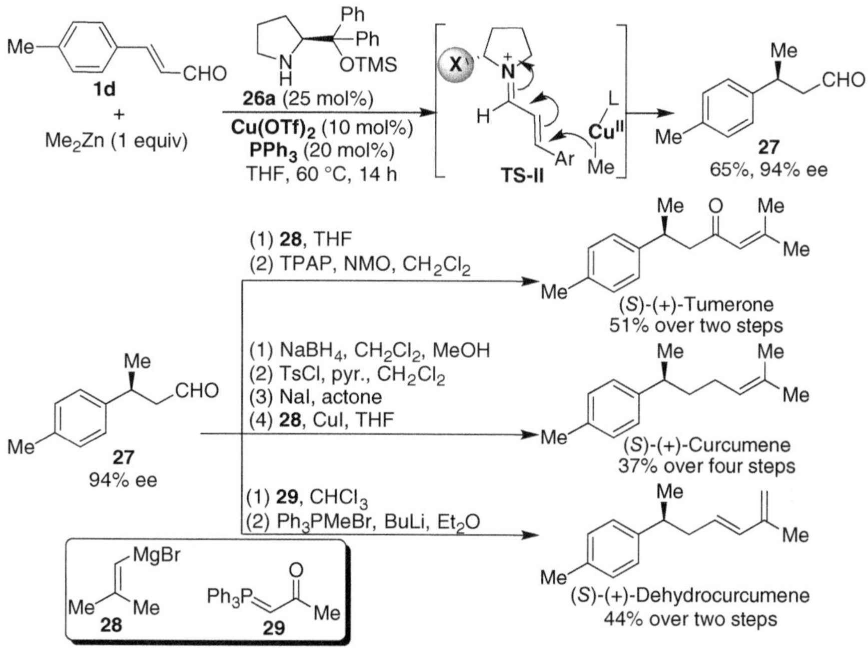

Scheme 10.8 Total synthesis of (*S*)-Tumerone, Curcumene, and Dehydrocurcumene. Source: Modified from Afewerki et al. [23].

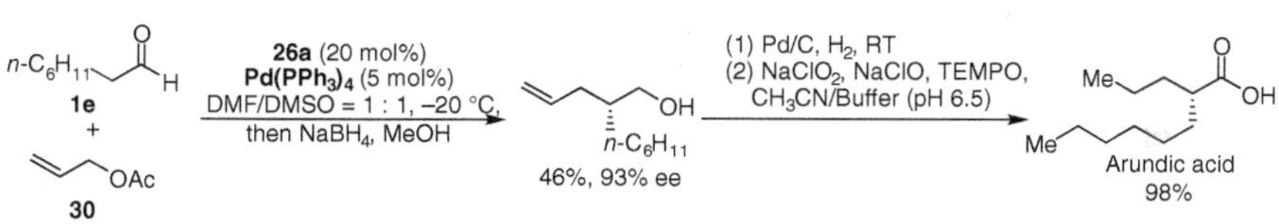

Scheme 10.9 Total synthesis of Arundic acid.

changing chiral secondary amine to ent-**26a** as a cocatalyst for the allylation step provides an access to (*R*)-(−)-Arundic acid.

(−)-Paroxetine, an antidepressant agent, is usually used in the treatment of depression, panic disorders, and obsessive-compulsive disorders. Carreira and coworkers established a stereodivergent α-allylation of linear aldehydes and allylic alcohols, which proceeds via the activation of the racemic allylic alcohols by the $[Ir(cod)Cl]_2$ complex leading to the allylic alkylation of π-allyl-iridium intermediates with enamines *in situ* generated [26]. In the presence of trichloroacetic acid, the enantioselective allylation of aldehyde **1f** and 4-fluorophenyl vinyl carbinol **31a** cooperatively catalyzed by iridium complex formed from $[Ir(cod)Cl]_2$ and ligand (*S*)-**32**, and chiral amine (*S*)-**26b**, affords an γ,δ-unsaturated aldehyde (*S, R*)-**33** in 64% yield, 6 : 1 dr, and >99% ee in gram scale (Scheme 10.10). After the reduction, Mitsunobu reaction, and hydroboration/oxidation, the corresponding alcohol **34** is obtained. A further deprotection and cyclization of **34** accomplish the concise synthesis of (−)-Paroxetine in 20% total yield over five steps. It is worthy to mention

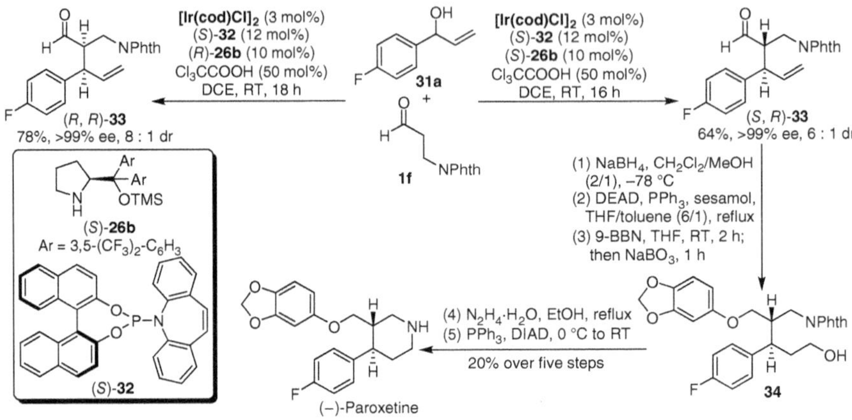

Scheme 10.10 Concise synthesis of (−)-Paroxetine.

that the stereoisomer (*R*, *R*)-**33** can also be prepared in optical pure form with the Ir[(*S*)-**32**]/(*R*)-**26b** dual catalytic system.

Similarly, all four stereoisomers of (−)-Δ^9-tetrahydrocannabinol, which was isolated from flowering tops of female plants of several *Cannabis sativa* L. varieties [27], were synthesized in an expeditious fashion using the same stereodivergent Ir/amine catalytic strategy reported by the same group [28]. The catalytic α-allylation of 5-methylhex-5-enal **1g** and allylic alcohol **31b** by using $[Ir(cod)Cl]_2$, ligand (*R*)- or (*S*)-**32** and chiral amine (*R*)- or (*S*)-**26b** as the catalytic system, provides all possible stereoisomers of γ,δ-unsaturated aldehyde **35** in moderate yields (54–62%) and with excellent stereoselectivities (15 : 1–20 : 1 dr, >99 : 1 er) with the assistance of 5 mol% $Zn(OTf)_2$ (Scheme 10.11).

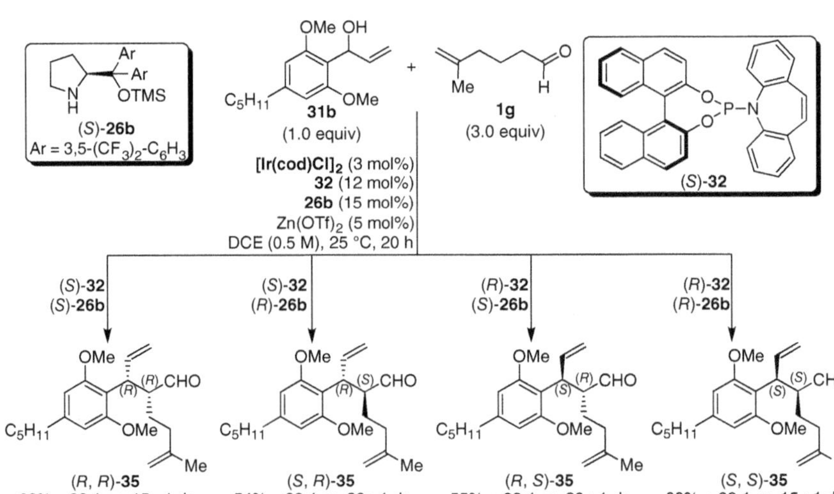

Scheme 10.11 Preparation of all stereoisomers of γ,δ-unsaturated aldehyde **35**.

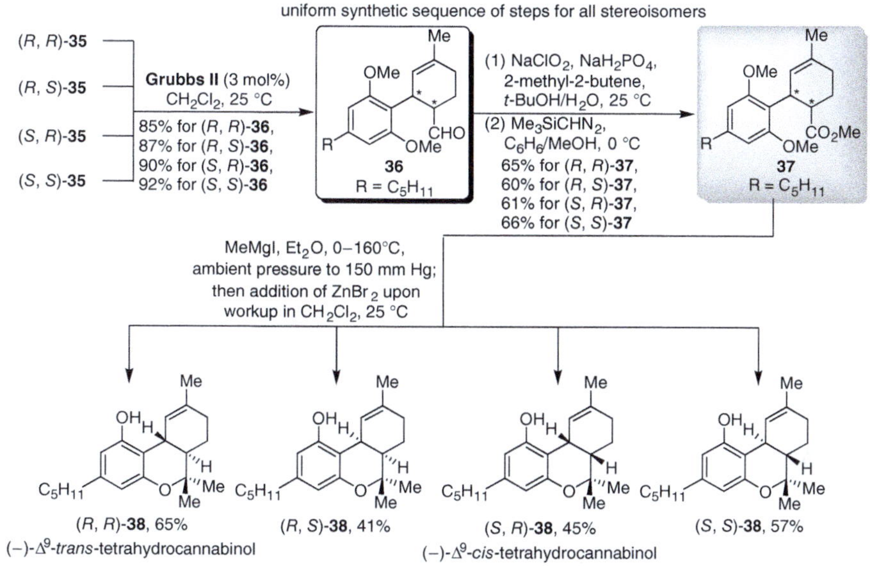

Scheme 10.12 Stereodivergent total synthesis of Δ^9-tetrahydrocannabinols.

As shown in Scheme 10.12, ring-closing metathesis of γ,δ-unsaturated aldehyde **35** using the second-generation Grubbs catalyst, followed by a Pinnick oxidation and methylation with Me_3SiCHN_2, gives the methyl ester **37** in 52–61% yield over three steps without noticeable erosion of the diastereoselectivity. Finally, Δ^9-tetrahydrocannabinol (**38**, all four stereoisomers) have been synthesized in 41–65% yield through the formation of tertiary alcohol and double methyl ether deprotection from ester **37** and followed by a subsequent intramolecular etherification.

Actinophyllic acid, isolated from the leaves of the Australian tree *Alstonia actinophylla* by Carroll et al. [29], is promising for the treatment of cardiovascular diseases. Distinct from the previous impressive strategy in the total synthesis of Actinophyllic acid [30], Yang and coworkers disclosed a catalytic asymmetric total synthesis of (−)-Actinophyllic acid [31]. Their synthesis begins with a catalytic allylation of 2-indolyl vinyl carbinol **31c** and aldehyde **1h** to give the corresponding enantiopure aldehyde **39** in 82% yield with 12 : 1 dr and >99% ee, in the presence of dimethyl hydrogen phosphate as the promoter along with (*S*)-**26b** and [Ir/(*R*)-**32**] as cooperative catalysts. Then the installation of the four contiguous chiral centers (C16, C15, C20, and C19) is completed through the aldol reaction of the chiral aldehyde **39** with *N*-Boc-protected 2-pyrrolidinone, providing **40** in 76% yield and with high diastereoselectivity (dr > 10 : 1). Subsequent protection and inversion of the configuration of C(19) deliver a *tert*-butyldimethylsilyl (TBS) silyl ether **41** in 64% yield over two steps. Followed by Lemieux–Johnson oxidation, Lindgren oxidation, and methylation, the tricyclic intermediate **42** is obtained in 60% yield, which is converted into the key tetracycle **43** through a selective Friedel–Crafts-like cyclization, with 77% overall yield and 6.5 : 1 dr. The following

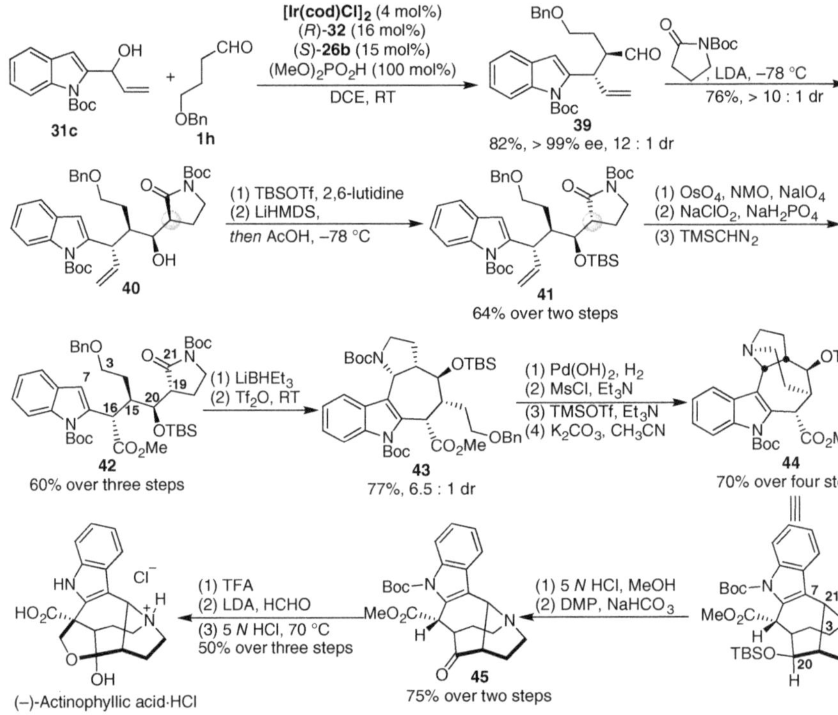

Scheme 10.13 Total synthesis of (–)-Actinophyllic acid.

conversions, including hydrogenolysis, mesylation, deprotection, and intramolecular substitution, transform the **43** into a pentacyclic building block **44**, which contains a 1-azabicyclo[4.2.1]nonane ring. Then, TBS deprotection of **44** followed by Dess–Martin oxidation gives ketone **45** in 75% yield, which can be transformed to (–)-Actinophyllic acid hydrochloride by three more steps (Scheme 10.13).

10.4 Application of Photocatalysis Combined with Organocatalysis

In 2010, MacMillan reported an enantioselective α-benzylation of aldehydes with electron-deficient aryl and heteroaryl methylene bromides under the irradiation of a 26 W compact fluorescent lamp, by employing the combined catalysis of chiral secondary amine **19b** and a commercially available iridium photoredox catalyst [32]. An enantioenriched α-heteroaryl methyl aldehyde **47**, derived from α-benzylation of aldehyde **1i** with 4-(bromomethyl)pyridine **46a**, is obtained in 93% ee. After subsequent three-step reaction sequence, including classical reductive amination and condensation with **48**, Angiogenesis inhibitor **49**, which has been used as a drug candidate for the treatment of diseases such as tumor proliferation and diabetic retinopathy [33], is accessed in 34% overall yield with maintained enantioselectivity (Scheme 10.14).

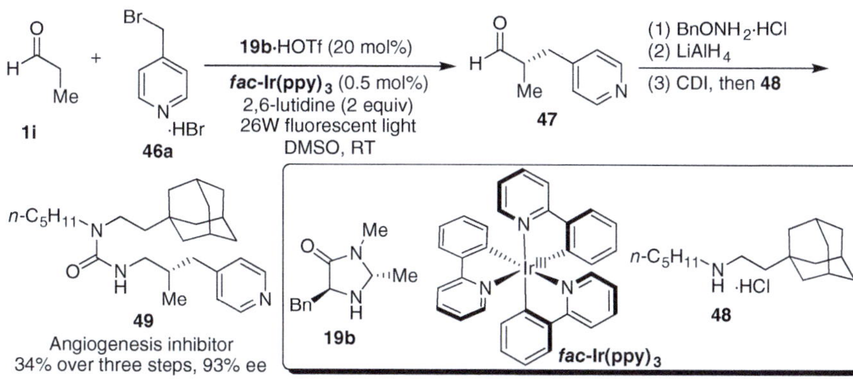

Scheme 10.14 Synthesis of Angiogenesis inhibitor.

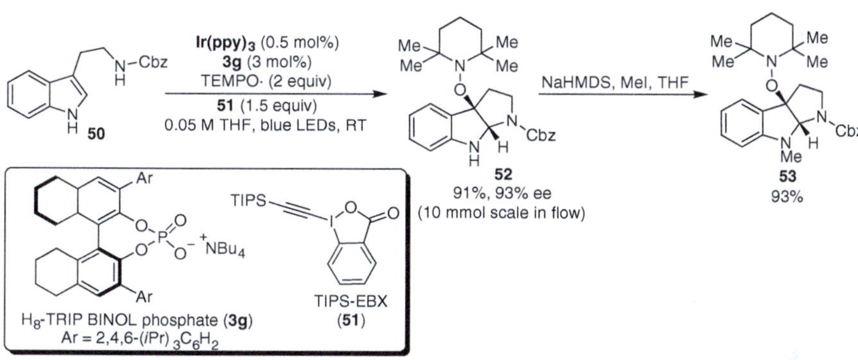

Scheme 10.15 Synthesis of the key pyrroloindoline intermediate.

Cyclotryptamine alkaloids, such as (–)-Calycanthidine and (–)-Chimonanthine, constitute a large family of natural products that show fascinating biological activities [34]. The construction of the key building block – bispyrrolidinoindoline has been the focus of sustainable synthetic interest for decades [35]. Recently, Knowles established an enantioselective synthesis of alkoxyamine-substituted pyrroloindolines with the stable nitroxyl radical 2,2,6,6-tetramethylpiperidine-1-oxyl (TEMPO), enabled by the combined catalysis of chiral phosphate anions (**3g**) and an iridium photoredox catalyst [36]. The tryptamine substrate **50** is converted to substituted pyrroloindoline **52** in 91% yield and 93% ee on 10 mmol scale under blue LED irradiation (Scheme 10.15). The treatment of **52** with MeI and sodium hexamethyldisilazide (NaHMDS) in tetrahydrofuran (THF) can also afford *N*-Me pyrroloindoline **53** in 93% yield.

As shown in Scheme 10.16, single-electron oxidation of the unprotected pyrroloindoline **52** with the iridium photoredox catalyst (**54**) generates a carbocation intermediate that can be captured by tryptamine **50** to form an adduct, which is subsequently reduced with Red-Al to provide the C_2-symmetrical dimer (–)-Chimonanthine with 24% overall yield and 93% ee in four steps from tryptamine **50**. Similarly, a sequential process involving the mesolytic cleavage, a Larock

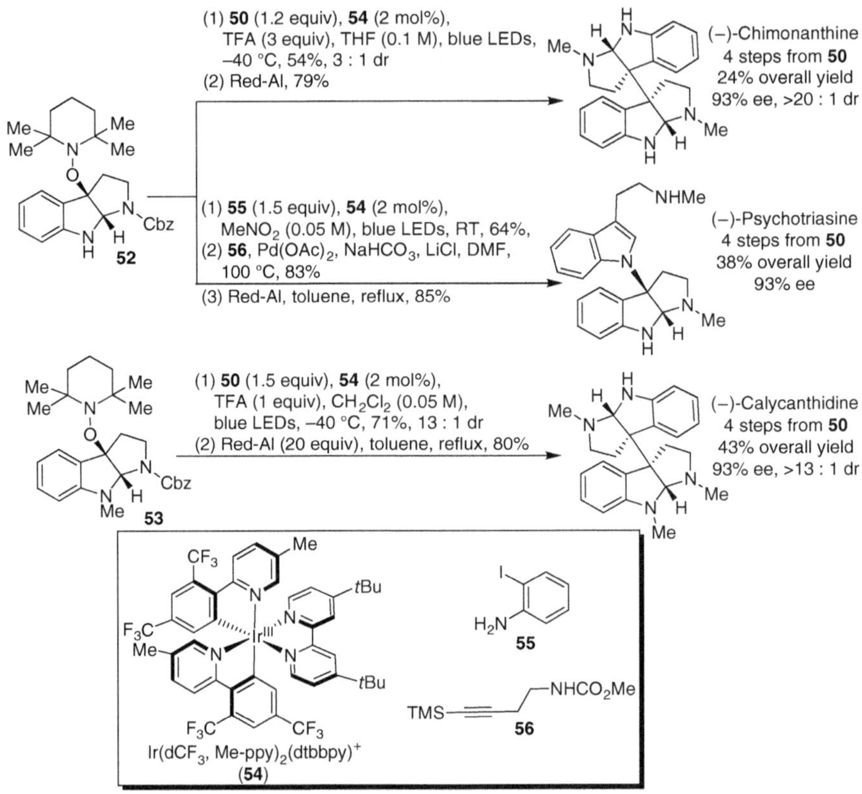

Scheme 10.16 Total synthesis of (–)-Chimonanthine, (–)-Psychotriasine, and (–)-Calycanthidine.

annulation and reduction, (–)-Psychotriasine is prepared in 38% yield over four steps from tryptamine **50** with maintained enantioselectivity. The reaction of *N*-Me pyrroloindoline **53** and tryptamine **50** under the mesolytic cleavage conditions followed by reduction provides the unsymmetrical dimer (–)-Calycanthidine in 43% yield over four steps with 93% ee and >13 : 1 dr.

10.5 Application of Lewis Base–Lewis Acid Cooperative Catalysis

Nelson and coworkers developed an enantioselective ketene-aldehyde cycloaddition of acid chlorides and aldehydes, enabled by cooperative catalysis of cinchona alkaloids (**58**) and Lewis acids, delivering 3,4-*cis*-disubstituted β-lactones with near-perfect stereoselectivity (>99% ee, ≥96% de) [37]. This efficient method was applied for the catalytic asymmetric total synthesis of (–)-Pironetin [38]. As shown in Scheme 10.17, the combination of $LiClO_4$ and *O*-trimethylsilyl quinidine **58a** allows the asymmetric cycloaddition of propionyl chloride **57** and 3-(*p*-methoxybenzyloxy)propionaldehyde **1j** to give β-lactone **59** in 70% yield

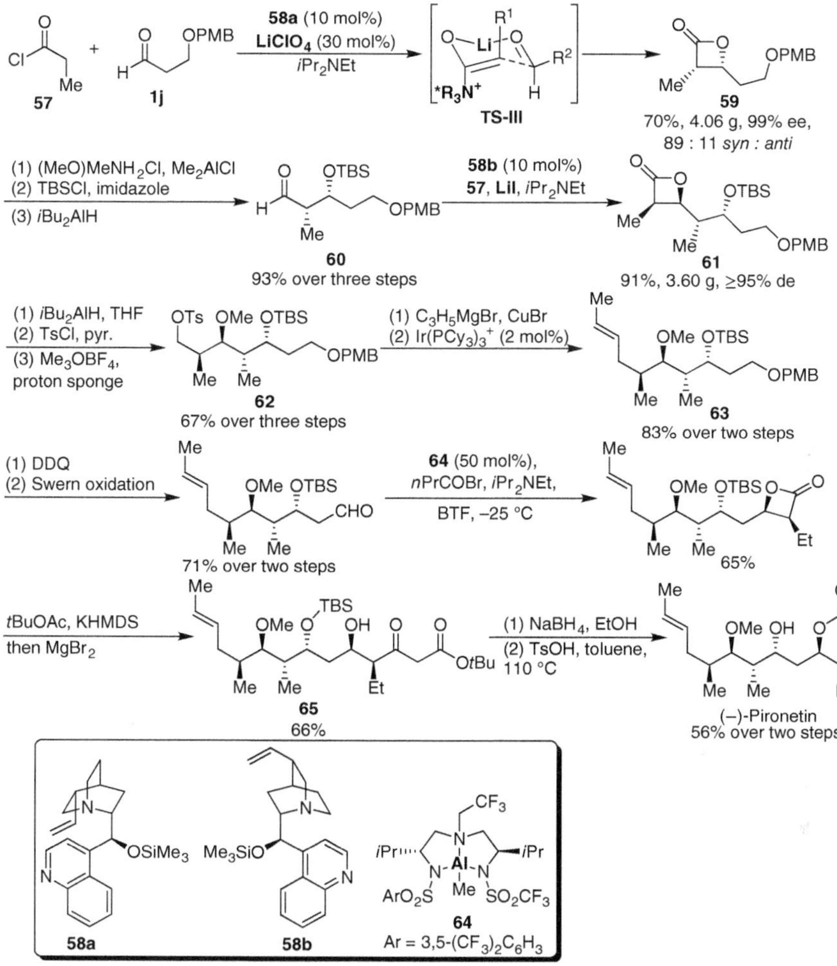

Scheme 10.17 Total synthesis of (−)-Pironetin.

and with 99% ee and 89 : 11 dr. In this case, the formation of a metal-stabilized ammonium enolate with lithium Lewis acid cocatalyst enables the enantioselective cycloaddition of enolate toward aldehyde via transition state **TS-III**. Following the three-step conversion, the *syn* aldehyde **60** is obtained in 93% overall yield. Then the aldehyde **60** is subjected to the *O*-trimethylsilyl quinine **58b** catalyzed cycloaddition reaction with the assistance of LiI, providing the *syn,anti,syn* β-lactone **61** in 91% yield with ≥95% de. After subsequent reduction, selective tosylation, and installation of C9 methyl ether, the protected tetraol **62** is accessed in 67% overall yield. The protected tetraol **62** could undergo the Cu(I)-mediated allyl Grignard substitution, followed by the Ir(I)-catalyzed olefin isomerization, furnishing the key synthon **63** in 83% yield. A sequential procedure including alcohol deprotection, oxidation, Lewis acid (**64**)-catalyzed cycloaddition, and ring-opening reaction, β-keto ester **65** can be prepared in 30% yield over four steps. Finally, the reduction with $NaBH_4$

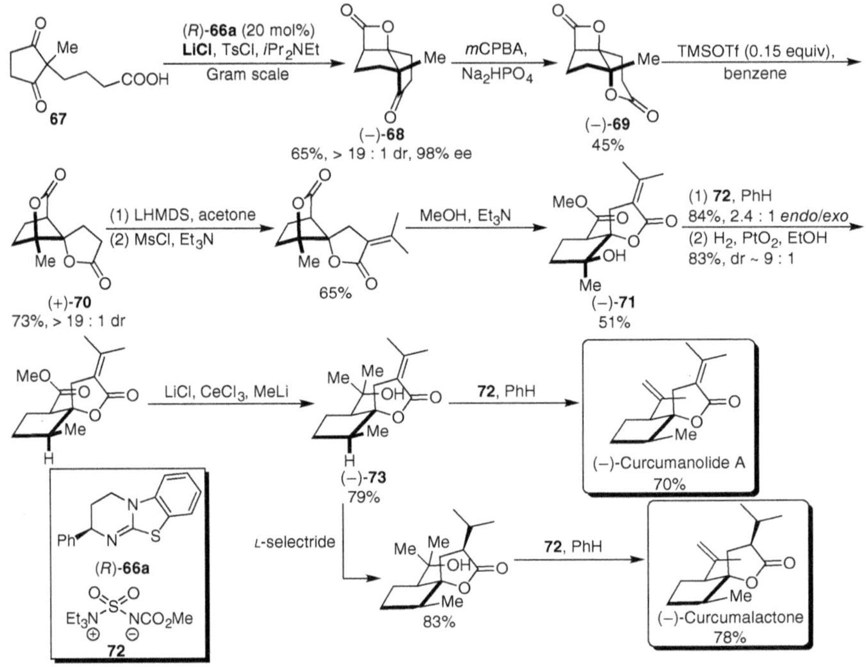

Scheme 10.18 Total synthesis of (−)-Curcumanolide A and (−)-Curcumalactone.

and the treatment of the resulting diol with TsOH leads to the cleavage of *tert*-butyl ester, lactonization, dehydration and the silyl ether removal, giving the desired (–)-Pironetin in 56% yield over two steps.

Romo and coworkers accomplished an asymmetric divergent synthesis of (–)-Curcumanolide A and (–)-Curcumalactone [39] based on the Lewis base-catalyzed aldol-lactonization of dione ketoacids [40]. The dione **67** undergoes the desymmetrization process under the cooperative catalysis of homobenzotetramisole **66a** and the mild Lewis acid LiCl to provide the tricycle β-lactone **68** with excellent enantioselectivity in gram scale (Scheme 10.18). Subsequent Baeyer–Villiger oxidation of **68** leads to the ring-expanded δ-lactone **69** in a moderate yield. The dyotropic rearrangement of **69** in the presence of substoichiometric TMSOTf produces the desired spiro-γ-lactone **70** in 73% yield with maintained enantioselectivity. A sequential procedure involving the aldol condensation with acetone and the methanolysis, the hydroxy ester **71** is given in 51% yield. The dehydration of the secondary alcohol of **71** with Burgess' reagent (**72**) and followed by selective hydrogenation over PtO_2 and a selective addition with $CeCl_3$/LiCl, delivers the tertiary alcohol **73**. The final dehydration toward (–)-Curcumanolide A is completed by employing Burgess' reagent (**72**) once again. Moreover, the synthesis of (–)-Curcumalactone can be accomplished via the diastereoselective reduction of **73** and followed by the dehydration with Burgess' reagent (**72**).

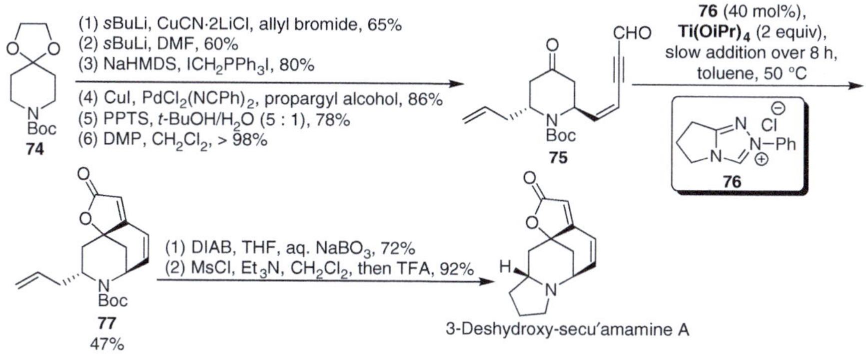

Scheme 10.19 Total synthesis of 3-Deshydroxy-secu'amamine A.

The Securinega alkaloids, which is considered as one class of the most important collections of secondary metabolites from plants of the *Euphorbiaceae* family, contain a unique bridged, tetracyclic substructure bearing an α,β,γ,δ-unsaturated lactone motif [41]. In 2013, Snyder and coworkers accomplished an efficient approach to access the bridged and tricyclic unsaturated lactone in racemic form by N-heterocyclic carbene (NHC)/Lewis acid cooperative catalysis [42]. The enynal **75**, derived from the Boc- and ketal-protected piperidone **74**, can undergo the cyclization under the cooperative catalysis of an NHC **76** and $Ti(O^iPr)_4$ to furnish the desired tricyclic compound **77** in a 47% yield. Then the 3-Deshydroxy-secu'amamine A is accessed in 66% yield through a chemoselective hydroboration/oxidation of **77** and followed by a one-pot methanesulfonate ester formation/*N*-Boc deprotection/cyclization sequence (Scheme 10.19).

The NHC/Lewis acid cooperative catalysis has also been applied to the creation of synthetic methods to access bridging butenolide core within [3.2.1]-bicycle. The enynal **78**, prepared from the 4-hydroxy-L-proline **21b**, can undergo [3+2] cyclization enabled by the synergistic activation of $Ti(O^iPr)_4$ and the carbene **76** to deliver the desired [3.2.1]-bicyclic compound **79** in 31% yield, which has been found as privileged structure motifs in various Securinega alkaloids such as Securinine and Norsecurinine (Scheme 10.20).

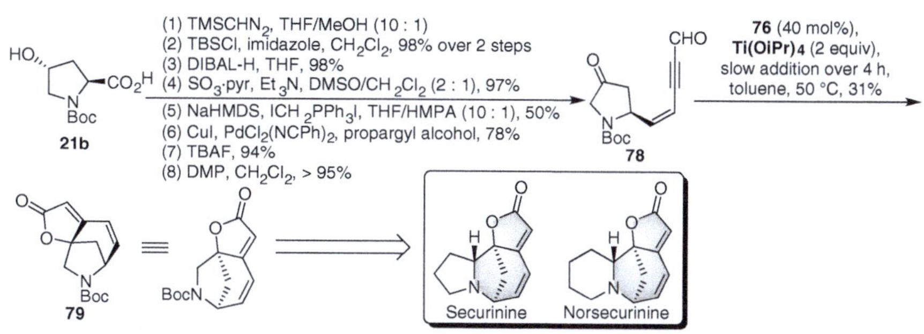

Scheme 10.20 The synthesis of the key [3.2.1]-bicyclic core.

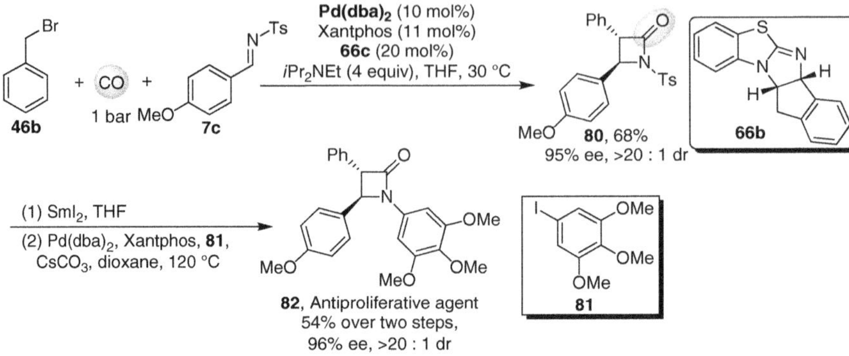

Scheme 10.21 Synthesis of antiproliferative agent.

10.6 Application of Lewis Base–Transition Metal Relay Catalysis

In 2019, Gong and coworkers established asymmetric formal [1+1+4] and [1+1+2] annulations using readily available halides and carbon monoxide as reaction components via relay catalysis of a palladium(0) complex and chiral Lewis base to afford chiral dihydropyridones and β-lactams, respectively [43]. The enantioselective [1+1+2] reaction of benzyl bromide **46b**, carbon monoxide, and *N*-tosylimine **7c** using the benzotetramisole (BTM, **66b**) as cocatalyst, provides the β-lactam **80** in 68% yield and with 95% ee and >20 : 1 dr (Scheme 10.21). Thus, the antiproliferative agent **82** [44] can be synthesized in good yield and excellent enantioselectivity from **80** by the sequential deprotection and Buchwald–Hartwig cross-coupling reaction with an iodobenzene derivative **81**.

10.7 Application of Metal-Brønsted Acid Relay Catalysis

Folicanthine, one of the Cyclotryptamine alkaloids, has been found to exhibit important biological activities [45]. Gong and coworkers [46], after finishing their study in the first catalytic enantioselective synthesis of (+)-Folicanthine, developed a more concise entry to (−)-Folicanthine based on the asymmetric three-component reaction of indole, diazooxindole **5f** and nitroethylene enabled by relay catalysis of $Rh_2(OAc)_4$ and chiral bifunctional squaramide **83** (Scheme 10.22) [47]. The resultant chiral oxindole derivative **84** featuring a 3,3′-bisindole skeleton in 88% ee is oxidized and followed by a subsequent Michael addition to nitroethylene catalyzed by $Mn(OAc)_2$ to afford the Kanai–Matsunaga intermediate **85** in 53% yield over three steps [48]. Finally, the synthesis of (−)-Folicanthine is finished in 14.5% overall yield according to the literature's procedures [48, 49].

Eupomatilone-6 and its analogs were isolated from the Australian shrubs *Eupomatia bennettii* by Carroll and Taylor in 1991 [50]. A relay catalytic intramolecular hydrosiloxylation of arylacetylenes/Mukaiyama aldol reactions

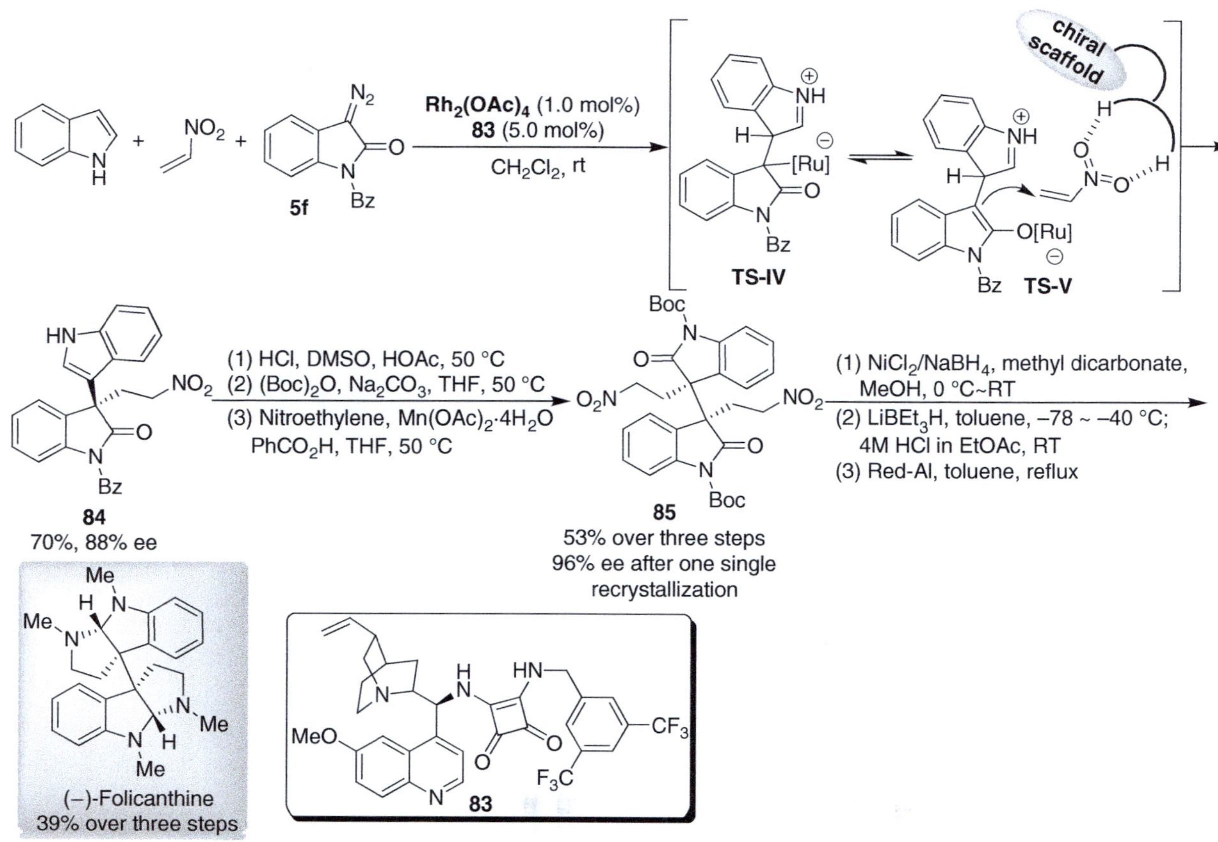

Scheme 10.22 Total synthesis of (−)-Folicanthine. Source: Modified from Chen et al. [47].

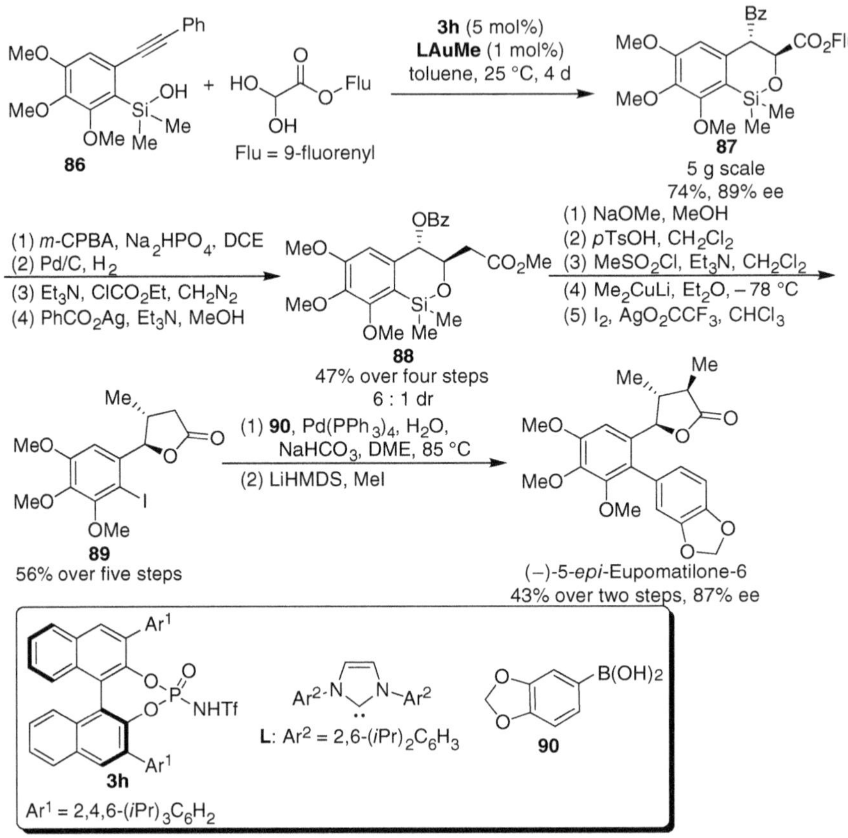

Scheme 10.23 Total synthesis of (−)-5-*epi*-Eupomatilone-6. Source: Modified from Wu et al. [51].

of 2,3,4-trimethoxy-6-(phenylethynyl)phenyl dimethylsilanol **86** with fluorenyl-glyoxylate yields a chiral building block for the enantioselective synthesis of (−)-5-*epi*-Eupomatilone-6 (Scheme 10.23) [51]. The cascade reaction that is conducted in a 5 g scale in the presence of a gold complex (LAuMe) and chiral Brønsted acid **3h** delivers the intermediate **87** in 74% yield and with 89% ee and 10 : 1 dr. The Baeyer–Villiger oxidation of **87**, and followed by hydrogenolysis, formation of α-diazoketone, and Arndt–Eistert reaction, affords the ester **88** in 47% yield over four steps with 6 : 1 dr. A sequential process involving methanolysis, lactonization accompanying with the removal of silyl group, elimination, conjugate addition with dimethyl copper lithium, and iodination provides the aryl iodide **89** in 56% overall yield. The following Suzuki–Miyaura coupling and alkylation steps deliver (−)-5-*epi*-Eupomatilone-6 with 87% ee.

In 2017, Gong and coworkers described a highly diastereo- and enantioselective synthesis of both *ent*-Calyxolane A and Calyxolane B, the diastereomers of marine natural products (Scheme 10.24) [52]. An enantioselective aza-ene-type reaction of enamide **91** and α-diazoester **5e** enabled by the relay catalysis of a gold complex and

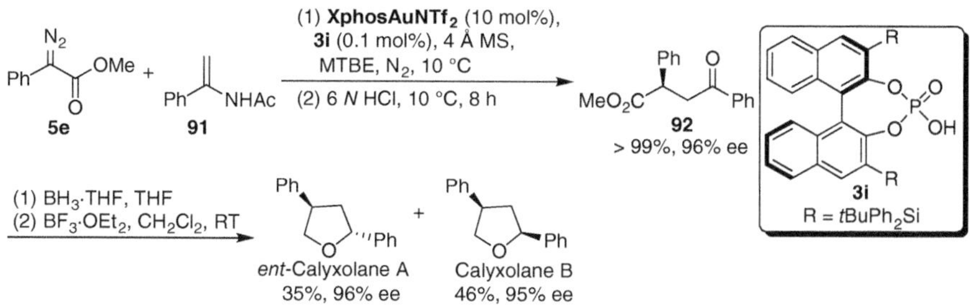

Scheme 10.24 Total synthesis of *ent*-Calyxolane A and Calyxolane B. Source: Modified from Zhao et al. [52].

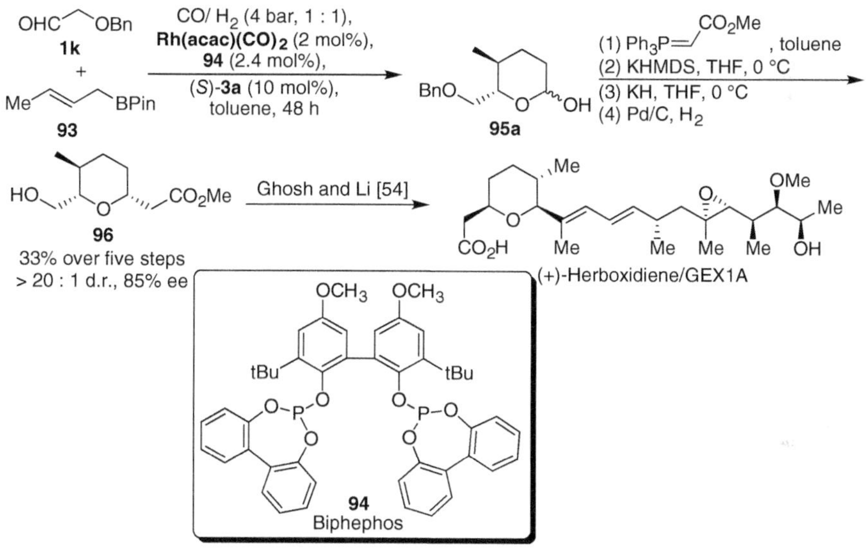

Scheme 10.25 Formal synthesis of (+)-Herboxidiene/GEX1A. Source: Modified from Afewerki et al. [23].

chiral phosphoric acid **3i** gives a γ-keto ester **92** in 96% ee. After the reduction of **92** with borane, followed by a cyclization promoted by boron trifluoride, the natural products *ent*-Calyxolane A and Calyxolane B are obtained in 35% yield and 46% yield, respectively, without the erosion of enantioselectivity.

By means of Rh^I/chiral phosphoric acid (**3a**) relay catalysis, an efficient cascade reaction involving an asymmetric allylboration of aldehyde (**1k**) with allylboronate (**93**) and hydroformylation with syngas and hemiacetalization provides a chiral hemiacetal **95a**, which is converted to the tetrahydropyran **96** in 33% yield over five steps through a Wittig reaction, intramolecular hetero-Michael addition and deprotection sequence (Scheme 10.25) [53]. The key intermediate **96** can be applied for the formal synthesis of (+)-Herboxidiene/GEX1A [54].

The key tetrahydropyran core **97**, which has been used for the synthesis of Leucascandrolide A [55], can be facilely prepared via the enantioselective cascade reaction

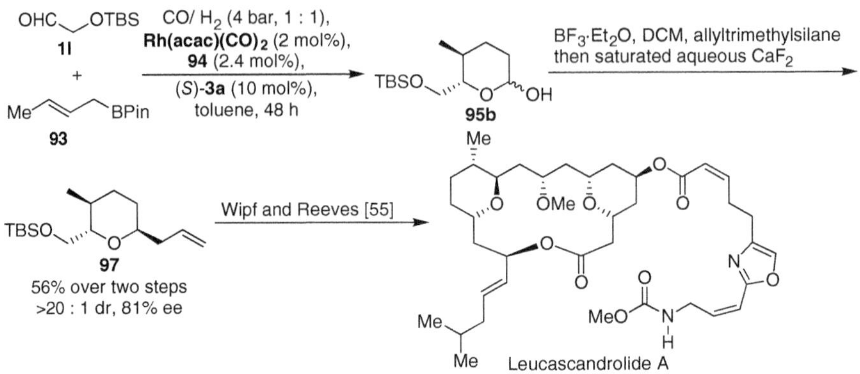

Scheme 10.26 Formal synthesis of Leucascandrolide A.

of aldehyde **1l** with allylboronate (**93**), followed by a diastereoselective allylation (Scheme 10.26).

10.8 Conclusion

Asymmetric organocatalysis combined with metal catalysis provides robust strategies to enable highly enantioselective processes, some of which have been applied to the stereoselective and efficient synthesis of natural products, pharmaceuticals, and related structurally diverse molecules. However, the repertoire of organo/metal combined catalysis methods for the asymmetric total synthesis has much less been displayed. The growth of asymmetric organo-metal combined catalysis will inspire the design of new synthetic routes to access natural products and biologically active molecules. In turn, the continuing fascination with the enantioselective synthesis of natural products and chiral core structures steadily serves as a driving force for the exploration of new efficient organo/metal combined catalysis.

References

1 (a) Breuer, M., Ditrich, K., Habicher, T. et al. (2004). *Angew. Chem. Int. Ed.* 43: 788–824. (b) Jacobson, E.N., Pfaltz, A., and Yamamoto, H. (eds.) (1999). *Comprehensive Asymmetric Catalysis*. Heidelberg: Springer-Verlag. (c) Zhou, Q.-L. (ed.) (2011). *Privileged Ligands and Catalysts*. Weinheim: Wiley-VCH. (d) Halpern, J. and Trost, B.M. (2004). *Proc. Natl. Acad. Sci. U.S.A.* 101: 5347–5347.

2 (a) Dalko, P.I. and Moisan, L. (2001). *Angew. Chem. Int. Ed.* 40: 3726–3748. (b) Dalko, P.I. (2007). *Enantioselective Organocatalysis: Reactions and Experimental Procedures*. Weinheim: Wiley-VCH. (c) List, B. (2007). *Chem. Rev.* 107: 5413–5415.

3 (a) Shao, Z.-H. and Zhang, H.-B. (2009). *Chem. Soc. Rev.* 38: 2745–2755. (b) Du, Z.T. and Shao, Z.-H. (2013). *Chem. Soc. Rev.* 42: 1337–1378. (c) Allen,

A.E. and MacMillan, D.W.C. (2012). *Chem. Sci.* 3: 633–658. (d) Fogg, D.E. and dos Santos, E.N. (2004). *Coord. Chem. Rev.* 248: 2365–2379. (e) Grossmann, A. and Enders, D. (2012). *Angew. Chem. Int. Ed.* 51: 314–325. (f) Patil, N.T., Shinde, V.S., and Gajula, B.A. (2012). *Org. Biomol. Chem.* 10: 211–224. (g) Han, Z.-Y., Wang, C., and Gong, L.-Z. (2012). *Science of Synthesis: Asymmetric Organocatalysis*, vol. 2 (ed. K. Maruoka), 697. Stuttgart: Georg Thieme Verlag. (h) Chen, D.-F., Han, Z.-Y., Zhou, X.-L., and Gong, L.-Z. (2014). *Acc. Chem. Res.* 47: 2365–2377. (i) Inamdar, S.M., Shinde, V.S., and Patil, N.T. (2015). *Org. Biomol. Chem.* 13: 8116–8162. (j) Gong, L.-Z. (2018). *Acta Chim. Sin.* 76: 817–818.

4 Enzell, C. and Erdtman, H. (1958). *Tetrahedron* 4: 361–368.

5 For selected examples, see: (a) Subba Rao, H.N., Damodaran, N.P., and Dev, S. (1968). *Tetrahedron Lett.*: 2213–2214. (b) Abad, A., Agulló, C., Arnó, M. et al. (1996). *J. Org. Chem.* 61: 5916–5919. (c) Fuganti, C. and Serra, S. (1999). *J. Org. Chem.* 64: 8728–8730. (d) Aavula, B.R., Cui, Q., and Mash, E.A. (2000). *Tetrahedron: Asymmetry* 11: 4681–4686. (e) Grainger, R.S. and Patel, A. (2003). *Chem. Commun.*: 1072–1073.

6 Mukherjee, S. and List, B. (2007). *J. Am. Chem. Soc.* 129: 11336–11337.

7 Reetz, M.T., Westermann, J., and Kyung, S.-H. (1985). *Chem. Ber.* 118: 1050–1057.

8 (a) Hu, W., Xu, X., Zhou, J. et al. (2008). *J. Am. Chem. Soc.* 130: 7782–7783. (b) Xu, X., Zhou, J., Yang, L., and Hu, W. (2008). *Chem. Commun.*: 6564–6566. (c) Guo, Z., Shi, T., Jiang, J. et al. (2009). *Org. Biomol. Chem.* 7: 5028–5033.

9 Qian, Y., Xu, X., Jiang, L. et al. (2010). *J. Org. Chem.* 75: 7483–7486.

10 Qian, Y., Jing, C., Liu, S., and Hu, W. (2013). *Chem. Commun.* 49: 2700–2702.

11 Xu, B., Zhu, S.-F., Zuo, X.-D. et al. (2014). *Angew. Chem. Int. Ed.* 53: 3913–3916.

12 Di Santo, R., Costi, R., Roux, A. et al. (2005). *J. Med. Chem.* 48: 4220–4223.

13 Kumaraswamy, G. and Pitchaiah, A. (2011). *Tetrahedron* 67: 2536–2541.

14 (a) Junquero, D., Oms, P., Carilla-Durand, E. et al. (2001). *Biochem. Pharmacol.* 61: 97–108. (b) José, J.Z.-L., Ruth, F.-S., Antonio, J.L.F. et al. (2006). *Pharmacol.* 48: 128 and references cited therein.

15 Xu, B., Zhu, S.-F., Zhang, Z.-C. et al. (2014). *Chem. Sci.* 5: 1442–1448.

16 Taga, T., Akimoto, N., and Ibuka, T. (1984). *Chem. Pharm. Bull.* 32: 4223–4225.

17 Hartrampf, N., Winter, N., Pupo, G. et al. (2018). *J. Am. Chem. Soc.* 140: 8675–8680.

18 Stork, G., Brizzolara, A., Landesman, H. et al. (1963). *J. Am. Chem. Soc.* 85: 207–222.

19 (a) Tsuji, J., Minami, I., and Shimizu, I. (1983). *Tetrahedron Lett.* 24: 1793–1796. (b) Tsuji, J., Shimizu, I., Minami, I. et al. (1985). *J. Org. Chem.* 50: 1523–1529. (c) Tsuji, J. and Minami, I. (1987). *Acc. Chem. Res.* 20: 140–145.

20 (a) Felker, I., Pupo, G., Kraft, P., and List, B. (2015). *Angew. Chem. Int. Ed.* 54: 1960–1964. (b) Pupo, G., Properzi, R., and List, B. (2016). *Angew. Chem. Int. Ed.* 55: 6099–6102. (c) Monaco, M.R., Pupo, G., and List, B. (2016). *Synlett* 27: 1027–1040.

21 Simmons, B., Walji, A.M., and MacMillan, D.W.C. (2009). *Angew. Chem. Int. Ed.* 48: 4349–4353.

22 (a) ApSimon, J. (1983). *The Total Synthesis of Natural Products*, vol. 5, 35. New York: Wiley. (b) Kitahara, T., Furusho, Y., and Mori, K. (1993). *Biosci. Biotechnol., Biochem.* 57: 1137–1140. (c) Tanaka, K., Nuruzzaman, M., Yoshida, M. et al. (1999). *Chem. Pharm. Bull.* 47: 1053–1055. (d) McEnroe, F.J. and Fenical, W. (1978). *Tetrahedron* 34: 1661–1664. (e) Damodaran, N.P. and Dev, S. (1968). *Tetrahedron* 24: 4113–4122. (f) Fuganti, C. and Serra, S. (2000). *J. Chem. Soc., Perkin Trans. 1*: 3758–3764. (g) Wright, A.E., Pomponi, S.A., McConnell, O.J. et al. (1987). *J. Nat. Prod.* 50: 976–978. (h) Fusetani, N., Sugano, M., Matsunaga, S., and Hashimoto, K. (1987). *Cell. Mol. Life Sci.* 43: 1234–1235. (i) ElSayed, K.A., Yousaf, M., Hamann, M.T. et al. (2002). *J. Nat. Prod.* 65: 1547–1553. (j) Cichewicz, R.H., Clifford, L.J., Lassen, P.R. et al. (2005). *Bioorg. Med. Chem.* 13: 5600–5612.

23 Afewerki, S., Breistein, P., Pirttilä, K. et al. (2011). *Chem. Eur. J.* 17: 8784–8788.

24 (a) Tateishi, N., Mori, T., Kagamiishi, Y. et al. (2002). *Blood Flow Metab.* 22: 723–724. (b) Mori, T., Town, T., Tan, J. et al. (2006). *J. Pharmacol. Exp. Ther.* 318: 571–578. (c) Hasegawa, T., Kawanaka, Y., Kasamatsu, E. et al. (2003). *Org. Process Res. Dev.* 7: 168–171. (d) Fernandes, R.A., Dhall, A., and Ingle, A.B. (2009). *Tetrahedron Lett.* 50: 5903–5905. (e) Sorbera, L.A., Castaner, J., and Castaner, J.M. (2004). *Drugs Future*: 441.

25 Afewerki, S., Ibrahem, I., Rydfjord, J. et al. (2012). *Chem. Eur. J.* 18: 2972–2977.

26 Krautwald, S., Schafroth, M.A., Sarlah, D., and Carreira, E.M. (2014). *J. Am. Chem. Soc.* 136: 3020–3023.

27 (a) Mechoulam, R. and Gaoni, Y. (1964). *J. Am. Chem. Soc.* 86: 1646–1647. (b) Taylor, E.C., Lenard, K., and Shvo, Y. (1966). *J. Am. Chem. Soc.* 88: 367. (c) Smith, R.M. and Kempfert, K.D. (1977). *Phytochemistry* 16: 1088–1089. (d) Turner, C.E., Elsohly, M.A., and Boeren, E.G. (1980). *J. Nat. Prod.* 43: 169–234.

28 Schafroth, M.A., Zuccarello, G., Krautwald, S. et al. (2014). *Angew. Chem. Int. Ed.* 53: 13898–13901.

29 Carroll, A.R., Hyde, E., Smith, J. et al. (2005). *J. Org. Chem.* 70: 1096–1099.

30 (a) Martin, C.L., Overman, L.E., and Rohde, J.M. (2008). *J. Am. Chem. Soc.* 130: 7568–7569. (b) Martin, C.L., Overman, L.E., and Rohde, J.M. (2010). *J. Am. Chem. Soc.* 132: 4894–4906. (c) Granger, B.A., Jewett, I.T., Butler, J.D. et al. (2013). *J. Am. Chem. Soc.* 135: 12984–12986. (d) Granger, B.A., Jewett, I.T., Butler, J.D., and Martin, S.F. (2014). *Tetrahedron* 70: 4094–4104. (e) Cai, L., Zhang, K., and Kwon, O. (2016). *J. Am. Chem. Soc.* 138: 3298–3301. (f) Yoshii, Y., Tokuyama, H., and Chen, D.Y.-K. (2017). *Angew. Chem. Int. Ed.* 56: 12277–12281. (g) Xue, F., Lu, H., He, L. et al. (2018). *J. Org. Chem.* 83: 754–764. (h) Vaswani, R.G., Day, J.J., and Wood, J.L. (2009). *Org. Lett.* 11: 4532–4535. (i) Zaimoku, H., Taniguchi, T., and Ishibashi, H. (2012). *Org. Lett.* 14: 1656–1658. (j) Galicia, I.Z. and Maldonado, L.A. (2013). *Tetrahedron Lett.* 54: 2180–2182. (k) Mortimer, D., Whiting, M., Harrity, J.P.A. et al. (2014). *Tetrahedron Lett.* 55: 1255–1257.

31 Liang, X., Zhang, T.-Y., Meng, C.-Y. et al. (2018). *Org. Lett.* 20: 4575–4578.

32 Shih, H.W., Wal, M.N.V., Grange, R.L., and MacMillan, D.W.C. (2010). *J. Am. Chem. Soc.* 132: 13600–13603.

33 Matsuoka, H., Nishimura, K., Seike, H. et al. (2008). Preparation of pyridylalkylurea derivatives as angiogenesis inhibitors. US Patent 2008/0161270 A1.

34 (a) Cordell, G.A. and Saxton, J.E. (1981). *The Alkaloids: Chemistry and Physiology*, vol. 20 (ed. R.G.A. Rodrigo), 3–295. New York: Academic Press. (b) Anthoni, U., Christophersen, C., and Nielsen, P.H. (1999). Naturally occurring cyclotryptophans and cyclotryptamines. In: *Alkaloids: Chemical and Biological Perspectivese*, vol. 13 (ed. S.W. Pelletier), 163–236. Pergamon: Oxford.

35 Steven, A. and Overman, L.E. (2007). *Angew. Chem. Int. Ed.* 46: 5488–5508.

36 Gentry, E.C., Rono, L.J., Hale, M.E. et al. (2018). *J. Am. Chem. Soc.* 140: 3394–3402.

37 Zhu, C., Shen, X., and Nelson, S.G. (2004). *J. Am. Chem. Soc.* 126: 5352–5353.

38 Shen, X., Wasmuth, A.S., Zhao, J. et al. (2006). *J. Am. Chem. Soc.* 128: 7438–7439.

39 Leverett, C.A., Purohit, V.C., Johnson, A.G. et al. (2012). *J. Am. Chem. Soc.* 134: 13348–13356.

40 Leverett, C.A., Purohit, V.C., and Romo, D. (2010). *Angew. Chem. Int. Ed.* 49: 9479–9483.

41 (a) Finefield, J.M., Sherman, D.H., Kreitman, M., and Williams, R.M. (2012). *Angew. Chem. Int. Ed.* 51: 4802–4836. (b) Ohsaki, A., Ishiyama, H., Yoneda, K., and Kobayashi, J. (2003). *Tetrahedron Lett.* 44: 3097–3099. (c) Zhao, B.-X., Wang, Y., Zhang, D.-M. et al. (2012). *Org. Lett.* 14: 3096–3099.

42 ElSohly, A.M., Wespe, D.A., Poore, T.J., and Snyder, S.A. (2013). *Angew. Chem. Int. Ed.* 52: 5789–5794.

43 Li, L.-L., Ding, D., Song, J. et al. (2019). *Angew. Chem. Int. Ed.* 58: 7647–7651.

44 O'Boyle, N.M., Carr, M., Greene, L.M. et al. (2011). *Eur. J. Med. Chem.* 46: 4595–4607.

45 (a) Crich, D. and Banerjee, A. (2007). *Acc. Chem. Res.* 40: 151–161. (b) Sevenet, T. and Pusset, J. (1996). *The Alkaloids: Chemistry and Pharmacology*, vol. 48, 1–73. New York: Academic Press.

46 Guo, C., Song, J., Huang, J.-Z. et al. (2012). *Angew. Chem. Int. Ed.* 51: 1046–1050.

47 Chen, D.-F., Zhao, F., Hu, Y., and Gong, L.-Z. (2014). *Angew. Chem. Int. Ed.* 53: 10763–10767.

48 Mitsunuma, H., Shibasaki, M., Kanai, M., and Matsunaga, S. (2012). *Angew. Chem. Int. Ed.* 51: 5217–5221.

49 Wada, M., Murata, T., and Oikawa, H. (2014). *Org. Biomol. Chem.* 12: 298–306.

50 Carroll, A.R. and Taylor, W.C. (1991). *Aust. J. Chem.* 44: 1615–1626.

51 Wu, X., Li, M.-L., and Wang, P.-S. (2014). *J. Org. Chem.* 79: 419–425.

52 Zhao, F., Li, N., Zhang, T. et al. (2017). *Angew. Chem. Int. Ed.* 56: 3247–3251.

53 Li, L.-L., Su, Y.-L., Han, Z.-Y., and Gong, L.-Z. (2018). *Chem. Eur. J.* 24: 7626–7630.

54 Ghosh, A.K. and Li, J. (2011). *Org. Lett.* 13: 66–69.

55 Wipf, P. and Reeves, J.T. (2002). *Chem. Commun.*: 2066–2067.

Index

Asymmetric Organo-Metal Catalysis: Concepts, Principles, and Applications, First Edition. Liu-Zhu Gong.
© 2022 WILEY-VCH GmbH. Published 2022 by WILEY-VCH GmbH.

b

C